T0324051

VIRUSES

SECOND EDITION

VIRUSES

From Understanding to Investigation

SECOND EDITION

SUSAN PAYNE

*Department of Veterinary Medicine and Biomedical Sciences, Texas A&M University,
College Station, TX, United States*

ILLUSTRATOR, MARCY EDELSTEIN

ELSEVIER

ACADEMIC PRESS

An imprint of Elsevier

Academic Press is an imprint of Elsevier
125 London Wall, London EC2Y 5AS, United Kingdom
525 B Street, Suite 1650, San Diego, CA 92101, United States
50 Hampshire Street, 5th Floor, Cambridge, MA 02139, United States
The Boulevard, Langford Lane, Kidlington, Oxford OX5 1GB, United Kingdom

Copyright © 2023 Elsevier Inc. All rights reserved.

No part of this publication may be reproduced or transmitted in any form or by any means, electronic or mechanical, including photocopying, recording, or any information storage and retrieval system, without permission in writing from the publisher. Details on how to seek permission, further information about the Publisher's permissions policies and our arrangements with organizations such as the Copyright Clearance Center and the Copyright Licensing Agency, can be found at our website: www.elsevier.com/permissions.

This book and the individual contributions contained in it are protected under copyright by the Publisher (other than as may be noted herein).

Notices
Knowledge and best practice in this field are constantly changing. As new research and experience broaden our understanding, changes in research methods, professional practices, or medical treatment may become necessary.

Practitioners and researchers must always rely on their own experience and knowledge in evaluating and using any information, methods, compounds, or experiments described herein. In using such information or methods they should be mindful of their own safety and the safety of others, including parties for whom they have a professional responsibility.

To the fullest extent of the law, neither the Publisher nor the authors, contributors, or editors, assume any liability for any injury and/or damage to persons or property as a matter of products liability, negligence or otherwise, or from any use or operation of any methods, products, instructions, or ideas contained in the material herein.

ISBN: 978-0-323-90385-1

For Information on all Academic Press publications
visit our website at https://www.elsevier.com/books-and-journals

Publisher: Stacy Masucci
Acquisitions Editor: Kattie Washington
Editorial Project Manager: Susan E. Ikeda
Production Project Manager: Fahmida Sultana
Cover Designer: Mark Rogers

Typeset by MPS Limited, Chennai, India

Contents

Contents

About the author

Dr. Susan Payne is an associate professor emeritus in the Department of Veterinary Pathobiology at the Texas A&M University, United States. During her academic career, she mentored graduate and undergraduate students at three universities and taught virology to undergraduate, graduate, medical, and veterinary students. Those courses are the basis for this textbook. She also had an active research program and authored more than 40 peer-reviewed articles. She is retired and lives in Caldwell, Texas with her husband and too many animals to count.

Preface

This book, *Viruses: From Understanding to Investigation*, was inspired by a long career of teaching and research. My students include undergraduate, graduate, medical, and veterinary students. As regards the book title, my intent is to lead students of virology from a basic understanding to an interest in the investigations that have provided the information contained herein. The focus of this textbook is on animal and human viruses, only because these have been the focus of my research and teaching for many years. The viruses of plants, fungi, bacteria, and single-celled organisms are certainly no less interesting.

There is a huge amount of information about viruses available online, in journals, books, websites, and blogs. So what is the need for another virology textbook? My intent was to organize and present a thoughtful, understandable, and up-to-date summary of the volumes of information available for consumption elsewhere. While every textbook, including this one, contains many facts, I have tried to emphasize general concepts.

With more than 40 chapters in the second edition, this book contains more than enough material for a semester-long course in introductory virology. The book is geared toward students with some background in cell biology, microbiology, immunology, and/or biochemistry, and I hope that it will be useful for both undergraduate and beginning graduate students. I also hope that no instructor will try to cover all of the material contained herein during a single semester. The book is organized into two parts, the first 10 chapters cover topics including an introduction to viruses (containing information on replication cycle, diversity, taxonomy, and outcomes of virus infection); structure; interactions with the host cell; methods for studying viruses; immunity to viruses; antiviral agents and introductions to viral epidemiology; evolution; and pathogenesis. There are also chapters that serve as introductions to RNA and DNA viruses. I imagine that this will be more than enough information for many instructors and students.

The remaining chapters present viruses by family, with information about structure, genome organization, replication strategies, and disease. I have tried to

be up-to-date and include virus families that are relatively new (hence, these chapters are short). While each chapter contains basic information about a particular virus family, I am fond of narratives that tie the molecular basis of virus replication to pathogenesis and have provided examples from a variety of animals, including human animals. The inclusion of "animal diseases" specifically serves as a reminder that companion and food animals play integral roles in human health and well-being. (As do plant and bacterial viruses, but those are subjects for other authors to address.)

I encourage instructors to review the material on virus families and choose a handful of these chapters to use in their courses. Positive-strand RNA viruses are presented first followed by negative and dsRNA viruses. The DNA viruses are presented from the smallest to the largest. Last, but certainly not least, are chapters covering the reverse transcribing retroviruses and hepadnaviruses. I have included some taxonomic information in each chapter, sometimes more, sometimes less. I have tried to include the most up-to-date information available from the International Committee on the Taxonomy of Viruses while still focusing on virus "families" to organize this book. The readers should be aware of the fact that the huge number of new viruses being identified using nucleic acid sequencing strategies has resulted in the need for an expansion of viral taxonomy, the details of which are, for the most part, not provided herein.

Throughout the book, I have included brief discussions of both a historical nature (e.g., oncogenic retroviruses and an account of the discovery of hepatitis C virus) and current issues [such as the initiative of the World Health Organization and the World Organization for Animal Health (OIE) to collaborate to reduce human deaths by rabies virus in underdeveloped countries]. In the mix are also topics relevant to basic research such as use of vesicular stomatitis virus G protein for pseudo-typing and lymphocytic choriomeningitis virus as a model for pathogenesis. I hope that the readers will find at least a few topics of interest and will use this textbook as a jumping-off point for more in-depth investigations into the world of viruses.

Acknowledgments

This book was beautifully and carefully illustrated by my sister, Marcy R. Edelstein, for which I thank her most sincerely. If a picture is worth a thousand words, this book is quite long indeed! Thank you, dear sister. Many thanks and much love to my husband Ross, for his patience during this revision. He built a lot of fence, cleared many acres, and even weeded my garden as I worked on this project. I also gratefully acknowledge the dedicated and imaginative researchers who work to unravel the complex and beautiful world of viruses. I wish I was a graduate student again!

1

Introduction to animal viruses

After studying this chapter, you should be able to:

- Provide a meaningful definition of a virus.
- Explain the difference between cell division and virus replication.
- Explain the correct usage of "virion" versus "virus."
- Describe the basic steps in a virus replication-cycle.
- Draw, label, and describe each part of a "one-step" growth curve.
- List possible outcomes of a virus infection (1) at the level of the individual cell and (2) at the level of the host animal.
- Define the term "host range" as regards viruses.

What is a virus?

Most of us are familiar with the term "virus" and understand that viruses are disease-causing agents, transmitted from one person or animal to another. We know that many colds and the "flu" are caused by viruses. Beginning in 2019, severe acute respiratory syndrome coronavirus 2 (SARS-CoV-2), the causative agent of Coronavirus Disease 2019 (COVID-19) spread worldwide causing the worst viral pandemic in a century. We might also be aware that viruses can be used to deliver genes to cells for the purposes of gene therapy or genetic engineering. How is it that viruses can

be pathogens causing millions of deaths as well as useful tools? To answer this question we must understand the fundamental nature of viruses; we can then begin to investigate how viruses shape our world. What are viruses? Major points to understand are listed below:

- Viruses are infectious (able to colonize and replicate within a host organism) agents that are *not* cellular in nature.
- Viruses have nucleic acid genomes that are surrounded by protein coats called capsids. Capsids protect genomes from environmental hazards and are needed for efficient delivery of viral genomes into new host cells. Some viruses have lipid membranes, called envelopes, that surround the capsid (Fig. 1.1).
- Viruses are structurally much simpler than cells.
- Viruses do not increase in number by cell division; instead they assemble from *newly synthesized* protein and nucleic acid building blocks. As viruses are not cells, they have none the organelles associated with cells (Fig. 1.2).
- A sample of purified virus particles (virions) has no metabolic activity.
- Viruses must enter a living host cell in order to replicate (thus *all* viruses are obligate intracellular parasites).
- Synthesis of proteins and genomes for new virus particles requires an energy source (ATP), building

© 2023 Elsevier Inc. All rights reserved.

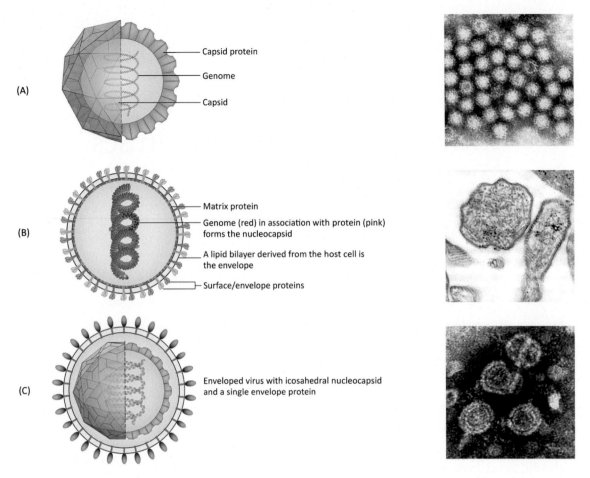

FIGURE 1.1   Basic features of virions. (A) On left, simple diagram of an unenveloped virus with icosahedral symmetry; on right, electron micrograph of calicivirus (CDC/Dr. Erskine Palmer CDC Public Health Image Library Image #198). (B) On left, simple diagram of enveloped virus with a helical nucleocapsid; on right, electron micrograph of mumps virus, a paramyxovirus (CDC/Dr. F. A. Murphy. CDC Public Health Image Library Image #8758.). (C) On left a simple diagram of an enveloped virus with an icosahedral nucleocapsid; on right, electron micrograph of hepadnavirus (CDC/Dr. Erskine Palmer CDC Public Health Image Library Image #270).

materials (amino acids and nucleotides), and protein synthesis machinery (ribosomes) supplied by the host cell. The cell also provides scaffolds (microtubules, filaments) and membranes on which viruses replicate their genomes and assemble. Thus the cell is a factory providing both working machinery and raw materials. The infected cell may or may not continue normal cellular processes (host cell mRNA and protein synthesis) during a viral infection.

- Virus particles or virions are essentially packages designed to deliver nucleic acids to cells.
- Viral genomes are excellent examples of "selfish genes."
- Viruses are not a homogeneous group of infectious agents; while sharing certain basic attributes (Box 1.1) viruses are extremely diverse as regards size, genome type, and evolutionary

history. The preceding list might suggest that viruses are uninteresting, inanimate particles but in fact, viruses provide windows into dynamic and diverse molecular, cellular, and evolutionary processes.

Diversity in the world of viruses

The following list highlights some important points about diversity in the world of viruses. Understanding viral diversity is important as it helps explain why no single antiviral drug or mitigation strategy will suffice when confronting these infectious agents. It also explains why some viruses have become essential tools in the fields of molecular biology, cell biology, and medicine among others.

- All viruses have nucleic acid genomes, but some utilize DNA as genetic material, while others have RNA genomes. Viral genomes are not always double-stranded molecules; there are many viruses with single-stranded RNA or DNA genomes. There are viral genomes that consist of a single molecule

of nucleic acid, but some viral genomes are segmented. For example, reoviruses package 11–12 different pieces of double-stranded RNA and each genome segment codes for a different protein.
- Some viruses have lipid envelopes in addition to a genome and protein coat. Viral envelopes are not homogeneous. Different types of host membranes may be utilized to form a viral envelope, and their specific lipid and protein components can differ.
- Virions range in size from 10 to 1000 nm (Fig. 1.3).
- Viral genomes range in size from 3000 nucleotides (nt) to over 1,000,000 nucleotide pairs.
- Outcomes of viral infections are as diverse as are their particles and genomes. Infection does not always result in cell or host death (or even disease).

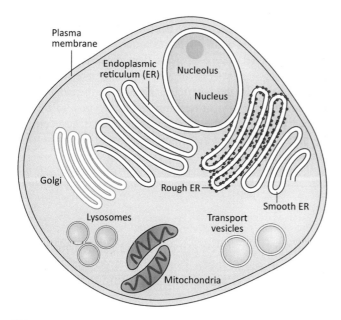

FIGURE 1.2 Simple schematic of a eukaryotic cell identifying some major organelles.

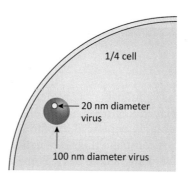

FIGURE 1.3 Relative sizes of an animal cell and virions.

BOX 1.1

Evolution of a Definition: What Is a Virus?

The word "virus" derived from Latin and referred to poisons or noxious liquids. However, by the late 19th century, the term "virus" was used to describe infectious agents that could pass through filters designed to remove bacteria from liquids. Thus early definitions of viruses focused on size: Viruses were biological entities that were "smaller than bacteria." We now know that some viruses that are larger than bacteria, so the trend has been to drop "smallness" from the definition. Another part of the definition of a virus is that they are all obligate intracellular parasites. This is certainly true of all viruses, but there are also bacteria and protozoan parasites that are obligate intracellular parasites. When the biochemical nature of viruses was discovered, it became clear that viruses lack many of the complex structures common to cells. This resulted in definitions of viruses based on comparisons to cells. While these comparisons emphasize the many ways

that viruses are different from cells, they do not help us understand these unique infectious agents. So, what are viruses? They are biological entities comprised of genomes packaged within a protein coats. Viral genomes may be RNA or DNA and may be large or small. The protein coats may be simple or complex but they all function to deliver the viral genomes into permissive host cells. Once inside cells, the viral genomes direct production of new viral genomes and proteins. New viral proteins plus new viral genomes assemble to form new particles or virions, and so the cycle continues. The distinction between the terms "virus" and "virion" can be confusing. In this book I use the term "virus" to generally describe any processes involving the biologic agent (i.e., virus replication-cycle). "Virion" specifically refers to a virus particle, an assembly of proteins and nucleic acids with no metabolic activity.

BOX 1.2

Viruses Can Be Beneficial!

Viruses were first recognized and studied as disease-causing agents. It is only within the last couple of decades that scientists have come to realize that viruses have shaped evolution of all life in fundamental ways and that they can, in fact, be beneficial. One example of the role of viruses in evolution involves the co-option of retroviral envelope proteins in the development of the placenta. The envelope glycoproteins of enveloped viruses are essential for viral entry into host cells. They bind to susceptible cells and induce fusion of the virion envelope with the cell plasma membrane; they are proteins that specialize membrane fusion. Genomic studies have revealed that placental mammals have "domesticated" the envelope glycoproteins of retroviruses to facilitate implantation of embryos into the maternal endometrium, a process central to formation of the placenta. Thus it appears that the evolution of placental mammals required retroviruses!

Another example of the benefits of viral infection involves plants. In a 2007 paper (Márquez et al., 2007) the authors analyzed a type of grass that tolerates high soil temperatures (up to 65°C) in Yellowstone National Park. They showed that the heat tolerant plants were colonized by a fungus that was in turn infected by a virus. In this seminal paper the authors clearly demonstrated that both the fungus AND the virus were necessary to confer heat resistance to the plant.

In fact, some cellular genes are derived from viruses and have played key roles in evolution. Some viruses are highly beneficial to plants (Box 1.2).

- Some viruses complete their replication-cycles in minutes while others take days. Some viruses are transiently associated with an infected host (days or weeks) while others (e.g., herpesviruses) are life-long residents.

Given the diversity described above we might naturally wonder how viruses arose. Did all viruses derive from one ancestor and which came first, the virus or the cell? It is well accepted that there is no single viral ancestor although there certainly are lineages of related viruses. Viruses with RNA genomes certainly evolved independently from DNA viruses. There are three general scenarios for virus evolution that have long been proposed:

- Retrograde evolution: A process in which independent organisms first became intracellular parasites with some metabolic activities, but then lost all ability for independent metabolism, keeping only those genes necessary for replication. Poxviruses are very large complex viruses that *may* have evolved in this manner.
- Origins from cellular DNA and RNA components: This scenario posits that nucleic acids acquired protein coats, and along with them, the ability to be transmitted from cell to cell. Some viral DNA genomes resemble plasmids or episomes. Perhaps such DNA elements acquired protein coats and the ability to be transferred efficiently from cell to cell.
- Descendants of primitive precellular life forms: This scenario posits that viruses originated and evolved along with primitive, self-replicating molecules.

Are viruses alive?

Viruses parasitize every known form of life on this planet and they have both short-term and long-term impacts on their hosts. *But are viruses alive?* This question is the subject of ongoing debate, but the answer does not change the *nature* of a virus. As we discuss and describe viruses it is easy to assume that they are alive. They replicate to increase in number and the terms "virus replication-cycle" and "virus *life-cycle*" are often used interchangeably. Viruses also evolve (change their genomes), sometimes very rapidly, to adapt to new hosts and environments. In contrast, the virion (the physical package that we view with an electron microscope) has no metabolic activity. Some virions can be generated simply by mixing purified genomes and proteins in a test tube. The genomes may have been synthesized by machine and the viral proteins may have been produced in bacteria. If those component parts combine under suitable conditions, a fully infectious virion can be produced. To avoid the question of living versus nonliving, the term "infectious agent" is both appropriate and descriptive. We can then speak of *infectious* virions that are *capable of entering a cell and initiating a replication-cycle*, or *inactivated* virions that cannot complete a replication-cycle. As we will see in later chapters, the difference between an infectious and a noninfectious virion may be as small as the cleavage of a single peptide bond.

Basic steps in the virus replication-cycle

Let's now discuss the basic steps in the virus replication-cycle. The first step is attachment (or binding) to a host cell (Fig. 1.4). Attachment results from very specific interactions between viral proteins and molecules (proteins, lipids, or sugars) present on the surface of the host cell. The interactions are usually hydrophobic and ionic in nature (not involving the formation of covalent bonds). Thus attachment is influenced by environmental conditions such as pH and salt concentration. Attachment becomes stronger as many copies of a viral surface protein (either capsid proteins or envelope proteins) interact with multiple copies of the host cell receptor molecules.

The next step in the virus replication-cycle is penetration of the viral genome into the host cell cytoplasm or nucleoplasm. After penetration, there may be a further rearrangement of viral proteins to release the viral genome, a process called uncoating. Penetration and uncoating are two distinct steps for some viruses, while for others the viral genome is uncoated during the process of penetration. The processes of penetration and uncoating are irreversible, the infecting virion cannot reassemble.

The next phase in the virus replication-cycle involves synthesis of new viral proteins and genomes. These are complex processes that require transcription (synthesis of mRNA), translation (protein synthesis), and genome replication to generate the parts that will assemble into new virions. Synthesis of viral proteins and genomes occurs in close association with, and depends upon, many host cell proteins and structures. The great diversity among viruses will be evident as we examine processes that regulate transcription, translation, genome replication, and the specific virus—host cell interactions that shape these processes.

The next step in the virus replication-cycle is assembly of new virions. New particles assemble from the genome and protein components that accumulate in the infected cell. Viruses are assembled at different sites in host cells; sometime large areas of the cell become "virus factories," concentrated regions of viral proteins and genomes from which host cell organelles are excluded.

The final step(s) in the virus replication-cycle are release from the host cell and maturation of the released virions. Virion release may occur upon cell rupture or lysis. Many enveloped viruses acquire their envelopes from cellular membranes in a process called budding. Some enveloped viruses bud through the plasma membrane, but budding can occur at other intracellular membranes such as nuclear, endoplasmic reticulum, or Golgi membranes. The budding process can, but does not always, kill the host cell. Viruses that obtain their lipid envelopes by budding into cellular vesicles are released when these vesicles fuse with the plasma membrane in a process called exocytosis.

Maturation is the term used to describe changes in virus structure that occur after a virus is released from the host cell. Maturation may be required before a virus is able to infect a new cell. Maturation may involve cleavage or rearrangement of one or more viral proteins when the virion is in the extracellular environment. The necessity for maturation can be explained as follows: Virions assemble within cells under conditions of favorable energy but when the released virions encounter new cells they must be able to disassemble (a process called uncoating). Maturation events that occur after virion release *set the stage* for a productive encounter with the next cell. Maturation processes are well understood for several important animal viruses and examples will be presented in future chapters.

It is important to stress that each step in a virus replication-cycle requires specific interactions between viral proteins and host cell proteins. Some viruses can infect many different cell types and organisms because they interact with proteins found on, and in, many cell types. These viruses are said to have a broad host range. Other viruses have a very narrow host range

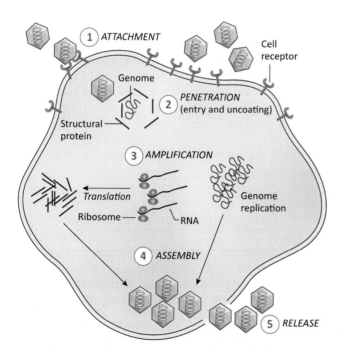

FIGURE 1.4 The basic virus life-cycle is shown in a generic cell. (For simplicity no cellular organelles are shown but the processes of virus replication are in fact intimately associated with cell organelles and structures.) The basic virus life-cycle begins with: (1) Attachment of the virion to receptors on a cell. (2) The genome is delivered into cytoplasm (penetration). (3) Viral proteins and nucleic acids are synthesized (amplification). (4) Genomes and proteins assemble to form new virions. (5) Virions are released from the cell.

due to their need to interact with specific cellular proteins that are expressed only in a few cell types. Factors that impact virus replication include the presence or absence of receptors on the host cell surface, the metabolic state of the cell and the presence or absence of any number of intracellular proteins required to complete the virus replication-cycle.

Another way to view the replication-cycle of a virus is the one-step growth curve (Fig. 1.5). This is a graphical representation of the events occurring when virions are added to susceptible cells in a flask, and a single replication-cycle is allowed to proceed. The one-step growth curve illustrates the concept that penetration of a virus into the host cell is not reversible. The x-axis of the graph depicts time after infection and the y-axis shows the number of extracellular virions at a given time. Soon after virions are added to cells they begin to attach to receptors and the number of free virions *decreases*. As virions enter cells we observe the so-called eclipse phase when infectious virions *cannot be detected*, even if cells are broken open (lysed). There are virtually no infectious particles to be found during the eclipse phase! At this point in the virus replication-cycle the capsid proteins have dissociated from the viral genome. Even though no infectious virions are present, the virus is actively being replicated. Viral proteins and genomes are being synthesized to high levels and can be detected in the cells using various molecular techniques. In the presence of sufficiently high levels of viral proteins and genomes the assembly process begins. In the case of some viruses, infectious particles can be released by experimental lysis of the cells. This is depicted by the blue dashed line in

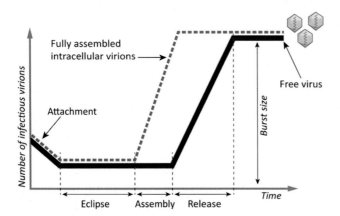

FIGURE 1.5 One-step virus growth curve. The red curve represents infectious virions released from the infected cells. The blue curve represents infectious virions released if the cells are lysed. Key to understanding the one-step growth curve is to note that after attachment, the number of virions detected in media and within cells *decreases*. These virions have penetrated cells and their genomes have uncoated, thus they are no longer "infectious." New virions are detected only after amplification and assembly.

Fig. 1.5. The burst size shown on the graph is a measure of the number of infectious virions released from an infected cell.

Propagating viruses

Viruses are obligate intracellular parasites; they replicate only within living cells. Thus in the laboratory, susceptible cells or organisms are required to study virus replication. For the virologist, ideal host cells are easily grown and maintained in the laboratory. Animal virologists often use cell and (less often) organ cultures. To culture animal cells, tissues or organs are harvested and disrupted (using mechanical and enzymatic methods) to obtain individual cells. Cells circulating in the blood, such as lymphocytes, can be obtained directly from animal blood samples. Often cultured cells are derived from tumors. If cells are provided with the appropriate environmental conditions (growth media, temperature, pH, and CO_2), they will remain metabolically active and may undergo cell divisions.

Historically the best-studied viruses are those that have been adapted for robust growth in a culture system. However, cell or organ cultures are quite different from the natural environment of a human or animal host. The biggest difference is that the cultured cells lack the many antiviral defenses encountered in an organism. It is not uncommon for a virus that is highly adapted to cultured cells to perform poorly when used to infect an animal. In fact, propagation in culture is a common method for producing attenuated (weakened) live viral vaccines. Attenuated viruses replicate in a host, but do not cause disease. When considering experiments with viruses, it is very important to understand both the host system and the origins of the virus being studied.

Categorizing viruses (taxonomy)

The most widely accepted general method to group viruses is by the type of nucleic acid (RNA or DNA) that serves as the viral genome. Within this scheme, there are three major groups of viruses:

- DNA viruses: These viruses package DNA genomes that are synthesized by a DNA-dependent DNA polymerase.
- RNA viruses: These viruses package RNA genomes that are synthesized by an RNA-dependent RNA polymerase (RdRp).
- The third group of viruses use the enzyme reverse transcriptase (RT) during the replication-cycle. RT is an RNA-dependent DNA polymerase that

synthesizes a DNA copy of an RNA molecule. Reverse transcribing viruses (examples are the retroviruses and hepadnaviruses) use both RNA and DNA versions of their genomes (at different times) during their replication-cycle.

DNA and RNA viruses are further differentiated by the physical makeup of their genomes (single stranded, double stranded, unsegmented, segmented, linear, circular). The importance of genome type, and how it influences virus replication will be covered in upcoming chapters. In addition to genome type, other physical traits were historically used to subdivide viruses into smaller groups. Some viruses have lipid envelopes (enveloped viruses) while others do not (naked viruses). Capsids also come in different shapes and sizes.

A formal viral taxonomy has been developed and curated by groups of expert virologists from around the world who volunteer to serve on the International Committee on the Taxonomy of Viruses (ICTV). Anyone can visit the ICTV website at https://ictv. global/taxonomy/ to find the most recent virus classification schemes. The site also provides a helpful history of virus names. The goal of viral taxonomy is to group and categorize viruses in a manner that focuses on their evolutionary relationships. For many years the ICTV, borrowing nomenclature from the Linnaean classification system, used the categories: order, family, genera, and species (Fig. 1.6). Orders contain two or more related families, and families are subdivided into multiple genera. A genus is further subdivided into species (or strains). Viruses in the same family are considerably more closely related to one another than to viruses from different families. Placement of viruses into families was often accomplished by examining shared characteristics such as genome type, presence or absence of an envelope, shape of the capsid, arrangement of genes on the viral genome, etc., and viruses within a family share a core set of properties. Thus if one knows the major characteristics of any single member of the family *Picornaviridae* (e.g., poliovirus), one knows the genome type, general genome organization, approximate size, and shape of all picornaviruses. One needs only to learn the characteristics of a handful of virus families, rather than thousands of individual viruses. The later chapters of this textbook are organized by virus family.

Before it was possible to generate genome sequences quickly and cheaply, classifying viruses was often done using phenotypic traits such as host range, or tissue tropism. Now it is standard practice to use genome sequences to categorize or classify viruses. Genome sequences provide detailed and objective criteria to subdivide viruses into related groups. Genome sequences from

Viral Species
May contain **strains** that differ by up to 10% at the nucleotide sequence level

Viral Genus
Group of similar species constitutes a genus. Members share nucleotide homology and are likely serologically related

Virus Family
Members share a core set of properties including genome type, presence/absence of an envelope and genome organization

Orders
Related families may be grouped into orders

FIGURE 1.6 Viral taxonomy is based on groups of characteristics such as genome type, genome organization, capsid structure, presence/absence of an envelope. The virus family is often considered the focal point of virus taxonomy. Viruses in a family share genome type, overall genome organization, size, and shape. Related families can be grouped into orders. Families are also subdivided into smaller groups of more closely related viruses (genera) within the family. A genus can contain a number of different species or strains. These may differ by up to 10% at the nucleotide sequence level. Closely related strains may sometimes be quite phenotypically distinct.

many different viruses can be compared to generate phylogenies that provide a visual "map" of relationships among viruses (Fig. 1.7). In some cases, many thousands of viral genome sequences are compared in order to generate detailed phylogenies. Such is the case with the human immunodeficiency virus (HIV). The recent explosion in viral genome sequence data has necessitated extensive taxonomic changes in some virus families. For example, until a few years ago the site of infection (respiratory vs enteric) was used as a criterion to define genera within the family *Picornaviridae*. However, a phylogeny based on genome sequences does not split the picornaviruses cleanly along these lines. Thus the family *Picornaviridae* still contains the genus *Enterovirus*, but there is no longer a genus *Rhinovirus*, although you will see frequent reference to it in older literature. Genome sequencing is also expediting the discovery of hundreds of new viruses each year. Many of these have not been propagated in the laboratory but their genomes have all the hallmarks of replication competent viruses and they are included in ICTV taxonomy. The discovery of so many new viruses has enabled and prompted the ICTV to expand its taxonomic system beyond the initial groups mentioned above. In 2021 the ICTV released a report (International Committee on Taxonomy of Viruses

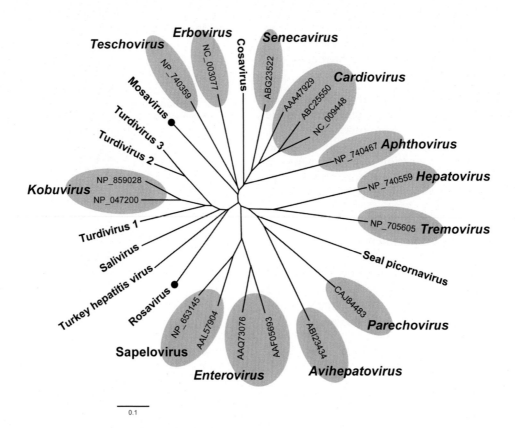

FIGURE 1.7 Phylogeny of the family *Picornaviridae* based on amino acid sequences of complete 3D proteins of all taxonomic genera in the family. In some cases a genus contains only one virus isolate or strain. Source: *From Phan, T.G., Kapusinszky, B., Wang, C., Rose, R.K., Lipton, H. L., et al., 2011. The fecal viral flora of wild rodents. PLoS. Pathog. 7 (9), e1002218. doi:10.1371/ journal.ppat.1002218.*

Executive Committee, 2020) that provides 15 ranks (8 principal and 7 derivative). The principal ranks include: realm, kingdom, phylum, class, order, family, genus, and species. The rationale for the new ranks is to allow an examination of deep evolutionary relationships among viruses. The ICTV currently recognizes six realms, based on similarities of their genes and proteins. The members of each realm are hypothesized to have arisen from a different common ancestor. In 2020 the ICTV recognized the following realms: *Riboviria, Duplodnaviria, Monodnaviria, Varidnaviria, Adnaviria,* and *Ribozyviria.* The characteristics that define each realm are briefly described in Table 1.1.

Alternatives to ICTV taxonomy are sometimes used to group viruses that share common phenotypic characteristics. Hepatitis viruses are so named because they share the phenotype of replicating in the liver. However, human hepatitis A virus (HAV), human hepatitis B virus (HBV), and the human hepatitis C virus (HCV) belong to three different virus families, and vaccines and antiviral treatments developed for one are not necessarily effective for treating, or preventing, infection with the others. Another common phenotypic grouping is use of the term arbovirus (meaning arthropod-borne virus) to describe viruses

that are transmitted by insects. Members of many different virus families can properly be called arboviruses; the term does not imply genetic relatedness among the diverse members of this "group."

You might ask if it is useful to generate or understand, phylogenies of viruses. The answer is a resounding yes. For example, the origins of a disease outbreak can be determined using detailed genetic information. Information from genome sequencing can be used to analyze past outbreaks and track the transmission of viruses from one person or animal to another in order to determine the best methods to curb virus transmission during an epidemic. Finally, understanding the deep evolutionary relatedness among seemingly diverse virus families may provide a window into the development of novel, broad acting antiviral compounds.

A final note on viral taxonomy is important. Throughout this book we will describe particular viruses, showing their overall genome organization and describing the physical structures of their genomes and capsids. However, just as there is genetic variation within a species of animal or plant, there is genetic variation within a given species of virus. We will discuss HIV in some detail later in this text and it will become clear that the term HIV does not describe a single genome sequence, but rather a somewhat diverse group of genomes that are highly related to one another. An experienced virologist can look at the overall sequence and organization of a genome and place it within a particular genus and species, even though its genetic sequence might not be identical to other members of that genus and species. If you find this a confusing concept, simply look around at the individuals in your classroom, and it should become clear.

Outcomes of viral infection

Viral infection impacts individual cells and these cellular changes may or may not noticeably influence the health and fitness of an organism. To understand the possible outcomes of a viral infection we start by examining infection at the level of an individual cell. There are four general possible outcomes when a virus encounters a cell:

- Productive or permissive infection. Viral proteins and nucleic acids are synthesized and virions are assembled and released. The released virions can in turn infect additional cells.
- Nonpermissive infection. The cell is completely resistant to infection. (See Box 1.3 for more insight into the concept of permissive vs nonpermissive cells.)
- Abortive or nonproductive infection. The virus enters the cell, but replication becomes irreversibly blocked at some step before particles are produced.

TABLE 1.1 Six major virus lineages described by the ICTV.

Realm name	Description	Number of recognized families[a]	Number of recognized species
Riboviria	Viruses that use an RNA-directed polymerase to replicate	103 (Including Reoviridae, Hepeviridae, Togaviridae, Flaviviridae, Bornaviridae, Filoviridae, Paramyxoviridae, Pneumoviridae, Rhabdovirdae, Arenaviridae, Hantaviridae, Orthomyxoviridae, Coronaviridae, Artiriviridae, Caliciviridae, Picornaviridae, Astroviridae, Hepadnaviridae, Retroviridae)	3850
Duplodnaviria	Double-stranded DNA viruses whose genomes encode a major capsid protein containing the HK97 fold (HK97-MCP). Viruses of this realm also share a portal protein and terminase complex.	17 (Including Herpesviridae)	2944
Monodnaviria	Single-stranded DNA viruses encoding a rolling-circle replication endonuclease of the HUH superfamily (or if it is derived from such a virus).	17 (Including Circoviridae, Parvoviridae, Polyomaviridae, Papillomaviridae, Anelloviridae Genomoviridae)	1416
Varidnaviria	DNA viruses that encodes a major capsid protein containing the vertical jelly-roll fold and forming pseudohexameric capsomers. (Note: Includes viruses demonstrated to have evolved from members of Varidnaviria that have lost the major capsid protein.)	17 (Including Adenoviridae, Asfarviridae Poxviridae)	268
Adnaviria	Filament-shaped viruses that infect archaea and share a shell structure	3	31
Ribozyviria	Small RNA viruses that have ribozyme activity. Includes hepatitis delta virus (HDV)	1	15

[a]Families discussed in this book.

BOX 1.3

Permissive or Not?

Some viruses replicate very poorly when first introduced into cultured cells. There may be no visible signs of virus infection, but upon prolonged incubation or "blind passage" (often over a period of weeks or months) the virus will adapt to the new environment. This "cell culture adapted" virus now grows well in cultured cells. Therefore the initial virus infection was permissive, although very poorly so. After becoming adapted to cell culture conditions, the virus may be attenuated (replicate poorly or become incapable of causing disease) in the animal host.

Latent infection. Describes a situation where a viral genome is present in the cell, but no or only a few viral proteins are produced. Latency implies that the virus can productively replicate given the right conditions (Box 1.4).

Note that the outcomes listed above focus on the fate of the *virus*, not the *cell*, as both productive and nonproductive infections can impact the cell in different ways. At the cellular level, the effects of infection can range from no apparent change to cell death. Infected cells may not die, but may take on altered morphology. Another virally induced change to a cell can be transformation or immortalization of the cell. Transformation is accompanied by major changes in the morphology of cells and/or their ability to divide. Immortalized cells divide continuously given the appropriate media and conditions. Many of the cells frequently used in biomedical research are immortal (though not all immortalized cells arise from virus infection). In the most extreme cases transformation may allow cells to form tumors in infected hosts.

Productive infections often result in cell death (often called lytic or cytopathic infections), but this is not always the case. Some viruses can productively replicate *without* damaging the cell. This is called an inapparent infection as no effects are seen on the cultured cells in which the virus is replicating. Viruses that cause inapparent infections are often produced in small amounts for the life of the cell. Sometimes an inapparent infection results from latency. A much less frequent outcome of infection is transformation or immortalization that allows the cell to divide without restriction. Immortalized cells may be productively infected (virus is released) or the condition may result from a nonproductive infection.

In the preceding sections we learned that cells can be inapparently infected by a virus. Inapparent infection also occurs at the level of the animal host as viruses may replicate in hosts without causing any disease. After all, the "job" of a virus is replicate and infect another host; disease is not a required side effect. Until very recently it was hard to find viruses that caused inapparent infections. But many inapparent infections are now being identified through large-scale sequencing of host nucleic acids. It turns out that viruses that cause obvious diseases of their hosts may just be "the tip of the iceberg."

Disease is defined as damage to tissues or organs. Many viral infections do cause disease, and diseases can be described as acute, chronic, or latent (Fig. 1.8). Acute disease has a rapid onset, lasts from days to months, and the virus is either controlled or cleared, or causes death of the host. There are many examples of acute viral diseases, the common cold being one and COVID-19 being another. From a public health standpoint, it is important to know that virus replication and spread may begin well before symptoms develop and virus may be shed for days or weeks after symptoms have resolved. The peak of clinical signs and symptoms may or may not correspond to peak virus titers, or the time of maximum transmissibility from one host to another. SARS-CoV-2 and COVID-19 disease will be discussed in detail elsewhere but it is important to note here that one reason this virus spread so widely and destructively is that early in the pandemic it was not understood that asymptomatic individuals could efficiently spread this virus.

Chronic viral infections have a slower progression and the time to resolution is years to a lifetime. These viral infections may, but do not always, lead to death of the host. Chronic infections are also called persistent infections. Virus is produced and shed continuously (albeit sometimes at very low levels). Examples of viruses that may cause chronic or persistent infections of humans are HCV, HBV, and HIV. It should be noted that a chronic viral infection can be without symptoms (inapparent) for years.

Latent infection describes the maintenance of a viral genome without the production of detectable virus.

BOX 1.4

Latent versus Chronic Infections: Where Is the Boundary?

A latent infection is one in which viral genomes are present in cells but virions are not produced. The term chronic infection describes one where virions can be routinely detected. Thus the sensitivity of the assays used for virus detection becomes an important factor in the distinction. As virus detection methods become more sensitive, the distinction between latency and chronic infection has become blurred. Consider genital herpes, caused by human herpesviruses 1 and 2 (HHV1 and 2). These viruses are abundant in visible lesions but also can be transmitted when there are no visible lesions. So is the infection latent or is it chronic? How often are the HHVs found on the skin in the absence of lesions? How often must a latent virus reactivate before an infection is considered chronic? From a public health standpoint calling genital herpes, a chronic infection might better convey the fact that herpesvirus can be transmitted in the absence of lesions.

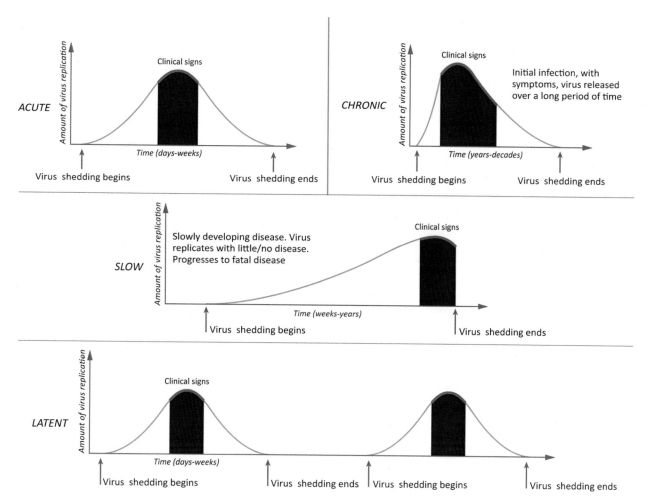

FIGURE 1.8 Outcomes of viral infection at the level of the animal host are quite variable. The red areas under the curves depict periods of clinical disease. In the examples depicted here, virus shedding begins before the onset of symptoms and ends after symptoms have resolved. Note that periods of virus shedding vary. Shedding may begin at the time of onset of clinical symptom and may end prior to the resolution of disease. During latent infection, there may be intermittent virus shedding without clinical symptoms.

Herpesviruses are a good example of viruses that cause latent infections. The chickenpox/shingles virus, formerly known as varicella-zoster virus, but recently renamed human herpesvirus 3 (HHV3) is an instructive example. Prior to 1995 chickenpox was a common childhood infection in the United States. Chickenpox infection is usually mild, characterized by blister-like pustules that resolve in about a week. However, HHV3 remains in the body long after the pustules have disappeared. HHV3 genomes are silently maintained in neurons, for decades. Shingles, a very painful and debilitating disease of adults, occurs when HHV3 exits latency and travels down neurons to the skin to produce blister-like lesions. These lesions contain infectious virus, thus a person with shingles can transmit chickenpox to a nonimmune person. HHV3 reactivates (breaks out of latency) when the host's immune system is impaired (by advancing age or stress, for

example). Shingles vaccines are now available that can boost immune responses to HHV3, reducing the likelihood of virus reactivation.

Introduction to viral pathogenesis

Viral pathogenesis is defined as the mechanism by which viruses cause disease. A simple view of viral pathogenesis is that viruses replicate and kill cells, thus causing disease. For example, death of liver cells (hepatocytes) causes hepatitis, death of enterocytes may cause diarrhea, death of respiratory epithelial cells may cause severe respiratory tract disease. However, loss of cell function, without cell death, can also produce disease. During HIV infection, immunodeficiency is not simply caused by cell death; the virus also alters the functions of cells needed to maintain a healthy immune system.

Signs and symptoms of disease can also result from tissue damage caused by host immune responses. Inflammation, killing of virus-infected cells by the immune system, or deposition of immune complexes are examples. Of course, like any biological event, disease is often a complex combination of direct damage by virus in concert with host immune responses. Understanding viral pathogenesis, the mechanism by which disease develops, is an important consideration in developing effective treatments. For example, SARS-CoV-2 has been shown to cause generalized and damaging inflammatory responses, thus in cases of severe COVID-19 one treatment is low-dose dexamethasone. Dexamethasone is a corticosteroid used in a wide range of conditions for its antiinflammatory and immunosuppressant effects.

Introduction to viral transmission

How are viruses transmitted from one animal to another? Common routes of infection include:

- fecal–oral,
- respiratory droplets,
- contact with contaminated fomites,
- exchange of infected bodily fluids, tissues, or organs,
- airborne,
- insect vectors.

Fecal–oral transmission occurs via ingestion of contaminated food or water. Virus enters the body through epithelial cells or lymphoid tissues in the gastrointestinal tract. Examples include rotaviruses and the Norwalk-like viruses (noroviruses). Noroviruses have caused notable outbreaks on cruise ships, sickening hundreds of guests and crew in a matter of days. Human HAV is also transmitted by the fecal–oral route via contaminated produce or uncooked shellfish. Fomites (objects contaminated with infectious organisms) can also play a role in fecal–oral transmission.

Respiratory transmission occurs when viruses are released from the respiratory tract as droplets or very fine particles. Virus may be inhaled, or infection may occur through contact with fomites contaminated by respiratory secretions hence the advice to wash your hands often! Viruses expelled from the respiratory tract may also be transmitted via mucosal surfaces such as the eye. This is why some heath care workers wear face shields when caring for patients with infectious diseases. Examples of viruses that can be spread by the respiratory route are influenza viruses, rhinoviruses (one of the common cold viruses), and SARS-CoV-2.

A few viruses, such as foot-and-mouth disease virus of livestock, can be transmitted over long distances through the air, a process called airborne transmission. Measles virus and SARS-CoV-2 are also known for airborne transmission. Simply sitting in a room with a measles-infected individual can lead to infection! It should be noted that airborne transmission is distinct from aerosol transmission. In airborne transmission, particle sizes are very small and the particles remain suspended in the air for long periods. The importance of understanding the distinction between these two types of transmission is exemplified by the 2014 Ebola virus epidemic. Ebola virus is most often transmitted through contact with body fluids of an infected individual. Transmission occurs when the patient is clearly symptomatic and virus titers are highest. Ebola virus can also be transmitted via large respiratory droplets or aerosols, but not by the tiny droplets that remain airborne for long periods, thus Ebola is not considered an "airborne" virus. Those not in close proximity to the infected patient are not at high risk of infection.

Transmission of viruses via exchange of bodily fluids can result from blood transfusions, use of dirty needles, trauma (bleeding), organ or tissue transplantation, sexual contact, or artificial insemination. HIV, HBV, and HCV are all transmitted via contaminated blood. But these viruses can also be transmitted through contact with other bodily fluids such as semen or saliva. HIV can also be transmitted via breast milk. Rabies virus is transmitted by saliva, often as the result of a bite.

Many viruses (e.g., West Nile virus, the equine encephalitis viruses, dengue virus, chikungunya virus, and zika virus) are transmitted from one host to another primarily via an *insect* intermediary. Blood-feeding insects such as mosquitos, ticks, and midges are common vectors. Viruses transmitted by insect vectors are collectively called arboviruses.

It should be emphasized that a virus can be transmitted by more than one route. SARS coronavirus 1 (SARS CoV-1), considered primarily a respiratory virus, is also transmitted by the fecal–oral route.

Blood transfusions, organ transplants, or even dirty needles may facilitate transmission of viruses usually spread by other routes. Mucosal surfaces, such as the eye, can be entry points for transmission of virus in present in blood or other bodily fluids. Some mosquito-vectored viruses (West Nile, chikungunya, yellow fever, and equine encephalitis viruses) require special precautions to avoid transmission in a research setting, where these viruses can be transmitted via aerosols.

Finally, a discussion of virus transmission should also include brief mention of virus transmissibility, a topic in the news during the SARS-CoV-2 pandemic that started in 2019. Transmissibility is the ease of virus spread from one host to another. Measles virus is highly transmissible by the airborne route, and outbreaks can quickly become widespread in nonimmune populations. As we have seen recently, SARS-CoV-2 is highly transmissible and variants with increased transmissibility have emerged during the pandemic. Increased transmissibility may occur when a virus binds more tightly to receptors, evolves to bind new receptors or has gained the ability to replicate to higher levels in an infected patient. However, transmissibility is not related to the ability of a virus to cause disease (virulence). A virus may be relatively difficult to transmit, but highly virulent if transmission does occur. It is easy to overestimate the transmissibility of a highly virulent virus.

In this chapter we have learned that:

- Viruses are infectious agents (but are not cells).
- Viruses are obligate intracellular parasites that require host cells for their replication.
- Virions are the packages that contain the viral genome.
- Virions assemble from viral proteins and genomes synthesized within the infected cell.
- In the laboratory viruses are grown in cell or organ cultures.

- Viruses can change or adapt to new growth conditions.
- Viruses have different genome types, capsid types, routes of infection, and diverse interactions with host cells.
- Virus infection may, but does not always, lead to cell death or host disease.
- Virus infection may be relatively short lived (acute infections) or may be life-long (chronic or persistent).
- Different viruses are transmitted by different routes (respiratory, fecal oral, exchange of bodily fluids).

References

International Committee on Taxonomy of Viruses Executive Committee, 2020. The new scope of virus taxonomy: partitioning the virosphere into 15 hierarchical ranks. Nat. Microbiol. 5 (5), 668–674. Available from: https://doi.org/10.1038/s41564-020-0709-x. Epub 2020 Apr 27. PMID: 32341570; PMCID: PMC7186216.

Márquez, L.M., Redman, R.S., Rodriguez, R.J., Roossinck, M.J., 2007. A virus in a fungus in a plant: three-way symbiosis required for thermal tolerance. Science 315 (5811), 513–515 Available from: https://doi.org/10.1126/science.1136237.

2

Virus structure

After studying this chapter, you should be able to:

- Define "capsid" and explain its functions.
- Define "nucleocapsid," "envelope," and "envelope protein."
- Describe the differences between "structural" and "nonstructural" proteins.
- Be able to distinguish between icosahedral and helical capsids.
- Indicate the twofold, threefold, and fivefold axes of symmetry on an icosahedral capsid.
- Describe the functions and activities of envelope glycoproteins.
- Describe the location and a common function of matrix (MA) proteins of enveloped viruses.

Anatomy of a virus

The simplest viruses consist of a genome packaged in a protein shell or capsid (Fig. 2.1). Capsids are assembled from many copies of a single, or a few types of capsid proteins. Some viruses have a lipid bilayer, called the envelope, surrounding their capsids (Fig. 2.1). Envelopes may be derived from the cell plasma membrane, nuclear membrane, or other intracellular membranes. All enveloped viruses encode proteins that are associated with the lipid bilayer. Envelope proteins are usually glycosylated (thus are envelope *glyco*proteins) and often contain transmembrane (TM) anchoring domains. They often project out from the surface of the envelope forming distinct spikes. Many enveloped viruses have a matrix protein positioned inside, and associated with the envelope. Matrix proteins may interact directly with the lipid membrane or may bind to the cytosolic tails of envelope glycoproteins. Matrix proteins often form a link between the membrane and the viral nucleocapsid (Fig. 2.1). The term nucleocapsid refers to a complex of viral nucleic acid and protein. The term is most often used to refer to the assemblage of protein and nucleic acid within an *enveloped* virus. If viral envelopes are gently lysed, nucleocapsids are released.

The proteins that assemble to form the virion (the extracellular particle) are called *structural proteins*. Additional proteins may be encoded by a virus, but may be absent from the virion. These so-called *nonstructural proteins* have a variety of functions in the virus replication cycle. For example, nonstructural proteins may block cell and host antiviral responses; others are enzymes such as proteases or polymerases. It is important to note: nonstructural does not mean nonfunctional, or unimportant. Most nonstructural proteins are in fact essential for virus replication.

Viruses.
DOI: https://doi.org/10.1016/B978-0-323-90385-1.00008-X

© 2023 Elsevier Inc. All rights reserved.

Capsid structure and function

In the simplest terms, viral capsids are protein packages that protect the genome. However capsids should not be considered static boxes, as they are dynamic structures that have other important functions. In addition to simply providing protective "packaging,"

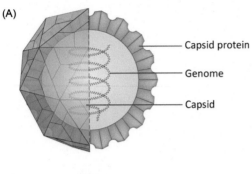

(A)

— Capsid protein

— Genome

— Capsid

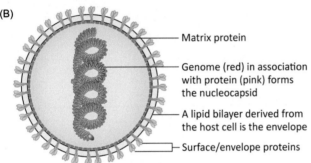

(B)

— Matrix protein

— Genome (red) in association with protein (pink) forms the nucleocapsid

— A lipid bilayer derived from the host cell is the envelope

— Surface/envelope proteins

FIGURE 2.1 Basic structures of an unenveloped (naked) virus (panel A) and an enveloped virus (panel B).

the capsids of unenveloped viruses mediate attachment to, and penetration into, the host cell. Capsids must also be able to specifically package viral genomes and direct budding or release from cells.

Capsids come in two basic shapes: helical (rod shaped) or icosahedral (spherical). The simplest capsids are assemblies of many copies of a single protein (often called the capsid protein). As capsids assemble, they are stabilized by the repeated interactions (largely electrostatic) between capsid protein building blocks. It should come as no surprise that there are few covalent bonds between these building blocks, because the genome must be released from the capsid at a later time!

The repeated occurrence of similar protein—protein interfaces leads to construction of a symmetrical capsid. The simple helical capsid of tobacco mosaic virus (TMV), a plant virus, is assembled from many copies of a single capsid protein. Each capsid protein forms identical side-to-side and top to bottom interactions with its neighbors, as indicated in Fig. 2.2. In addition to interacting with neighboring proteins, each TMV capsid protein interacts with three nucleotides of the viral (RNA) genome. The capsid proteins are tightly packed around the RNA and form a rigid rod whose length is determined by the length of the genome. Not all helical capsids are rigid rods, many enveloped animal viruses have very flexible, helical nucleocapsids surrounded by an envelope.

Viruses with spherical capsids have icosahedral symmetry. An icosahedron is a closed cube with twofold, threefold, and fivefold axes of symmetry (Fig. 2.2). The simplest icosahedron can be assembled from 20

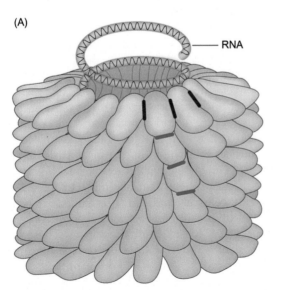

(A)

— RNA

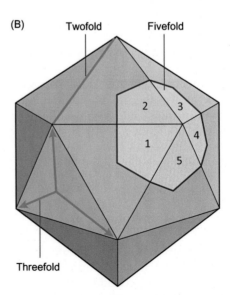

(B)

Twofold   Fivefold

2   3

1   4

5

Threefold

FIGURE 2.2 Capsids come in two basic shapes: helices (rods shown on left) and spherical particles with icosahedral symmetry (shown on right). (A) The red and blue bars on the helical virus indicate sites of repeating contacts between subunits. Red bars indicate side-to-side contacts. Blue bars indicated top to bottom contacts. (B) The twofold, threefold, and fivefold axes symmetry are indicated for this simple icosahedral capsid.

equilateral triangles. More complex, and larger, icosahedra can be built by assembling more than 20 triangular subunits.

Capsids that vary from the definitions of a helix or an icosahedron are sometimes called "complex." For example, there are viruses of bacteria (bacteriophage) with icosahedral "heads," rod like (helical) "tails" and long "fibers" extending from the tails.

Capsids are built using many copies of one or a few types of protein

Biological constraints require that capsids be assembled from multiple copies of one, or a few, small (usually in the range of 20–60 kDa) proteins. A key consideration of capsid assembly is the relative size of an amino acid and the triplet codon for that amino acid. The average size of an amino acid is 110 daltons (Da) while the average size of a triplet codon is $\sim 330\,\text{Da} \times 3$, or nearly 1000 Da. Thus a polypeptide is always physically smaller than the gene that encodes it. Luckily, translation produces many copies of protein from each mRNA. Thus it is easy to produce dozens of capsid proteins that can assemble into a shell to package a viral genome.

Another biological constraint faced by capsid proteins is their ability to fold. Small polypeptides can fold tightly while larger ones are unable to fold without leaving gaps in the structure. Tightly folded proteins are better at protecting their nucleic acid cargo as gaps could render the viral genome susceptible to environmental damage. Larger proteins also need chaperones to help them fold correctly. Indeed, there are some viral capsids that assemble with the help of chaperones.

Fidelity of protein synthesis is another issue that constrains the size of a capsid protein. The observed error rate of protein synthesis is $\sim 10^{-4}$, or 1 mistake per 1000 amino acids polymerized. Thus on average, a 1000 amino acid long protein would contain one mistake. In contrast, synthesis of 10 copies of a 100 amino acid protein would produce, on average, 9 perfect proteins and 1 with a mistake that could be excluded from the virion.

Simple icosahedral capsids

The most common icosahedral capsids can be envisioned as an arrangement of three capsid proteins into a triangle (Fig. 2.3A). As the simplest icosahedron can be assembled from 20 equilateral triangles, the simplest icosahedral capsid can be drawn with 20 triangular faces, each containing 3 capsid proteins on each face. This structure is called a $T=1$ capsid, and T is the triangulation number (Box 2.1). A $T=1$ icosahedron is an assembly of 20 triangular faces with 3 capsid proteins per face for a total of 60 capsid proteins in the shell. A $T=3$ icosahedron has 3×20 or 60 triangular faces (with 3 capsid proteins per face for a total of 180 capsid proteins), a $T=4$ icosahedron has 4×20 or 80 triangular faces (built from 240 capsid proteins) and so on. The T number can be used to determine the *theoretical* number of capsid proteins required to assemble a shell, if each triangular face is constructed from three capsid proteins. It follows that larger viruses have higher T numbers, using more total capsid proteins for shell assembly. A $T=1$ virus [e.g., a parvovirus (Family *Parvoviridae*)] has a shell assembled from 60 capsid proteins. Caliciviruses (Family *Caliciviridae*) have

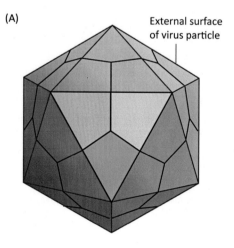

(A)

External surface of virus particle

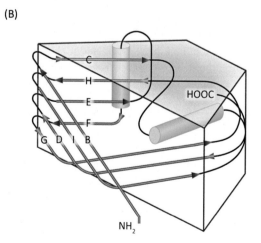

(B)

FIGURE 2.3 (A) The icosahedral capsids of most animal viruses are assembled from proteins that have a common shape: an eight-stranded jelly roll β-barrel motif. (B) Each triangular "face" of the capsid is assembled from three copies of an eight-stranded jelly roll β-barrel motif.

$T = 3$ capsids, assembled from 180 molecules of a capsid proteins. Togaviruses (Family *Togaviridae*) have $T = 4$ capsids are assembled from 240 capsid proteins (Table 2.1). Thus larger capsids are assembled by using a larger number of building blocks (Fig. 2.4).

In the previous section the icosahedron was described as an assembly of equilateral triangles. However, capsids are not assembled from "triangular" or even symmetrical, proteins. Among animal viruses, the most common type of capsid protein structure is called an eight-stranded jelly roll β-barrel motif. This structure consists of four pairs of antiparallel β sheets (Fig. 2.3B). Three of these units can be envisioned to form one triangular face of an icosahedral capsid. While many types of viruses use many copies of a single type of capsid protein, there are variations. Picornaviruses (Family *Picornaviridae*) assemble capsids from 180 proteins. However, they use 60 copies each, of three different capsid proteins (with similar structures) to form a shell. There are additional variations that will be discussed below but perhaps the most important "take home" message is that capsids are stabilized by repeating interactions among many copies of one or a few capsid proteins.

How do capsid proteins assemble to form a simple icosahedral shell? Is this an orderly process? The answer is yes, the process is stepwise and sometimes the substructures can be identified. As was noted above, $T = 1$ capsids assemble from 60 capsid proteins. A general model is that three capsid proteins assemble to form a capsomere. Five capsomers form a pentamer and 12 pentamers assemble to form the closed shell. $T = 3$ and $T = 4$ capsids are built with pentamers and hexamers (hexamers are assemblies of six capsid proteins). $T = 3$ capsids assemble from 12 pentamers and 20 hexamers while $T = 4$ viruses assemble from 12 pentamers and 30 hexamers.

BOX 2.1

Triangulation Numbers and Structural Subunits

The T number was first described by Caspar and Klug in 1962 (Caspar and Klug, 1962) to explain the structural basis for icosahedral capsids of different sizes. In their models, the T number predicted the number of capsid proteins in a structure, and their predictions proved true for many small viruses. However, as more icosahedral capsids were analyzed it became apparent that T numbers more often predicted numbers of "structural subunits" rather than numbers of polypeptides in a capsid. The most common structural subunits of icosahedral animal virus capsids are β-barrels (eight-stranded jelly roll β-barrel motifs). This common structure is shown in Fig. 2.3. Capsid proteins of different viruses do not have conserved amino acid sequences; however, they do assume this conserved structure. The distinction between a structural subunit and a polypeptide in building capsids can be demonstrated by comparing the capsids of caliciviruses, picornaviruses, and cowpea mosaic virus. All three virus groups have T = 3 capsid architecture. The calicivirus capsid is assembled from 180 copies of a single capsid (C) protein while picornaviruses are assembled from 60 copies each, of three capsid proteins (VP1, VP2, and VP3). The three picornavirus capsid proteins have different primary amino acid sequences but fold to assume similar structures. Thus the capsids of picornaviruses are assembled from 180 equivalent structural subunits. The T = 3 capsids of picornaviruses are sometimes called pseudo T = 3 or P = 3 capsids as they do not strictly adhere to the original predictions put forth by Caspar and Klug. In the case of cowpea mosaic virus the capsid is assembled from two different polypeptides, 60 copies each of a large (L) and a small (S) capsid protein. However, a close look at capsid architecture shows that L folds into two independent β-barrels connected by a hinge region. Thus the cowpea mosaic virus capsid is assembled from 180 "structural" domains, where the structural unit is the β-barrel.

TABLE 2.1   Construction of simple icosahedral capsids.

T number	Triangular faces/ icosahedron	Number of proteins (structural subunits)/capsid	Number of pentamers + number of hexamers
1	20	60	12 + 0
3	60	180	12 + 20
4	80	240	12 + 30

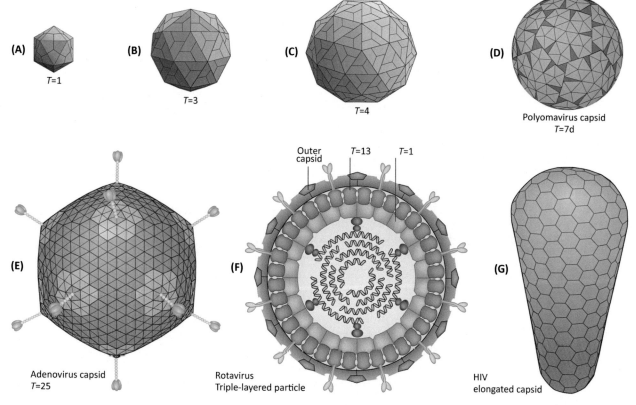

FIGURE 2.4 Examples of large and small icosahedral capsids. (A) T = 1 icosahedral capsid; (B) T = 3 icosahedral capsid; (C) T = 4 icosahedral capsid; (D) T = 7d polyomavirus capsid; (E) T = 25 adenovirus capsid; (F) Triple layer particle of rotavirus contains a T = 1 core surrounded by T = 13 shells. (G) HIV capsid. Elongated icosahedral capsids are formed by adding rows of hexamers to the middle of the structure.

Larger icosahedral capsids

Large icosahedral capsids can also be envisioned as assemblages of triangular faces. For example, one could construct a $T = 7$ icosahedron using 140 equilateral triangles assembled from 420 structural subunits or a $T = 13$ icosahedron from 260 equilateral triangles (780 structural subunits). However, cryo-electron microscopic visualization of viruses with larger icosahedral capsids reveals that some of the larger capsids (described below) are not simply larger versions of a $T = 1$ or $T = 3$ virus.

Polyomavirus and papillomavirus capsids

Based on the predictions of Caspar and Klug, (Caspar and Klug, 1962) the $T = 7$ capsids of polyomaviruses (Family *Polyomaviridae*) and for papillomaviruses (Family *Papillomaviridae*) would contain 420 structural subunits assembled into 12 pentamers and 80 hexamers, but this is not exactly the case. Instead the capsids of polyomaviruses and papillomaviruses are assembled from 360 structural subunits organized as 72 pentamers (Fig. 2.4). These capsids structures are referred to at $T = 7$ d (rather than $T = 7$).

Adenovirus capsids

The large (~95 nm diameter) icosahedral capsids of adenoviruses (Family *Adenoviridae*) contain 12 different polypeptides. Their capsid structure is described as pseudo $T = 25$. According to the predictions of Caspar and Klug the adenovirus capsid should be assembled from 1500 copies of a β-barrel motif type of capsid protein. However, this is not the case. Instead, 720 copies of the major capsid protein (MCP, also called the hexon protein) assemble to form 240 trimers. It turns out that each MCP has 2 structural domains; thus a trimer of MCP proteins has 6 structural domains. Adenovirus capsids also contain 12 pentamers or vertices (Fig. 2.4E). Although pentons and hexons are assembled from different types of capsid proteins the math adds up! There are 240 trimers of MCP that contribute 1440 structural units and 12 pentamers that contribute 60 structural units.

Reoviruses

Reoviruses (Family *Reoviridae*) have multilayered icosahedral capsids. Viruses in the genus *Rotavirus* have triple-layered particles (Fig. 2.4F). The smallest, innermost layer is a $T = 1$ capsid formed from 60 copies of viral protein 2

(VP2). The middle layer is a $T = 13$ capsid assembled from the ~45 kDa VP6. The outmost layer of the rotavirus capsid is assembled from two additional structural proteins, VP4 and VP7. Members of the genus *Reovirus* have two capsids layers, an inner $T = 1$ core surrounded by an outer $T = 13$ shell.

Herpesvirus capsids

Herpesviruses (Family *Herpesviridae*) are large, enveloped viruses with icosahedral capsids. The major capsid protein or MCP is quite large (~1300 amino acids) and folds to assume a structure that is distinct from the eight-stranded jelly roll β-barrel motif of other animal viruses. The MCP assumes a shape very similar to the capsid proteins of double-stranded (ds) DNA bacteriophage such as P22 or HK97. Herpesviral capsids are assembled from a total of 162 capsomers (150 hexons and 12 pentons). The MCP forms all of the hexons and 11 of 12 pentons. The 12th penton is formed from a different herpesvirus protein called the portal protein to form the portal "complex." The portal complex is cylindrical and contains a channel. Bacteriophage portal complexes actively transport DNA into and out of bacteriophage capsids. Herpesvirus portals likely serve a similar function. The structure of herpesvirus capsids suggests an evolutionary relationship between herpesviruses and dsDNA phage. Based on capsid protein structure, herpesviruses are contained in the realm *Duplodnaviria* (ds DNA viruses whose genomes encode a major capsid protein containing the HK97 fold) along with numerous tailed DNA bacteriophage.

Surface structures of capsids

Finally, the surfaces of icosahedral capsids can be quite variable. Some capsids are relatively smooth, while others have prominent projections and/or deep canyons or pits (Fig. 2.5). Electron micrographs of unenveloped icosahedral viruses are shown in Fig. 2.6. Unenveloped viruses are sometimes referred to as "naked" viruses. Attachment proteins of some naked viruses project out from the capsid to engage a receptor (upon examining adenovirus capsids it is not surprising to learn that the long fiber proteins engage host cell receptors). In other cases attachment sites are inside a canyon or pit in the capsid, serving as receptacles for long, thin host cell proteins.

Viral envelopes

Many important human and animal pathogens are enveloped viruses. Their helical or icosahedral capsids are enclosed within a lipid bilayer. Enveloped viruses come in a variety of shapes, ranging from the long, filamentous ebolaviruses (Family *Filoviridae*) to the icosahedral togaviruses (Family *Togaviridae*), to the large, brick-shaped poxviruses (Family *Poxviridae*). Most enveloped viruses acquire their lipid bilayer by budding through a cellular membrane [e.g., plasma membrane, endoplasmic reticulum (ER), Golgi, or nuclear membranes]. However, the poxviruses assemble their membranes from crescent-shaped lipid fragments associated with protein scaffolds.

Enveloped viruses (with the exception of poxviruses) have envelope proteins that are anchored into or across the lipid bilayer. Thus they can be categorized as type I or type II membrane proteins depending on their orientation in the membrane (Fig. 2.7). Types I and II membrane proteins span the lipid bilayer a single time, via an α-helical TM domain. Type I proteins are inserted into membranes with their amino-terminal domain to the outside and their carboxyl-terminal domain to the inside of the cell or

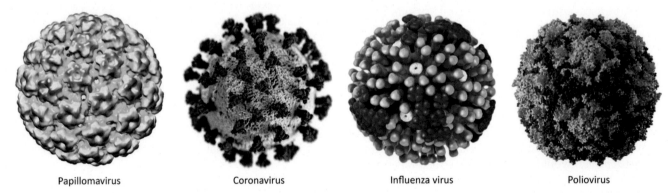

Papillomavirus   Coronavirus   Influenza virus   Poliovirus

FIGURE 2.5   Molecular models of viruses. Papillomavirus, an unenveloped icosahedral virus (https://ftp.wwpdb.org/pub/emdb/structures/EMD-5993/images/emd_5993.tif); coronavirus, an enveloped virus (CDC PHIL 23312, content providers CDC/Alissa Eckert, MSMI; Dan Higgins, MAMS, photo credit Alissa Eckert, MSMI, Dan Higgins, MAMS); influenza virus (CDC PHIL 19013, content provider CDC/Douglas Jordan, photo credit Dan Higgins); poliovirus, an unenveloped icosahedral virus (CDC PHIL 22498, content provider CDC/Sarah Poser, photo credit, Meredith Boyter Newlove).

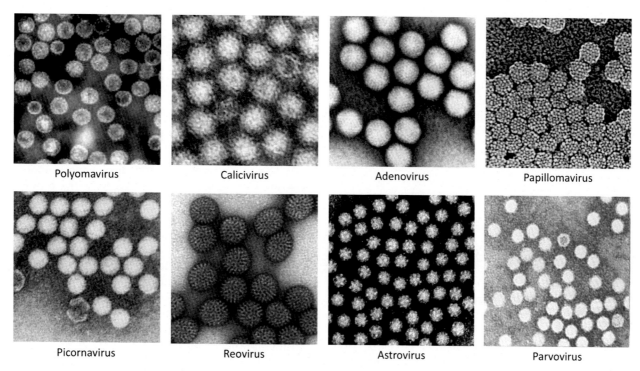

FIGURE 2.6 Examples of unenveloped viruses. Polyomavirus (CDC PHIL 5629, content providers CDC/Dr. Erskine Palmer); calicivirus (CDC PHIL 198, content providers: CDC/Charles D. Humphrey); adenovirus (CDC PHIL 237 content providers CDC/Dr. G. William Gary, Jr.); papillomavirus (From: (Belnap et al., 1996); picornavirus (CDC PHIL 235, content providers CDC/J. J. Esposito; F. A. Murphy); reoviruses (CDC PHIL 197, content providers CDC/Dr. Erskine Palmer); astovirus (Graham Beards at English Wikipedia); parvovirus (CDC PHIL 20040407, content providers: CDC/R. Regnery; E. L. Palmer).

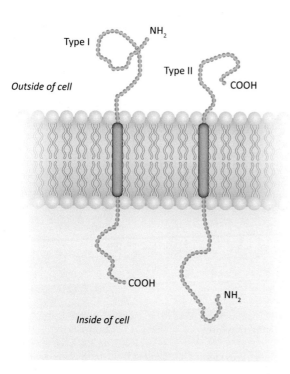

FIGURE 2.7 Membrane arrangement of type I and type II membrane proteins.

virion. Type II membrane proteins have the opposite polarity with respect to the membrane. Envelope proteins often take the form of long spikes. Spikes are usually homo- or heterodimers or trimers and many have distinct globular heads. Examples of viral type I membrane proteins include: influenza virus (Family *Orthomyxoviridae*) hemagglutinin (HA), rabies virus (Family *Rhabdoviridae*) G protein, paramyxovirus (Family *Paramyxoviridae*) fusion proteins, and hepatitis C virus (Family *Flaviviridae*) E1 and E2. Paramyxovirus (Family *Paramyxoviridae*) attachment proteins are type II membrane proteins.

Virus envelope proteins must carry out at least two functions: *receptor binding and membrane fusion*. These activities may be carried out by a single protein (influenza virus HA, rabies virus G) or by two distinct proteins in the case of the paramyxoviruses. In the case of influenza virus, the HA glycoprotein is cleaved by cellular proteases to generate a fusion-active version of the protein but the two halves of HA (HA1 and HA2) remain associated via disulfide bonds. Rabies virus G protein also has both attachment and fusion activities; however, rabies virus G remains uncleaved. Human immunodeficiency virus (HIV), a retrovirus (Family *Retroviridae*), encodes a single glycoprotein precursor that is cleaved during synthesis to produce two distinct structural proteins. The amino-terminal domain of the HIV envelope precursor is the surface unit (SU) protein, which functions in attachment. The carboxyl-terminal half of the HIV envelope precursor is the TM protein. As its name implies, it has a membrane-spanning domain, anchoring it into the envelope. TM is the fusion protein. Paramyxoviruses produce two distinct envelope glycoproteins. Both are membrane-anchored proteins. The attachment protein is a type II integral membrane protein while the fusion protein is a type I integral membrane protein.

Glycosylation

The ectodomains of envelope proteins are usually glycosylated (they have polysaccharides linked to the protein backbone). Carbohydrates can be linked to the peptide backbone via the nitrogen atom on an asparagine side chain (so-called N-linked) or to the oxygen atoms of serine or threonine side chains (O-linked). Glycosylation occurs during transit of envelope proteins through the ER and Golgi. The amount of carbohydrate decorating envelope proteins varies. Some viral envelope proteins have one molecule of polysaccharide per polypeptide chain. In the case of the HIV

SU, the molecular mass of carbohydrate equals that of the polypeptide backbone. HIV SU protein has an apparent molecular weight of 120 kDa (as determined by gel electrophoresis) but the calculated mass of the amino acid backbone is <50 kDa.

Other membrane proteins

A few enveloped viruses also have membrane-associated proteins with other roles in the replication cycle. Influenza viruses have a receptor-destroying protein called neuraminidase (NA) that is required for efficient release of virions from the surface of the infected cell. Drugs that block the enzymatic activity of NA do not block virion formation, but virions remain tightly associated with the infected cell, thereby limiting their ability to infect other cells.

Another role for viral membrane-associated proteins is ion transport. Influenza viruses encode a small integral membrane protein called M2. M2 assembles into a low pH-activated ion channel that facilitates transport of H^+ ions across the viral envelope thereby acidifying the core during virus entry. Fig. 2.8 shows micrographs of some enveloped viruses.

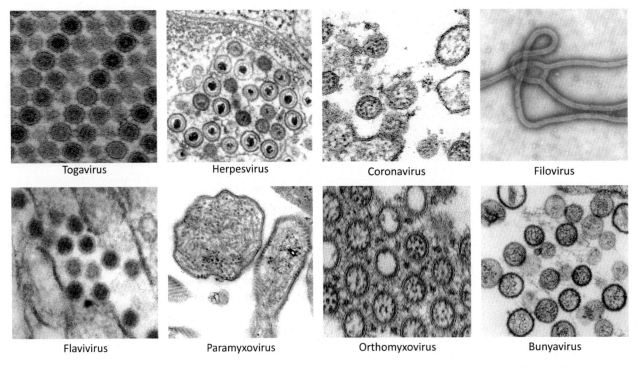

Togavirus   Herpesvirus   Coronavirus   Filovirus

Flavivirus   Paramyxovirus   Orthomyxovirus   Bunyavirus

FIGURE 2.8   Examples of enveloped viruses. Togavirus (CDC PHIL 17369, content providers CDC/Cynthia Goldsmith); herpesvirus (CDC PHIL 10260, content providers CDC/Dr. Fred Murphy; Sylvia Whitfield); coronavirus (CDC PHIL 23336, content providers CDC/Cynthia S. Goldsmith and A. Tamin); filovirus (CDC PHIL 10228 content providers CDC/F. Murphy, Sylvia Whitfield); paramyxovirus (CDC PHIL 8758, content providers CDC/F.A. Murphy); orthomyxovirus (CDC PHIL 11746, content providers CDC/Cynthia Goldsmith); bunyavirus (CDC PHIL 14340, content providers CDC/Brian W.J. May, Luanne H. Elliot).

Matrix proteins

The term matrix protein is used to describe a protein that forms layer on the inside of the viral envelope (Fig. 2.1). Matrix proteins play important roles in virus assembly, as they form links or bridge between nucleocapsids/cores and the envelope. Retroviral matrix (MA) proteins are fatty-acylated, allowing them to interact tightly with cellular membranes to form budding sites. The influenza virus matrix protein (M or M1) first makes contact with nucleocapsids in the nucleus of the infected cell and may participate in nucleocapsid assembly; influenza virus M is required for transport of nuclecapsids from the nucleus to the plasma membrane. Additional M is found at the plasma membrane where it associates with the cytoplasmic domains of influenza virus envelope glycoproteins. Thus M organizes both glycoproteins and the nucleocapsids. The matrix (M) proteins of the paramyxoviruses are highly basic proteins and paramyxovirus M is the most abundant protein in the virion. Paramyxovirus M associates with nucleocapsids and the plasma membrane (probably via interactions with cytosolic domains envelope glycoproteins) and is likely the driving force in budding.

Nucleocapsid structure

When discussing enveloped viruses, the term nucleocapsid is commonly used instead of capsid. The nucleocapsid refers to the assembly of protein and nucleic acid (the genome) that remains after the viral envelope is removed. In the case of the negative-strand RNA viruses, the main structural protein of the nucleocapsid is an RNA-binding nucleoprotein (N) but it also contains several molecules of the viral RNA-dependent RNA polymerase. Hepadnaviruses (Family *Hepadnaviridae*) and retroviruses have icosahedral capsids surrounded by an envelope. These structures are often called "cores."

In this chapter we learned that:

- Capsids are protein coats that package viral genomes. Simple capsids are either rods (helical) or spheres. Spherical viruses have icosahedral symmetry. An icosahedron is a shell defined by its symmetry. An icosahedral shell has twofold, threefold, and fivefold axes of symmetry.

- Capsids are not static packages. They have important functions such as genome packaging, attachment, and entry. Structural flexibility of capsids allows them to carry out multiple functions in a highly regulated manner.

- Virions are either naked (unenveloped) or have a lipid envelope. Naked virions use capsid proteins to mediate attachment and entry.

- Most enveloped viruses obtain their lipid membranes during budding. Viral envelopes may be derived from plasma membranes, nuclear membranes, or other internal membranes such as Golgi, ER, or transport vesicles.

- Enveloped viruses encode one or more envelope proteins with TM-domain anchors. Envelope proteins often form spikes extending from the surface of the virion and most are glycosylated. Envelope proteins mediate attachment and fusion.

- Matrix proteins are found beneath the lipid membrane of enveloped viruses. They often function to drive virion assembly and budding.

- The assembly of genome (nucleic acid) and capsid proteins found inside of the viral envelope is often called the nucleocapsid (particularly in the case of negative-strand RNA viruses). In the case of retroviruses, hepadnaviruses, and more complex viruses, the structure is often called the core. Gently lysing the envelope will release nucleocapsids or cores. Viral polymerases may also be present in nucleocapsids/cores.

- The group of proteins associated with the extracellular virus particle or virion are collectively called *structural proteins*. All viruses encode one or more structural proteins.

- The term *nonstructural protein* describes any virus proteins produced in the infected cell, but not packaged in the virion. Nonstructural proteins have many critical roles in the virus replication cycle.

References

Belnap, D.M., Olson, N.H., Cladel, N.M., Newcomb, W.W., Brown, J. C., Kreider, J.W., et al., 1996. Conserved features in papillomavirus and polyomavirus capsids. J. Mol. Biol. 259, 249.

Caspar, D., Klug, A., 1962. Physical principles in the construction of regular viruses. Cold Spring Harbor Symposia on Quantitative Biology 1—24.

3

Virus interactions with the cell

After studying this chapter, you should be able to:

- List conditions that impact virus attachment.
- Explain how to set up a synchronized infection in the laboratory.
- Define "penetration" and "uncoating" as regards the virus replication cycle.
- Describe the cellular machinery that viruses use to move through the cell.
- Describe the general structure of a eukaryotic gene, including definitions of promoter, intron, and exon.
- Explain the differences between DNA-dependent DNA polymerases, RNA-dependent RNA polymerases (RdRp), and reverse transcriptase (RT).
- Explain why DNA virus replication is often linked to cell cycle.

This chapter examines the major steps in virus replication within the context of cellular structures and processes. The emphasis is on virus interactions with eukaryotic, primarily animal, cells. Recall that the major steps in virus replication are: (1) attachment, (2) penetration and genome uncoating, (3) synthesis of new viral genomes and proteins, (4) assembly of new virions, and (5) virion release and maturation. In the following sections, we will briefly review the major structures and activities of the eukaryotic cells to provide framework on which to outline the major steps in virus replication.

Virion attachment

The extracellular space

The first step in the virus replication cycle is generally described as "attachment to plasma membrane (PM)-associated molecules." However, we are beginning to appreciate that some important virus—host interactions occur within extracellular spaces. The extracellular matrix (ECM) is a three-dimensional network occupied by fluid and ions that helps to form the structure of tissues and organs. The components of the ECM are unique to each organ but key to ECM structure are proteoglycans and fibrous proteins as collagen, elastin, fibronectin, and laminin. Proteoglycans are composed of

Viruses.
DOI: https://doi.org/10.1016/B978-0-323-90385-1.00025-X

© 2023 Elsevier Inc. All rights reserved.

unbranched, negatively charged, polysaccharide chains linked to protein cores. Proteoglycans are hydrophilic and adopt extended conformations. The ECM also contains various low molecular weight molecules with functions in metabolism as well as enzymes, such as a variety of extracellular metalloproteases. Some viruses exploit extracellular proteases to perform critical viral protein cleavages necessary for productive infection. In addition, viral infection of some organs, such as the lung and liver, have the effect of causing a remodeling of the ECM leading to pathogenic processes such as pneumonia or liver fibrosis.

The plasma membrane

The PM is both a barrier to, and required for, virus attachment and entry. The PM is a selectively permeable lipid bilayer that contains many proteins; it is a complex and dynamic structure. Small molecules such as carbon dioxide and oxygen cross the PM by diffusion. Sugars and amino acids cross the PM using protein channels or transporters. Larger molecules must be endocytosed in order to enter the cell. PM-associated proteins include receptors, signaling molecules, enzymes, and adherence proteins. As shown in Fig. 3.1, some proteins associated with the PM have cytoplasmic, membrane, and extracellular domains while others are embedded entirely with the lipid bilayer. PM-associated proteins are mobile; they move laterally through the membrane and can reorganize and form complexes as the result of signaling. PM proteins are often glycosylated (carbohydrate chains are attached to the protein backbone). The lipids of the

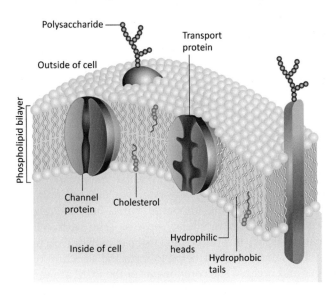

FIGURE 3.1 The PM is a selectively permeable lipid bilayer. Proteins associated with the PM include receptors, signaling molecules, enzymes, and adherence proteins.

outer leaflet of the PM can also be glycosylated. Some cell types, such as epithelial and endothelial cells have a microscopically visible layer of carbohydrate-rich molecules on their luminal surfaces that form the so-called glycocalyx or perinuclear matrix. In the lungs, the glycocalyx is comprised mainly of heparin sulfate and chondroitin sulfate and plays critical roles in lung architecture and function. The glycocalyx can serve as a barrier to infectious agents, but some viruses use the glycocalyx to their advantage by interacting with its negatively charged polysaccharides.

Adding more complexity to the PM, the lipid bilayer is not homogeneous. Preferential association of certain lipids, for example, cholesterol and sphingolipids, results in the formation of discrete regions or microdomains called "lipid rafts." Specific proteins are often associated with these lipid rafts. Rafts are thought to be dynamic and transient, and may be shaped by processes such as cell signaling, membrane trafficking, and even virus assembly and budding. In summary, the PM is not a homogenous or static structure; like the rest of the cell, the PM is a structure that may differ in protein and lipid composition, depending on cell type and metabolism, as well as in response to viral infection.

Attachment

Attachment requires the interaction of receptors on a host cell with attachment proteins presented by the infecting virus. Attachment is often mediated by protein—protein interactions but some viral attachment proteins bind to the sugar residues found on glycoproteins or glycolipids. Influenza virus is an example of a virus that attaches to carbohydrate receptors. Naked (unenveloped) virions use capsid proteins for attachment while enveloped viruses use envelope-associated proteins.

Virus attachment is achieved by interactions of small subdomains of molecules. Interaction faces usually comprise just a few amino acids or sugar residues, and the interactions are usually electrostatic in nature. Thus the initial contacts between the viral attachment protein and a receptor are weak and reversible. However, as multiple viral attachment proteins interact with multiple receptor molecules, binding becomes strong and irreversible. Thus it follows that cells with a higher density or number of receptors are more readily infected (Fig. 3.2). Because virus attachment is electrostatic, it can be affected by pH, ion concentration, and types of ions in the extracellular space or culture media. When we propagate viruses in the laboratory, the type of media used, and its pH, are important factors to consider. Attachment does not require energy and can take place in the cold (4°C). In the laboratory, a virus infection can be synchronized by allowing attachment to occur in the cold. Over a period of hours, virions diffuse through the media to the PM and attach to their receptors. The culture is then quickly warmed up and the attached virions penetrate at the same time.

The presence or absence of receptors is a major host range determinant, as absence of receptors excludes viruses from a cell. Different cells within an organism display different surface molecules, thus one cell or tissue type may be permissive for virus attachment while others are not. Epithelial cells are an example of polarized cells. They display different molecules on their apical (facing the lumen) and basolateral (facing the inside) surfaces (Fig. 3.3). Receptor molecules may be expressed on only one surface of the polarized cell, such that passage through those cells is a one-way process.

The receptors for many human and animal pathogens have been identified, but it is important to note that viruses adapted to cell cultures may use different receptors than those used during a natural infection. Even within an organism, a virus may utilize a variety of receptors. Attachment can also require interaction of

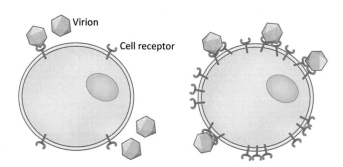

FIGURE 3.2 Receptor density and attachment. Cells with a higher density or number of receptors are more readily infected.

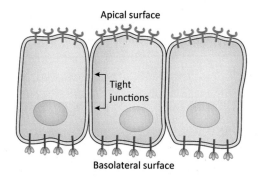

FIGURE 3.3 Polarized cells have discrete apical and basolateral surfaces. Apical surfaces face the outside (e.g., the airspace in the lung or the lumen of the intestine) while basolateral surfaces face the inside of the body. Polarized cells display distinct proteins on their apical versus basolateral sides. Materials, such as viruses, can be transported directionally through polarized cells.

the virus with more than one type of receptor molecule. An example is human immunodeficiency virus (HIV). The surface unit (SU) glycoprotein of HIV (also called gp120) attaches to the CD4 protein present on helper T-lymphocytes, macrophages, and dendritic cells. After initial interactions between SU and CD4, SU then binds to a second receptor, one of several chemokine receptors on the cell surface. The chemokine receptors are the coreceptors for HIV.

Virus penetration and uncoating

After attaching to a cell, a virus or its genome, must cross the plasma membrane to gain access to internal cell compartments; there are a variety of mechanisms that can be utilized. The process may occur at the PM or a virus particle may be taken into the cell in a membrane bound vesicle by phagocytosis, pinocytosis, or endocytosis. Recent studies using a variety of microscopic techniques have shown that initial sites of virus attachment may be distant from the eventual site of penetration. Receptors may diffuse laterally through the PM or more strikingly, attached viruses may engage actin-based structures such as filopodia or microvilli. In a process called virus surfing, actin-dependent mechanisms propel an attached virion along toward the cell body. The process has been observed for retroviruses, herpesviruses, and papillomaviruses, among others. Viral surfing is an energy-dependent process and requires functional myosin motors.

Many viruses also exploit cell processes designed to bring cargo into cells via membrane bound vesicles. These processes include phagocytosis, macropinocytosis, and receptor-mediated endocytosis. There are different mechanisms by which endocytosis can occur, such as clathrin-dependent or caveolin-1-dependent pathways (Fig. 3.4). The environment inside of endosomes is distinct from the cell cytosol. Of importance to our discussion of viruses, endosomes become acidified as they mature and low pH often triggers events that lead release of viral nucleic acids from endosomes.

Virus penetration and uncoating of unenveloped viruses

How do the relatively large capsids of unenveloped viruses breach a cell membrane? The well-studied picornaviruses generate a channel or pore through the cell membrane by inserting hydrophobic domains of capsid proteins into the membrane to form a pore that allows the RNA genome to enter the cell. This process can occur at the PM or after uptake into an endocytic vesicle where it is triggered by low pH. Picornaviruses are an example of viruses that perform penetration and genome uncoating in a single step (Fig. 3.5). Reoviruses have presented a mystery to virologists as an entire capsid crosses the PM using an undiscovered mechanism. Reoviruses are also unique in that they never uncoat their segmented RNA genomes. Instead, mRNAs are transcribed within a capsid structure and are extruded into the cytoplasm through pores in the capsid.

Virus penetration and uncoating of enveloped viruses

Enveloped viruses must transport their capsids and/or genomes across two sets of lipid membranes, both the viral envelope and a cell membrane. The process can either occur at the PM (Fig. 3.6) or after endocytosis of the viral particle (Fig. 3.7). Paramyxoviruses are an example of enveloped viruses that mediate membrane fusion at PM. If fusion occurs at the PM,

viral envelope proteins remain on the cell surface. Whether at the PM or from an endosome, membrane fusion must overcome unfavorable energetic barriers to bring two negatively charged lipid membranes into close proximity. This is accomplished by specialized viral fusion proteins (Box 3.1). Membrane fusion results in lipid mixing or formation of a pore through both the viral envelope and a host cell membrane. To initiate fusion, hydrophobic portions of a viral protein are inserted into the target cell membrane. This triggers protein rearrangements that pull the two membranes into close proximity (within a few angstroms) (Fig. 3.8). Some viruses use discrete domains of a single protein to accomplish attachment and fusion. Influenza viruses have attachment and hydrophobic fusion domains in different regions of a single molecule, the hemagglutinin (HA) protein. Other viruses (i.e., the paramyxoviruses) encode distinct attachment and fusion proteins. After fusion, the viral nucleocapsids or cores are released into the cytosol.

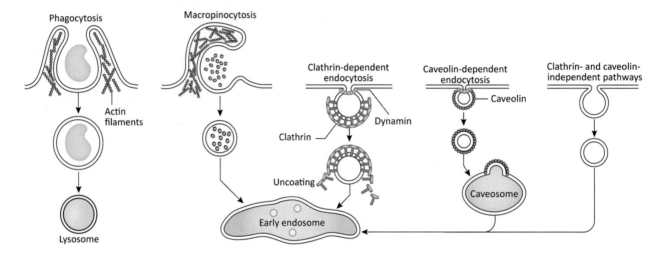

FIGURE 3.4   Cellular uptake of macromolecules by phagocytosis, macropinocytosis, and various endocytic pathways.

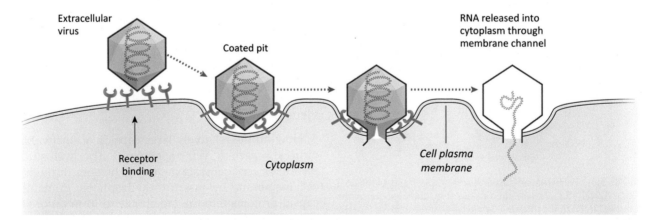

FIGURE 3.5   Illustration depicting the penetration of a viral genome across the PM. In this example, the RNA genome of a picornavirus crosses the PM through a channel or pore formed by capsid proteins.

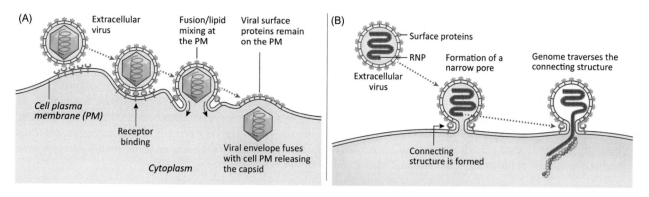

FIGURE 3.6   Illustration depicting the penetration of a viral genome after fusion of an enveloped virus with the PM. Panel A shows a process in which the viral envelope and PM become contiguous. Panel B shows a process in which a small pore is formed through the PM.

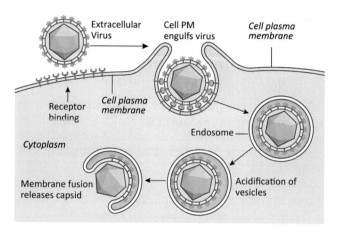

FIGURE 3.7   Many viruses, both enveloped and unenveloped, are brought into cells by endocytosis. The low pH environment in the endosome triggers molecular rearrangements of capsid or envelope proteins. In this example, an enveloped virus is fusing with an endosomal membrane to release the capsid into the cytosol.

Movement through the cell, interactions with the cytoskeleton

The cytosol is a dense network of filaments, organelles, and molecular assemblies (see Box 3.2). It is a highly viscous environment that restricts diffusion of molecules larger than 500 kDa or particles larger than 20 nm. (Recall that most viruses are larger than 20 nm.) Cells have evolved highly regulated processes to move and sort cargo in order to maintain the highly organized and complex environment of the cell. These processes can potentially inhibit virus movement and replication but are often exploited by viruses to enhance their replication. For example, viruses use actin filaments and microtubules to move through the cell as well as from cell to cell. To aid in their movement, viruses highjack the cellular molecular motors associated with actin filaments and microtubules. Specific interactions with the cell cytoskeleton are important in every step of the virus replication cycle.

As mentioned above, viruses engage actin to move along filopodia toward the cell body; sites of endocytosis and cytoskeletal elements are also involved in endocytic processes. Other viral process that utilize cytoskeletal elements include migration of viral structures within cells, formation of specialized sites of replication, assembly of new virions as well as virion budding, release and movement between cells.

Amplification of viral proteins and nucleic acids in the context of the infected cell

After penetration and uncoating have been achieved, the next major events in the viral replication cycle are the synthesis (amplification) of viral proteins and genomes. In the absence of successful amplification, the proteins and genomes needed to assemble new virions are not made. To better appreciate the processes by which viruses synthesize mRNAs, genomes, and proteins, let us review some of the most basic aspects of these processes in the eukaryotic host cell.

A short review of transcription in the eukaryotic cell

Synthesis of cellular mRNAs (transcription) occurs in the nucleus. The enzyme that synthesizes mRNAs is RNA polymerase II (RNA polII). RNA polII is a DNA-dependent RNA polymerase (an enzyme that uses DNA as the template for synthesis of an RNA product). RNA polII forms protein complexes with other cellular proteins in order to be directed to specific genes. The regions of a eukaryotic gene that interact with the transcription complex are the promoter/enhancer sequences (Fig. 3.9). The promoter/enhancer regions of a gene define the conditions under which a gene product will be synthesized. Cellular promoters can be quite long and complex to allow graded cell

BOX 3.1

Fusion Proteins

Fusion of a viral membrane to a cellular membrane requires that the high kinetic barrier to membrane fusion be breached. Fusion proteins serve as the catalysts in this process. A hallmark of fusion proteins is that they undergo structural changes as a result of attachment and/or changes in pH (during endocytosis). Often these structural changes expose a hydrophobic segment called the fusion loop or fusion peptide whose function is to engage the target cell membrane. The fusion protein becomes a bridge between the two membranes, drawing them together. Viral fusion proteins are suicide enzymes that function only once.

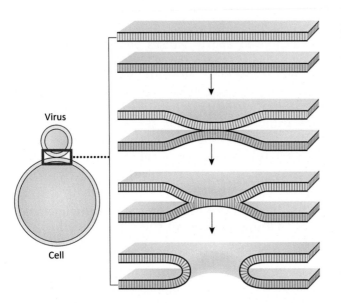

FIGURE 3.8   General illustration of the membrane fusion process. Membrane fusion or pore formation require membranes to be in very close proximity.

nt downstream of the polyadenylation signal sequence, followed by addition of the polyA tail by an enzyme called poly(A) polymerase. The poly(A) tail controls degradation of the mRNA from the 3' end and is also involved in initiating translation. The mature mRNA is now ready to be transported from the nucleus. In summary:

- Most eukaryotic genes have nontranscribed promoter/enhancer sequences.
- The 5' end of the mRNA is modified (capped) by addition of a methylated guanosine, using a 5' to 5' linkage.
- Most eukaryotic genes have introns that are removed from RNA by splicing.
- The 3' ends of mRNAs are cleaved and polyadenylated.
- Capped, polyadenylated, spliced mRNAs are transported out of the nucleus.
- The open reading frame or coding region of the mRNA is usually flanked by 5' and 3' nontranslated regions.
- For the most part, one mRNA encodes one protein.

responses to a variety of stimuli. Promoter-enhancer sequences themselves are not transcribed.

Shortly after initiation of a transcript, the newly synthesized RNA is modified by addition of a 5'-methyl guanosine "cap." The cap protects the 5' end of the mRNA from degradation, and also serves as the assembly site of the ribosome. Most eukaryotic genes have long regions, called introns that are not found in the mature mRNA. Instead, they are removed by a process called RNA splicing. Splicing is accomplished by spliceosomes (assemblies of small cellular RNAs and proteins). Recall that the regions of the mRNA that are not removed (the protein coding regions) are called exons.

An additional modification to the eukaryotic transcript is addition of nontemplated adenosines near the 3' end, to produce the poly(A) tail. A short RNA sequence (usually AAUAAA), called the polyadenylation signal, serves as the binding site for the protein complex that carries out polyadenylation. The transcript is cleaved 10–30

Transcription of viral mRNAs

Do viruses follow the rules of gene organization, transcription, and RNA processing used by the eukaryotic cell? Some do, but many do not. As we will see throughout this text, viruses use a variety of unique strategies to control synthesis of their mRNAs, as the abundance of an mRNA can directly impact the amount of protein product produced. Why have many viruses adopted unique strategies to regulate mRNA and protein synthesis?

- The coding capacity of many viruses is small because their genomes are small (compared to the host cell genome), thus they cannot have long complex promoter/enhancer regions and/or long nontranscribed introns.
- Most RNA viruses transcribe their mRNAs in the cell cytoplasm, therefore have no access to spliceosomes or cellular RNA polII as these are found in the nucleus.

BOX 3.2

Cell Cytoskeleton

The cytoskeleton is a dynamic and interconnected network of filaments. There are three major types of filaments. Actin filaments control cell shape, locomotion, and cytokinesis (separation of daughter cells after cell division). Intermediate filaments provide mechanical strength to cells. Microtubules function in intracellular transport and chromosome segregation. Microtubules also control the position of organelles. The cytoskeleton is not static. It is constantly reorganizing via the assembly and disassembly of filaments. Regulation of the cytoskeleton is quite complex, requiring many molecular players. Molecular motors are key to moving materials along filaments. Members of the myosin superfamily of proteins are actin-based motors. For example, myosins facilitate movement of cargo along tracks of actin filaments and also move actin filaments relative to the PM.

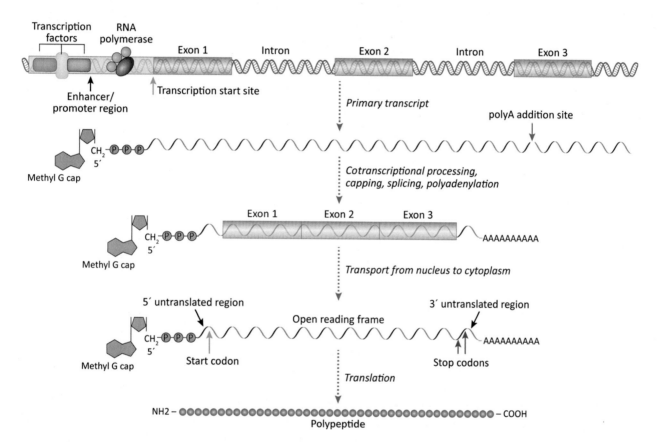

FIGURE 3.9 Organization of a eukaryotic gene showing a promoter region followed by introns and exons. The primary transcript is produced, followed by splicing to remove introns. After export to the cytoplasm, the capped and polyadenylated transcript is translated to produce the protein product of the gene.

- Many viruses inhibit host cell transcription and/or translation, therefore must use alternatives to "normal" cellular processes to produce viral proteins.
- An infecting virus usually brings one copy of its genome into a very crowded cell. Viral mRNA synthesis and protein synthesis must be very efficient in order to compete for building materials.

Thus the organization and expression strategies of viral genes often differ from host genes:

- Many viral genes have no introns.
- Some viral mRNAs do not have 5' caps.
- Some viral mRNAs do not have poly(A) tails.
- Some viral mRNAs have overlapping open reading frames such that more than one type of protein can be produced from a single transcript.

- Viral transcripts with introns can be alternatively spliced to generate multiple, different mRNAs. (Obviously these must be viruses that are replicating in cell nucleus where the splicing machinery is present.)
- A very few viral mRNAs are not exact copies of the genome! In a process called RNA editing or pseudotemplated transcription, the polymerases of a few RNA viruses add nucleotides not present in the genome sequence. The term pseudotemplated suggests that there are some specific signals in the genome that instruct the addition of these extra residues.

A short review of translation in the eukaryotic host cell

Let us review a few basics of translation in the cell, focusing on initial interactions of the translation apparatus with the mRNA. Both the 5′ cap and the 3′ poly(A) tail are involved in translation initiation. The poly(A) tail is bound by poly(A)-binding protein (PAPB). A complex of initiation proteins (eIF4F complex) binds to the 5′ cap (Fig. 3.10) and interacts with PAPB. This is followed by association with a preinitiation complex that includes the 40S ribosomal subunit, tRNA^{met}, and other initiation factors. The preinitiation complex moves or scans along the untranslated region of the mRNA until it encounters an AUG codon with the appropriate surrounding sequences (Kozac sequence). Now the 60S subunit binds at the AUG codon to generate the 80S initiation complex with tRNA^{met} in the A site. A second tRNA enters the P site and a peptide bond is formed. The ribosome moves along the mRNA (translocates) one codon (3 nt) at a time. At the end of an open reading frame, a ribosome will encounter stop codons that trigger termination. In eukaryotes transcription termination is facilitated by two release factors (eRF1 and eRF3). eRF1 recognizes stop codons (UAA, UGA, or UAG) in the A site.

Translation of viral proteins

We noted above that most eukaryotic mRNAs code for a single protein. But one hallmark of viruses is their ability to efficiently exploit a small genome. Thus it is not uncommon for individual viral mRNAs to encode different versions of a protein, or two or more

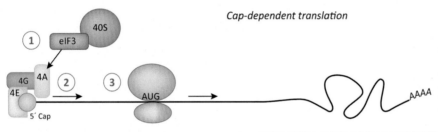

Cap-dependent translation

Cap-dependent translation begins with (1) recruitment of the 40S ribosomal subunit/eIF complex to the cap-binding complex (eIF4E, 4G and 4A). (2) The complex scans the mRNA to find the start codon. (3) The 60S subunit is recruited.

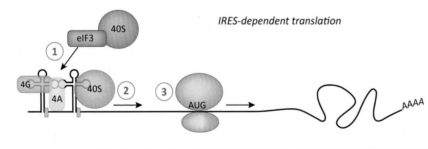

IRES-dependent translation

IRES-dependent translation begins with (1) formation of an initiation complex on the internal ribosome entry site (IRES). Initiation does not require the cap binding protein. (2) The IRES positions the 40S ribosomal subunit at the correct initiation codon. (3) the 60S subunit is recruited.

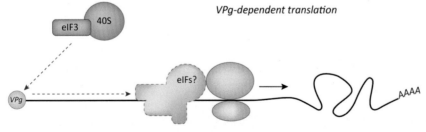

VPg-dependent translation

VPg-dependent translation has been described for some plus strand RNA viruses. It appears to begin with formation of an initiation complex by interactions with VPg. Other details of the process are not known.

FIGURE 3.10   Methods of translation initiation. Most cellular transcripts have a 5′ cap and a 3′ poly(A) sequence that are key players in ribosome assembly. Many types of viruses adhere to this host cell strategy. However, several positive strand RNA viruses use cap-independent translation. Picornaviruses and some flaviviruses use an RNA structure called the internal ribosome entry site (IRES) to direct ribosome assembly. Another strategy is ribosome assembly directed by a viral protein (VpG) covalently linked to an RNA transcript.

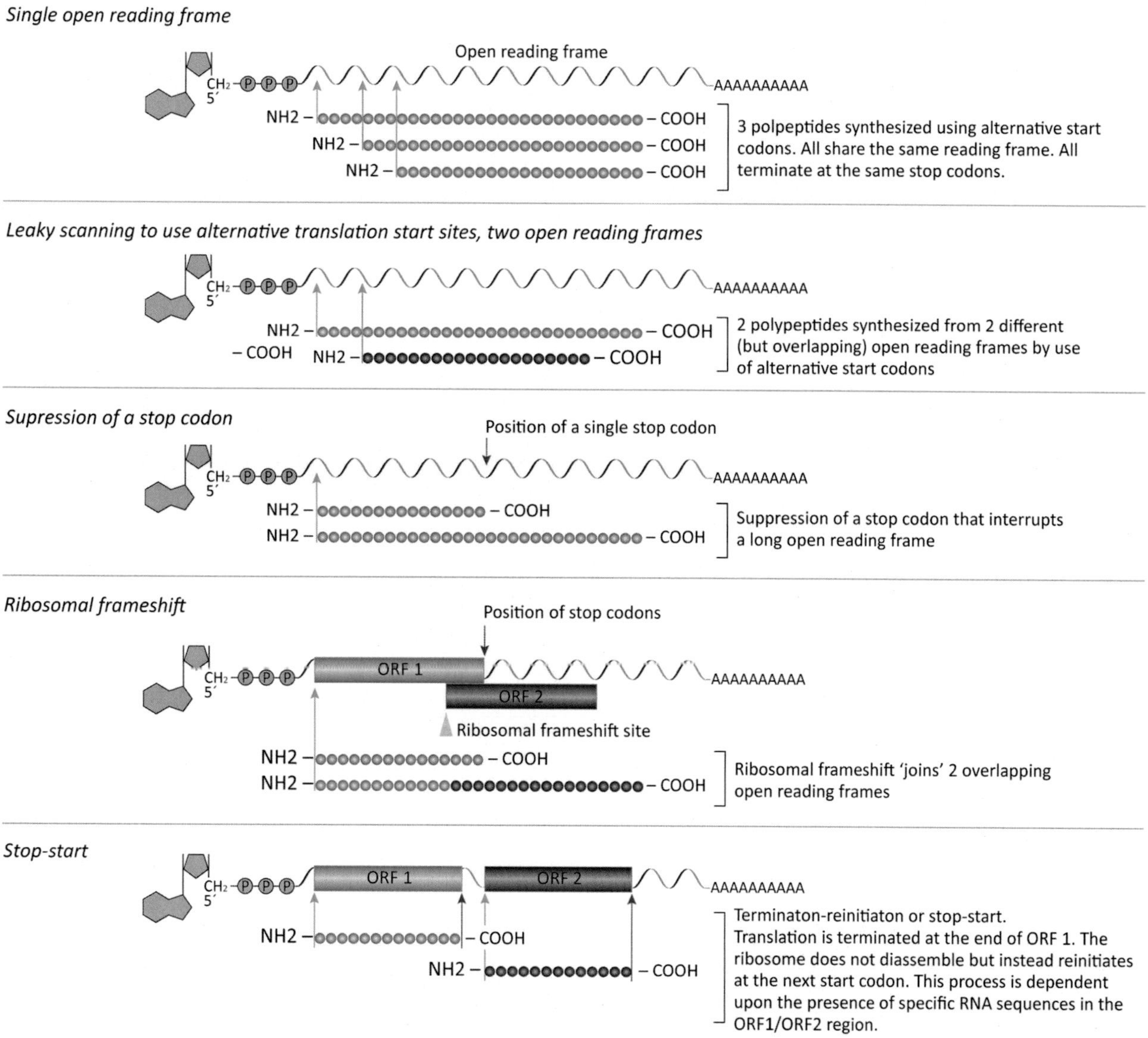

FIGURE 3.11   Viruses use a number of strategies to produce more than one protein from a transcript.

completely different proteins. Common strategies are illustrated in Fig. 3.11 and include:

- Use of alternative start codons, a process often called leaky scanning.
- Suppression of stop codons to produce a longer protein product.
- Frame-shifting moves a ribosome to another reading frame to produce an alternative protein product.
- Termination-reinitiation or stop-start, a process in which the ribosome completes synthesis of a protein then reinitiates synthesis of a second protein, without disassembly from the mRNA.

These mechanisms all function at the level of the ribosome and serve to control relative amounts of protein products. For example, use of alternate start codons will result in production of a greater amount of the protein with the best Kozak sequence and lesser amounts of proteins that initiate at alternative start codons. In the case of ribosomal frame-shifting, folding of the mRNA into structures called pseudoknots modifies the translation process (Fig. 3.12; Box 3.3).

Many RNA viruses produce polyproteins that are cleaved by viral proteases to generate smaller, functional proteins (Fig. 3.13). The process is exemplified by picornaviruses. The picornaviral genome contains a single, long open reading frame that is translated to produce a large precursor polyprotein. Protease domains within the polyprotein are enzymatically active immediately after being translated and work to quickly cleave the large precursor into mature products. The proteolytic cleavages occur in an ordered sequence. Intermediate cleavage products may have activities distinct from those of the final cleavage products, allowing one coding sequence to have multiple functions.

As viruses must use the host cell translational apparatus, their viral mRNAs must complete with cellular mRNAs for ribosomes and amino acids. One way to compete is simply to inhibit synthesis of host cell transcription and/or translation. (This also limits the ability of the cell to respond to infection.) Another strategy is to efficiently compete for the translational machinery. Some viruses use a combination of both methods. But how can a virus inhibit translation of host proteins without effecting viral protein synthesis? An example is provided in Box 3.4.

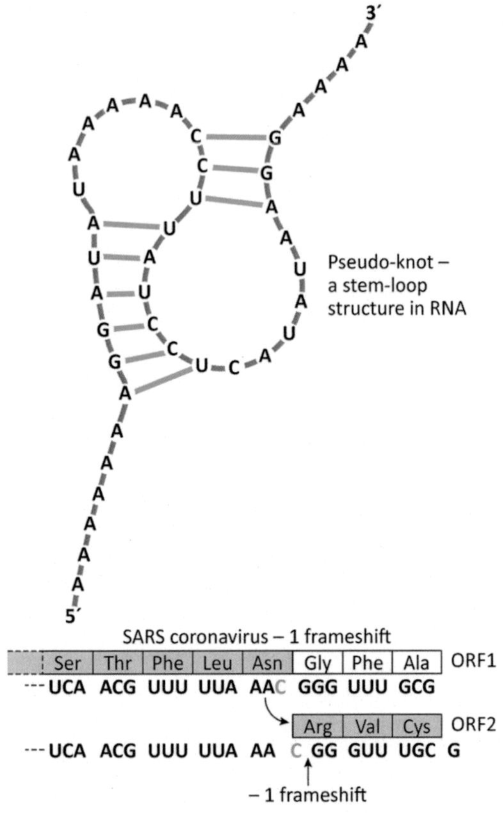

Synthesis of viral genomes

Viruses can be placed into one of three major groups based on genome type and replication strategy. The groups are (1) DNA viruses, (2) RNA viruses, and (3) viruses that use RT.

There are many families of DNA viruses. The genomes of DNA viruses range from 3000 nt to well over 1 million base pairs. Some DNA viruses have circular genomes, others linear. But all DNA viruses use a DNA-dependent DNA polymerase to synthesize additional genome copies. Some DNA viruses use a cellular DNA polymerase but others encode their own DNA polymerases for genome synthesis. There is also obviously a need for a sufficient pool of dNTPs. Eukaryotic DNA synthesis is a highly regulated process. Some types of animal cells divide regularly (e.g., epithelial cells) but others are quiescent, seldom dividing except in response to damage [e.g., hepatocytes (liver cells)]. Nondividing cells have limited DNA replication machinery and very limited dNTP pools. Strict regulation of cell division causes a problem for some DNA viruses. Some can only replicate in mitotically active cells. But others can stimulate quiescent cells to divide. Some large DNA viruses encode enzymes required for dNTP synthesis, thereby increasing the cellular pools of these building blocks in nondividing cells. Animal DNA viruses (with the exception of the poxviruses and

FIGURE 3.12 An RNA pseudoknot directs ribosomal frame-shifting.

BOX 3.3

Ribosomal Frame-Shifting and Stop Codon Suppression

Ribosomal frame-shifting is common among retroviruses and coronaviruses. It is a process whereby the ribosome moves from one reading frame (protein 1) to a second reading frame to produce a different protein (protein 2). Ribosomes frame-shift when they stall during protein synthesis. They stall at specific sites on viral mRNAs called "pseudoknots." The pseudoknot forms as a result of base pairing within the mRNA. In most cases, the pseudoknot causes frame-shifting about 20% of the time. A surprisingly efficient process!

Suppression of a stop codon is a process whereby a ribosome fails to terminate protein synthesis at a stop codon. Most eukaryotic genes terminate with multiple stop codons, but if there is a single stop codon, an amino acid can be inserted into the growing polypeptide and translation continues. This is estimated to occur as often as 20% of the time. Thus a protein coding sequence downstream of a pseudoknot or a single stop codon will be synthesized at a lower quantity ($\sim$20%) than the protein encoded when the signal is ignored ($\sim$80% of the time).

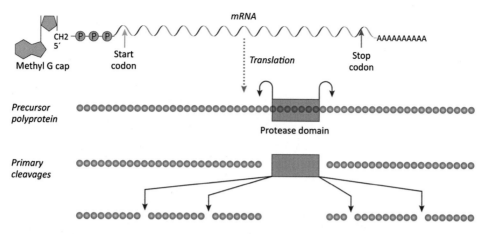

FIGURE 3.13   Proteolytic cleavage of a viral polyprotein.

BOX 3.4

Poliovirus (PV) and Cap-Independent Translation

PV inhibits cap-dependent translation in the infected host cell. It does so by cleaving the eukaryotic initiation factor 4G (eIF-4G). A function of eIF-4G is to interact with both the 5'-cap on the mRNA and the 40S ribosomal subunit, thereby bringing mRNA and ribosome together. A protease encoded by PV cuts eIF-4G into two pieces. One piece binds to eIF-3 on the 40S subunit and the other piece binds to the cap-binding protein eIF-4E. As the two business ends of eIF-4G are now separate polypeptides, the 40S subunit cannot bind the cap. PV RNA is not capped so the cleavage does not interfere with viral protein synthesis. PV uses a cap-independent mechanism for ribosome assembly onto its mRNA.

African swine fever virus) replicate their genomes in the nucleus.

RNA viruses were originally defined as viruses with an RNA genome packaged within the virion. But more critical to the definition is that the RNA genome found in the virion is used as the template for the synthesis of additional RNA genomes. With the exception of hepatitis D virus, animal RNA viruses encode an RdRp for genome synthesis. The RdRp is also called the "replicase" and is used both for genome synthesis and transcription. As all living cells continuously produce RNAs for "housekeeping" purposes, the synthesis of viral RNA genomes can occur in dividing or nondividing cells.

Members of the families *Retroviridae* and *Hepadnaviridae* use a polymerase called RT to synthesize their genomes. Retroviruses package an RNA genome while hepadnaviruses package a DNA genome. Both virus families use the enzyme RT to make a DNA copy of an RNA molecule. The RNA genome of a reverse transcribing virus is an mRNA, transcribed from the viral DNA genome by host cell RNA polII. (In the case of retroviruses, the DNA form of the viral genome is integrated into the cell DNA.) Comparison of RT with RdRps of RNA viruses suggests that these enzymes are highly related. Thus in the most current viral taxonomy, the realm *Riboviria* contains viruses that encode an RdRp as well as viruses that encode RT.

The endgame: virus assembly, release, and maturation

Assembly

Viruses assemble in the infected cell when the local concentrations of structural polypeptides and genomes are sufficiently high. Sometimes assembly takes place in discrete regions of the cell called "virus factories" containing high concentrations of capsid proteins and genomes. Cytoskeletal components and cellular organelles are often intimately involved in both virus assembly and release, although some very simple viruses can be assembled (albeit inefficiently) in the test tube.

There are two general ways to build a simple icosahedral capsid. In one scenario, capsid proteins form an empty shell into which the genome is inserted. Picornaviruses use this strategy and empty capsids can be seen in virus preparations. Empty particles are less dense than complete virions and can be separated from them by density centrifugation. In the second scenario, an encapsidation signal on the genome interacts with one or more capsid proteins, followed by recruitment of additional capsid proteins.

Most animal viruses with helical nucleocapsids are RNA viruses. They usually encode a basic protein [often called the nucleocapsid (N) protein] that interacts with viral RNA. Initial interactions are specific, such that cellular RNAs are not packaged indiscriminately. Genome packaging signals often include sequences at both the 5′ and 3′ ends of the viral genomic RNA. This is a mechanism whereby genomic RNA can be "distinguished" from subgenomic mRNAs. In some cases, the N protein surrounds the viral RNA but in others (i.e., influenza viruses) it appears that the RNA winds around a protein core.

Virion release

Mechanisms for virus release from cells include cell death (lysis), budding, and exocytosis. The cytoskeleton can present a barrier to release and some unenveloped viruses encode proteins that disrupt the cytoskeleton to allow dispersal of newly assembled virions. Most enveloped viruses obtain their envelope by a budding process (poxviruses and African swine fever virus are exceptions). A viral nucleocapsid interacts with a region of host cell membrane into which glycosylated viral envelope proteins have been inserted. The nucleocapsid "finds" the proper place to bud from the cell by forming specific interactions with the cytoplasmic tail(s) of the envelope proteins. Cell membranes that can serve as sites of budding include the PM, endosomal, and nuclear membranes. Viruses released

by budding from the PM (e.g., the HIV) are released individually (Fig. 3.14). Viruses with envelopes derived from endosomal or nuclear membranes may bud into vesicles that traffic to the PM and fuse, releasing their cargo of virions by a process called exocytosis (Fig. 3.15).

It is easy to assume that release always results in free virions escaping into the extracellular environment. And while this certainly does happen, virions can also be transmitted directly from one cell to another. Vaccinia virus (family *Poxviridae*) uses actin tails to move between cells. We can visualize this process using cells with fluorescent-tagged actin and virions tagged with a different fluorescent tag (Fig. 3.16).

Virion maturation

Maturation cleavages of virion proteins may occur after particle release, and these may be required to produce infectious virus. Why do some viruses mature after release from a cell? The virion has two very different roles in the replication cycle: to assemble under conditions of favorable energy and to disassemble during the processes of attachment and penetration into a new cell. Maturation cleavages prepare the newly released virion to attach to a new cell, penetrate, and disassemble. For example, picornaviruses include in their capsids, an uncleaved precursor protein. The precursor is cleaved within the assembled capsid. If this cleavage is blocked, the capsid is unable to deliver the genome to a new cell. Retroviruses require a set of cleavages that convert long precursor polyproteins into individual structural proteins. The cleavages are made by a retroviral protease and inhibitors of HIV protease activity are powerful antiviral drugs. Protease

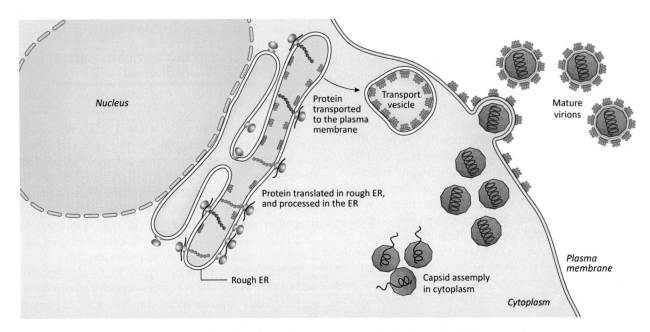

FIGURE 3.14   Illustration showing the process of assembly and budding of a virus particle from the PM.

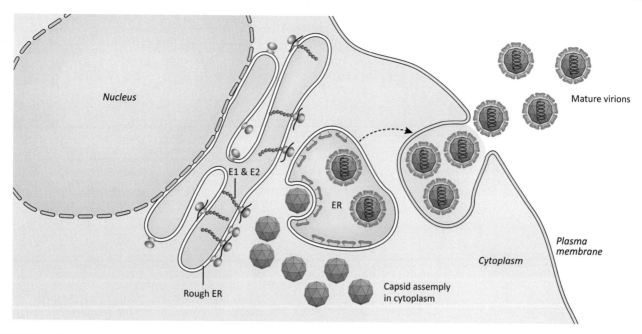

FIGURE 3.15   Illustration showing release of enveloped virions by exocytosis.

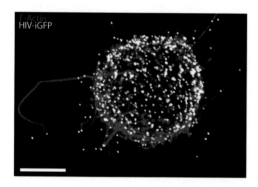

FIGURE 3.16   Showing human immunodeficiency virus particles (white) on filopodia of infected dendritic cells. Source: *From Aggarwal, et al., 2012. Mobilization of HIV spread by diaphanous 2 dependent filopodia in infected dendritic cells. PLoS Pathog. 8, e1002762.*

inhibitors are small molecules that interact with the HIV protease to prevent it from cleaving the polyproteins in the immature particle. For many enveloped viruses, maturation involves cleavage of glycoproteins that mediate fusion of the viral envelope with a host cell membrane.

In this chapter, we have learned that:

- Virus attachment requires specific host receptors and the process is influenced by receptor density, pH, and ions.
- Attachment does not require energy, thus can occur in the cold. Adding virions to cells in the cold is a way to synchronize a viral infection. Penetration is an energy-requiring step that occurs only after the culture has been warmed to physiologic temperature (37°C for most animal cells).
- In order to penetrate into the host cell, a virion or genome must cross a lipid membrane. This can occur at the PM or after endocytosis.
- Uncoating is a process that can occur during or after penetration. Uncoating provides the viral genome access to cytoplasm or nucleoplasm.
- Enveloped viruses use fusion proteins to overcome the kinetic barriers to membrane fusion. Fusion proteins are suicide enzymes that catalyze membrane fusion.
- The cell has a complex and dynamic cytoskeletal system that is extensively utilized by viruses.
- Eukaryotic genes are large. They are preceded by untranscribed promoter regions and most have large introns. RNA polII transcribes cellular mRNAs. Cellular mRNAs are highly processed in the nucleus, undergoing capping, splicing (removal of introns), and polyadenylation.
- Viruses maximize use of small genomes by minimizing the size of promoters and encoding multiple proteins from a single mRNA. Some viruses make extensive use of alternative splicing. Some viruses modulate the process of translation by inducing ribosomal frame-shifting and suppression of stop codons.
- DNA viruses use cellular or virally encoded DNA-dependent DNA polymerases for genome replication. Most DNA viruses require that the cellular DNA synthesis machinery be active to provide adequate dNTP pools.
- RNA viruses use virally encoded RdRp to transcribe mRNAs and to replicate genomes. RNA viruses can replicate in dividing or nondividing cells.

4

Methods to study viruses

This chapter describes methods for growing, purifying, counting, and characterizing viruses. It also presents general principles of diagnostic virology. After studying this chapter, you should be able to:

- Describe general requirements for culturing cells and tissues.
- Describe differences between primary cells and transformed cells.
- Describe how centrifugation is used to purify viruses.
- Understand the types of information provided by negative staining electron microscopy (EM), thin sectioning EM, electron cryomicroscopy (cryo-EM), and confocal microscopy.
- Understand what is being measured by each of the following techniques: plaque assays, polymerase chain reaction (PCR), enzyme-linked immunosorbent assays (ELISA), hemagglutination, and hemagglutination inhibition assays.

many early studies of viruses were done in bacteria or plants. Tobacco mosaic virus (TMV) was an early "model virus" as it replicates in a variety of plants, at levels sufficient for biochemical analysis and imaging. Growing TMV is as simple as applying virus to abraded leaves of a susceptible plant. The earliest studies of animal viruses were limited to using whole animals. When possible, pathogens of large animals (including humans) were adapted to small animals such as mice, rats, and rabbits. These small animal models provided a means to study viral pathogenesis and develop vaccines. Fertile chicken and duck eggs were, and continue to be, widely used for propagating viruses. In the 1940s and 1950s development of robust cell culture techniques revolutionized the study of animal viruses. Today, most animal viruses are grown in cultured cells.

Growing viruses

Viruses replicate only within living cells, thus prior to development of methods for culturing animal cells,

Generating cell cultures

A primary cell culture consists of cells taken directly from an animal. The replication capacity, or life span, of primary cell cultures is limited. The following steps

Viruses.
DOI: https://doi.org/10.1016/B978-0-323-90385-1.00022-4

© 2023 Elsevier Inc. All rights reserved.

describe an overall strategy for generating primary cell cultures. It is of utmost importance that all work is done under sterile conditions (Fig. 4.1):

1. The desired tissue is removed from the animal and is chopped or minced.
2. Tissue fragments are treated with enzymes such as collagenase to degrade the extracellular matrix and release single cells and small aggregates of cells.
3. Larger pieces of tissue are removed by filtration, cells are pelleted by centrifugation and resuspended in buffered saline or cell culture media.
4. Additional centrifugation steps may be performed to separate single cells from cell aggregates.
5. Cells and growth media are added to culture dishes and are maintained in a humidified incubator (typically 37°C, 5% CO_2). Cells attach to the bottom of the dish where they grow and divide to form a monolayer.
6. The adherent cells can be removed with trypsin, washed, and divided among new culture plates or dishes. This is called a passage, and is done to increase cell number.

Primary cells can be propagated for only a limited number of passages before the cells undergo a crisis and the culture dies. Embryonic cells can be passaged many more times than cells taken from adults. Some types of cells (e.g., fibroblasts) divide more readily than do cells that are normally nondividing in the adult animal (e.g., neurons). Tumors provide another source of cells for virus culture.

Tumors can be used to generate immortalized cell lines. However this is not a simple process; establishing cell lines is time consuming, with an overall low rate of success. Therefore well-established lines are widely shared among researchers. Once established, tumor-derived cells can often be passaged indefinitely and these so-called "immortalized cells" are excellent tools for the virologist. As they are relatively easy to grow, many types are commercially available and they can be genetically modified. Multiple genes can be introduced, mutated, or deleted to generate an unlimited supply of "designer" cells.

Purifying viruses

Viruses grown in cultured cells can be purified, quantitated (enumerated), imaged, and biochemically analyzed. The higher the initial virus concentration the easier it is to purify virus away from cell debris and media components. If a virus kill cells (is cytopathic) it is present among cell debris and culture media. In contrast, if the virus is highly cell associated, the cells must be gently lysed to release the virus. Viruses that bud from cells without killing them are found in the culture media and are minimally contaminated with cell debris. In any of these cases, mixtures of virus and cell debris are initially subjected to low-speed ($\sim 5000 \times g$) centrifugation to pellet cells or cell debris, but not the much smaller virions. At the end of the low-speed centrifugation, the liquid supernatant, containing the virus, is saved and the pellet (containing the cell debris) is discarded. The process of separating larger from smaller molecules is called differential centrifugation (Fig. 4.2).

To concentrate and further purify virions, the virus-containing supernatant can be centrifuged at a much higher speed ($\sim 30,000-100,000 \times g$). After the centrifugation

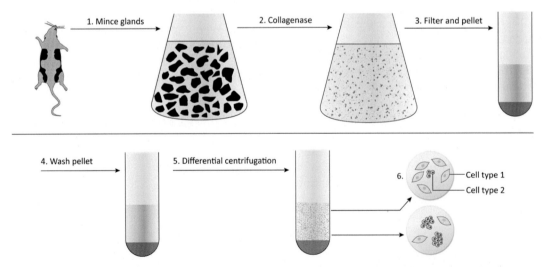

FIGURE 4.1   Generating cell cultures begins with removing tissues (normal or tumor) from an animal. Tissues are minced and treated with enzymes to degrade the extracellular matrix. Centrifugation is used to pellet the cells. Cells are resuspended in media and placed in culture vessels.

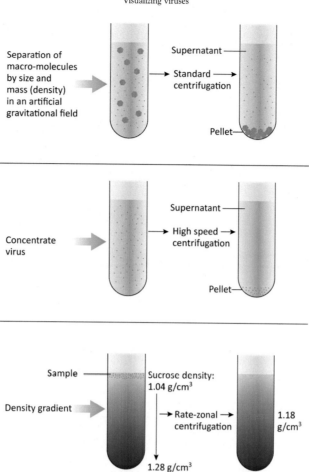

FIGURE 4.2 Centrifugation is a common technique used to purify viruses away from cell debris and other small molecules. Low-speed centrifugation ($\sim 5000 \times g$) is used to remove cells and large cell debris, leaving virions in the supernatant (top panel). Virus can be further purified and concentrated by centrifugation at higher speeds (middle panel). A gradient of sucrose or glycerol can be used to achieve further purification as materials in the sample will separate into layers depending on their buoyant density (lower panel).

is completed, the supernatant is discarded and the virus pellet is saved. If desired, the virus can be further purified by centrifugation through a density gradient. Sucrose and glycerol gradients are commonly used; a sucrose gradient might range from 40% sucrose at the bottom of the tube to 5% sucrose at the top of the tube. The gradient is prepared and the virus sample layered carefully onto the top. During centrifugation, the components of the sample will separate into layers depending on their buoyant density. By using a gradient, one can achieve a finer separation of the macromolecules (including virions) present in a sample. If sufficient virus is present in the sample, a visible band forms and can be carefully removed from the tube.

Visualizing viruses

Optical microscopy

Light microscopes use visible light (400−700 nm wavelengths) to image objects. They are seldom used to visualize viruses, as they lack sufficient magnifying or resolving power to do so. However, the largest known viruses can be seen with a light microscope. Examples of these very large viruses include mimivirus, pithovirus, megavirus, and pandoravirus, all of which infect amoebas. Even though many viruses are far too small to be seen with the light microscope, one can often observe virally induced changes to infected cells. A classic example is the Negri body, once used as a diagnostic test for rabies virus infection. The Negri body is not "a virus" but is a structure (a type of inclusion body) that is seen in rabies virus-infected neurons (Fig. 4.3). Most inclusion bodies are densely packed virus factories that form at discrete locations in the infected cell.

Fluorescence is the emission of light by a substance that has absorbed light (or other electromagnetic radiation). The basic function of a fluorescence microscope is to irradiate a specimen with high-energy excitation light and detect the much weaker emitted fluorescence. The fluorescence microscope is designed so that only

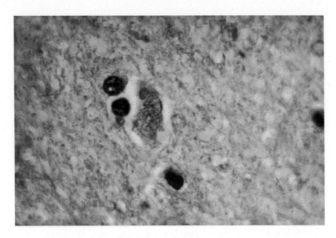

FIGURE 4.3   Hematoxylin-eosin stain of brain tissue showing Negri bodies, darkly staining area that contain high concentrations of rabies virus proteins. (CDC PHIL 14559 content providers CDC/Dr. Daniel P. Perl.)

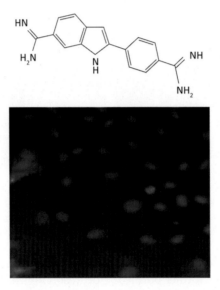

FIGURE 4.4   Chemical structure of DAPI (top) a fluorescent stain used to stain nuclei. Bottom panel shows a field of cells with DAPI (blue) stained nuclei. Other cell structures are not visible.

the emission light (fluorescence) reaches the eye or detector, allowing fluorescent structures to be seen with high contrast against a dark background. This is achieved by use of filters of specific wavelength. The advantage of fluorescence microscopy is that a single fluorescein molecule can emit as many as 300,000 photons before it is destroyed, thus making it theoretically possible to detect a single molecule in a field of view.

Fluorochromes (also termed fluorophores) are stains used in fluorescence microscopy. As in light microscopy, the most useful of these stains attach themselves to specific targets in the cell. For example, DAPI (4′,6-diamidino-2-phenylindole) is a fluorescent stain that

binds strongly to A-T rich regions in DNA. DAPI (Fig. 4.4) can pass through an intact cell membrane thus can be used to stain both live and fixed cells. It is used extensively in fluorescence microscopy to label the cell nucleus. Different fluorochromes are excited by specific wavelengths of irradiating light and emit light of defined wavelength and intensity. Thus multiple fluorochromes can be used simultaneously to identify different target molecules within a cell.

Alternatively, fluorochromes can be chemically attached to molecules, such as antibodies, to specifically label their ligands (Fig. 4.5). Alexa dyes are a series of fluorescent molecules that are widely used as fluorochromes. Among the Alexa dyes, there are choices of

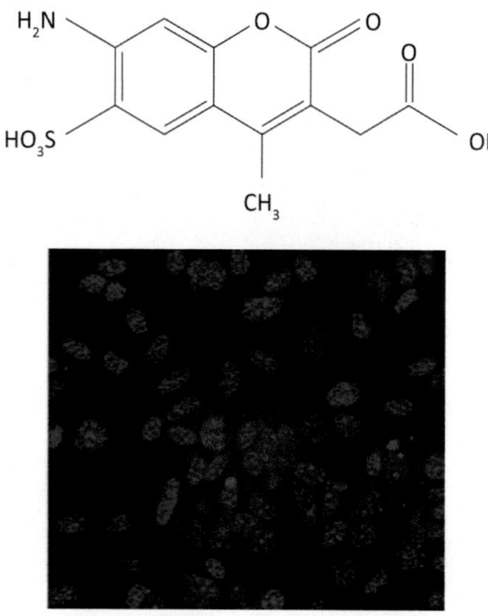

FIGURE 4.5 Chemical structure of Alexa dye (top). Bottom panel: Alexa Fluor 568 goat antimouse IgG staining of hepatitis A virus (HAV) infected BSC-1 cells. Uninfected cells have blue, DAPI-stained nuclei. Infected cells show red staining of HAV (Counihan et al., 2006). *Source: From Counihan, N.A., Daniel, L.M., Chojnacki, J., Anderson, D.A., 2006. Infrared fluorescent immunofocus assay (IR-FIFA) for the quantitation of non-cytopathic and minimally cytopathic viruses. J. Virol. Methods 133, 62.*

molecules with a variety of excitation and emission spectra. If different Alexa dyes are attached to different molecules, cell images can be collected using filters specific for each color. The relationship of one macromolecule to another can be determined by overlaying the images. For example, if a green image and a red image are superimposed, yellow pixels are seen where the tagged macromolecules colocalize in the cell.

A viral diagnostic assay called an immunofluorescence assay (IFA) uses tagged antibodies to detect viral proteins in infected cells. IFAs often include DAPI to stain cell nuclei. The result is that all cells have blue DAPI-stained nuclei and infected cells glow green or red. If cells have fused together due to virus infection, one may see a large cell (syncytium) containing many blue nuclei.

Another way to add a fluorescent tag to a protein is to use genetic engineering (cloning) to add the coding sequence for a fluorescent protein to the coding sequence of a protein of interest. Green fluorescent protein (GFP), isolated from a jellyfish, is a 238-amino acid protein (Fig. 4.6) that folds to create a fluorescent center that emits green light. Genes encoding cellular or viral proteins can be modified by addition of the GFP gene to create a "tagged" protein. If the engineered gene is introduced into a cell, the protein product is visible when cells are exposed to ultraviolet (UV) light. This is a very powerful imaging technique; however, one drawback is that the addition of a relatively large protein tag can alter the trafficking, localization, or enzymatic activity of the protein of interest.

Confocal microscopy is a type of fluorescent microscopy that offers several advantages. The method provides the ability to collect serial *optical* sections through a thick specimen by collecting and measuring the light intensity originating from numerous thin focal areas in a cell. Confocal microscopy can image either fixed or living cells that have been labeled with one or more fluorescent probes. However, the excitation energies required to reach deep into the cells are very high and specimen damage can result. In contrast *deconvolution methods* employ conventional widefield fluorescence microscopes for image acquisition. Excitation intensities can be kept low, resulting in less damage to the specimen. Deconvolution is often used to image monolayers of living cells. Deconvolution methods have high computational demands as the images are computationally processed.

Electron microscopy

The development of EM in the 1930s allowed individual viruses to be seen for the first time. EM uses an electron beam in place of light, and the beam is focused by electromagnets, rather than by glass lenses. EM has great resolving power and can magnify objects by up to ~10,000,000 times. Disadvantages of EM are that fixed and processed samples are dead, and biological samples are heavily damaged by the electron beam even as they are being imaged (sample damage results from the interaction of electrons with organic

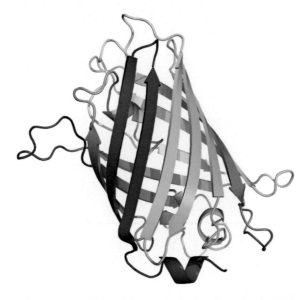

FIGURE 4.6   Ribbon diagram showing the structure of the GFP protein. Source: *By Zephyris, CC BY-SA 3.0, https://commons.wikimedia.org/w/index.php?curid = 2718822.*

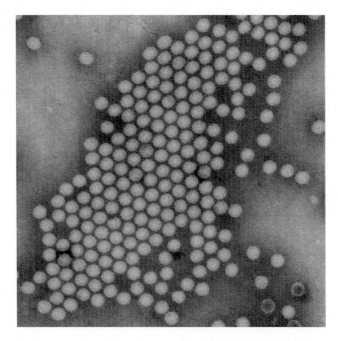

FIGURE 4.7   Transmission electron microscopic (TEM), negative stain image, reveals some of the ultrastructural features exhibited by a grouping of icosahedral-shaped polio virus particles. CDC Public Health Image ID #1837. Content Providers(s): CDC/ J. J. Esposito; F. A. Murphy.

matter). However, the simple structures and crystalline nature of viruses were first revealed by EM and it remains a common tool of the virologist. In a process called negative staining, heavy metals are used to coat the surface of virus particles to make a "cast" or "relief map" (Fig. 4.7). The heavy metal coating is much less sensitive to radiation damage than the biological sample, thus the coated surface structures are well preserved. However, information about internal structures is poor.

Thin sectioning is a method that reveals structures inside viruses and cells. Samples are fixed, embedded in resin, and sliced with a diamond knife and a microtome. The sections are carefully placed onto small carbon grids and can be processed in a variety of ways. Samples can be treated with electron dense stains such as uranyl acetate or lead citrate to provide contrast. Samples may also be incubated with gold-labeled antibodies. Use of gold-labeled antibodies is a powerful technique that allows the localization of specific molecules within a virus or cell. Micrographs showing the process of virus budding from a cell are produced by thin sectioning (Fig. 4.8).

Cryo-EM uses ultralow temperatures (liquid nitrogen or liquid helium) to preserve biological samples. Cryofixation of samples produces a glass-like (vitrified) material rather than crystalline ice. Cryo-EM allows the observation of specimens that have not been chemically stained or fixed, thus showing them in their native aqueous environment. Low temperatures provide for better images with less specimen damage, but cryo-EM micrographs have less contrast than stained images. Cryo-EM

images are more complex than negative stains because the entire particle (including the interior) is seen in the image. However, thicker frozen specimens can be cut into thin sections using a cryoultramicrotome. Images generated by cryo-EM are often enhanced by collecting and computationally averaging multiple images. This allows use of lower and less damaging doses of electrons. Tomography is a process whereby a series of images are taken with the specimen tilted at different angles relative to the direction of the electron beam. The images are computationally combined to create 3D images (tomograms) of the sample. Extremely high-resolution (<4 Å) images have been achieved for icosahedral viruses because they are constructed from many repeating subunits that can be computationally averaged. Cryo-EM, while a powerful technique, requires expensive and specialized equipment and highly trained personnel (Fig. 4.9).

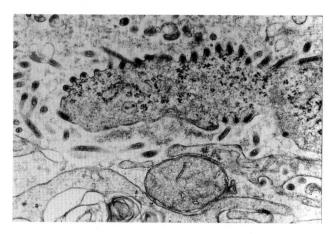

FIGURE 4.8 Transmission electron microscopic image depicting some ultrastructural details of Ebola virus particles are budding from the cell surface. CDC Public Health Image Library ID# 12794. Content providers CDC/Cynthia Goldsmith.

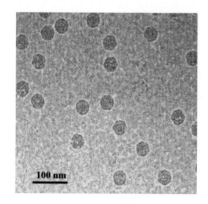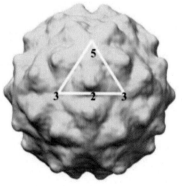

FIGURE 4.9 Cryo-EM structure of Flock house virus, virus-like particle (VLP). Left panel: electron cryomicrograph of frozen hydrated VLPs. Right panel: 3D reconstruction using image processing and viewed down the twofold axis of symmetry. The icosahedral asymmetric unit is highlighted by the white triangle, showing the fivefold, twofold, and threefold axes of symmetry (Bajaj et al., 2016). Source: *From Bajaj, S., Dey, D., Bhukar, R., Kumar, M., Banerjee, M., 2016. Non-enveloped virus entry: structural determinants and mechanism of functioning of a viral lytic peptide. J. Mol. Biol. 428, 3540.*

Counting viruses

Methods of counting or quantitating viruses fall into two discrete categories. Infectivity assays, as the name implies, measure virions that can successfully infect a cell to produce infectious progeny. Inactivated (noninfectious) virions are not counted. The second group of techniques measures specific virion components, often a specific viral protein, or the viral genome. These techniques are *chemical/physical measures* of virus quantification and they include serologic assays, PCR, and hemagglutination assays (HA). Negative staining EM can also be used as a chemical/physical assay to detect and count virus particles. Chemical/physical assays do not distinguish between infectious and inactivated (noninfectious) virions. For example, if one were to take two identical samples and expose one to UV radiation (thus damaging viral genomes), the amount of viral protein measured would be the same for both samples, even though the irradiated sample contains no infectious virus.

Infectivity assays

Infectivity assays require cell cultures, embryonated eggs, or animals (or plants, bacteria, fungi, etc.). An infectivity assay measures particles capable of replicating in a particular cell type or organism. *Plaque assays* are a common type of infectivity assay, used to count discrete "infectious centers." Samples containing virus are serially diluted and aliquots of each dilution are added to a dish of cultured cells (or a plant leaf in the case of a plant virus) (Fig. 4.10). Each dilution is usually tested in triplicate. The process is as follows:

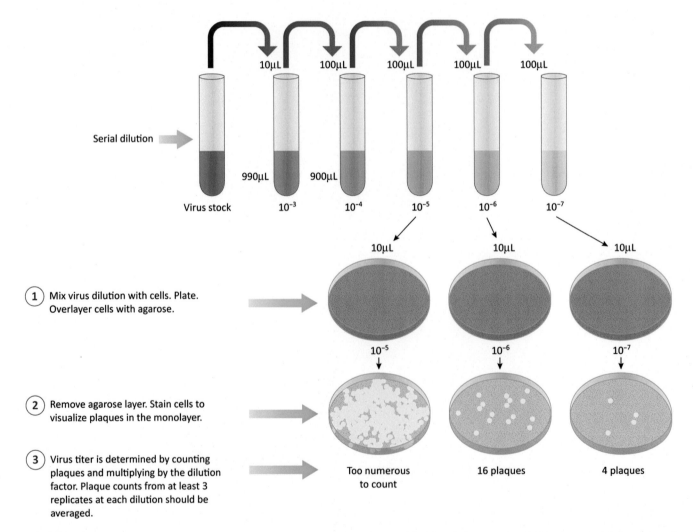

1. Mix virus dilution with cells. Plate. Overlayer cells with agarose.

2. Remove agarose layer. Stain cells to visualize plaques in the monolayer.

3. Virus titer is determined by counting plaques and multiplying by the dilution factor. Plaque counts from at least 3 replicates at each dilution should be averaged.

FIGURE 4.10   Plaque assays are used to count infectious particles. Samples are diluted and aliquots of each dilution are added to cultured cells. The cells are covered with an agarose overlay. Virus produced from an infected cell can infect nearby cells. If infected cells are killed, a region free of cells (the plaque) develops.

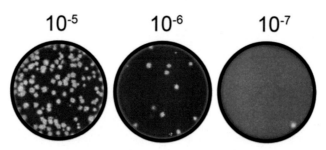

FIGURE 4.11 Stained cell monolayer with plaques. Living cells are stained purple leaving clear areas (plaques) where cells are absent (Jureka et al., 2020). Source: *From Jureka, A.S., Silvas, J.A., Basler, C.F., 2020. Propagation, inactivation, and safety testing of SARS-CoV-2. Viruses 12, 622.*

- An individual virion infects a cell.
- The virus replicates, producing progeny virions.
- Second generation virions infect and kill surrounding cells to generate a hole or plaque in the cell monolayer that is visible to the naked eye (or at low magnification).
- Live cells are stained, providing a dark background for the clear plaques (Fig. 4.11).

To determine a virus titer the number of plaques is multiplied by the dilution factor. Plaque titers are usually expressed at plaque-forming units (PFU)/mL. Thus for the assay shown in Fig. 4.10, the titer would be calculated as 16 plaques times 10^6 (the reciprocal of the dilution factor) divided by 0.01 mL (as 10 μL was put on the plate) for a titer of 16×10^8 PFU/mL in the virus stock.

The purpose of making and testing serial dilutions is to achieve a "countable" number of plaques in the cell monolayer. If too many infectious particles are present, individual cells are likely to be infected with more than one virion or individual plaques may merge to form large areas lacking cells. Plots of the number of plaques versus the dilution factor are linear if a single virion is sufficient to initiate a replication cycle ("single hit" kinetics). Results of plaque assays are reported as PFU per volume of measure (usually a mL). The term PFU acknowledges that we cannot know, with certainty, that the visualized plaques were formed by infection with a single virus of interest. Use of the term PFU should also remind us that variables such as type of host cell, type of culture media, ion concentration, or pH might affect the apparent concentration of virus. For example, a sample with a titer of 10^3 PFU/mL on mouse cells might have a titer of 10^6 PFU/mL on more permissive human cells.

What if the virus of interest does not kill infected cells? Can plaque assays be performed? The answer is yes, if there is a method for counting groups of infected cells. Groups of infected cells are called foci and the assays are focus-forming assays. Results are reported as focus-forming units (FFU). A common method of detecting foci of infected cells is to use a cell-based fluorescent antibody assay. Highly specific antibodies tagged with fluorescent molecules can be used to detect the presence of viral antigens in cells, as shown in Fig. 4.5. Focus-forming assays can be "direct" (using labeled primary antibody) or "indirect" (using labeled secondary antibody) as illustrated in Fig. 4.12. The technique allows one to distinguish infected from uninfected cells, thus can be used to perform focus-forming assays or endpoint dilution assays (described below). Foci of infected cells can also be detected using antibodies that are tagged with enzymes that cleave an uncolored substrate to produce a colored product. There are several variations on the focus-forming assays but all have in common the identification of small, discrete groups of infected cells in a monolayer culture (Figs. 4.13 and 4.14).

Endpoint dilution assays are another type of infectivity assay. In these assays, serial dilutions of a virus sample are made and are added to susceptible cells. The cells are incubated for period of time and then an assay is used to determine how many cultures are infected. The general protocol is as follows:

1. Prepare serial dilutions of the virus stock.
2. Use aliquots of each dilution to infect three to six test units. A test unit might be a culture dish, a single well on a multiwell plate, a leaf on a plant, or an animal.
3. Incubate test units to allow virus replication.
4. After a predetermined period of time, check each test unit and determine if virus has replicated. If virus can be detected, the unit is scored positive. If no virus is detected, the unit is scored negative. Virus detection assays might include PCR to detect viral genomes, serologic assays including ELISA or cell-based immunofluorescence, protein assays or enzyme assays (e.g., growth of a retrovirus is often detected by measuring the reverse transcriptase enzyme).
5. Count the total number of infected and uninfected units at various dilutions.

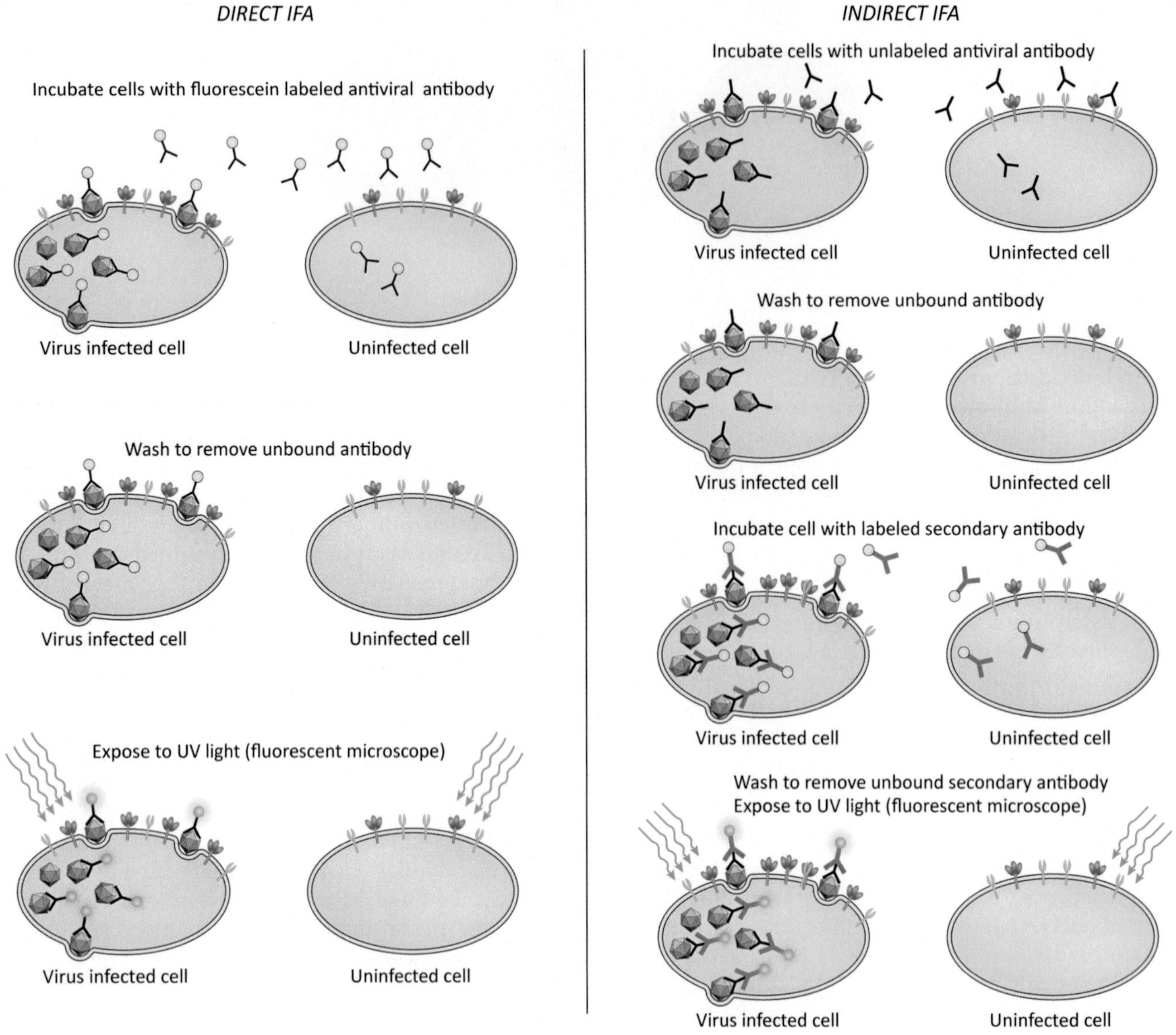

DIRECT IFA

Incubate cells with fluorescein labeled antiviral antibody

Virus infected cell   Uninfected cell

Wash to remove unbound antibody

Virus infected cell   Uninfected cell

Expose to UV light (fluorescent microscope)

Virus infected cell   Uninfected cell

INDIRECT IFA

Incubate cells with unlabeled antiviral antibody

Virus infected cell   Uninfected cell

Wash to remove unbound antibody

Virus infected cell   Uninfected cell

Incubate cell with labeled secondary antibody

Virus infected cell   Uninfected cell

Wash to remove unbound secondary antibody
Expose to UV light (fluorescent microscope)

Virus infected cell   Uninfected cell

FIGURE 4.12   Cell-based IFAs use tagged antibodies to mark infected cells. Direct immunofluorescence uses tagged primary (antiviral) antibodies. Indirect immunofluorescence uses a secondary antibody that is tagged.

FIGURE 4.13   Red blood cells (RBCs) binding to coronavirus-infected cells. Bovine coronavirus-infected cells are not killed but do express the viral envelope protein on their surface. The viral glycoprotein binds to RBCs. The RBCs serve to "mark" infected area in this electron micrograph. Source: *Courtesy HR Payne.*

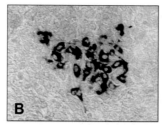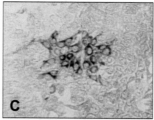

FIGURE 4.14 Focus assays for detection of hepatitis C virus. Left panel showing staining with an alkaline phosphatase-conjugated secondary antibody and right panel showing staining with a horseradish peroxidase-conjugated secondary antibody (Kang and Shin, 2012). Source: *From Kang W., Shin E.-C., 2012. Colorimetric focus-forming assay with automated focus counting by image analysis for quantification of infectious hepatitis C virions. PLoS One, e43960.*

When the endpoint dilution assay is done in cell cultures, the titer is reported in tissue culture infectious doses 50% ($TCID_{50}$) where one $TCID_{50}$ is the amount of sample that will infect 50% of the test units. For example, if 1 mL of virus sample is added to each of six units, and three of those units are positive at the end of the assay, the $TCID_{50}$ is 1. However, the results of endpoint dilution assays are seldom "perfect" and precise calculations are based on summing all of the data from a series of dilutions. Endpoint dilution assays can be performed using animals and results are reported as infectious dose 50% (ID_{50}) or lethal dose 50% (LD_{50}), if death is the endpoint. For a given virus sample, the $TCID_{50}$ titer is often higher than the ID_{50} titer as achieving successful infection of an animal usually requires more than one infectious particle to overcome host defenses. In turn, the ID_{50} is often greater than LD_{50} titer. Note that virus titers are not absolute values, but rather depend on the type of cells or animals used for the assay.

Chemical/physical methods of virus quantitation

Chemical/physical methods of virus quantitation measure the amount (or relative amount) of a viral protein, genome, or enzyme in a sample. Types of chemical/physical methods for enumerating viruses include the following:

- Direct visualization of virions by EM.
- HA.
- Serological assays (based on antigen–antibody interactions, see Box 4.1). Examples include ELISA, cell-based immunofluorescence assays (using tagged antibodies), and Western blots.
- Genome detection and quantification by PCR.

The readouts from chemical/physical assays provide no information about the amount of *infectious* virus in a sample, but they are often convenient, quick, and quite reproducible. They can often be correlated back to infectivity assays as a quick way to *estimate* the

---

BOX 4.1

Serologic assays

Serologic assays use the power of antibody–antigen interactions (the ability of an antibody to bind its cognate antigen with high specificity). There are many types of serologic assay in a variety of formats (i.e., ELISA, IFA, Western blot, lateral flow assays, and precipitation assays) and their use is not limited to viral antigens. The terms primary and secondary antibody are often used when describing serologic assays. A primary antibody is one that binds a specific antigen. If HIV is the antigen, primary antibodies are those that bind to HIV proteins. A secondary antibody is one that binds to a primary antibody. If the primary antibody is human-anti-HIV (e.g., from an infected patient), the secondary antibody recognizes the human antibody. The secondary antibody might be rabbit, mouse, or goat-antihuman IgM or IgG. When secondary antibodies are used, they are tagged with enzyme or fluorescent tags. The benefit of using a secondary antibody is that a single aliquot of secondary antibody can be used to detect many different human antibodies, regardless of what antigen those antibodies bind.

infectivity of a sample. However, if a virus sample is prepared or stored incorrectly, the protein or genome concentration might remain unchanged despite a significant decrease in the infectivity titer.

Direct visualization by electron microscopy

The number of virions in a sample can be determined using EM. While simple in concept, there are a number of technical issues. The concentration of virus must actually be relatively high as the sample size that can be applied to an EM grid is quite small (microliters). A virus at a concentration of 10^4 virions per mL would only yield an average of 10 virions per microscope grid. It would require considerable time to carefully scan the entire grid to find them. In addition, virus preparations are seldom completely pure and it can be extremely difficult to distinguish between cell debris and virions. However, if the virus concentration is sufficiently high, a sample can be mixed with a known volume of electron dense beads. One can then count, and compare, the number of beads and virions in a field of view.

Hemagglutination assay

Some viruses bind to the sialic acid residues on the surface of red blood cells (RBCs). A single virion can bind to several different RBCs, and an RBC can be bound by multiple virions to form a large network, or web, of cell and virus that is easily visualized. This process is called hemagglutination as the RBCs form a clump (Fig. 4.15). The HA is fast and inexpensive and does not require either sophisticated instrumentation or extensive training. It is done by preparing serial dilutions of a virus sample. An aliquot of each dilution is added to RBCs in a microtiter plate well or test tube. One well contains RBCs and saline (negative control) and another contains a known positive reference sample of virus. The samples are gently mixed and allowed to sit at room temperature. In the negative wells, the RBCs will slide down to form a tight button at the bottom of the tube. In positive wells, the RBCs and virions will bind to each other to form a mesh of cells on the bottom of the tube. The reciprocal of the highest dilution of virus that gives a positive HA is the HA titer. It should be noted that a particular virus may only hemagglutinate RBCs from a specific animal species.

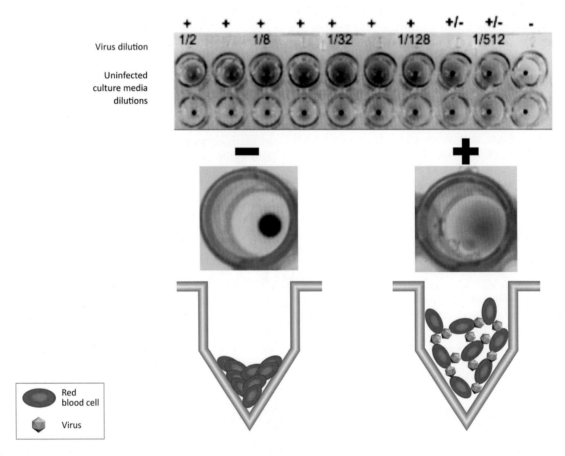

FIGURE 4.15   Some viruses bind to RBCs causing the cells to form a lattice. Note that positive hemagglutination is the presence of a lacey layer of RBCs. If virus is not present, the RBCs slide to the bottom of the tube to form a tight "button."

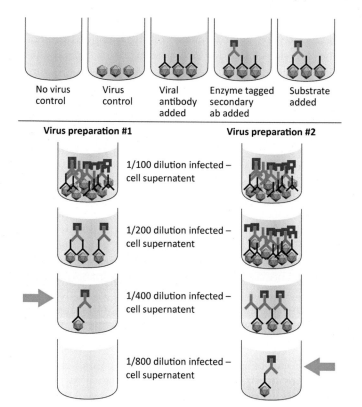

FIGURE 4.16 An ELISA can be used to determine the relative titer of a virus stock. The titer is expressed as the reciprocal of the highest dilution that produces a positive color.

Serological assays

Viruses can be quantitated using a variety of immunologic or serologic assays. A serologic assay is one that uses antibodies to detect an antigen (in our case a viral protein) in a sample. Most serologic assays used "tagged" antibodies. The tag can be an enzyme that cleaves a substrate (colorless substrate turns colored upon cleavage), a radioactive isotope, or a fluorescent molecule. Serologic assays are commonly used in research and diagnostic laboratories.

The cell-based immunofluorescence assays described above as a means of determining a virus titer are also serological assays.

Enzyme-linked immunosorbent assays

ELISAs (Fig. 4.16 and Box 4.2) require that either antigen or antibody be adsorbed onto a plate or tube (often plastic). As proteins readily bind to many types of glass or plastic, the adsorption part of the assay is straightforward. The enzyme used for detection is

---

BOX 4.2

Variations of the ELISA or IFA can be used for antigen or antibody detection

Direct ELISA or IFA. Primary antibodies (i.e., antiinfluenza virus antibodies) are chemically linked to enzymes such as horseradish peroxidase (HRP) or alkaline phosphatase (AP). These reagents facilitate detection of viral antigens in a sample.

Indirect ELISA or IFA. The primary antibody is untagged, but a tagged secondary antibody is used to detect the presence of primary antibody. For example, if the primary antibody is obtained from a human patient the secondary antibody is antihuman IgG (perhaps obtained from mice immunized with human IgG). Indirect ELISAs are common because it is cheaper to tag

one batch of antihuman IgG than to tag specific antibodies directed against hundreds of different viral (or other) proteins.

Sandwich ELISA. The antigen of interest is "sandwiched" between two antibodies. An antibody or a serum sample is used to coat a dish or tube. A sample containing an antigen is added and allowed to bind to the antibody. A second antibody is added to the wells. It will bind to the antigen (if present). In this example the antigen must be able to interact with two antibodies. Requiring that two different antibodies react with the same antigen can improve the specificity of the assay.

often either horseradish peroxidase (HRP) or alkaline phosphatase (AP). These enzymes are relatively stable, cheap, and easy to purify, and they can be chemically linked to antibodies to aid in detection of an immune complex. There are inexpensive substrates that change from colorless to colored when cleaved by HRP or AP. ELISAs can be used to detect either antigen (e.g., a virus) or antibody (from a potentially infected individual). Antigen detection ELISAs can be used to quantitate the amount of virus in a cell culture supernatant or patient sample. In this case the primary antibody is a purchased and/or standardized reagent. Antibody detection ELISAs demonstrate the presence of antibodies to a specific pathogen in patient sera. In this case the antigen (usually a viral protein) is the standardized reagent. The presence of antibody signifies that the patient has either been infected (or vaccinated) with the agent in question.

Western blots

Western blots are labor intensive and expensive, but provide a method of detecting specific viral antigens that may be present in a mixture. Western blots are done by electrophoresis of antigen (i.e., purified virus) by sodium dodecyl sulfate (SDS) polyacrylamide gel electrophoresis (PAGE). The proteins in a mixture are separated by size in the gel and are transferred from the gel to a solid (paper-like) substrate called a membrane. The membrane is then incubated with a source of antibody. The antibody may be tagged or conversely, untagged antibody may be detected using a tagged secondary antibody as described for the ELISA assay. If dilutions of a sample are prepared and used for electrophoresis, the Western blot may be a semiquantitative assay.

Western blots are also important diagnostic tools that are used to screen for specific antibodies in a patient. In this case, a patient's serum is incubated with a membrane containing viral antigens. A tagged secondary antibody is then used to detect any patient antibodies bound to the membrane. Western blots are highly specific as one can determine the molecular weights (sizes) of the proteins recognized by patient sera. False positive reactions can be identified when immunoreactive protein bands do not correspond in size to known viral proteins. For diagnostic purposes, a positive Western blot may require that more than one viral protein be bound by patient antibodies.

Genome detection and quantitation by PCR

PCR can be used to identify and/or quantitate viral genomes in a sample. PCR is a very sensitive method and uses oligonucleotide primers designed to detect specific viruses. The general process of PCR is shown in Fig. 4.17. PCR assays require knowledge of viral genome sequences to allow design of primers that will hybridize specifically to the virus of interest. A heat

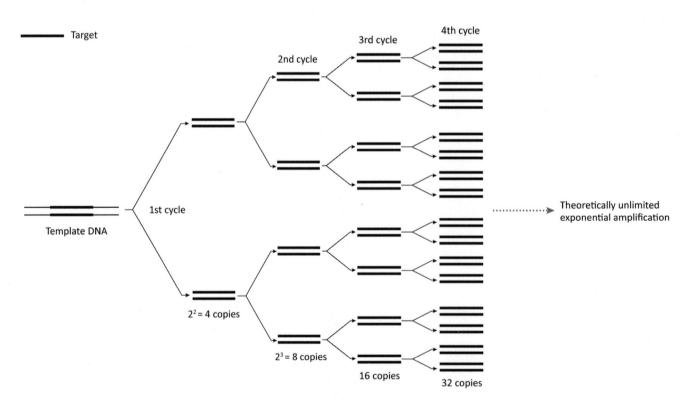

FIGURE 4.17   Overview of the PCR process to generate specific pieces of DNA from a sample.

stable DNA polymerase is used to synthesize a copy of the target nucleic acid. The sample is then heated, to denature the DNA, and the primers are allowed to rehybridize before the next round of DNA synthesis. The number of molecules doubles after each cycle of PCR. In a "conventional" PCR assay, the product is analyzed at the end of many cycles (generally 30–40). Analysis is done by electrophoresis to detect a DNA band of a specific size. The power of PCR is the ability to highly amplify, and identify, as little as a single DNA molecule present in a complex mixture.

Conventional PCR assays are not usually quantitative, but can be designed to be semiquantitative. A semiquantitative assay would involve making dilutions of a test sample, performing PCR on each dilution and, at the end of the assay, running the samples on an agarose gel to detect a DNA band of a predicted size. The density of the DNA band could be compared to known standards. The equipment required for conventional PCR is relatively low tech and inexpensive.

"Real time" PCR uses the same set of enzymatic reactions as conventional PCR, but the method for detecting the PCR product requires expensive equipment. Real time PCR works by detecting PCR product (amplified DNA fragments) at the completion of each cycle of the reaction and this version of a PCR assay can be quantitative. There are two general methods that are commonly used. The first is to use a nonspecific fluorescent dye that intercalates with double-stranded DNA (similar to staining a gel-containing DNA). The dyes are minimally fluorescent until they interact with double-stranded DNA. As the concentration of PCR product increases, the fluorescent signal increases. The second detection method makes use of a fluorescent reporter bound to the specific oligonucleotides used in the reaction. A fluorescent signal is only produced when the oligonucleotide hybridizes with the target DNA. Fig. 4.18 shows the results of a theoretical real time PCR assay. The X-axis is the cycle

number and the Y-axis is a measure of fluorescence. As DNA product is being produced, fluorescence increases until a plateau is reached. Two important parameters of real time PCR are the threshold value and cycle threshold (Ct). The threshold value is the fluorescence value that exceeds the background level. *The Ct is the cycle at which the background fluorescence value (the threshold) is exceeded.* Ct values are inversely proportional to the starting amount of target nucleic acid in the sample. The higher the Ct value, the lower the amount of starting material in the sample. Real time PCR assays are quantitative when known amounts of target nucleic acids are included in the assay to generate a standard curve.

PCR is a process that amplifies (by synthesizing new molecules) a particular target DNA. However, many viruses have RNA genomes so how can PCR be used to detect them? The solution is simply to use the enzyme reverse transcriptase (RT) in the first cycle of the PCR reaction. RT synthesizes a DNA copy of an RNA template. Once a DNA copy has been synthesized, the standard heat stable DNA polymerase in the reaction goes to work. This process is sometimes called reverse transcription PCR (RT-PCR). It is easy to confuse reverse transcription PCR with real time PCR when this shorthand is used.

The most recent iteration of PCR is digital PCR (dPCR). The same reagents are used as for real time PCR but this process incorporates the power of microfluidics. A single PCR reaction is assembled but then is separated into tens of thousands of partitions (e.g., nanoliter-sized drops) and the PCR reaction is carried out individually in each partition. After multiple rounds of amplification each drop is checked for the presence of product (detection of a fluorescent signal). Droplets without fluorescence are scored 0 and those with fluorescence are scored 1; absolute measurement of target DNA is determined using the Poisson distribution method. With dPCR there is no need to run standard

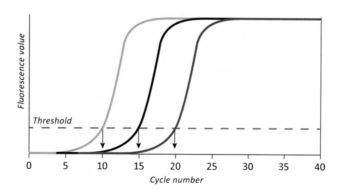

FIGURE 4.18 Theoretical real time PCR result. The three curves represent three concentrations of a target DNA. The light blue curve (left) has the highest starting concentration of target and therefore the signal (of specific DNA product) crosses the threshold at cycle 10. The green curve (right) has the lowest concentration of target DNA, it crosses the threshold at cycle 20.

curves and the method appears to offer increased sensitivity.

There are several advantages of PCR (any of the methods discussed above) as a virus detection tool. These include: (1) primer sets can be designed to recognize sequences common among groups of related viruses; (2) primer sets can designed to detect a specific member of a virus group; (3) multiple primer sets can be included in a single reaction to screen for more than one suspect virus in a sample; (4) PCR product (DNA) can be rapidly sequenced providing verification of a positive signal and genetic information about the virus; (5) assays can be quantitative; (6) much of the process can be automated; and (7) hundreds of samples can be processed at one time.

PCR assays are also very sensitive, but sensitivity can be a disadvantage as well as an advantage as it is quite easy to introduce contaminants into the reactions (e.g., PCR products from a previous assay or other DNAs, such as plasmids, that are present in the laboratory). Thus when performing PCR for diagnostic purposes, it is essential that every precaution be taken to avoid contaminating patient samples. This often requires separate equipment and work areas. For example, a "clean" area to process the patient sample, a second area to set up the assays and a third area where the PCR products are synthesized and analyzed. It is also important to test all purchased reagents for the presence of contaminating nucleic acids. This requires multiple negative controls. For example, one might apply sterile buffer or water to a nucleic acid purification column to check the column for contamination with nucleic acids that might have been introduced during the manufacturing process.

Basic principles of diagnostic virology

The purpose of diagnostic virology is to identify the agent most likely responsible for causing disease in a human or animal patient. Virus identification is important and can be used to:

- Determine treatment strategies (although only a handful of antiviral medications are available).
- Predict disease course and expected outcome.
- Predict the potential for virus spread.
- Allow identification of, and vaccination of, susceptible individuals.
- Trace the movement of a virus through a community, or world-wide.

For the medical practitioner, methods for identifying the virus in an infected patient ideally should be sensitive, specific, and rapid, as once a patient has recovered (or died) diagnosis has less practical value (although a diagnosis could benefit family and community members). On the other hand, epidemiologic studies may include hundreds or thousands of samples requiring use of low cost, high throughput modalities such as PCR.

Common targets for diagnostic tests are viral proteins (antigens), viral genomes, and/or antiviral antibodies. Some diagnostics are designed to detect viral proteins, enzymes, or genomes directly from a patient sample while other diagnostic methods require that the patient sample be added to susceptible cell cultures to allow for propagation of the virus. An indirect method of detecting current or prior virus infection is to assay the patient's blood for antibodies.

Methods to directly detect viruses in patient samples include some of those described above. ELISA assays are very sensitive and can be designed to detect viral proteins. EM is very rarely used, but can be used to detect virions in stool samples. PCR (highly sensitive and highly specific) is widely used to detect viral nucleic acids.

An increasingly common type of diagnostic assay is the lateral flow immunoassay that uses the process of diffusion to move a small liquid sample across a test chamber. These assays are very rapid and require no specialized equipment. The liquid sample contacts various dried reagents as it flows through the chamber (Fig. 4.19). Lateral flow immunoassays provide rapid diagnostic capabilities in medical/veterinary settings. Tests can be designed to detect either antigen or antibody from a patient sample. An example of an *antigen capture assay* is shown in Fig. 4.19. The test strip contains three key assay reagents (dried on the strip). Closest to the sample addition chamber is an antiviral antibody. In our example the antibody is conjugated to gold nanoparticles. When the liquid from the patient sample is applied, it flows, by capillary action, across the slide and will encounter the labeled antibody. The labeled antibody is picked up in the flowing liquid and will bind to any viral antigen present in the sample. The liquid continues to move across the slide (by capillary action) until it hits the test (T) strip. In our example the T strip contains antibody to the virus in question. If the labeled antibody is bound to virus, it will be stopped (captured) at the T strip. If the labeled antibody is not bound to virus, it will move past the T strip and reach the control (C) strip. Antiimmunoglobulin antibodies are bound at the C

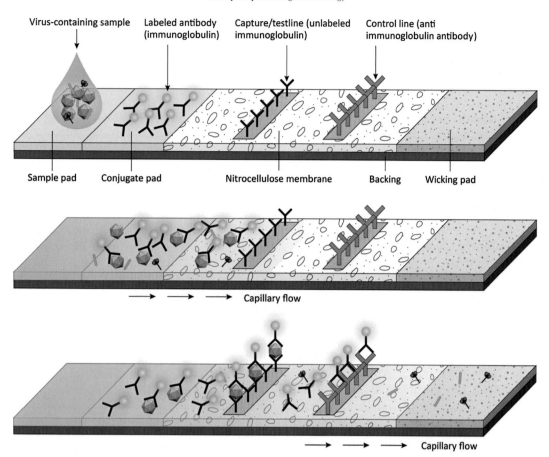

FIGURE 4.19   An antigen capture, lateral flow immunoassay. The test strip contains three key assay reagents. In this example, a gold-conjugated antibody binds to virus in the test sample. The liquid moves across the slide by capillary action to the test (T) strip. The T strip contains membrane bound antiviral antibody. It will bind to capture the virus particles and their associated gold-tagged antibodies to generate a visible signal at the T strip. If no virus is present in the sample, the gold-labeled antibody travels past the T strip and binds to antiimmuno-globulin antibodies bound at the C strip. A positive sample must show color reactions at both the T strip and the C strip. A negative sample must show a color reaction at the C strip (to validate that the test worked correctly).

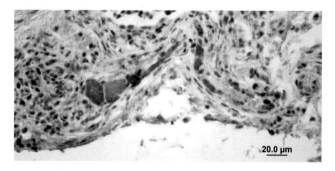

FIGURE 4.20   Immunohistochemistry is a technique used to detect viral antigens in tissue sections. In this example, a brain section has been treated with antibodies to a bornavirus. The antibodies are tagged with an enzyme, producing a brown precipitate that indicates bornavirus-infected cells. Source: *Courtesy I. Tizard.*

strip and will capture any labeled immunoglobulin in the liquid. Thus if the sample is positive for virus, col-ored lines should appear at the T strip and the C strip (because there is excess labeled antibody in the system). If the sample is negative for virus, the T strip will remain colorless but the C strip will be positive. Lateral flow immunoassays are used to detect a variety of proteins in patient samples and some are designed for home use. They are available to detect viruses such as influenza, SARS-CoV-2, rotavirus, and Ebola. A drawback of these tests can be lack of sensitivity.

Immunohistochemistry is a method for detecting protein antigens in tissues or cultured cells. Like the ELISA, this assay uses tagged antibodies, but they are incubated with cells or tissue sections. The antibodies can be tagged with enzymes or with fluorescent molecules. Immunohistochemistry is a powerful technique as it allows one to examine individual virus-infected cells in a tissue section (Fig. 4.20). Patient samples (biopsies) are often preserved in formaldehyde or are stored frozen at ultracold temperatures. If these samples are archived, they can be tested for the presence of viral antigen even after years or decades. In a research setting, this technique is often used in studies of viral pathogenesis as it allows the investigator to examine, in detail, the sites where virus has replicated in the infected organism.

Often there is not enough virus present in the infected host to allow for *direct detection* in a patient sample. In that case, the sample may be sent to a diagnostic laboratory for inoculation into cultured cells (or fertile eggs) to generate higher concentrations of virus. The infected cultures are closely observed for visible changes, such as cell killing (cytopathic effects), changes in cell morphology, or formation of syncytia (fused cells). Any of these changes in the infected cell cultures (as compared to uninfected controls) provides an indication that an agent is replicating, and that further testing is warranted to identify the agent. Growing viruses is labor intensive, takes days to weeks, and presents biosafety issues. Patient samples are often inoculated into several different types of cells in the hopes that at least one type will be susceptible. However, infected cultures can provide abundant material for detection of viral antigens by ELISA, IFA, immunohistochemistry, or Western blots. They can also provide material for PCR or direct sequencing, or EM can be used to look for the presence of virus particles in samples.

Screening for antibodies in patient samples is also an important diagnostic tool. Recent or past exposure to viruses (or immunization) can be determined by detecting antiviral antibodies in patient samples. This type of assay requires a panel of viral antigens against which antisera can be tested. It is possible to detect not only antiviral antibody, but also to determine the class of antibodies in the sample. Detection of IgM indicates recent or acute infection, as these are the first antibodies produced in response to infection. Detection of antiviral IgG indicates that the patient is, or has been, infected with a particular virus; however, IgG may also be present due to prior immunization. Measuring IgG levels can be helpful for diagnostic purposes if *two* patient samples are available: one collected early in the disease process (acute serum sample) and one collected after recovery (convalescent serum sample). If the IgG titer in the convalescent serum is higher (> fourfold higher) than in the acute sample, a diagnosis can be made. These data are not necessarily helpful to the patient, but are helpful in determining what viruses are circulating in a population. Methods to determine antibody titers include ELISA, cell-based indirect immunofluorescence, and Western blot assays. The key here is that a *specific* virus is used to capture antibody that might be present in the patient's serum.

A virus *neutralization assay* is a method to detect antibodies that are capable of inhibiting virus replication. Neutralization assays are a modification of the infectivity assays described above. Neutralization assays specifically detect antibodies capable of *inhibiting virus replication* (or in other words, antibody that can neutralize virus infection). Virus neutralization is a specialized type of immunoassay because it does not detect all antigen—antibody reactions. It only detects antibody that can block virus replication. This is important because related groups of viruses may share common antigens, but only a fraction of these antigens are targets of neutralizing antibody. A virus *serotype* is usually based on virus neutralization. For example, there are three major serotypes of poliovirus (Types 1, 2, and 3). In order to protect against poliovirus infection, a successful vaccine must induce neutralizing antibodies to all three serotypes. When developing vaccines, those that induce high levels of neutralizing antibodies are considered superior.

Deep sequencing for virus discovery and characterization

In the sections above, we learned about traditional methods for culturing and identifying viruses. These methods have been used for many decades to identify and characterize viral pathogens. However, in the past few years, the development of powerful DNA sequencing methods and computational strategies have provided new tools for virus discovery, characterization, and even diagnostics. Deep sequencing refers to sequencing a sample of DNA hundreds or

BOX 4.3

Common biochemical and molecular techniques for investigative virology (1)

Virologistists have a wide array of molecular and biochemical tools that can be used to study virus replication and/or pathogenesis. The scope of this text does not allow a complete discussion of these techniques but short descriptions of some common techniques are provided.

Electrophoresis. Charged molecules can be separated in an electric field. Often the electric field is applied to a semisolid gel (agarose or polyacrylamide) so the components in a sample can be separated by charge and/or size. Addition of SDS to a protein sample coats the proteins with a uniform negative charge so they are separated based on size alone. Nucleic acids have a uniform net negative charge, thus are easily separated by size using agarose or polyacrylamide gels. Gels are treated with specific stains to detect the presence of proteins or nucleic acids. Materials in the semisolid gel can be transferred to a solid support, or membrane, for additional manipulations, such as incubation with antibodies (Western blot).

Chromatography is the collective term for a group of techniques used to separate molecules in a mixture. A liquid mixture is applied to material packed in a column (column chromatography), or layered onto plate (thin layer chromatography). Separation is based on the relative ability of the components in the mixture to move through or across the support materials. Chromatography can be applied to any mixture of molecules but the research virologist most often uses column chromatography to separate viruses from contaminating cellular material or to separate individual viral proteins.

Size exclusion chromatography employs beads with pores of defined sizes. Larger molecules are excluded from the beads, thus move quickly through the column. Smaller molecules enter the pores in the beads and this slows down their passage through the column. The result is that larger molecules are eluted from the column before smaller molecules. Thus molecules (i.e., different viral proteins) are separated by size.

Ion exchange chromatography uses charged materials in a column to separate macromolecules based on their charge. If the material in the column is negatively charged, molecules with a positive charge will be retained in the column. By changing buffer conditions (pH and ion concentrations), macromolecules can be separated on the basis of charge.

The basis of *affinity chromatography* is the very specific interaction between two molecules (e.g., the interaction of an antibody to its cognate antigen). If an antibody is attached to a solid support, it can be used to capture antigen from a dilute solution. After washing away unbound material, the antigen can be released from the column by changing salt and or pH of the buffer. Proteins A and G are bacterial proteins that bind to immunoglobulins. They are used to affinity purify antibodies and antibody/antigen complexes.

even thousands of times over. This approach allows researchers to detect rare sequences comprising as little as 1% of the original sample. A strength of deep sequencing is that no prior knowledge of viral sequences is needed and there are few biases involved in the data collection process. The details of these tools are beyond the scope of this book but there are a few points to keep in mind as we read about new discoveries based on deep sequencing. One is that the methods used for acquiring sequences will affect the outcome. For example, if DNA is sequenced one will obviously not identify RNA viruses in a sample. Likewise, some techniques target mRNAs in a sample (by use of poly(A) primers), and this method will not reveal the genomes of unpolyadenylated RNA viruses. Selection of analysis techniques may also bias discovery results. Finally, it can be challenging to link a disease to the presence of a viral sequence; the simple presence of a viral sequence in a patient is not sufficient to ascribe causation. But with those caveats, deep sequencing techniques are now in the forefront of virus identification and characterization. Use of unbiased sequencing has resulted in an explosion of new viruses from humans, animals, and environmental samples. The current challenge is to develop an understanding of which viruses might be threats and which are part of our normal viral flora.

A few biochemical and molecular techniques commonly used by virologists are described in Boxes 4.3 and 4.4.

In this chapter we have learned:

BOX 4.4

Common biochemical and molecular techniques for investigative virology (2)

Flow cytometry. Cells are suspended in a stream of fluid and flow past an electronic detection apparatus. The number of cells passing the detector is counted. Thousands of cells per second can be counted. Often cells are labeled with fluorescent dyes (e.g., tagged antibodies) and are not only counted, but are separated into different collection tubes based on labeling patterns. The process is often used to get pure populations of cells from a mixture. Immunologists take advantage of the fact that different types of proteins are found on the surfaces of different types of lymphocytes. Thus labeled antibodies can be used to quantitate and separate different lymphocyte subsets from a blood sample. Not only can the presence of a molecule on the cell surface be detected, but the relative amounts can be determined as well. Flow cytometry is also used for viral diagnostics. Cells infected with an unknown agent can be incubated with panels of antibodies.

Reverse genetics. Virologists study viral genes to determine their functions. Historically this was accomplished by finding and purifying mutant viruses to determine how they differed from their "wild-type" parents. With the development of gene cloning technologies, virologists are able to clone entire viral genomes, manipulate them in the laboratory (introduce specific mutations), and examine the effects on virus replication or disease. Introducing specific mutation is the "reverse" of traditional genetic method of identifying a mutant phenotype and then determining the genetic cause. In order to do reverse genetics, the virologist usually starts with a cloned viral genome that can be introduced into cells to produce infectious viruses that can be analyzed in great detail.

Designer cells and designer animals. In addition to mutating viral genes, virologists manipulate the hosts as well. It is relatively easy to add, delete, or modify the genetic makeup of cultured cells. Animal genomes can be modified as well. Systems to create designer mice are reasonably efficient and hundreds of types of genetically modified mice are commercially available. What is gained by modifying the host? At every step in the virus life cycle, viral proteins interact with cellular proteins and a single viral protein may interact with dozens of cellular proteins. To analyze the effects of just one type of interaction, it may be preferable to modify the host genome. It is also possible to genetically alter a resistant host to render it susceptible to a virus. For example, the receptor for a human virus could be added to mouse cells to generate a tractable animal model. It can be challenging for a student of virology, new to reading scientific literature, to determine the types of virus or host mutations that have been used to obtain data for a specific study. But this is critical to understanding and evaluating experimental results.

- A general method for obtaining and culturing cells. Primary cultures contain cells obtained from an animal and they have a limited lifespan. Tumors are a source of immortalized or transformed cells that can divide continuously.
- A basic method for virus purification using centrifugation.
- Methods to visualize viruses including negative staining EM, thin sectioning EM, cryo-EM, and confocal microscopy.
- Methods to count infectious virus particles (plaque assays and endpoint dilution).
- The basic features of common techniques used to detect viruses and antibodies.

References

Bajaj, S., Dey, D., Bhukar, R., Kumar, M., Banerjee, M., 2016. Non-enveloped virus entry: structural determinants and mechanism of functioning of a viral lytic peptide. J. Mol. Biol. 428, 3540.

Counihan, N.A., Daniel, L.M., Chojnacki, J., Anderson, D.A., 2006. Infrared fluorescent immunofocus assay (IR-FIFA) for the quantitation of non-cytopathic and minimally cytopathic viruses. J. Virol. Methods 133, 62.

Jureka, A.S., Silvas, J.A., Basler, C.F., 2020. Propagation, inactivation, and safety testing of SARS-CoV-2. Viruses 12, 622.

Kang, W., Shin, E.-C., 2012. Colorimetric focus-forming assay with automated focus counting by image analysis for quantification of infectious hepatitis C virions. PLoS One 7, e43960. Available from: https://doi.org/10.1371/journal.pone.0043960.

5

Virus transmission and epidemiology

After studying this chapter, you should be able to:

- List modes of virus transmission.
- List some factors that impact virus transmission.
- Define or explain the terms virulence, pathogenicity, prevalence, incidence, incubation period, and infectious period.
- Explain how serological surveys provide information needed to determine the pathogenicity of a virus.
- List some of the factors that impact the spread of a virus.
- List some of the factors that impact the outcome of a viral infection.

Infectious disease epidemiology (which includes the epidemiology of viral diseases) is the study of the complex relationships between hosts and infectious agents. Epidemiologists are interested in spread or transmission of infections agents, with or without disease. Epidemiologists also try to predict the potential for development of epidemics and a very important part of their job is to define the kinds of interventions that could prevent or control an outbreak. Veterinarians are often concerned with threats to food animals (how a disease of food animals might be spread, or be introduced into a disease-free area). In this chapter, we focus on the epidemiology of viruses but some principals apply across all manner of infectious organisms. However, a feature that is

unique to viral epidemiology is that *all* viruses are obligate intracellular parasites, unable to replicate in the absence of susceptible host cells. Thus in order to model virus transmission, epidemiologists must try to account for a variety of factors involving both virus *and* host. Factors that can impact virus transmission and spread include the following:

- Mode or method of transmission of the viral agent.
- Duration of the infection and the window of transmissibility.
- Prevalence of the viral agent within a population. Numbers of susceptible and nonsusceptible individuals in the population.
- Population density.
- Patterns of travel or associations (e.g., schoolchildren and their families form interconnected networks).
- Living conditions.
- Climate and/or season.
- Presence/absence of insect vectors.

Predicting the course of an outbreak is particularly challenging if a new or novel virus (such as SARS-CoV-2) is involved, as there may be inadequate information about modes of virus transmission, window of transmissibility, stability of the virus in the environment, duration of infection, and even the proportion of infections that lead to overt disease. Predictions can be further complicated by additional factors that may impact the outcome

© 2023 Elsevier Inc. All rights reserved.

of infection. For example, differences in age, gender, nutrition, and genetic susceptibility of the host are important factors in the outcome of infection. Thus some infected individuals may have an asymptomatic infection while others develop severe, life-threatening disease.

categories: respiratory routes, food and water borne, exchange of body fluids, contact with contaminated fomites (objects such as toys, utensils, clothing, door knobs, medical or veterinary instruments), and insect vectored (Fig. 5.1).

Modes of transmission

Mode or mechanism of transmission is a key player in infectious disease epidemiology. Mechanisms of virus transmission can be divided into five broad

Respiratory routes

Human viruses that primarily replicate in the respiratory tract include human influenza viruses, rhinoviruses, respiratory syncytial virus, human coronaviruses (SARS, MERS, SARS-CoV-2, 229E, NL63, OC43, and HKU1) and

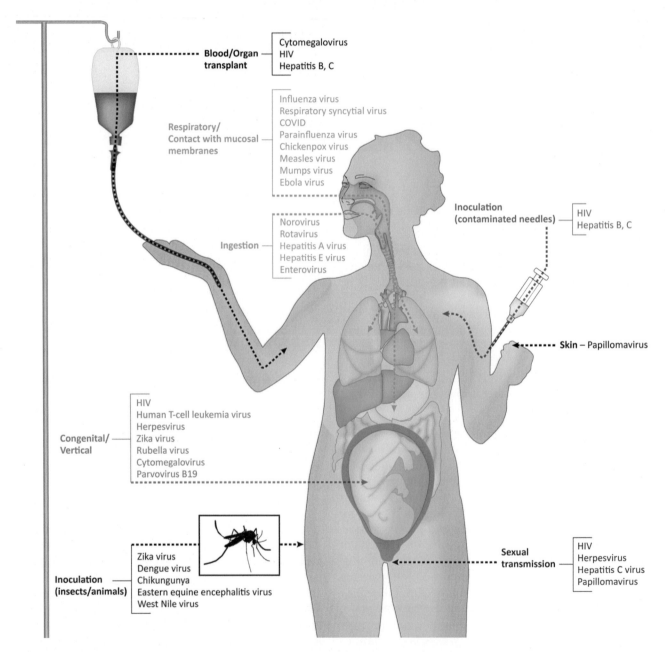

FIGURE 5.1   Major routes of virus transmission.

some adenoviruses and they are spread primarily via respiratory routes. However, other viruses that are not considered "respiratory" viruses are also spread by this route. Examples include measles virus, chickenpox virus [human herpesvirus 3 (HHV3) or varicella zoster virus], mumps virus, and Ebola virus.

The recent COVID-19 pandemic has generated intense interest in better understanding and defining the details of respiratory transmission of viruses. At first glance, respiratory transmission seems simple enough: it occurs when viruses replicating in the respiratory tract are expelled as droplets or aerosols and are inhaled by a permissive host. However, there are distinct mechanisms by which viruses can be transmitted by the respiratory route, and preventing spread requires a more nuanced understanding. Viruses can be released from the respiratory tract as relatively large droplets or as fine aerosols (Fig. 5.2). The definitions of "large droplet" and "aerosol" are inexact but large droplets are often characterized as greater than 5−20 μm. Large droplets stay in the air for a few minutes, travel less than 1 m, and rapidly fall out of the air column; coughing and sneezing release large droplets. Large droplets are likely to contaminate surfaces around the infected individual, leading to transmission through fomites. This occurs when a person touches a contaminated surface and then touches mucus membranes, such as their nose or eyes. Frequent hand washing can reduce this route of infection by influenza virus. Respiratory transmission of Ebola virus seems to occur most often in hospital settings, and is believed to occur via large droplets.

Aerosols are often described as infectious droplets of less than 5−20 μm that can remain in the air for an extended period of time and travel over longer distances. Particles less than 3 μm can remain airborne indefinitely. The COVID-19 pandemic has underscored the importance of fine aerosols in virus transmission. Virus replicating in the respiratory tracts of asymptomatic individuals can be released with each breath and inhaled by people in the shared space.

Masking, social distancing, improvements in ventilation, and moving activities to outdoor settings during the COVID-19 pandemic are all strategies implemented to reduce respiratory transmission of SARS-CoV-2. Social distancing and use of cloth masks are likely most effective at preventing spread by larger droplets, but not as effective for preventing spread by finer aerosols. N95 respirators or similar masks are effective at reducing virus spread by fine aerosols (Box 5.1). Measles virus, one of the most

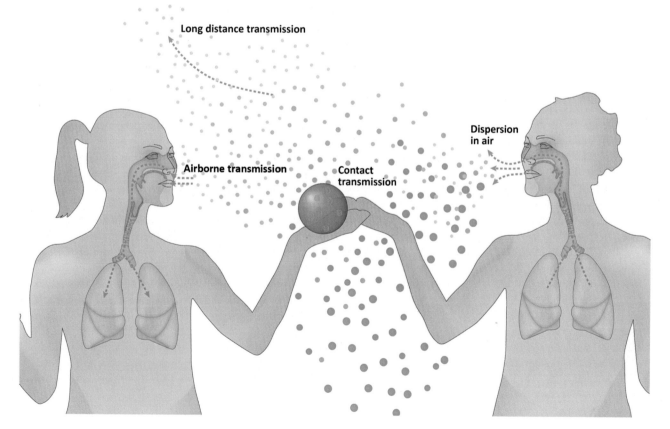

FIGURE 5.2   Respiratory droplets and aerosols are released by coughing, sneezing, and breathing. Larger droplets fall the ground within a few feet. The smallest aerosols can stay in the air for extended periods of time and travel much further.

BOX 5.1

Masking for infection control

Masking has become more common since the beginning of the COVID-19 pandemic. But in the context of infection control, what is the history of masking and for what purpose was it originally used? How effective is masking for infection control? What types of masks are best for that purpose?

As far back as the 1600s European plague doctors adopted "protective" clothing for use during outbreaks of bubonic plague. The outfits included masks that were shaped like long bird beaks. The beaks had air holes near the tip, and were filled with incense or fragrant herbs to mask the smell of the dead. At that time the prevailing theory was that disease was caused by bad odors, or miasmas. As we now know, bubonic plague (caused by the bacterium *Yersinia pestis*) is transmitted by fleas, thus plague masks would have provided little protection.

The mid to late 1800s saw the robust development of the science of microbiology in Europe. Pasteur had demonstrated that microorganisms were responsible for the "souring" of milk and wine, and correctly postulated they might also be the cause of human and animal diseases. A better understanding of the role of microbes in disease processes soon led to the use of masks during surgery, with the aim of preventing infection via surgical incisions. The first surgical masks were simple cloth coverings that prevented droplets (from coughing or sneezing) from entering wounds. By 1935 most surgeons in the United States and Europe routinely wore them. Surgical masks serve the same purpose today.

Two events at the beginning of the 20th century saw masking adopted by the general public for infection control: the Manchurian plague of 1910–11 and the influenza pandemic of 1918–19. The Manchurian plague was an outbreak of pneumonic plague largely centered in the extreme Northeast of China. Caused by the same bacterium, *Y. pestis*, as the bubonic plague, the pneumonic version is spread by the respiratory route from person to person. The death toll of that outbreak is estimated to have reached between 50,000 and 60,000 people, with an astounding mortality of 100%. An international team of doctors assembled in China to discuss control measures and a University of Cambridge (United Kingdom) trained doctor named Wu Lien-teh promoted the wearing of cloth face masks as one measure to stem the outbreak.

Numerous photographs taken during the influenza pandemic of 1918–19 depict the cloth masks worn by health care workers, police, and the general public. In some US communities, masks were mandated, not without controversy. A historical analysis of mask use during the influenza pandemic suggests that mandates, per se, did not ensure correct and consistent use of masks and certainly did not guarantee that masks were properly constructed of the best materials. Additionally, then as now, masks likely worked best as one part of a larger, more comprehensive strategy for controlling infection.

Important developments in medical masks took place in the mid to late 20th century. First was the general movement away from reusable masks that could be repeatedly sterilized, to disposable paper masks. Another was the development of newer nonwoven synthetic fibers purported to have superior filtering capacity that could be shaped to fit snugly on the face. These so-call respirator masks were designed to filter out incoming air, thus protecting the actual wearer of the mask. An important improvement on the design of early respirator masks was made by an American materials scientist, Dr. Peter Tsai in the 1990s. His group at the University of Tennessee found a way to add an electrostatic charge to mask materials. Viruses adhered to the charged material, providing significant improvement in comfort and performance. The charged material could have a larger pore size (to allow breathability), while still filtering out 95% of infectious agents from the air. Dr. Tsai's technology is used to manufacture today's N95 respirator masks.

The emergence of COVID-19 in 2019 generated a renewed interest in medical masking. First and foremost, the rapidly spreading virus revealed the reality of a fragile and inadequate supply chain for personal protective equipment (PPE) used by medical workers and other first responders. Masks of all types were in short supply, and those most knowledgeable about infectious diseases were particularly upset by the lack of N95s. The shortage of masks led to the necessity of their reuse, development of methods for sterilizing disposable masks, and the growth of a cottage industry in cloth mask manufacture. The shortage of medical grade masks also resulted in government agencies, such as the US Centers for Disease Control and Prevention (CDC), recommending *against* the use of medical masks by the general public. This advice changed when the CDC recommended use of cloth masks by the public to prevent spread of COVID-19 from infected to uninfected persons. In 2022, the CDC updated and clarified its mask guidelines to account for both the increased availability of high quality masks and the emergence of highly transmissible COVID-19 variants, such as Omicron. As of this writing, key messages from the US CDC include: the importance of masking as a public health tool; consistent use of the most protective mask

---

BOX 5.1 *(cont'd)*

that you, as an individual, can access and will wear; the importance of highly effective respirator masks in certain high risk situations; and use of the most effective masks by persons at increased risk for severe disease. Other information from the CDC on masking suggests use of snugly fitting cloth masks made from multiple layers of materials and/or double masking with a disposable surgical mask covered by a cloth mask. The US CDC publication (https://www.cdc.gov/coronavirus/2019-ncov/your-health/effective-masks.html) also includes additional valuable information, including descriptions of respirator masks.

---

transmissible viruses of humans, is easily spread by aerosol transmission.

Another factor that impacts transmission of respiratory viruses is the site of virus replication. The upper and lower respiratory tracts are discrete environments. The cells lining the upper versus lower respiratory tracts have different cell surface molecules thus play host to distinct viruses. In addition, the temperature of the upper respiratory tract is lower than the lungs. Finally, the fact that viruses transmitted by the respiratory route are released into the environment means that factors such as air temperature, humidity, and UV light play roles in virus survival and transmission.

A final example that illustrates the complexity of respiratory transmission of pathogens involves transmission of viruses from animal to humans via dried feces or urine (Fig. 5.3). Viruses that move between humans and animals are termed *zoonotic viruses*. Many hantaviruses infect rodents without causing significant disease, but occasionally jump into humans, causing severe disease, such as fatal pneumonia. Humans become infected by inhaling dust from rodent droppings. An outbreak of severe pulmonary disease in the Southwestern United States in 1993 was caused by a hantavirus named sin nombre virus. The outbreak was linked to an explosion in the deer mouse population that year. Since that time over 600 cases of hantavirus pulmonary syndrome have been reported in the United States. In 2012 an outbreak occurred among visitors to Yosemite National Park, United States. The outbreak resulted in 10 cases with 3 deaths. Most of the victims lodged at a single location in the Park (the Signature Tent Cabins) and the outbreak was linked to high rodent activity in the area. Avoidance of human hantavirus infection requires appropriate rodent control and cleaning methods. Rodent-infested areas should be sprayed and mopped with disinfectant as it is important to avoid generating dust that can be easily inhaled.

Food and water

Many viruses can be transmitted by ingestion of contaminated food or water. Virus enters the body through epithelial cells or lymphoid tissues in the gastrointestinal

FIGURE 5.3 Hantavirus transmission. The normal hosts of hantaviruses (family *Hantaviridae*) are rodents. They excrete virus in urine and feces. When rodent activity is high, structures may become contaminated. Humans can be infected by inhaling contaminated dust.

tract. Some viruses transmitted in this way are primary residents of the gastrointestinal tract and are collectively referred to as *enteric viruses*. Large amounts of these viruses are shed via feces into the environment. Enteric viruses can also be transmitted via contact with contaminated fomites. Vomiting and toilet flushing can aerosolize enteric viruses. Transmission via fomites is of particular concern in settings such as hospitals, day care centers, and nursing homes. Examples of enteric viruses of humans include rotaviruses (family *Reoviridae*) and the Norwalk-like viruses or noroviruses (family *Caliciviridae*).

Noroviruses have caused notable outbreaks on cruise ships, sickening hundreds of guests and crew in a matter of days. Many enteric viruses cause gastroenteritis but some cause hepatitis [e.g., human hepatitis A virus (HAV)] or neurological disease (e.g., polio virus); many enteric viruses induce silent infections. Most enteric viruses are unenveloped, and as might be expected, are quite stable in the environment. Adequate wastewater treatment is key to minimizing transmission by water consumption but enteric viruses can persist in recreational waters. Among the foods that are common transmitters of enteric viruses are shellfish (particularly if eaten raw). Shellfish, such as muscles, clams, and oysters are filter feeders: they concentrate pathogens present at low levels in water. Other foods that may be contaminated with enteric viruses include soft fruits and salad vegetables. These may be contaminating during cultivation (primary contamination) if irrigated with contaminated water or at any point during processing or preparation (secondary contamination).

Exchange of body fluids

Transmission of viruses via exchange of bodily fluids among humans can result from kissing, sexual contact, trauma (bleeding), blood transfusions, sharing dirty needles, organ or tissue transplantation, or artificial insemination. Veterinary procedures can enable virus transmission among animals, and animal to human transmission of viruses such as Rift Valley fever virus (family *Phenuiviridae*) occurs via contact with animal blood or milk. The term vertical transmission is used to describe various modes of transmission from mother to offspring that occur by exchange of bodily fluids. Vertical transmission can be across the placenta, during birth or via colostrum or milk. Among human viruses, human immunodeficiency virus (HIV), hepatitis C virus (HCV), and hepatitis B virus (HBV) may all be transmitted via contaminated blood, but may also be transmitted through contact with other bodily fluids such as semen or saliva. HIV can be transmitted across the placenta, in the birth canal or via breast milk. Rabies virus (family *Rhabdoviridae*) is unique in that it is naturally transmitted via saliva through a bite. Rabies has also been transmitted via organ transplant. Many viruses that are not usually associated with transmission by exchange of bodily fluids can in fact be transmitted by blood transfusion or organ transplant.

Insect vectors

Some viruses *must* infect multiple hosts to be maintained in nature. Examples include the yellow fever virus (Box 5.2) and dengue virus (family *Flaviviridae*). Both

---

BOX 5.2

Mosquitos in the water barrels. How yellow fever reached the Western Hemisphere

Author: I. Tizard, Distinguished Professor, Texas A&M University.

The yellow fever virus originated in the rain forests of Central Africa where it was a disease of monkeys. It was transmitted between these monkeys by mosquitos of the genus *Aedes*, most notably *Aedes aegypti*. Provided the monkeys were left in peace, the virus could circulate among them. Occasionally local hunters would wander into the deep forest and some, the unlucky ones, would be bitten by an affected mosquito and develop severe disease. This was probably not a common occurrence, and, given the low human population density in the rain forests, the likelihood of transmission between two humans was very low and major disease outbreaks were rare.

This changed with the development of the transatlantic slave trade. Africans were taken from interior Africa, marched to the coast, and loaded onto slave ships. Some of these victims were probably infected with the yellow fever virus. The slave ships had to carry food and water for their cargo. Water barrels were filled at any convenient lake or river. Some of this water would have contained mosquito larvae. Thus eventually, the two major

components required for disease transmission, multiple potential human victims and infected mosquitos came together as they were transported across the Atlantic.

The first yellow fever epidemic in the new world (as opposed to single cases in slaves) likely occurred in 1647 on the English colony of Barbados. The population here was sufficiently dense (about 200 persons per square mile) to sustain the epidemic. From Barbados the disease spread throughout the Caribbean, associated with the spread of the mosquito vector, *A. aegypti*. The virus reached Mexico, Florida, and northern South America within a year. The disease eventually established itself (became endemic) in much of Central America, the Caribbean islands, the southern United States, especially New Orleans, and occasionally reached as far north as the cities of Charleston, New York, and Philadelphia (cities that engaged in trade with the West Indies). It persisted in these urban areas because of the presence of *Aedes* mosquitos and readily available susceptible hosts.

Yellow fever is an acute hepatitis. As a result of liver damage, patients develop jaundice and thus appear yellow. Eventually it results in hemorrhagic disease, and bleeding

BOX 5.2 (cont'd)

into the gastrointestinal tract. Hence the Spanish name is "vomito negro" or black vomit. It is a lethal disease with 40% mortality. For example, in 1793, 44% of British soldiers based on Santo Domingo died within 3 months. The soldiers were based in the West Indies as part of the ongoing struggle between the British, Spanish, and French, for what were then exceedingly valuable island colonies, the source of most of the world's sugar at that time.

The French established a colony on the Western side of Haiti and this became a major sugar-producing site supported by a very large number of African slaves. When wars broke out between Britain and France, the Atlantic was dominated by the British Royal Navy who effectively cut Haiti off from France. As a result, a slave revolt led by Francois l'Overture succeeded in declaring Haiti's independence and it became the second independent state in the Western Hemisphere. In 1801 a brief truce between the British and French permitted Napoleon to attempt to reassert French sovereignty. He sent 20,000 soldiers under the command of his brother-in-law General LeClerc to retake the colony. This was rapidly achieved, but within a few months the French were dying of yellow fever at a rate of 50 per day. LeClerc himself died within a few months.

By 1803 only 3000 French soldiers were left alive, and the war with Britain had resumed. Napoleon recognized that he could no longer retain possession of colonies in the New World so rather than let the British capture them, he sold them to the United States. The Louisiana Purchase Treaty was signed in 1803. Given the importance of this to the United States, it is no exaggeration to say that yellow fever can be considered the most significant disease in American history.

For most of the 19th century, yellow fever continued to be a scourge throughout the Caribbean basin and sporadic outbreaks occurred regularly in New Orleans and around the US gulf coast. At this time, however, it was still not known how the disease was transmitted. But by the 1880s, there was a growing recognition that some diseases, malaria for example, could be transmitted by mosquitos. A Cuban physician, Dr. Carlos Findlay went so far as to propose that yellow fever was mosquito borne but was unable to prove this. His theory was correct but his timing was wrong! He left insufficient time for the virus to grow in the mosquito after it fed on a yellow fever victim before allowing it to bite a healthy human volunteer.

In 1898, the Spanish–American war broke out as a result of Spanish attempts to suppress a revolt in its colony of Cuba. The United States resolved to expel the Spaniards from Cuba and did so in a short campaign conducted during the dry season. Unfortunately once the fighting was over, US troops remained in the country during the rainy season and soon began to experience losses from yellow fever. Seriously alarmed, in 1900, the US Army established the "Yellow Fever Commission" under Major Walter Reed and sent it to Havana, Cuba. Their task was to determine the cause and seek to prevent the disease.

After first demonstrating that yellow fever was not a bacterial disease as some claimed, the commission began to consider seriously the theories of Carlos Findlay, namely that it was a mosquito-borne disease. They paid soldiers to volunteer for studies. Their first experiment allowing mosquitos to feed on yellow fever victims and then immediately letting them feed on healthy volunteers did not work. Nobody developed yellow fever. One of the Commission members, however, James Carroll, allowed himself to be bitten by a mosquito that had fed on a yellow fever victim 12 days earlier. Carroll developed yellow fever but survived. Because Carroll had been in contact with a yellow fever case a few days previously, the experiment was repeated on two more volunteers who had not been in contact with such cases. They both developed yellow fever and one, Jesse Lazear, died. The Commission now conducted a series of elegant experiments that confirmed conclusively that yellow fever was transmitted by the *A. aegypti* mosquito.

The next step was to control or eliminate the disease. William Gorgas, the chief sanitary officer in Havana organized an aggressive mosquito eradication campaign. By eliminating standing water, the sites where mosquito larvae developed, the mosquito population was reduced to such an extent that the number of yellow fever cases in the city fell from 1400 in 1900 to none in 1902.

The French engineer Ferdinand de Lesseps built the Suez Canal in 1869. This encouraged him to attempt to build a Panama Canal. The French began digging in 1882 but gave up after several years as a result of several factors, the most important of which was yellow fever. Not knowing how yellow fever was transmitted, there was much standing water in the construction zone and mosquitos thrived. In the first year of construction 400 workers died, in the second, 1300. When construction was abandoned 7 years late it is estimated that more than 22,000 workers had died. de Lesseps and his company went bankrupt. The United States decided to complete the construction of the canal and this recommenced in 1907. Prior to this, however, William Gorgas, the eliminator of yellow fever from Havana was called in and pursued aggressive mosquito control policies in the canal zone. Removal of standing water and screening of workers housing worked well. The death rate from yellow fever was reduced to insignificance and the last case of yellow fever in Panama occurred in 1906.

BOX 5.2   (cont'd)

In the years following, research into the cause of yellow fever demonstrated that it was caused by a virus. Eventually Max Theiler, working at the Rockefeller Foundation in New York, showed that it would infect mice when inoculated into the brain. By passing the virus many times through mice, he showed that the virus lost its virulence for monkeys. This attenuated 17D strain could not induce disease in human volunteers but did induce protective immunity. The 17D strain has proved to be an incredibly effective vaccine and has been used to protect millions of people. As a result, the virus is now confined to the remoter areas of rain forests in Amazonia and Central Africa where it causes occasional disease in unvaccinated visitors.

viruses are human pathogens but natural transmission is seldom directly from human to human. Instead the natural transmission cycle is from infected mosquitos to humans (Fig. 5.4). Insects known to transmit viruses include mosquitos, midges, and ticks. Viruses transmitted by insects are collectively called *arboviruses* (from arthropod borne). West Nile virus is another example of an arbovirus, but in this case the virus is largely maintained in a transmission cycle between mosquitos and birds (Fig. 5.4). Humans can be infected, but usually are dead end hosts because we seldom transmit virus back to mosquitos (and human—human transmission is very rare). Transmission from an animal to insect vector requires that sufficient virus be present in the blood.

Once ingested by the insect, most arbovirus replicate in the insect to facilitate spread back to the human, animal, or avian host. Many are maintained for long periods in their insect hosts. For example, mosquitos can transmit virus to their offspring (transovarial transmission). Thus even in the absence of susceptible vertebrate hosts, these viruses persists in nature. Virus transmission via insect vectors depends on a number of variables including the species of insect vector, the host preference of the blood-feeding vector, and environmental factors (temperature, rainfall) that may limit or facilitate vector spread. Vesicular stomatitis virus (family *Rhabdoviridae*), a virus of livestock, is thought to be maintained in infected midges or mosquitos but once an outbreak begins, animal to animal transmission occurs via virus-filled blisters. There is also a known example of *mechanical transmission* of a virus via insects: equine infectious anemia virus (family *Retroviridae*) is naturally transmitted among horses by biting horseflies. The mouth parts of these large flies become contaminated with blood after biting one horse and then transmits virus when biting a nearby horse. The virus does not persist, nor does it replicate, in the fly.

Role of the host specificity in virus transmission

Another important consideration in virus epidemiology is the number of different hosts that can support

Simple transmission cycle

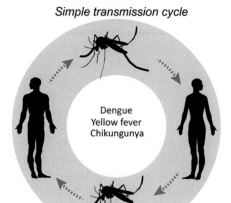

Complex transmission cycle

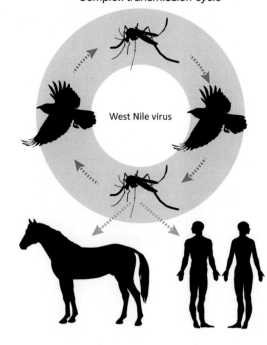

FIGURE 5.4   Transmission of viruses by insects. The top cycle depicts a simple transmission cycle involving humans and mosquitos. The bottom cycle shows human/horse infections via a complex transmission cycle. In this process the virus is normally maintained in birds and mosquitos. The virus may occasionally infect humans or horses but is usually not transmitted further.

replication of a virus. Some viruses are very host specific, infecting only a single species. For example, polio, measles mumps, rubella, and chickenpox viruses infect only humans. Some viruses (e.g., influenza virus) infect multiple species and spread across species boundaries. Some influenza viruses are relatively well adapted to multiple species. For example, influenza viruses can be transmitted between humans and pigs on farms, as well as at county fairs and petting zoos. Spread is not always from pigs to humans, spread from humans to pigs has been well documented. Avian influenza viruses can be transmitted to humans involved in the slaughter and processing of poultry. As mentioned above, rodent hantaviruses can occasionally infect humans. The natural hosts for Ebola viruses are bats, but can on occasion be transmitted to humans. A number of arboviruses that routinely infect birds and mosquitos occasionally jump to humans or horses (e.g., eastern equine encephalitis virus and West Nile virus). An important consideration in the epidemiology of viruses is whether or not a virus can easily spread within multiple hosts. West Nile virus can be transmitted to humans via mosquitos but once in the human host, the virus is seldom transmitted further. Viruses that infect only humans (or any single species) can potentially be eradicated (e.g., smallpox virus). In contrast, viruses that exist widely in nature (e.g., influenza viruses) will continuously evolve as they move among various hosts.

Transmissibility

An important goal of infectious disease epidemiology is to provide useful models of virus transmission. Such models can be used to determine the types of interventions (e.g., masking, hand washing, vaccination, quarantine, drug therapy) that will most effectively reduce transmission and limit the scope of an outbreak. Understanding the mode of virus transmission is only one step in developing robust models. Another important factor is to understand the transmissibility of a pathogen.

The ability of a virus to move between hosts is termed *transmissibility*. Transmissibility is dependent on a number of different factors. A virus might become more transmissible if mutations allow it to bind more tightly to a receptor or to use new receptors. Avian influenza viruses are poorly transmissible among humans because they bind receptors deep in the lungs. Mutations that would allow them to bind to receptors in the upper respiratory tract of humans would potentially increase their transmissibility among humans. Transmissibility might also increase if a virus can replicate more robustly (produce more new virions)

during an infection, such that larger numbers of infectious particles are released from each host.

Transmissibility can be described quantitatively. Epidemiologists use the expression *basic reproductive rate* (R_0) to describe transmissibility. R_0 is defined as the average number of new infections initiated by a single individual in a completely susceptible population during that individual's infectious period. Basic reproductive rate is a function of both virus characteristics and contact patterns within a community. Measles virus is highly infectious with an R_0 of 12−18. The R_0 of the Alpha variant of SARS-CoV-2 is two to three, while the more transmissible Delta variant is six to eight. The Omicron variant appears to be even more transmissible, with initial estimates of two to three times higher than Delta.

R_0 is calculated by assuming that every contact is susceptible. However, prior infection and/or vaccination reduces the number of susceptible individuals in the population. The effective or net reproductive number (R) is the actual average number of secondary cases and it equals the product of R_0 and the proportion of susceptible individuals in the population. R is smaller than R_0 because not all individuals in real populations are susceptible. What is the value of calculating R? When R is greater than 1, we expect the incidence of infection to increase, but when R is less than 1, the infection burns itself out. When R equals 1, the number of infections is constant.

It should be noted that transmissibility is not related to the ability of a virus to cause disease. A virus may be relatively difficult to transmit, but highly virulent if transmission does occur. It is easy to overestimate the transmissibility of a highly virulent virus.

Describing viral infection and disease

Infection versus disease

Another important concept in infectious disease epidemiology is the difference between infection and disease. While lay persons might use these terms interchangeably, they have very different meanings to infectious disease professionals. Many viral infections are in fact completely without disease. The infected individual is without clinical symptoms, but is harboring infectious, transmissible virus. The ability to survey healthy individuals for the presence of viral genomes has, in recent years, greatly expanded to the number of known viruses and increased appreciation of silent infections. Viruses that always, or usually, produce observable disease are easier to control. A key reason for the successful eradication of the smallpox virus was that over 95% of infections resulted in observable, distinct, disease. The virus was not able to hide or spread silently.

In contrast, poliovirus has proven much more difficult to eradicate. One reason is that there is less than one case of observable disease per 100 infections, but all of the infected individuals are shedding virus! Models of virus spread, as well as containment strategies can differ greatly, depending on whether or not individuals without disease can spread a virus.

A method to determine if there are, or have been, silent infections in a population is the *serological survey*. These are used to determine the *case infection ratio*, the number of *observed* cases per 100 infections (Fig. 5.5). During, or after a disease outbreak, blood samples are collected from a population and are tested for the presence of antiviral antibodies. The presence of antiviral antibodies indicates an infection occurred even in the absence of symptoms.

It is important to note that a serologic assay may not be able to differentiate between natural infection and immunization. Recently there has been increased interest in developing vaccines designed to allow for discrimination between infection and vaccination (DIVA). Most so-called DIVA vaccines are *subunit vaccines*. For example, the mRNA vaccines licensed for COVID-19 in the United States are based on the spike protein of SARS-CoV-2. A vaccinated individual develops only antispike protein antibodies, while natural infection elicits antibodies to other SARS-CoV-2 proteins, such as nucleocapsid.

Pathogenicity and virulence

Pathogenicity and virulence are two terms that are quite easily confused and are often, incorrectly, used interchangeably. *Pathogenicity is a measure of the proportion of infections resulting in overt disease.* Measles virus is highly pathogenic as over 95% of infected individuals will experience an observable disease episode.

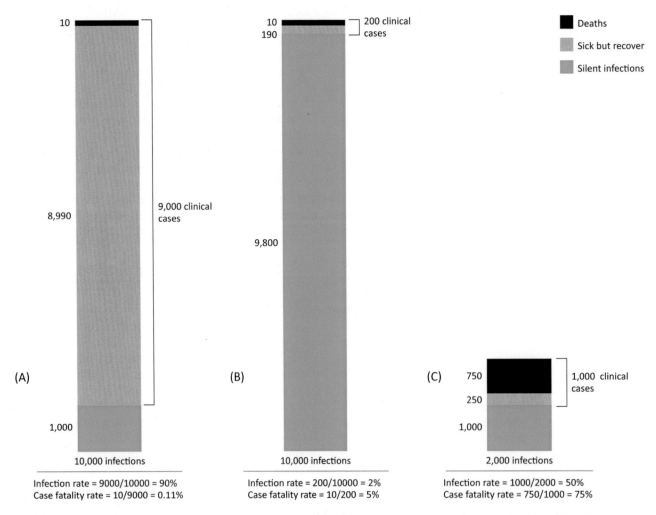

FIGURE 5.5 Infection ratio versus fatality ratio. The infection ratio is the number of clinical cases per case of infection. The case fatality ratio is the number of deaths per clinical cases. Three scenarios are presented. The virus in panel A has high pathogenicity but low virulence. The virus in panel B has low pathogenicity. The virus in panel C has high virulence.

Pathogenicity is technically not a measure of severity of disease. The disease in question may be a simple rash, or the common cold.

A related, but distinct term used to describe disease is *virulence*. *Virulence is a measure of the severity of disease and is often measured as the number of deaths*. An infectious agent with high virulence has a *high case fatality ratio*. The case fatality ratio is defined as *the number of deaths per 100 clinical cases*. It is important to note that a virus can have low pathogenicity (few infections lead to a disease episode) but high virulence (when disease does occur it is severe) (Fig. 5.5). This is the case with poliovirus, where 99% of infectious are silent but the 1% of clinically apparent cases present as severe neurological disease. Alternatively, the pathogenicity of a virus might be high, while its virulence (the case fatality ratio) is low. Measles virus provides an example of this scenario: the case infection ratio is 95% while the case fatality ratio varies from 0% to 25% depending on the overall health of the affected population. Outbreaks of novel viral diseases such as COVID-19 can be very stressful for clinicians, public health workers, and the population at large because we may not have enough information to distinguish between pathogenicity and virulence. We cannot assume that every infection results in disease.

Another important concept, particularly as regard viruses, is that very similar viruses can vary greatly in pathogenicity, virulence, and transmissibility. This is because very small genetic differences can have huge biological effects. For example, influenza viruses can vary greatly in their pathogenicity, virulence, and transmissibility, based on just one or very few amino acid differences.

Acute versus persistent infection

Acute viral infections are those in which host defenses are triggered upon infection and succeed in clearing the virus. Acute viral infections are typically characterized by rapid onset of disease and a relatively brief period (days or weeks) of symptoms. Acute infections may be mild (e.g., common colds caused by rhinoviruses) or severe (Ebola). Clinical symptoms may persist after viral clearance due to residual tissue damage caused by the virus or the immune system. For example, influenza virus is usually cleared from the body within 7 days, but weakness and other symptoms may persist much longer.

Persistent viral infections are those that last for many years, often for the life of the infected individual. Immune responses may control the virus, but are unable to completely eliminate it. The factors involved in maintaining viral persistence are complex, and

involve intimate interactions of host and virus. During some persistent infections, virus replicates and is detectable throughout. Examples include hepatitis B, hepatitis C, and (untreated) HIV infections; such infections are described as *chronic*. Herpesviruses commonly cause so-called *latent* infections. These are persistent infections in which the virus is often undetectable. Human herpesvirus-1 (HHV-1), which causes cold sores, replicates productively in mucosal epithelial cells but retreats to neurons in the face of an immune response. The herpesvirus genome persists in neurons for years, but if an individual's immunity wanes (perhaps due to stress) virus replication commences (the virus is activated from latency), travels back to epithelial cells and replicates robustly therein. Persons infected with the chickenpox virus (varicella-zoster virus, a herpesvirus) as children become latently infected for life. If their immunity wanes, as it often does in old age, a painful rash called shingles can develop. The shingles consists of fluid-filled vesicles that contain large amounts of infectious virus.

As mentioned above, outcomes of infection can depend highly on host factors. The possible outcomes of infection with HBV are complex and provide an example of how an individual's immune response can alter the course of infection. A robust immune response to HBV can lead to severe, acute disease, and the virus is eliminated. However, a more modest immune response can lead to long-term persistent infection without obvious symptoms.

Incubation, latent, and infectious periods

Development of epidemiologic models requires knowledge about the number of susceptible individuals in the population, the proportion of *infectious individuals*, those that can transmit virus, and the rate of contact between the two groups. In the case of an acute virus, a patient may be infectious for days or weeks, before and after the peak of clinical symptoms. Recovered patients usually have detectable antibody and no longer participate in the transmission chain. Vaccinated individuals may or may not be links in a transmission chain. (One hope of high vaccination rates is that this will provide protection for unvaccinated individuals, a concept called *herd immunity*. In reality, most vaccines are developed to protect against severe disease, rather than infection.) Transmission of persistent viral infections often occurs in the presence of antiviral antibody. Persistently infected individuals are capable of transmitting infection over many years or decades.

The time between acquiring an infection and the ability to transmit the agent is the *latent period*. The time between acquiring an infection and the onset of illness is the *incubation period*. The period during which the infected person can potentially transmit the infectious agent to others is the *infectious period*. If the incubation period is longer than the latent period, asymptomatic individuals unknowingly transmit virus. In a persistent infection, such as HIV, a long incubation period provides months to years during which an active infection may go undetected. Following the Ebola virus outbreak of 2014−15, it came as a surprise that some survivors continued to shed virus (e.g., in semen) for many weeks. Understanding the distinction between clinical illness and the ability to transmit virus is key to developing effective quarantine measures.

Incidence versus prevalence

Two terms used by epidemiologists are "incidence" and "prevalence." These terms have very specific meanings and are not interchangeable. *Incidence rate* (also called the *attack rate* for acute infectious diseases) is a count of the number of *new* infections during a specific time period. A specific population and a time frame are defined, and the number of new cases is counted in order to arrive at the numerator of the ratio. The denominator of the ratio includes both the size of population and time frame, and is often expressed as person-years.

Prevalence refers to the total number of cases present or counted. In the case of a persistent infection such as HIV, the numerator includes both patients infected recently, and those infected years or decades earlier. Prevalence is expressed as a ratio such as "cases per million." Note that there is no time parameter in this ratio.

For an acute viral infection, such as measles, the incidence rate or attack rate may be similar to the prevalence of the virus, given that virus is present and transmissible for a relatively short period of time. In contrast, when considering persistent or chronic infections, such as those caused by HIV or HCV incidence and prevalence may actually be quite different. In the case of HIV, education about safer sex practices has decreased incidence (new infections), while the availability of powerful antiviral drugs has resulted in significantly longer survival times for patients, thus increasing prevalence (Fig. 5.6). Interestingly, HIV drug treatment also decreases transmission (leading to fewer new cases).

Determination of incidence rates and/or prevalence of a virus depends on the ability to count infections and the values obtained depend on the methods used for counting. These rates might be determined by clinical symptoms, by testing for antibodies to an infectious agent or by using methods (e.g., PCR testing) designed to directly detect the agent. The entire set of criteria used in making a decision as to whether an individual has a disease is called the *case definition*. Changes in criteria for defining a case can dramatically change estimates of disease incidence.

The deliberate investigation of disease cases is called *active case detection*. Such investigations seek to classify an illness, determine the causative organism, assess the extent of an outbreak, stop the outbreak, and inform the public. *Passive surveillance* is the method by which healthcare workers report cases of disease among their patients. Many viral diseases are designated as "reportable" and in theory, each and every case should be reported. The list of reportable viral diseases includes: HIV, hepatitis A, B, and C, measles, mumps, rubella, rabies, and poliomyelitis among others. The weak link in passive surveillance is the

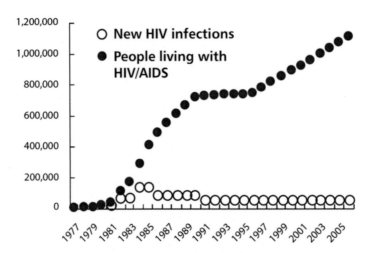

FIGURE 5.6   Incidence versus prevalence of HIV infections. The number of new HIV infections per year in the United States has remained steady but most infected individuals live years to decades with treatment, hence the yearly increase in prevalence. (U.S. CDC.)

filing of the initial report, as in practice only a small fraction of cases of common, reportable viral diseases are actually reported (reporting frequency may be as low as 10%−15%). Even so, passive surveillance is useful for monitoring infectious disease trends.

In summary, modeling the epidemiology of viral outbreaks requires knowledge of the particular virus, the host, the place (geography), and the time (season). Host factors include: age, nutrition, immunity, personal conduct, occupation, and interaction networks. Geography includes important factors such as temperature and humidity but also includes factors such as types of dwellings, water sources, and sanitary infrastructure. Many viral outbreaks are seasonal and the factors that influence seasonality include both host factors and geography. The most important reason for developing models of virus transmission is development of infection control programs. Controls might include quarantine, masking, social distancing, vaccination, water treatment, and/or insect control. However, we must also consider the human factor: the ability and willingness of a population to adhere to recommended control methods and guidelines.

In this chapter we have learned that:

- Modes of virus transmission include respiratory, food−water, exchange of bodily fluids, and insect vectors. Some viruses are transmitted by more than one route.
- A variety of environmental, biological, and societal/behavioral factors impact virus transmission.
- Viruses (even closely related viruses) differ in their virulence, pathogenicity, and transmissibility, and these attributes are not linked. A virus may be highly transmissible but have low pathogenicity. A virus may have high pathogenicity but low virulence. Recall that pathogenicity is a measure of how many clinical cases are seen among infected individuals. Virulence is a measure of the severity of disease.
- Serological surveys are important for gathering information about the total number of infections within a group, including those instances where infection was inapparent.
- Viral infections can be either acute or persistent; in rare cases, host factors may determine the outcome.
- Persistent viruses may be either *chronic* (virus produced constantly) or *latent* (virus produced only rarely).
- The term *incidence* is used to describe the rate of recent infections; *prevalence* describes the total number of individuals infected with a virus.
- The period between infection and development of symptoms is the *incubation period*. Virus transmission can occur during the incubation period, or even after clinical symptoms have resolved.
- A variety of factors impact the outcome of an infection. These include the age, sex, immune status, nutritional status, and overall health of the individual or animal.

6

Immunity and resistance to viruses

After studying this chapter, you should be able to answer the following questions:

- What are the major differences between innate and specific immune responses?
- What kinds of unique molecules (proteins, nucleic acids, carbohydrates) are associated with pathogens?
- How do cells detect the presence of viral pathogens?
- What are the first responses to viral infection?
- What are interferons (IFNs)?
- How are IFNs induced and what do they do?
- Describe at least one IFN-induced antiviral pathway.

- What is intrinsic viral immunity? Provide an example.
- What are the two main types of specific immune responses?
- What is antigen presentation?
- Describe three methods by which viruses evade immune responses.

What is immunity?

Immunity is the defense of the body against microbial invaders such as bacteria, viruses, parasitic

Viruses.
DOI: https://doi.org/10.1016/B978-0-323-90385-1.00016-9

© 2023 Elsevier Inc. All rights reserved.

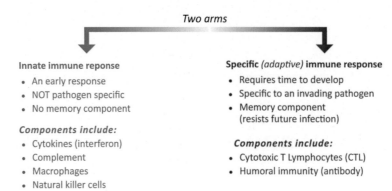

FIGURE 6.1   Overview of immune responses. A general comparison of innate and specific immune responses to pathogens.

protozoa, parasitic worms, and cancer cells. Immunity includes physical barriers to infection such as mucus, skin, and normal flora. *Innate immunity* involves cells and signaling pathways that are triggered by large groups of infectious agents. *Specific immune responses* use cells and antibodies that recognize very specific pathogens. There is a memory component to specific immunity that allows an animal to respond faster and more robustly the second (or third, fourth...) time an infectious agent is encountered. This type of immunity is often called *adaptive immunity*. In this chapter we review some basic information about innate and specific immunity (Fig. 6.1), with a focus on viral infections. If you are not familiar with the basic principles of immune responses, it would be helpful to read about the immune system in a general microbiology text. The following terms should be very familiar: antigen, antibody, lymphocyte, macrophage, T-cell, and B-cell. Some helpful definitions are provided in Box 6.1 and a brief description of the major cellular players in the immune response is provided in Fig. 6.2.

The immune system is complex, involving multiple cell types, protein activators, signaling pathways, and responses. Within a cell, *signaling pathways* connect activators to responses and the pathways intersect at many points. If signaling pathways are a roadmap, this chapter will introduce some highways and major intersections but will not consider the many smaller roads that may lead to the same destination. When reviewing this material, try to keep the "big picture" in mind by considering the following questions:

- If the immune system recognizes molecules unique to pathogens, why do mutations not arise that allow the pathogens to escape? (Sometime they do!)
- If pathogens change in response to the body's defenses, what is the response of the host? (Develop new defenses!)
- Can host and infectious agent reach a balance? (Often, but not always.)

Innate immunity

The foundation of innate immunity is the ability to detect and respond to foreign macromolecules, especially those common to groups of pathogens (Fig. 6.3). When a virus breaches physical barriers and infects a cell, hard-wired, innate responses to infection are immediately triggered. The virus is recognized as an invader because the body recognizes and responds to specific groups of molecules, collectively called *pathogen-associated molecular patterns* (PAMPs). This phenomenon is called *pattern recognition* and is accomplished by groups of proteins called *pattern recognition receptors* (PRRs). In the case of viruses, several different types of PRR recognize viral nucleic acids. When specific PRRs bind to viral nucleic acids, they initiate signaling cascades that change patterns of cellular gene expression, prompting cells to synthesize inflammatory and defensive molecules needed to controlling viral infections. Binding of PAMPs to PRRs not only triggers the innate immune response, but can also play roles in activating the development of specific immunity. The responses to viral PAMPs are diverse and varied, depending in part, on the cell type infected. Different cell types express different PRRs and signaling molecules. Sometimes multiple PRRs respond to a PAMP, or an initial response may trigger a positive feedback loop. Sometimes virally induced cell damage produces *damage-associated molecular patterns* (DAMPS) that engage PRRs. The following sections do not delve deeply into these intricately orchestrated responses, but provide a basic framework for understanding the incredible complexity of innate immune responses.

Pattern recognition receptors

Several different groups or families of PRRs have been discovered and described; they are named and grouped based on *homology of protein domains*. Probably

---

BOX 6.1

Definitions of key terms relating to innate immune responses

Antiviral State: Cells that have been stimulated by interferons express a large set of interferon-stimulated genes. The protein products have a variety of activities that help these cells block viral infection.

Cytokines: A large group of small proteins (~5−20 kDa) whose main function is cell signaling. They play a major role in immune responses. Subsets of cytokines include interferons, interleukins, lymphokines, tumor necrosis factors, and chemokines. In the context of infection, cytokines are released from cells such as macrophages, T- and B-lymphocytes, mast cells, endothelial cells, fibroblasts, and stromal cells. Once released from a cell, cytokines work by binding to receptor proteins on other cells. Binding of a cytokine to its receptor initiates a signaling pathway that can alter gene expression of the affected cell. *Cytokine storms* result from and excessive release of proinflammatory cytokines. The proinflammatory cytokines include: IL-1α, TNFα, and IL6. The signs and symptoms of a cytokine storm are varied but can include: fever, fatigue, swelling, muscle aches, nausea, and vomiting among others. More severe symptoms are increased blood clotting and very low blood pressure that can lead to organ failure.

Cytokine/Interferon Receptors: Plasma membrane-associated proteins that bind cytokines. Each type of cytokine has a specific receptor. Only those cells that express a receptor can respond to a cytokine.

Inflammation: A general term used to describe a body's response to any type of injury, including exposure to irritants (e.g., poison ivy or bleach), injuries that cause cell damage and infectious agents. The goal of inflammation is to eliminate the causative agent, remove damaged cells/tissue, and initiate repair and healing of affected tissues. Inflammation is a complex balancing act: too little can allow tissue destruction by pathogens while too much can cause excess damage to bodily tissues. Signs of inflammation include heat, pain, redness, swelling, and loss of function. An inflammatory response is quite complex and involves the vascular system, the immune system, and cells within an injured tissue. The initial response to a tissue injury involves movement of plasma and white blood cells into the injured area. Inflammation can be localized to the site of injury or may become systemic, involving many organs and tissues. Body-wide inflammatory responses result from circulation of proinflammatory cytokines.

Inflammasome: Cytosolic multiprotein complexes that assemble in response to PRR recognition of PAMPs. Activation and assembly of inflammasomes promote proteolytic cleavage and secretion of proinflammatory cytokines IL-1β and IL-18.

Interferons (IFN): A group of small signaling proteins (a subset of cytokines) that *interfere* with virus replication (hence the name). Viral infection stimulates cells to release IFNs that then bind to IFN receptors (IFNRs) on other cells. The signaling cascades initiated by IFNs result in expression of a large number of interferon-stimulated genes.

PAMP: Pathogen-associated molecular pattern. PAMPs are molecules shared by groups of related microbes that are essential for the survival of those organisms and are not found associated with host cells.

PRR: Pattern recognition receptors (PRRs) play a crucial role in the proper function of the innate immune system. PRRs are protein sensors that detect PAMPs. There are several groups of PRRs, some of which are displayed on the cell surface. Others are cytosolic or are associated with mitochondria. Binding of PRRs to PAMPs initiates signaling pathways.

---

the best known PRRs are the toll-like receptors (TLRs). Other families of PRRs include RIG-I like receptors (RLRs), NOD-like receptors (NLRs), pyrin and HIN200 domain-containing (PYHIN) proteins, and cytosol DNA sensing PRR cyclic GMP-AMP synthase (cGAS) (Fig. 6.3). PRRs recognize a variety of PAMPs from viruses, bacteria, fungi, and protozoa. PRRs that play major roles in innate responses to viruses recognize nucleic acids. A review of the PRR literature reveals a wide variety of proteins with roles in innate immunity, cancer, autoimmune diseases, and more. Here we will briefly describe PRRs importance for innate immunity to viruses.

Toll-like receptors

The word "toll," in German, means amazing or weird, and was used by geneticists to describe abnormal-looking fruit flies. The toll molecule was eventually found to be responsible for antifungal activities in Drosophila and when proteins with similar sequence, structure, and function were found in mammals, they were called "toll-like." Ten TLRs have been identified in humans. Some are found in the plasma membrane (PM) where they sense extracellular PAMPS, such as bacterial lipopolysaccharide, while others are associated with endosomes. TLRs 3, 7, 8,

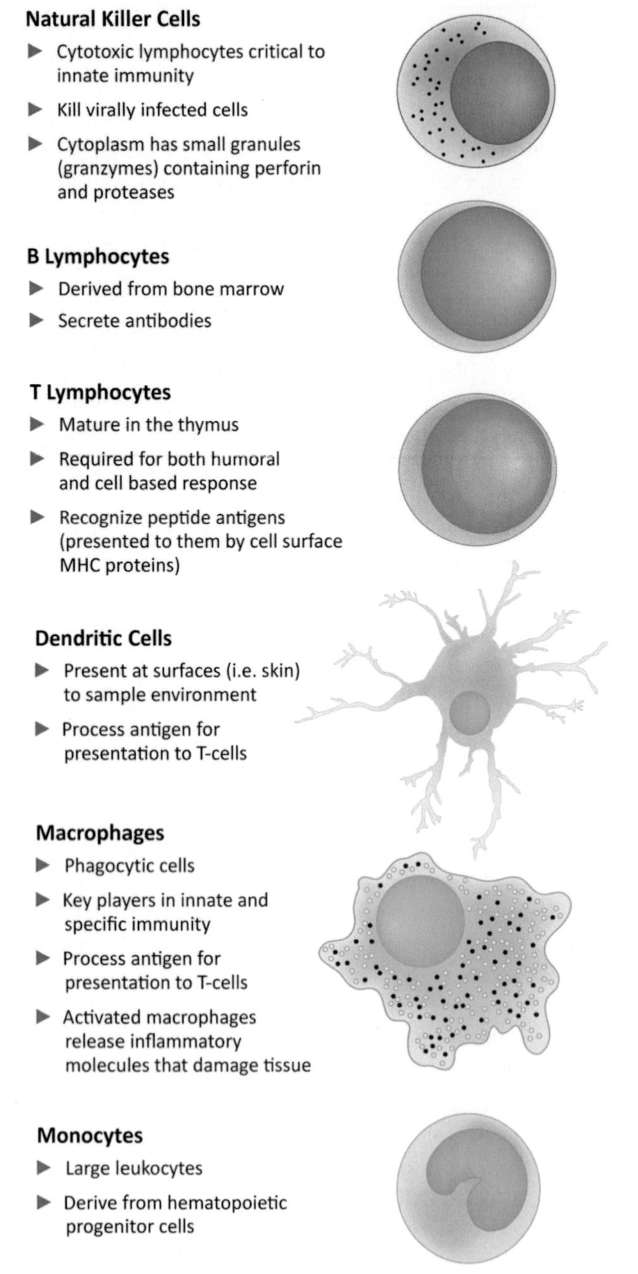

Natural Killer Cells
► Cytotoxic lymphocytes critical to innate immunity
► Kill virally infected cells
► Cytoplasm has small granules (granzymes) containing perforin and proteases
B Lymphocytes
► Derived from bone marrow
► Secrete antibodies
T Lymphocytes
► Mature in the thymus
► Required for both humoral and cell based response
► Recognize peptide antigens (presented to them by cell surface MHC proteins)
Dendritic Cells
► Present at surfaces (i.e. skin) to sample environment
► Process antigen for presentation to T-cells
Macrophages
► Phagocytic cells
► Key players in innate and specific immunity
► Process antigen for presentation to T-cells
► Activated macrophages release inflammatory molecules that damage tissue
Monocytes
► Large leukocytes
► Derive from hematopoietic progenitor cells

FIGURE 6.2 Cells of the immune system. Types of cells and their key roles in immunity to viruses are summarized.

and 9 are endosomal and are key players in antiviral responses (Fig. 6.4). TLR3 is expressed in many immune cells and is activated by double-stranded (ds) RNAs that are generated during replication of many RNA viruses. TLR3 also recognizes dsRNAs generated during the replication of some DNA viruses. TLR7 and TLR8 are highly related proteins that engage single-stranded (ss) RNA. TLR7 is expressed in some dendritic cells and B cells while TLR8 is expressed mainly in monocytes, macrophages, and dendritic cells. TLR7 and TLR8 bind to RNAs of several different RNA

PAMPS bind PRRs
TLRs
RLRs
cGAS, IFI16
MyD88
MAVS
STING
NfKb
MAPK
IRF3
IRF7
Into Nucleus
Genes for interferons and pro-inflammatory cytokines are expressed
Protein Synthesis
Interferons and pro-inflammatory cytokines are secreted to activate immune and antiviral responses

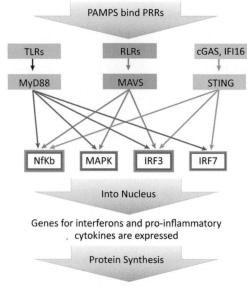

FIGURE 6.3 Signaling cascades are activated by PAMPs. PAMPs interact with classes of receptors (NODs, TLRs, RLRs) that in turn interact with major signaling molecules (MYD88, MAV, STING) that then activate specific transcription factors enabling genes encoding interferon and proinflammatory cytokines to be activated.

Cell membrane
Endosome
dsRNA
ssRNA
dsDNA
TLR7/8
TLR9
TLR3
TRIF
MYD88
MYD88
Cytoplasm
Active IRF3, IRF7, NF-kβ to the nucleus
Nucleus
Type I IFN
Type III IFN
Proinflammatory cytokines

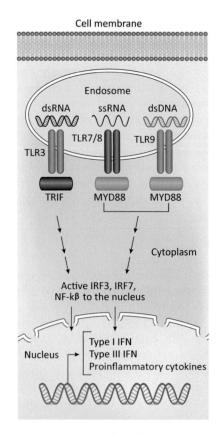

FIGURE 6.4 A subset of TLRs are nucleic acid sensors. When activated by viral nucleic acids they signal through TRIF and MYD88 to activate synthesis of IFNs and other proinflammatory cytokines.

viruses, but also seem to recognize RNAs produced during vaccinia virus (a DNA virus) infection. TLR9 is a DNA sensor and is highly expressed in some dendritic cells and B cells. TLR9 senses unmethylated CpG motifs in DNA of viral and bacterial genomes. TLR9 has been reported to detect DNAs of herpesviruses, papillomaviruses, and polyomaviruses.

After engaging with PAMPs, TLRs 3, 7, 8, and 9 activate signaling cascades. TLR3 initially interacts with a protein called tumor necrosis factor receptor-associated factor 3 (TRAF3) while TLRs 7—9 initially interact with the myeloid differentiation primary response 88 (MyD88) protein. Additional signaling proteins are recruited, eventually leading to nuclear localization of IFN regulatory factor 3 (IRF3), IFN regulatory factor 7 (IRF7), and/or nuclear factor kappa—factor beta (NF-κB). These transcription factors bind DNA to drive expression of proinflammatory cytokines and IFNs.

Retinoic acid-inducible gene-like receptors

Three proteins, retinoic acid inducible gene-1 (RIG-I), melanoma differentiation-associated protein 5 (MDA5), and laboratory of genetics and physiology 2 (LGP2) are members of the retinoic acid-inducible gene-like receptor (RLR) family. RLRs are cytosolic receptors for viral RNA. RIG-I and MDA5 have ATP-dependent RNA helicase domains as well as amino-terminal caspase activation and recruitment domains (CARDs) that function in signaling. LGP2 has a helicase domain but no CARD domain. The most intensively studied RLR to date is RIG-I. It is expressed in most cells at low levels, but increases in response to IFNs. RIG-I binds RNAs with a 5′ triphosphate group. Recall that cellular mRNAs have 5′-methyl guanosine caps, thus do not display a triphosphate at the 5′-end.

The genomes of some RNA viruses have complementary sequences at their 5′ and 3′ ends that form base paired (double stranded) structures and many plus strand RNA viruses form long dsRNAs during genome replication. MDA5 recognizes long dsRNA molecules. When RIG-I and MDA5 bind viral RNAs, their CARD domains are activated and bind the CARD domain of interferon-beta promoter stimulator-1 protein (ISP1, also called mitochondrial antiviral signaling protein, MAVS). This interaction initiates a multistep signaling pathway leading to activation of the transcription factor IRF3 and gene expression to produce IFNs (Fig. 6.5). The activities of LGP2 are not as clear; there is evidence for roles of LGP2 in both positive and negative regulation of RLR signaling.

Nucleotide-binding oligomerization domain (NOD)-like receptors

NLR proteins are characterized by three motifs: a C-terminal leucine rich repeat (LRR) domain, a central

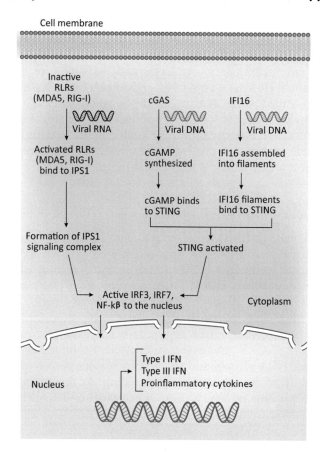

FIGURE 6.5 Cytosolic nucleic acid sensors include RLRs, MDA5, RIG-I, cGAS, and IFI16. They activate signaling pathways that activate IRF3, IRF7, and NF-κB, leading to the production of IFNs and other proinflammatory cytokines.

nucleotide-binding domain (NBD domain), and a variable N-terminal effector domain. Over 20 human NLRs have been identified. NOD1 and NOD2 were described early on, and were shown to detect specific structures within bacterial peptidoglycan. More recently NOD1 and NOD2 have been shown to have other activities. NOD2 is a sensor of viral ssRNA and interacts with ISP-1/MAVS, leading to activation of IRF3 and expression of type-I IFN. NOD2 activation not only drives of type-I IFN expression, type-I IFN also feeds back to increase expression of both NOD2 and NOD1.

Pyrin and HIN200 domain-containing proteins

Members of this protein family contain a pyrin domain (PYD) at the amino terminus and carboxyl terminal DNA-binding domains (called HIN200 domains). Proteins with PYDs are often involved in cell death processes, such as autophagy. Humans encode two PYHIN proteins with proposed roles in DNA sensing: absent in melanoma 2 (AIM2) and IFN- γ inducible 16 (IFI16). AIM2 is mainly expressed in intestinal epithelial cells, keratinocytes, and monocytes where recognition of dsDNA leads to the assembly of a large multiprotein complex termed the inflammasome, followed by release of proinflammatory cytokines and induction of cell death. IFI16 senses viral DNA and signals through a protein called stimulator of interferon genes (STING) to induce production of IFN-β and assembly of inflammasomes. The majority of IFI16 is located in the nucleus but small amounts are present in the cytosol. IFI16 also regulates transcription of some cellular genes.

Cyclic GMP-AMP synthase

Cyclic GMP-AMP (cGAMP) synthase (cGAS) recognizes cytosolic dsDNA thus is a sensor of many DNA viruses. Activated cGAS synthesizes the second messenger cGAMP. cGAMP then activates the STING protein, a major player in IFN signaling. Interestingly, the cGAS/STING pathway also seems to be important in regulating RNA virus replication. For example, mice lacking cGAS are more susceptible to infection by some RNA viruses and some RNA viruses encode proteins that can inhibit STING signaling. At this time the mechanisms by which cGAS is activated by RNA virus infection are not clearly defined.

Adapter proteins: next level players in innate immunity

As described above, there are an impressive array of PRRs that respond to pathogens. In the process of sensing diverse viral invaders, the next steps involve a few key signaling proteins downstream of PRRs. These include MYD88, TRIF, MAVS/ISP1, and STING (Fig. 6.1).

MyD88 and TRIF (TIR domain-containing adapter-inducing interferon-beta) are the primary signal transducers for TLRs. MyD88 interacts with all TLRs except for TLR3 and mice deficient in MyD88 show increased susceptibility to a wide variety of pathogens. TRIF is the adapter protein for TLR3 and TLR4. Engagement of MyD88 or TRIF begins a signaling cascade that eventually leads to activation of NF-κB and IRF3 and their translocation to the nucleus where they in turn regulate expression of type-I IFNs.

MAVS/ISP1 is a mitochondrial protein that activates NF-κB and IRF3 in response to viral infection through interactions with RIG-I and MDA5. Recall that binding of RIG-I or MDA5 to viral RNAs activates their CARD domains that in turn bind the CARD domain of MAVS/ISP1. This interaction initiates a multistep signaling pathway leading to activation of the transcription factors NF-κB and IRF3. STING is an endoplasmic reticulum membrane protein. STING appears to act as both a direct cytosolic DNA sensor protein and as an adapter for DNA sensors such as cGAS. When activated, STING dimerizes and activates downstream signaling pathways to ultimately induce type-I IFN. STING has also been reported to be an inhibitor of protein synthesis in response to RNA virus infection.

Inflammasomes

Inflammasomes are multiprotein complexes that assemble in response to pathogen sensing by NLRs or PYHIN proteins. Inflammasomes also have roles in cancer development, auto-inflammatory, neurodegenerative, and metabolic diseases. There are different types of inflammasomes, named for the PRRs that induce their assembly. For example, there are NLR and RIG-I specific inflammasomes. Inflammasomes are quite complex, but for our purposes their key function is the activation of caspases (a type of protease) that cleave and activate proinflammatory cytokines such as interleukin (IL)-1β and IL-18 (Box 6.1) initiating a type of proinflammatory cell death, termed pyroptosis.

Other proinflammatory molecules

A variety of proinflammatory molecules and IFNs are released from cells as a result of PRR activation. In addition to proteins (i.e., IFN and ILs) other proinflammatory molecules include nitric oxide (NO), produced by the enzymes nitric oxide synthase 2 and cyclooxygenase-2, and proinflammatory lipids such as prostaglandins and leukotrienes. These molecules increase local blood flow, attract defensive cells (e.g., neutrophils), and increase blood vessel permeability so that antimicrobial molecules and cells can reach the affected tissues.

Interferons

Signaling pathways activated by viral PAMPs result in the production of important antiviral molecules called IFNs. IFNs are a subset of cytokines, proteins important in cell signaling. The name interferon was derived from observations that infected cells released molecules that could *interfere* with many types of viral infection when added to other cells. There are three classes of IFNs, types I, II, and III.

Type-I interferons

Type-I IFNs are a group of structurally similar cytokines. IFN-α and IFN-β were the first discovered, but other type-I IFNs include IFN-ε, IFN-κ, IFN-ω, IFN-δ,

IFN-ζ, and IFN-τ. Type-I IFNs are found in all mammals and similar molecules have been found in birds, reptiles, amphibians, and fish. The importance of type-I IFNs in controlling viral infection is underscored by the presence of 13 IFN-α subtypes in humans. Type-I IFNs are secreted by most cell types in response to stimulation of PRRs by viral infection. The secreted type-I IFNs act in an autocrine (same cell), paracrine (nearby cells), or systemic fashion by binding to cells expressing the type-I IFN receptor (IFNAR) (Fig. 6.6). IFNAR is a heterodimer composed of two subunits, INFAR1 and INFAR2. The intracellular domain of IFNAR is associated with two tyrosine kinases, tyrosine kinase-2 (TYK2) and Janus kinase 1 (JAK1) to initiate a signaling cascade through the signal transducer and activator of transcription (STAT) protein, leading to the expression of about 300 IFN-stimulated genes (ISGs). Expression of ISGs leads to the so-called "antiviral state" of uninfected cells (Fig. 6.7).

ISGs have a wide variety of antiviral activities. The number and diverse antiviral activities of ISGs, as well as ability of viruses to circumvent them, underscore the truly ancient relationship between viruses and

cells. While the interaction of a PAMP with a PRR sets up a signaling cascade in a particular cell, release of type-1 IFNs by that cell has a global effect on the body. This is why a localized viral infection (e.g., influenza A virus in the respiratory tract) can result in a "whole body" illness typified by fatigue and achiness.

Type-II interferon

Interferon gamma (IFN-γ) is the sole member of the type-II IFN class. IFN-γ has roles in both innate responses and specific responses to viruses. It is produced primarily by specialized cells of the immune system: natural killer cells, subsets of T-cells and innate lymphoid cells. However, IFN-γ signals through the IFN-γ receptor (IFNGR) which is found in most cells. Therefore IFN-γ is produced by specialized cells, but can bind nearly all cell types to elicit a response. The key output of interaction of IFN-γ with its receptor is activation of the JAK-STAT pathway. One important role of IFN-γ is to bridge the innate and specific immune response pathways by playing a major role in establishment of cellular immunity. The genes induced by IFN-γ are a subset of those induced by the type-I IFNs.

Type-III interferons

The type-III IFN group consists of IFN-lambda (λ) 1 through 4. IFN-λ is induced by PRRs, such as RIG-I and is produced by a variety of cells. High levels of IFN-λ expression are seen in lung and liver cells; it may be the major IFN induced by respiratory virus infection. Certain dendritic cells also produce large amounts of IFN-λ. After release, IFN-λ binds to a heterodimeric IFN-lambda receptor (IFNLR), whose expression levels vary by cell type. Dimerization of the receptor leads to activation of JAK1 and TYK2 and phosphorylation and nuclear localization of STAT-1 and -2 to activate ISGs. In addition to functions in innate immunity to viruses, via expression of ISGs, IFN-λ also appears to play an important role in the development of specific immunity, controlling development of cytotoxic T-cell responses versus antibody responses.

JAK-STAT signaling

As described above, IFNs modulate cell activities through binding to their receptors. JAKS are key players in IFN signaling as they bind to the cytosolic domains of IFNRs. In the absence of a stimulus (the IFN ligand), the cytoplasmic domain of an IFNR is bound by an inactive JAK protein. Upon IFN binding, JAKs are activated and phosphorylate the IFNR. This leads to the repositioning and binding of STATs. Type-I IFNs bind to heterodimeric

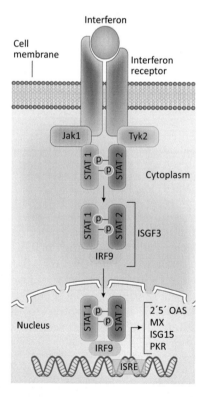

FIGURE 6.6   IFN signaling pathway. IFNs bind to receptors and signal through JAK STAT pathways. JAKS are found in association with IFN receptors. In the absence of a stimulus, JAK proteins are inactive. If IFN binds a receptor, the kinase domains of the JAKs phosphorylate the IFN receptor chains. This in turn activates STATs by exposing their nuclear localization signals. In the nucleus, STATs are transcriptional activators that drive expression of ISGs.

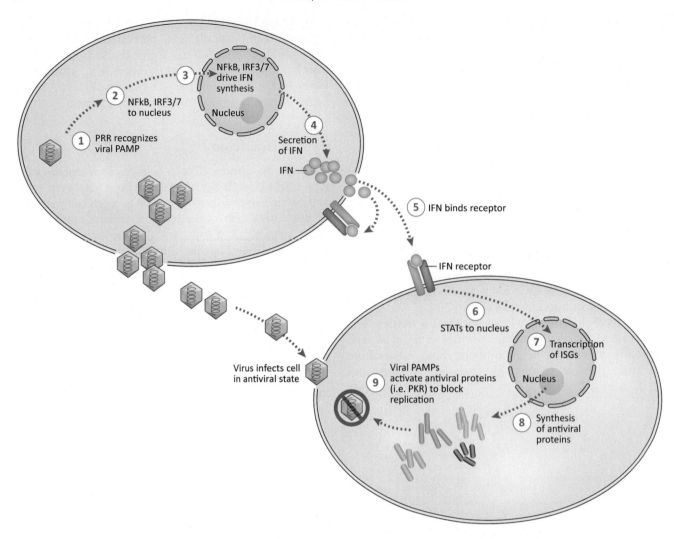

FIGURE 6.7   The antiviral state. Production and secretion of IFNs in response to viral infection leads to the development of an antiviral state in other cells. The response is engendered by binding of IFNs to their receptors, activating JAK-STAT pathways, and expression of a large number of ISGs. These include proteins that are ready to be activated, and can shut down replication of infecting virus.

IFNAR which is expressed by most cells. The intracellular domains of IFNAR and IFNLR are associated with TYK2 and JAK1. The IFNGR interacts with JAK1 and JAK2, also with the end result of homodimerization and phosphorylation of STAT1. As a result, STATs are phosphorylated, released from the receptor, and dimerized. There are seven different STAT proteins but STAT1 and STAT2 are the most important when it comes to signaling through IFNRs. STAT dimers have exposed nuclear localization signals. Once in the nucleus, STATs are transcriptional activators that drive expression of ISGs.

Interferon-stimulated genes and the antiviral state

There are over 300 ISGs and they encode proteins with a variety of activities. Some ISGs are regulators of

IFN signaling itself. Others have specific or general antiviral effects. Cells responding to IFN binding are said to be in an *antiviral state* (Fig. 6.7); they are refractory to infection by many viruses (not just the specific virus that originally triggered the IFN response). Some types of viruses are more sensitive to the effects of IFNs than others. The following sections describe the activities of a few ISGs.

Protein kinase RNA-activated pathway

The enzyme protein kinase RNA-activated (PKR) is produced in response to type-I IFNs. PKR remains in an inactive or latent state until, as its name suggests, it is activated, via autophosphorylation, upon binding to viral dsRNA (Fig. 6.8). Once active, PKR phosphorylates the eukaryotic translation initiation factor EIF2A.

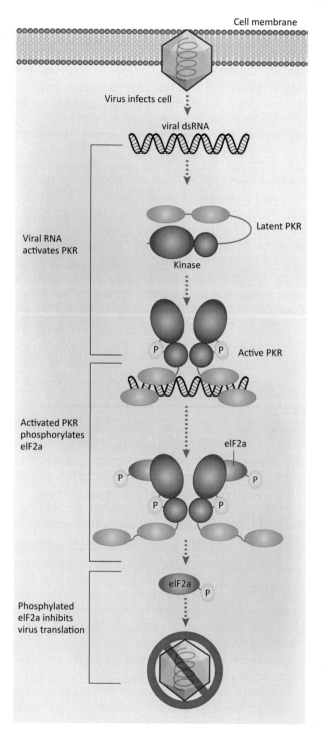

FIGURE 6.8　PKR pathway. Protein kinase RNA activated (PKR) is induced by IFN and activated, by autophosphorylation, upon binding dsRNA. PKR then phosphorylates eukaryotic initiation factor eIF2a, thereby inactivating it and inhibiting translation. Both host and viral translation are inhibited.

Phosphorylation of EIF2A inhibits mRNA translation, thereby preventing host and viral protein synthesis. Active PKR also induces cellular apoptosis, as a way to prevent further viral spread.

2′-5′ Oligo A synthase pathway

Two ISGs are key to this pathway: ribonuclease L (RNaseL) and 2′-5′ oligoadenylate synthase (OAS). RNaseL (L for latent) is inactive when synthesized, but when activated destroys RNA within the cell, leading to autophagy and apoptosis, two processes that are robust antiviral responses. In order for RNAseL to become active, it must bind to polymers of 2′-5′-linked oligoadenylate (2′-5′ oligo A) and these are synthesized by OAS. However, OAS is also produced in an inactive form and only becomes active upon binding to dsRNA (Fig. 6.9). The requirement for a multistep pathway helps insure that RNaseL is not activated in the absence of viral infection. To summarize the pathway: (1) RNaseL and OAS are ISGs; (2) if a cell expressing these proteins is infected by an RNA virus, long dsRNA accumulates in the cytoplasm; (3) OAS is activated and synthesizes 2′-5′ oligo A; (4) RNaseL binds to 2′-5′oligo A and is activated; and (5) active RNaseL cleaves viral and cellular RNA.

Other interferon-stimulated genes

Tetherin is an ISG that inhibits a variety of enveloped viruses, including human immunodeficiency virus (HIV), Ebola virus, Lassa virus, vesicular stomatitis virus, and Nipah virus. As the name implies, tetherin inhibits release of enveloped viruses by tethering them to the cell surface. Tetherin is a small glycosylated membrane protein (181 aa). It is found associated with cholesterol-rich microdomains in the PM, preferential budding sites of several enveloped viruses. The antiviral role of tetherin appears to be physically crosslink virions to the PM. A number of viruses, including HIV and Ebola virus, have evolved mechanisms to counteract tetherin.

The *IFN-inducible transmembrane protein (IFITM)* family of proteins inhibits a long list of enveloped viruses including influenza A and B viruses, dengue virus, hepatitis C virus (HCV), and Ebola virus among others. IFITMs contain two membrane-associated domains separated by a conserved intracellular loop. They are found on cytoplasmic and endosomal membranes and block viral replication by preventing fusion of the viral envelope to the host membrane. Imaging studies suggest that IFITMs trap infecting virions leading to their destruction by lysosomes and autolysosomes. The importance of IFITM-mediated virus restriction has been demonstrated in mice, where lack of IFITM3 increases susceptibility to influenza virus infection. A role for IFITM3 in human susceptibility to influenza virus infection has been suggested, based on evidence linking a minor human allele of IFITM3 with more severe disease.

Viperin (virus inhibitory protein, endoplasmic reticulum associated, IFN inducible, also known as RSAD2) is an

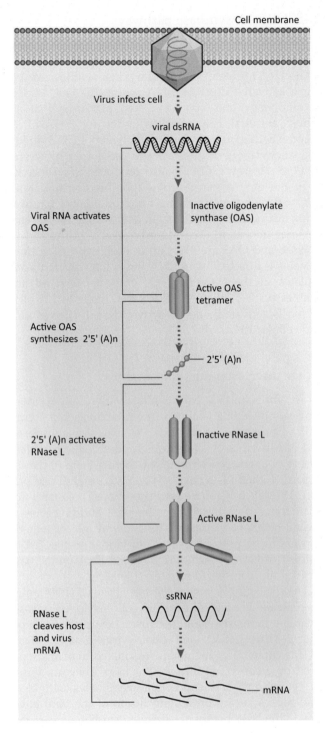

FIGURE 6.9   Oligo 5A pathway. Oligoadenylate synthase (OAS) is produced upon IFN stimulation of a cell. OAS is activated by dsRNA of an infecting virus and synthesizes 2′5 ′-A oligonucleotides. 2′5 ′-(A)n then serves to activate latent RNaseL. Active RNaseL cleaves host and viral RNAs.

ISG induced by type-I, type-II, and type-III IFNs. Viperin has broad antiviral activity against both RNA and DNA viruses including some herpesviruses, flaviviruses, retroviruses, orthomyxoviruses, paramyxoviruses, togaviruses, and rhabdoviruses. Viperin is expressed by a variety of animals, including mammals, fish, and reptiles. Viperin inhibits the release of some viruses but may have other means of inhibiting virus replication. At least one virus, human herpesvirus 5 (HHV-5) induces viperin expression to enhance viral infectivity; this is an example of how viruses can evolve to coopt an antiviral response.

Mx proteins are IFN inducible, large GTPases that belong to the *dynamin superfamily* of high molecular weight GTPases. Mx proteins are expressed by most vertebrates and they commonly localize to the cytoplasm. An exception is mice, whose Mx proteins are nuclear and cytoplasmic. Mx proteins assemble to form large, highly ordered oligomers that associate with intracellular membranes. Humans encode two Mx proteins, MxA and MxB, however only MxA has antiviral activities. In mice the Mx1 protein confers significant resistance to influenza viruses. The antiviral activities of Mx proteins seem to arise from their ability to bind to viral nucleocapsids to inhibit early steps in virus replication.

Intrinsic immunity

Intrinsic immunity is a type of innate immunity. However, the proteins that mediate intrinsic immunity usually restrict the replication of specific viruses. These so-called viral restriction factors may not be IFN inducible and they do not activate signaling cascades. Their presence renders certain cells nonpermissive to certain viruses. Examples of intrinsic factors important in species-specific immunity to retroviruses are described below.

Restriction factors

Tripartite motif-containing protein 5, alpha isoform (*TRIM5α*) is a retrovirus restriction factor. It is a 493 aa E3 ubiquitin ligase expressed by cells of most primates. Ubiquitin ligases direct ubiquitin conjugating enzymes to their target proteins. They assist in, or directly catalyze transfer of uniquin to those targets; the ubiquinated proteins are destroyed by proteasomes. The human genome encodes numerous E3 ligases that interact with a diversity of substrates. In humans TRIM5α is one of a cluster of eight TRIM genes on chromosome 11. There is evidence for positive selective pressure in the cluster and several TRIM genes have been identified as viral restriction factors. Depending on the viral protein targeted for ubiquitination, restriction can take place at different points in the virus replication pathway including entry, transcription, or release.

Apolipoprotein B mRNA editing enzyme catalytic polypeptide 3 (APOBEC3 or A3) catalyzes the deamination of cytidine to uridine in ssDNA. Primates encode 5—7 different A3 proteins (A3A to A3G) in a gene cluster on chromosome 20. The antiretroviral activity of A3 proteins requires they be packaged into virions to be present during reverse transcription of the retroviral genome in the newly infected cell. Thus A3 proteins do not interfere with virus production in an infected cell, but the released progeny viruses will be restricted in their ability to establish productive infections of new cells. Retroviruses can avoid the activity of A3 proteins by excluding them from virions. For example, the retrovirus murine leukemia virus (MLV) is not restricted by the mouse A3 protein because it fails to interact with MLV proteins, and is not packaged. Primate retroviruses encode VIF (virion infectivity) proteins that interact with A3 proteins, preventing their packaging into virions. Most simian immunodeficiency viruses (SIVs) do not target the human A3G for degradation therefore the SIVs are, for the most part, unable to replicate in human cells. A key adaptation allowing HIV to replicate in human cells is its ability to inactivate human A3G. There are a variety of SIVs, each adapted to a particular species of monkey, and for the most part an SIV VIF inactivates the A3G protein of the species in which it replicates.

Sterile alpha motif and histidine-aspartic domain (HD) containing protein 1 (SAMHD1) is another restriction factor that limits retroviral replication in certain cell types. SAMHD1 is a deoxynucleoside triphosphate phosphohydrolase that degrades the intracellular pool of deoxynucleoside triphosphates available during early retroviral reverse transcription. It seems to be particularly effective in limiting retrovirus replication in monocyte-derived macrophages and dendritic cells.

Retroviruses that can infect these cell types have evolved mechanisms to block SAMHD1 activity. For example, the VPX accessory protein of HIV-2 targets SAMHD1 for proteosomal degradation.

Autophagy

The autophagosome is a vesicle with a double membrane (Box 6.2). In normal cells the autophagosome contains cellular components targeted for degradation. The contents of the autophagosome are degraded when it fuses to a lysosome. During viral infection, autophagy can act as a surveillance mechanism that delivers viral antigens to endosomal/lysosomal compartments enriched in immune sensing molecules. Activated immune sensors can signal to activate autophagy. To evade this antiviral activity, many viruses actively block the autophagy pathway. Alternatively, some viruses subvert autophagy for their own benefit. Manipulated autophagy has been proposed to facilitate nearly every stage of the viral lifecycle in direct and indirect ways.

Natural killer cell responses to viral infection

Natural killer (NK) cells are a lymphocyte subset that play an important role in antitumor and antiviral responses. They make up about 5%—20% of the circulating lymphocytes in humans. They are large granular, cytotoxic cells. Their name, *natural* killer cells comes from their ability to kill virally infected cells without prior exposure to a virus. Thus they are key

---

BOX 6.2

Autophagy

Autophagy is a complex and highly regulated process whereby cellular components (e.g., protein aggregates or damaged organelles) or infectious agents (viruses) are engulfed by double-membrane vesicles in the cytosol. These vesicles, called autophagosomes, then fuse with lysosomes leading to degradation of their contents and providing building blocks for the cell. Not surprisingly, a general trigger for autophagy is nutrient deprivation.

The selective removal of intracellular infectious agents by autophagy is called xenophagy and the process has been shown to restrict virus titers. Autophagy also provides peptides for presentation to the adaptive immune system. In addition autophagy controls inflammasomes triggered by innate immunity signaling cascades, possibly preventing excessive immune activation. There are more than 30 autophagy-related gene products that control and drive these diverse processes.

Viral pathogens manipulate autophagy as an immune escape mechanism. Some viruses encode proteins that interfere with the formation of autophagosomes while others inhibit their fusion with late endosomes or lysosomes. There are also examples of viruses coopting autophagosomal membranes for use as platforms for genome replication and to facilitate their release from infected cells. For example, picornaviruses and flaviviruses seem to rely on autophagosomal membranes for optimum replication and release from cells.

players in innate immunity. NK cells provide rapid responses to virally infected cells, acting at around 3 days postinfection, well before development of adaptive immune responses. The importance of NK cells in controlling virus infection is underscored by the ability of some viruses to alter NK function.

Specific immunity

Induction of specific immunity to a virus is a critical way the body protects against viral disease. If an initial infection is survived, specific immune responses provide robust protection against future infection. Thus specific immunity involves pathogen-specific immune responses that gets better with time and experience (practice makes perfect). Vaccines take advantage of specific immune responses. The generation of specific immune responses is key to protection against all types of pathogens and the topic will only be reviewed briefly here. Two major types of specific immunity are antibody responses (humoral immunity) and cell-mediated immunity (CMI). These processes involve the activities of T- and B-lymphocytes. As summarized in Fig. 6.10, the general steps in activating specific immune responses are as follows:

- Antigen is trapped and processed.
- Antigen is recognized by T- and B-lymphocytes.
- Antigen is eliminated.
- Memory is established due to the presence of long-lived T- and B-lymphocytes.

Antibody responses

Antibodies are produced by B-lymphocytes in response to foreign antigens. Antibodies are found in the

blood, lymph, and mucosal secretions. Their role is to bind to foreign antigens, either inactivating them directly, or tagging them for removal by complement or phagocytic cells (e.g., macrophages). From a protective standpoint, the most important antiviral antibodies are those that *neutralize* viruses, preventing them from infecting cells. Neutralizing antibodies are measured by testing their ability to inhibit virus replication in cultured cells or animals. It should be noted that when virologists refer to viral serotypes, they are often referring to neutralizing, not simply binding, antibodies. Some types of neutralizing antibodies prevent viruses from binding to receptors, but others act later in the infection cycle, blocking penetration, and/or uncoating.

As a result of a virus infection, we produce many antibodies that bind to, but do not neutralize viruses. Detecting these so-called binding or nonneutralizing antibodies is very important for viral diagnostics. However, nonneutralizing antibodies also have a role in controlling a virus infection. They function in a variety of ways including antibody-dependent cell-mediated cytotoxicity, complement-mediated cytotoxicity, and antibody-dependent cell-mediated phagocytosis. These mechanisms can result in lysis or phagocytosis of virions. The Fc portions of immunoglobulins are key to these mechanisms as they are recognized by immune cells (NK cells, macrophage, neutrophils, and eosinophils) and proteins (such as complement proteins) thereby targeting these effectors to virions or virus-infected cells.

Cell-mediated responses

Cell-mediated responses (CMI) are directed at intracellular infectious agents or tumor cells. They are very important for control of virus replication. Cytotoxic T-lymphocytes (CTL) are key players in specific responses to viruses. CTL recognize and kill virus-infected cells often by triggering apoptosis. Key to the function of CTL is that they express T-cell receptors (TCR) that can recognize specific antigens. Another name for CTLs are CD8 + T cells, as they also display the protein, CD8, a glycoprotein on their PM. A CTL can recognize a virally infected cell if its receptor can recognize a viral peptide presented in the context of Class I major histocompatibility (MHC) proteins. The TCR recognizes the viral antigen while CD8 binds to the MHC protein. These interactions keep the two cells in close proximity. When activated, the T cell releases cytotoxins that damage the target cell. There are different classes of CTL. Those that have never encountered antigen are called naive. After a contact with a specific foreign antigen, the T cells become memory cells. When they encounter antigen again, they become stimulated to divide and have increased ability to kill.

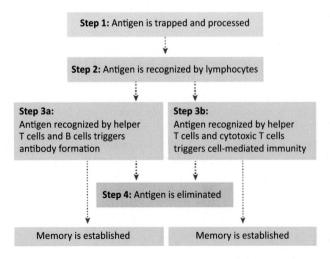

FIGURE 6.10   Overall steps in development of specific immunity.

Class I proteins

▶ Found on the surface of most mammalian cells

▶ Presented peptides derived from intracellular proteins

▶ Presented peptides represent a sampling of all the proteins synthesized in the cell (including viral proteins)

Class II proteins

▶ MHC Class II
  • B-cells
  • Macrophages
  • Dendritic cells
  • Activated T-cells

▶ Presented peptides are from extracellular proteins (that are taken up by cells)

FIGURE 6.11   MHC proteins bind and present both "self" and "foreign" peptides. Peptides bound to MHC can interact with specific T-cell receptors.

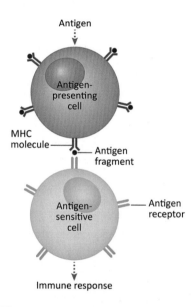

FIGURE 6.12   A general model of antigen presentation. In this figure the APC (top) might be a professional APC (dendritic cell, macrophage, or B-cell) or an infected fibroblast or epithelial cell. The antigen-sensitive cell (bottom) could be a T- or B-lymphocyte that becomes activated as a result of the interaction.

Antigen presentation

Antigen presentation is central to specific immunity. Extracellular antigens can bind to professional antigen presenting cells (APCs) (macrophage, dendritic cells, and B cells). Viral (or other) antigens produced *inside* of cells are proteolytically processed and are presented on the surface of the infected cell. Antigens are presented by a set of cell surface proteins called MHC proteins (Fig. 6.11). Their main function is to bind peptide fragments and display them on the cell surface for recognition by the appropriate T cells. All cells produce MHC class I molecules and that

bind fragments of cellular proteins as well as peptides derived from pathogens. Antigen presenting cells (APCs), such as macrophages and dendritic cells express MHC class II proteins (Fig. 6.12).

Viruses fight back

The previous sections provided a glimpse into the myriad immune responses to infectious agents. However, if even a fraction of those responses worked perfectly we would not be plagued by infections. The complex mix of immune responses has been driven by the evolution of infectious agents to counteract host defenses! In the previous sections we saw that how some retroviruses have evolved specific proteins to counteract specific retroviral restriction factors. However, there are also more general strategies, used by very diverse groups of viruses, to counteract immune responses.

Antigenic variation

Some viruses (e.g., influenza virus and HIV) defend against specific immune responses by mutation (Box 6.3). The mutated viruses display proteins that are not recognized by previously primed immune responses. For example, neutralizing antibodies may no longer bind the mutated viruses. Viruses may avoid preexisting immune responses by accumulating point mutations or by reassorting or recombining gene segments. Some viral proteins seem designed to excel at antigenic variation. The surface proteins of influenza viruses and HIV contain long, disordered segments that tolerate significant mutation. These regions serve to shield the more highly conserved receptor binding domains of the proteins.

BOX 6.3

Escape from neutralizing antibodies

Neutralizing antibodies are an important specific defense against viral invaders. Neutralizing antibodies not only bind to a virus, they bind in a manner that blocks infection. A neutralizing antibody might block interactions with the receptor, or might bind to a viral capsid in a manner that inhibits uncoating of the genome. Only a small subset of the many antibodies that bind a virus are capable of neutralization. After an infection, it can take some time for the host to produce highly effective neutralizing antibodies but these persist to protect against future encounters with the agent.

It is not surprising that some viruses have developed mechanisms to avoid antibody-mediated neutralization. For example, one obvious target for neutralizing antibodies are viral attachment proteins. If attachment proteins are bound to antibody, they cannot interact with the cell receptor. However, some viruses have developed mechanisms to avoid antibody-mediated neutralization. For example, one obvious target for neutralizing antibodies, a receptor binding domain (RBD) of a surface protein, may remain inaccessible to antibody. This is the case with the surface proteins of HIV and the HA protein of influenza virus. The RBDs of these surface proteins are shielded from immune recognition by nearby disordered protein domains. Often these domains are glycosylated, linked to polysaccharides, further

limiting access to the RBD by antibody. As the disordered regions or loops are not constrained by the need to maintain a specific structure they tolerate significant mutation. Thus if the host produces antibodies that successfully neutralize virus by binding to these loops, mutated viruses, that no longer bind such antibodies become predominant. In the case of HIV, this process happens within an infected individual. In the case of influenza viruses, we see the process occur on a population-wide basis. Virologists call this process *antigenic variation*.

Many picornaviruses, small unenveloped viruses, hide their RBDs in deep canyons or pits on the surface of their capsids. With their attachment domains hidden, other regions of their capsid proteins become targets for antibody, some of it neutralizing. Often the targets for neutralizing antibodies are surface loops or projections. These regions are not only good targets for antibodies, they are regions that can tolerate some amino acid sequence variation. While variation in picornaviral capsid proteins does not occur as rapidly as observed for influenza viruses or HIV, there are hundreds of picornaviruses that cause common colds; infection with one type does not confer protection against others. Many of these viruses use the same cell receptor, thus have conserved RBDs, but they have accumulated mutations in other regions of their capsid proteins.

Suppression of antigen presentation

Many viruses interfere with the ability of a cell to present antigens, thereby avoiding detection. Recall that in a virally infected cell, antigen presentation in presented in the context of MHC-class I proteins. The viral proteins must initially be digested by proteases. They then move to the endoplasmic reticulum (ER) with the help of heat shock proteins and the transporter associated with antigen processing (TAP), which translocates the peptides into the ER lumen. Once in the ER, chaperone proteins assist with loading the peptides onto MHC proteins. After binding peptides, MCH complexes exit the ER and are transported to the PM. Many of the steps required for antigen presentation are targeted by viruses, thereby interfering with the process.

Suppression of interferon responses

Viruses suppress IFN synthesis and innate immune responses by a variety of mechanisms. Here we will briefly describe some of the major mechanisms used by viruses to subvert the IFN response. Many viruses cause cytopathic infections, killing the infected cell in the process of their replication. These viruses often use the global strategy of inhibiting all host cell transcription and translation early in the infection process. Many viruses use additional, more targeted strategies to suppress the IFN response, starting with interfering with PAMP-PRR interactions and signaling. Influenza A viruses encode a protein, NS1, that binds ssRNA and RIG-I, inhibiting its ability to promulgate signals. Other viruses can sequester MDA-5, a downstream partner of RIG-I. Still other viruses encode proteins that block downstream signals; some inhibit the movement of transcriptions factors (e.g., IRF3/IRF7) into the nucleus.

Once IFNs are synthesized and secreted, viruses have mechanisms to inhibit other antiviral processes. For example, some paramyxoviruses encode proteins that interfere with JAK and/or STAT proteins, thereby suppressing the expression of ISGs. Finally, some viruses can inhibit the activity of antiviral proteins such as PKR by binding to and sequestering the dsRNAs that would activate the pathway.

Immunosuppression

Some viral infections cause overall disruptions in the immune system. For example, HIV replicates in, and kills, the CD4+ subset of T-lymphocytes. HIV also replicates in macrophages, disrupting their normal function. As CD4+ T-cells are key players in modulating both humoral and cell-mediated responses, the result is that untreated HIV infection leads to severe immunosuppression, opening the door for a variety of viral, fungal, and parasitic infections. Measles virus is another immuosuppressive virus. Mortality is usually low, but can reach 25% in children whose overall health is poor. Measles virus causes a transient immunosuppression that includes decreased lymphocytes numbers, cytokine imbalances, and decreased antibody levels, leaving the host open to infection by other pathogens. Eventually the measles virus is eliminated and the immune system recovers.

Poxviruses encode molecules that mimic cytokines, interferons, and their receptors

Poxviruses are large DNA viruses that encode proteins that mimic host defense molecules such as cytokines or their receptors. For example, poxviruses encode viroceptors that bind (and inactivate) host cytokines, preventing them from activating their true receptors. Virokines are viral cytokines that bind to receptors such that they can no longer response to the cell's own cytokines.

Hiding from the immune system

Some viruses replicate in tissue/organs that are not readily targeted by the immune system, for example, the brain. Others hide from immune recognition by entering latency. Herpesvirus have essentially perfected latency

with the result that they cause life-long infections. During latency genomes are maintained within cells but no (or very few) viral proteins are expressed. After years or decades of latency, the virus may emerge and productively replicate at times when the immune system is compromised (by advancing age, stress, other infections). A few viruses hide from the immune system by infecting a fetus, thereby allowing the viral antigens to be seen as self-antigens. An example is bovine viral diarrhea virus. Persistently infected bovids shed copious amounts of virus and are antibody negative.

Micro-RNAs

Micro-RNAs (miRNAs) are short (21–23 nt) host-encoded RNAs that bind to complementary sequences on mRNAs thereby blocking translation or inducing mRNA degradation. miRNA have many functions in host. However, some viruses have coopted cellular miRNAs to aid in their replication. Herpesviruses encode miRNAs that can interfere with immune responses, cell cycle control, and intracellular trafficking. For example, herpes simpex virus encodes a miRNA that can bind to the STING mRNA, inhibiting its translation (Box 6.4).

In this chapter we have learned that

- Innate immune responses are rapid responses to groups of pathogens. They are initiated by detection of PAMPs such as viral nucleic acids.
- PAMPS are recognized by PRRs (TLRs, RLRs, and NLRs) activating signaling pathways, many of which converge at common points.
- Innate responses to viruses include production of IFNs and proinflammatory cytokines.
- IFNs are released from virally infected cells and bind to receptors on other cells to initiate the

---

BOX 6.4

Micro-RNAs and viral infection

Micro-RNAs (miRNAs) are short (21–23 nt) host-encoded RNAs that bind to complementary sequences on mRNAs thereby blocking translation or inducing mRNA degradation. Thus miRNAs play an important role in regulating gene expression. Some viruses have coopted cellular miRNAs to aid in their replication. HCV replicates in human hepatocytes and requires the abundant, hepatocyte specific miRNA miR-122 to stabilize viral RNA and enhance replication. Other viruses, notably the herpesviruses, encode their own miRNAs; they target both host and viral mRNAs. By targeting host mRNAs, herpesviral miRNAs interfere with

processes such as immune response, cell cycle control, and intracellular trafficking.

In vertebrates, host miRNAs do not appear to play a major role in restricting virus replication. This is in contrast to *plants*, *nematodes*, and *arthropods* that utilize small RNAs as a major mechanism to inhibit virus replication. These small *interfering* RNAs are the major players in antiviral responses in these groups of organisms. They function by targeting viral mRNAs for cleavage by a process called antiviral RNA interference. That is not to say that miRNAs have no role in antiviral defenses in vertebrates, but the major antiviral players are most certainly the proteins expressed by ISGs.

antiviral state. The antiviral state results from synthesis of more than 300 ISG products such as PKR, OAS, tetherin, viperin, and Mx proteins.

- Some antiviral proteins (TRIM5α, APOBEC3G, and SAMHD1) are very species specific as regards host and virus and are players in the so-called intrinsic responses to viral infection. Many are directed against retroviruses and they serve as specific examples host−virus coevolution.
- Specific immune responses depend on antibodies and CTL to target specific pathogens. Specific responses are primed and aided by innate responses.
- Specific responses depend on processing and presentation of antigens by professional APCs.
- Neutralizing antibodies block viral infection.
- MHC proteins present peptide antigens on the surfaces of cells. MHC class I proteins are produced by all cells. MHC class II proteins are produced by APC.
- Viruses counteract host defenses of all types.
- Some viruses depend on host processes such as autophagy to support their replication.

7

Viral vaccines

After reading this chapter, you should be able to answer the following questions:

- What is the purpose of vaccination?
- What is an antigen?
- What is antigen recognition?
- What are some advantages of a killed/inactivated vaccine? What are some disadvantages?
- What are the advantages of a replication competent viral vaccine? What are the disadvantages?
- What is unique about nucleic acid vaccines?
- When considering vaccine development, what are "correlates of protection" and why is it helpful to determine them?

Vaccines provide a source of antigen

The purpose of a vaccine is to provide protection against a pathogen (in our case a virus). Early attempts at vaccine development were based on the observation that persons surviving a disease, for example, smallpox, were protected against further occurrences of that particular disease. Today we know that immune protection against a specific virus is provided by "specialized" (also called "adaptive") responses that inhibit virus replication and/or kill infected cells. The goal of vaccination is to stimulate the body to develop a specialized, protective immune response in the absence of disease.

Immune responses target one or more viral proteins. In the case of an infection, the immune response can potentially target every protein encoded by the virus. When discussing vaccines, the proteins targeted by the immune system are frequently referred to as antigens. An antigen is simply a molecule that is recognized by the immune system. A successful vaccine provides a safe source of viral antigens. The type of vaccine used dictates not only the specific antigens provided, but the *manner* in which those antigens are presented to the immune system.

Antigen recognition

Specialized immunity is achieved by selective proliferation of T- and B-lymphocytes that recognize specific foreign antigens. B-lymphocytes have receptors that can bind large, folded proteins but T-lymphocytes only recognize processed antigens that are bound to the major histocompatibility complex (MHC) proteins. Antigen processing takes place in two ways. Proteins produced within cells are digested and fragments (peptides) bind to MHC class I and are presented on the cell surface. Large proteins that are internalized by endocytosis are digested and peptides

Viruses.
DOI: https://doi.org/10.1016/B978-0-323-90385-1.00026-1

© 2023 Elsevier Inc. All rights reserved.

are presented by MHC class II molecules. Lymphocytes that bind antigen are stimulated to proliferate.

Development of a B-cell response by an infected or vaccinated individual results in production of antibodies that bind extracellular antigens, for example, viral capsids. In contrast, T-lymphocytes recognize antigens that have been processed and presented from within a cell. T-lymphocytes are key players in so-called cell-mediated immune (CMI) responses. In the context of a viral infection, CMI responses result in killing of infected cells. Most complex viral proteins contain sequences that stimulate B-cell responses (so-called B-cell epitopes) as well as regions that stimulate T-cell responses (so-called T-cell epitopes). B-cells can be stimulated by large extracellular protein antigens, while T-cells respond only to antigens (short peptides) that have been processed within a cell.

The purpose of recognizing a foreign antigen is to allow its removal by the various players of the immune system. Binding of a pathogen by antibodies may render it available to phagocytic cells such as macrophages, or may target it for destruction by complement-mediated lysis. In the case of viruses, the most protective antibodies are those that neutralize or block virus infection. These so-called *neutralizing antibodies* might: (1) prevent a virus from binding to a receptor; (2) prevent membrane fusion; or (3) prevent uncoating of the viral genome. There are many antiviral antibodies that bind, but do not neutralize, a virus. These so-called binding antibodies might enhance phagocytosis or promote complement-mediated lysis of the virus but would not be measured in a cell culture-based neutralization assay. In rare cases (e.g., dengue virus, family *Flaviviridae*) binding antibodies may even enhance or exacerbate infection and disease. In contrast to antibodies, cytotoxic T-lymphocytes recognize and kill virally infected cells thereby limiting the amount of virus produced.

As the development of specific immune responses depends on how antigen is presented, it follows that different approaches to vaccine development might be needed, depending on the type of immune response that is desired. There are a variety of ways that viral proteins can be delivered as vaccines. These include administration of: weakened (so-called attenuated) viruses that replicate in the host without causing disease; inactivated (killed) virions; purified protein products; nucleic acids (genes) to direct synthesis of desired proteins.

Classical versus engineered vaccines

Classical vaccines are those developed using strategies that were available prior to the advent of recombinant DNA techniques. Traditional or classical methods to produce a viral vaccine require large quantities of virus, thus require virus be replicated in cell cultures, embryonated eggs or animals. Classical vaccines consist either of killed virus preparations or attenuated (weakened) viruses (Fig. 7.1). Historically, attenuated viruses were obtained by repeated passage in animals or cultured cells. This is the method that Pasteur used to create a rabies vaccine (Box 7.1). Unfortunately the results of such passages were impossible to predict and some viruses were never adequately attenuated even after prolonged passage.

Vaccines developed using recombinant DNA technologies are often called recombinant or engineered vaccines. The ability to manipulate the viral genome provides an array of opportunities to achieve vaccine

---

BOX 7.1

Rabies vaccine success in 1885

Louis Pasteur was a talented and innovative microbiologist. His work (among others) was key to understanding that diseases are caused by infectious agents that can be identified and grown in the laboratory. Pasteur initially developed and tested animal vaccines for the bacterial diseases anthrax (*Bacillus anthracis*) and fowl cholera (*Pasteurella multocida*). His studies of rabies were more challenging as he could not grow agent on an agar plate. Instead Pasteur used monkeys and rabbits to grow and attenuate the rabies virus. He successfully used dried spinal cord material from infected rabbits to protect against rabies in dogs. He even determined that the vaccine could be administered to dogs *after* they were exposed to the virus. In 1885 Pasteur was asked to treat Joseph Meister, a 9-year-old boy who had been bitten multiple times by a rabid dog. No one doubted that without treatment the boy would die from rabies. Pasteur treated Joseph Meister with injections of rabbit spinal cord material and the boy survived. This was such a celebrated and important event that people from around the world made contributions to support Pasteur's work. Thus the Pasteur institute, dedicated to the study of biology, microorganisms, diseases, and vaccines, opened in 1888. To this day the institute continues to support cutting edge biomedical research.

CLASSICAL VACCINES

INACTIVATED
Grow virus and inactivate (i.e. heat or formalin).

REPLICATION COMPETENT (LIVE)
Attenuated Vaccines. Attenuation reduces viral virulence. Attenuation is an inexact process (i.e. propagate virus in cell culture or adapt to an animal other than normal host).

RECOMBINANT/ ENGINEERED VACCINES

INACTIVATED
Construct, grow and inactivate genetically modified virus. Engineered virus may be a vector expressing heterologous genes or may be a modified pathogen.

SUBUNIT
Viral antigens expressed in recombinant organisms (bacteria, yeast, cultured cells, plants). May be purified or crude extracts.

Viral antigens may be delivered as genes (DNA vaccine).

REPLICATION COMPETENT (LIVE)
Genetically modified virus (vector or pathogen) replicates in the vaccine recipient.

FIGURE 7.1   General types of viral vaccines.

goals. Engineered vaccines may consist of replicating virus that has been attenuated by removal of specific genes, may be preparations of purified viral antigens or even nucleic acids.

Fig. 7.2 provides a summary of the ways in which recombinant DNA technologies can be used in vaccine development and production. For example, highly virulent viruses can be weakened by deletion or modification of *specific* genes. Often the genes deleted are those required for the virus to grow robustly in the animal host, but which are not needed for virus replication in cultured cells. These are often genes encoding viral proteins that are involved in subverting immune responses. Another strategy is to move specific genes from a virulent virus (often genes for capsid or envelope proteins) to a safer virus. The recombinant virus now expresses one or more proteins from the pathogen and might be administered as a replicating virus or might be grown and inactivated. A different recombinant approach is to move specific genes from a virulent virus into bacteria, yeast, or cultured cells to produce large quantities of a particular viral protein. This approach has also been used in the development of vaccines for viruses that cannot be readily grown in culture, for example, papillomaviruses. After synthesis, the viral proteins must be highly purified and concentrated to produce a vaccine.

Another recombinant strategy that is gaining acceptance is the use of nucleic acids as vaccines. DNA molecules encoding specific proteins are generated and administered directly to the patient. Alternatively, mRNA can be produced from a cloned gene, assembled into lipid nanoparticles (Buschmann et al., 2021) and administered directly to the patient. An advantage of nucleic acid vaccines is the ability to standardize processes for synthesis and purification; regardless of the protein encoded, the nucleic acids have the same overall chemical composition. DNA and RNA vaccines also have the advantage of stimulating T-cell responses

because viral antigens are produced inside of living cells. The antigens can be processed and presented in the context of MHC proteins. Antigen released from cells can stimulate B-cell responses.

Replicating versus inactivated vaccines

Whether a vaccine is derived by classical techniques or genetic engineering, a key consideration is whether or not the vaccine contains a source of infectious material (a virus that will replicate in the recipient with the potential for transmission to another individual). A summary of the relative advantages of replicating versus inactivated vaccines is provided in Fig. 7.3. Vaccines containing replicating virus generally require smaller doses, as the virus is amplified in the host. Adjuvants (substances that modify the effects of other agents; these are added to vaccine to modify immune responses) are not usually needed and it may be possible to administer the vaccine orally or intranasally. However, there is the possibility that a replicating agent could be transmitted from a vaccinated person to someone else. In the case of classically attenuated viral vaccines, cases of reversion back to virulence have been documented.

Vaccines that do not contain any infectious material are often called inactivated or killed vaccines. However, this broad category also includes vaccines that consist of purified proteins. These vaccines typically require that larger doses of material be administered but they have many advantages, not the least of which is the inability to spread beyond the vaccine recipient.

Viral vaccines that contain infectious agents

Viral vaccines containing infectious agents can be developed by classical techniques (attenuation by

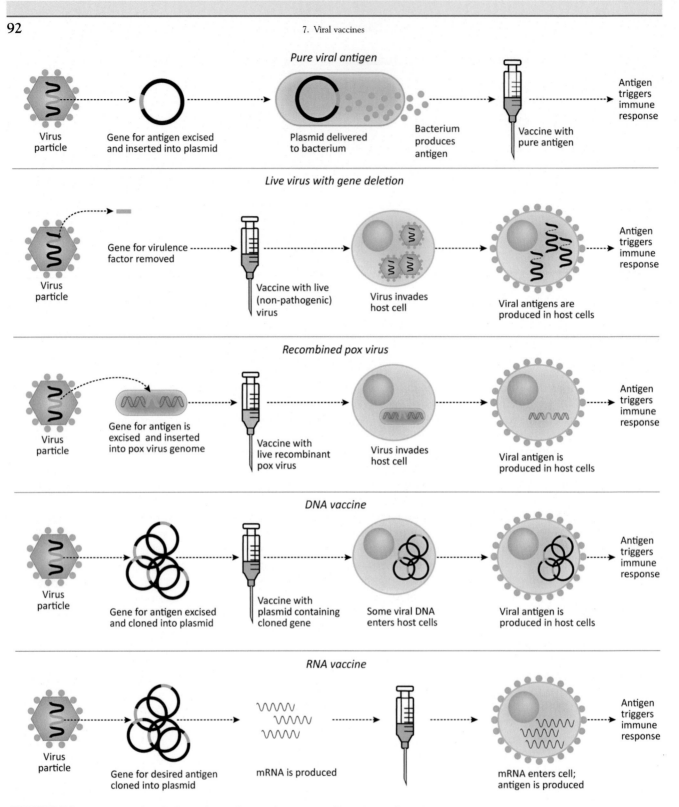

FIGURE 7.2   A variety of methods can be used to produce genetically engineered vaccines.

repeated passage in cultured cells) or by genetic engineering. The key point is that they contain viruses capable of *replicating* in the vaccine recipient. The "perfect" attenuated virus replicates enough to stimulate protective responses but not enough to cause harm. Production of an attenuated viral vaccine by classical methods requires that the virus be grown in cultured cells, or more rarely, in an animal. Genetically engineered attenuated vaccines use viruses specifically modified to reduce virulence. What sets these modern

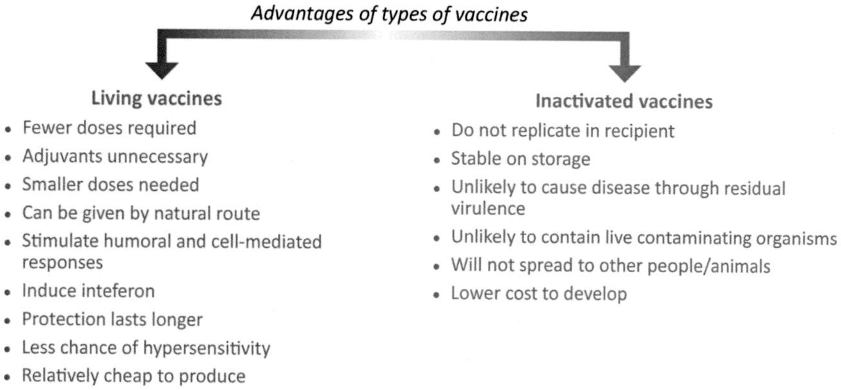

Advantages of types of vaccines

Living vaccines
- Fewer doses required
- Adjuvants unnecessary
- Smaller doses needed
- Can be given by natural route
- Stimulate humoral and cell-mediated responses
- Induce inteferon
- Protection lasts longer
- Less chance of hypersensitivity
- Relatively cheap to produce

Inactivated vaccines
- Do not replicate in recipient
- Stable on storage
- Unlikely to cause disease through residual virulence
- Unlikely to contain live contaminating organisms
- Will not spread to other people/animals
- Lower cost to develop

FIGURE 7.3   Relative advantages of killed and attenuated vaccines.

vaccines apart from classically attenuated viral vaccines is that *specific* genes are modified in a manner to prevent any possibility that the virus can revert to virulence.

Alternatively, a genetically engineered vaccine may utilize a nonpathogenic virus expressing proteins from a pathogen. For example, nonpathogenic poxviruses such as vaccinia virus have been modified to produce the rabies virus envelope glycoprotein. The recombinant poxviruses replicate to a limited extent in vaccine recipients, but produce enough rabies glycoprotein to elicit a protective response. Such vaccines have been successfully used to vaccinate wild animals (raccoon, skunk, coyote) against rabies. They are delivered in baits designed to be eaten. It should be noted that vaccine manufacturers often produce different types of vaccines for a particular agent. For example, the vaccine manufacturer Merial (Atlanta, GA, United States) produces a live recombinant vaccinia-virus-based rabies vaccine for use in wild animals but produces an inactivated recombinant vaccinia-virus-based rabies vaccines for pets and livestock. It should also be noted that a virus attenuated for one species may not be equally attenuated for all species. This is of particular importance in the practice of veterinary medicine.

Killed, inactivated, and subunit vaccines

Classical killed or inactivated vaccines are produced by growing large amounts of virus followed by treatment with heat or formalin. This is the process used for many years to produce influenza virus (flu) vaccines. Most flu vaccines are composed of virus that was grown in eggs and subsequently inactivated. This is why persons with egg allergies must avoid the classical flu vaccine.

Protein subunit vaccines contain one or a few viral proteins. While theoretically they could be produced

by purifying a specific protein from a batch of virus, this is highly impractical compared with cloning the gene for the desired antigen and introducing it into bacteria, yeast, plants, or cultured cells. The protein product is then purified and administered. This is the method used to produce a recombinant hepatitis B surface antigen vaccine, which is produced in yeast cells (Box 7.2). In the case of the cancer preventing papillomavirus vaccines, capsid proteins are produced in yeast cells and are then processed so that they assemble into capsid structures. These are very immunogenic and contain no viral DNA.

Many inactivated and subunit vaccines are highly effective, but one consideration is that they present a limited number antigens to the immune system. Recall that while a virus particle may be assembled from a few types of capsid/envelope proteins, there are *many* additional viral proteins produced in the infected cell (e.g., the nonstructural proteins required for genome replication and/or subversion of immune responses). These nonstructural proteins are not present in inactivated virions but might be important to elicit highly protective immune responses.

Inactivated/subunit vaccines typically require administration of a "large" dose of antigen. Manufacture and purification of the proteins may be relatively expensive. Inactivated/subunit vaccines are typically administered by injection. A variety of adjuvants might be included in the vaccine to increase immunogenicity and to protect the vaccine components from being degraded too rapidly. Adjuvants are substances that modify the effect of other agents, such as the immunogens found in vaccines. One common use of adjuvants is to reduce the amount of antigen required to elicit an immune response. Adjuvants may also be used to direct the immune response in a particular manner, for example, by specifically activating cytotoxic T-cells versus antibody-secreting B-cells.

DNA and mRNA vaccines can also be considered among the group of inactivated or subunit vaccines. In

---

BOX 7.2

HBV vaccine: first subunit vaccine, first recombinant vaccine

Neither the discovery of the hepatitis B virus (HBV) nor the development of a protective vaccine followed predictable paths. Hepatitis is an infection of the liver that can be caused by a variety of infectious agents. By the mid-20th century, hepatitis B or serum hepatitis was recognized as a significant health problem, leading to increased risk of hepatocellular carcinoma. It was assumed that hepatitis B was caused by a viral infection, and well known that blood transfusions greatly increased the risk of infection. But efforts to culture the virus were unsuccessful for many years. A major breakthrough occurred in 1965 when Dr. Baruch Blumberg (who was not studying hepatitis B!) identified an antigen in a blood sample that was eventually identified as an HBV protein. Following this discovery, plasma from chronically infected patients was used a source of antigen for developing a blood test (1971), and eventually for developing a successful vaccine (licensed in the United States in 1981). Vaccine production was possible because some patients have large quantities of the hepatitis B surface (S) antigen that can be harvested from their plasma. Heat treatment of the product inactivates infectious virus, leaving the S antigen intact and highly immunogenic. The vaccine was very effective, but use of human plasma as starting material was not without risk and not without its detractors. But the success of the plasma-derived vaccine indicated that whole virus was not needed to induce protection. Rather, protection was provided by immune responses that specifically targeted the S antigen. This knowledge dovetailed beautifully with new recombinant DNA technologies (gene cloning and protein expression). By 1986 an HBV recombinant S antigen vaccine (a subunit vaccine) had been developed, tested, and approved. During HBV vaccine development, another important discovery was made: in order to induce a protective response, the S antigen had to be correctly folded and presented to immune system in a manner that mimicked the antigen produced during a natural infection. The recombinant vaccine product quickly replaced the human plasma-derived product in the United States, due to its superior safety.

---

the case of a DNA vaccine, the desired genes are cloned, and the DNA is administered directly into a patient. If the DNA reaches the cell nucleus and is transcribed, a protein product can be synthesized. The DNA approach avoids the problem of purifying large amounts of recombinant proteins. One limitation of the DNA vaccine approach is the necessity to administer relatively large doses of DNA.

Alternatively, cloned DNA can be used to generate large amounts of messenger RNA (mRNA) for administration to patients. mRNA vaccines against SARS-CoV-2 have been used widely during the COVID-19 pandemic. An advantage of both DNA and mRNA vaccines is that the desired protein is produced within cells, therefore it can be processed and presented to T-lymphocytes to induce CMI responses.

In some cases, a viral pathogen is so virulent that any risk of improper inactivation is considered unacceptable. An example is rabies virus. Although traditional rabies vaccines used inactivated rabies viruses, modern vaccines are derived from a vaccinia virus vector expressing the rabies virus glycoprotein. The recombinant viruses are grown and then inactivated. Even if not properly inactivated, the recombinant vaccinia virus is relatively safe and certainly cannot cause rabies in the vaccine recipient.

Correlates of protection

An important consideration for vaccine development is to determine the type(s) of response that is (are) protective. Protection may require neutralizing antibodies, virus-specific cytotoxic T-cells, or both. Protection may require recognition of one or more viral proteins, or recognition of a specific epitope on a target protein. An obvious way to determine if a vaccine is protective is to compare the response of vaccinated versus unvaccinated animals to a viral challenge. (We anticipate that the unvaccinated animals will get sick but the vaccinated animals will remain healthy.) This direct approach is the "gold standard" but uses a lot of animals and can be quite costly. It is also possible that no animal model is suitable for a human virus. Thus during the vaccine development process, it is often helpful to determine the type and magnitude of responses that confer protection. In other words, what responses (both type of response and magnitude of response) *correlate* with the desired outcome, be it protection against infection, disease, or death. For example, a specific titer of neutralizing antibody (Box 7.3) may serve to protect against infection by a particular viral pathogen. Various combinations of antigens, doses, delivery strategies, and adjuvants can then be administered, followed by measurement of neutralizing

---

BOX 7.3

Neutralizing antibody

Antibodies are proteins made by the immune system in response to pathogens, or any foreign molecule. Antibodies bind to their targets facilitating their removal or degradation. For any given protein antigen, many different antibodies may be produced, binding at different locations and with different affinities. Virologists pay particular attention to antibodies that can bind and *inactivate* a virus. These so-called *neutralizing antibodies* prevent a virus from entering or replicating in a cell. They are measured by testing their ability to inhibit virus replication in cultured cells or animals. It should be noted that when virologists refer to viral serotypes, they are often referring to neutralizing, not simply binding, antibodies. Some types of neutralizing antibodies prevent viruses from binding to receptors, but others act later in the infection cycle, blocking penetration, and/or uncoating. When developing vaccines, particular attention is often paid to the levels of neutralizing antibodies that are produced. Higher levels of neutralizing antibodies generally correlate with better protection.

Neutralizing antibodies are measured using plaque assays or focus forming assays. A standardized amount of virus is mixed with dilutions of antibody and then applied to cells. The cultures are incubated and then assayed for the presence of virus. A reduction in the number of infected cells (e.g., as indicated by a reduction in the number of plaques) is indicative of the presence of neutralizing antibody.

---

antibody titers. Only those preparations that induce adequate levels of neutralizing antibody need be tested further. Another advantage of understanding correlates of protection is that vaccine recipients can be tested to determine their status (protected or not protected). For example, in the case of rabies vaccines, protection is well correlated with high titers of antibody to the G protein. Among veterinarians or other "at risk" persons, antibody titers can be measured to provide information about the need for a booster dose.

In the case of natural infections (e.g., Ebola virus), we may examine survivors to determine the types of specific responses that correlate with survival. We can then design vaccines that mimic the natural, protective response. But what happens when natural infection is not protective and our immune system is unable to clear or control the virus? The best (worst?) example of this is infection with the human immunodeficiency virus (HIV). An untreated HIV infection is almost universally fatal, despite the fact that patients do mount an immune response. The inability to control infection is due, in part, to the fact that HIV attacks the very immune cells we need to sustain a protective response. HIV also mutates rapidly and its surface proteins are designed to evade host responses. In short, there is little/no natural *protective* response, thus there are no well-defined correlates of protection to serve as the standard for vaccine design. (World-wide, a very few persons have been identified who seem to be naturally resistant to HIV. They are an important resource in the fight against HIV!)

Vaccines to protect against influenza virus infection have been used for many decades, but we still face vaccine challenges. Most human influenza virus infections are not fatal and infection with one particular strain of virus protects against future infection *with that strain*. However, we remain susceptible to other strains of influenza virus, and there are many of these as influenza viruses mutate readily. Thus new vaccines must be manufactured each year, in an attempt to keep one step ahead of this viral pathogen. The "perfect" influenza virus vaccine would provide protection against many different strains. It remains to be seen when, or if, such a vaccine will be developed.

Vaccine development and assessment of efficacy

Vaccine development can be complex and time consuming. There are many factors that must be considered. Some of these are:

- Safety: Does the vaccine produce major/minor side effects? Is there risk that an attenuated virus will revert and cause disease?
- Efficacy: How well does the vaccine work? How many vaccinated individuals are resistant to infection and/or disease?
- Development costs: How much time/effort/money is needed to produce a vaccine candidate?
- Manufacturing costs: How much money does it cost to produce each dose of vaccine?
- Costs of transport and storage: Must the vaccine be refrigerated or stored at ultralow temperature?
- Ease of administration: An injectable vaccine requires adequate supply of sterile needles.
- Number of doses required to achieve protection.
- Duration of immunity: How often are booster doses needed?

Vaccine efficacy is a measure of how well a vaccine works. For statisticians, efficacy is a measurement of how much a vaccine lowers the risk of an outcome. Thus in order to determine efficacy one must define "successful outcome." Does vaccine success means prevention of infection, disease, hospitalization, death, or prevention of virus spread? And in what percentage of vaccinated individuals? The definition of vaccine success has been front and center during the COVID-19 pandemic.

Once a particular measure of success has been defined, efficacy can be determined in vaccine trials. In trials of COVID-19 vaccines, an early measure of success was infection. Vaccine developers gave equal numbers of individuals either vaccine or a placebo. Then they monitored this cohort and observed how many individuals who were vaccinated actually contracted COVID. The placebo group was an important control as infections among these individuals were used as the comparison group. Efficacy is measured as the *difference in infections* among the two groups. For example, if 20% of the placebo group were infected during a trial, but 0% of the vaccinated group, the vaccine efficacy is 100%. However, if 10% of the vaccinated group were infected, the vaccine efficacy is 50%.

It should be noted that a vaccine that fails to prevent infection might in fact be very efficacious in preventing hospitalization or death. Vaccine efficacy can also vary depending on a variety of factors including: strain of the virus, age of a population, overall health or health care availability. Finally, one must consider the duration of the protective effect. As immunity wanes, protection may fail against infection but may remain high for protection against severe disease. It has become very clear, in the midst of the COVID-19 pandemic, that the public's understanding of vaccine efficacy is an important factor in vaccine acceptance. A vaccine that is considered a failure because it does not prevent infection 100% of the time may unnecessarily leave many persons vulnerable to severe disease and hospitalization.

COVID-19 vaccines

SARS-CoV-2, the etiologic agent of COVID-19, emerged as a human pathogen in late 2019 in China and quickly spread world-wide. A respiratory pathogen, the virus spreads relatively easily and causes severe disease and death in a subset of infected individuals. SARS-CoV-2 is also a somewhat stealthy pathogen, causing inapparent infection in many individuals, facilitating its spread. Vaccine developers around the world quickly began work and within months, vaccines were granted emergency use authorization in multiple countries. This was astoundingly fast considering that vaccine development historically took years to decades (Fig. 7.4). However, development and production of COVID-19 vaccines were achieved on a foundation of decades of research, by thousands of research organizations and companies. To those outside of the medical and scientific research fields much of that work was completely unknown. The work was done, and funded, on the expectation that a viral pandemic of the scale of the 1918−1919 influenza virus pandemic would occur one day. Relatively small outbreaks (compared to COVID-19) of novel human viruses underscored the need to respond quickly to the next threat. By July 2021 there were 18 COVID-19 vaccines approved for emergency use, 105 in clinical development and an additional 184 vaccine candidates in preclinical development (Ndwandwe and Wiysonge, 2021). Many different vaccine platforms were being employed, ranging from the classical method growing and inactivating the virus to many of the recombinant technologies outlined in Fig. 7.2. Somewhat surprisingly, mRNA technologies emerged among the earliest approved vaccines.

mRNA vaccines are simple in concept: first, a gene of interest is cloned or synthesized. The gene is transcribed in a test tube and the resulting mRNA is mixed with lipids.

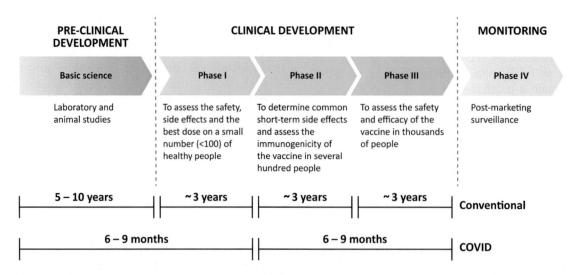

FIGURE 7.4   Timeline for vaccine development. Source: *Adapted from Ndwandwe, D., Wiysonge, C.S., 2021. COVID-19 vaccines. Curr. Opin. Immunol. 71, 111. doi:10.1016/j.coi.2021.07.003.*

When the mRNA−lipid mixture is added to cells, the lipids fuse with the cytoplasmic membrane and release the mRNA into the cytosol where it is translated. The resulting protein is modified and processed, for example, glycosylated, by the host cell. Some of the foreign protein is proteolytically cleaved and presented in the context of MHC on the cell surface. Some protein may be released and recognized by B-lymphocytes. Eventually the administered mRNA will be degraded by the host cells. While simple in concept, working with RNA is incredibly difficult as it is very unstable and subject to degradation by abundant and ubiquitous ribonucleases. The success of COVID-19 mRNA vaccines was the result of multifaceted

collaborations, over many years. The earliest indications that mRNA might have medical applications emerged in the 1960s and 1970s just as the structure and function of mRNAs were being elucidated (Dolgin, 2021). By the 1980s, mRNA was being synthesized in the laboratory and lipids used to deliver it to cells. Development of scalable methods for generating mRNA−lipid nanoparticles pushed the mRNA vaccine platform into high gear in the 2000s (Dolgin, 2021). The general structure of the nanoparticles that make up mRNA vaccines is shown in Fig. 7.5. Key to assembly of these particles is the interaction of negatively charged mRNA with positively charged lipids. Studies indicate that mRNA vaccines directing synthesis of

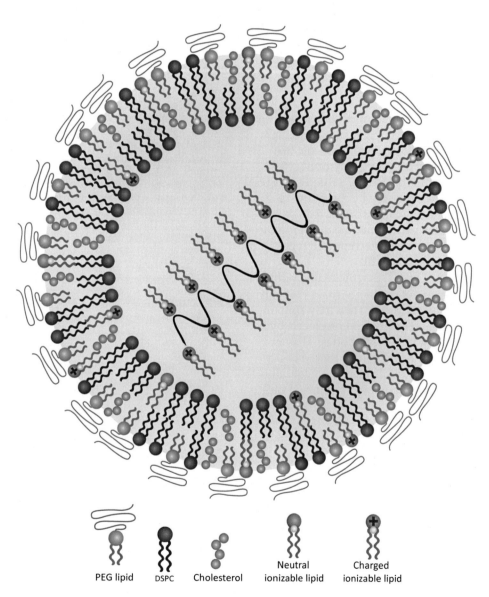

FIGURE 7.5   The structure of an mRNA lipid vaccine showing interaction of negatively charged mRNA with positively charged lipids. *Source: Adapted from Buschmann, M.D., Carrasco, M.J., Alishetty, S., Paige, M., Alameh, M.G., Weissman, D., 2021. Nanomaterial delivery systems for mRNA vaccines. Vaccines 9, 65. https://doi.org/10.3390/vaccines9010065.*

the SARS-CoV-2 spike protein are over 90% efficacious in preventing death (Bajema et al., 2021; Nanduri et al., 2021).

A final word about safety

Vaccine safety is very important. In particular, the risk of vaccine-related adverse events (complications) versus risk of natural infection must be considered. If the risk (or perceived risk) of natural infection is extremely low, relatively minor side effects such as fever, headache, or a sore arm may be unacceptable. Complicating this is the occurrence of random events that just happen to coincide with administration of millions of doses of vaccine. It is intellectually and emotionally difficult not to blame the vaccine if we become ill soon after having an injection, although there may be no connection at all. It has been said that vaccines are victims of their own success: *the more they successfully prevent disease, the easier it is to underestimate their value.*

In this chapter we have learned that:

- The purpose of vaccination is to protect against specific diseases. This is accomplished by administering antigens (usually proteins) that are processed and presented to T- and B-lymphocytes.
- Viruses encode a few to hundreds of proteins. Some of these are targeted successfully by the immune system to confer protection against repeated infection. An important part of vaccine development is to determine correlates of protection (the types immune responses are protective).
- The common feature of killed/inactivated/subunit vaccines is that they do not contain viruses that can replicate in the vaccine recipient. They may contain a few or many antigens. They can be produced by classical or recombinant DNA techniques. They are relatively inexpensive to develop, but can be costly to produce (high cost per dose). They are generally safe and can be given to immunocompromised or pregnant recipients.

- Some vaccines contain infectious viruses (capable of replicating in the recipient). They can be produced by classical methods (attenuated by repeated passage in cultured cell) or by genetic engineering (altering specific viral genes). Classically attenuated viruses may revert to virulence or may be spread from a vaccinated individual to other contacts. In theory, genetic engineering can be used to generate very safe attenuated vaccines. Some live vaccines cannot be given to immunocompromised or pregnant individuals. A virus attenuated for one species may not be attenuated for all species.
- The greatest challenges for vaccine development are those diseases for which *natural infection* does not result in life-long immunity.

References

Bajema, K.L., Dahl, R.M., Prill, M.M., Meites, E., Rodriguez-Barradas, M.C., et al., 2021. Effectiveness of COVID-19 mRNA vaccines against COVID-19-associated hospitalization - Five Veterans Affairs Medical Centers, United States, February 1-August 6, 2021. MMWR Morb. Mortal. Wkly. Rep. 70, 1294. Available from: https://doi.org/10.15585/mmwr.mm7037e3.

Buschmann, M.D., Carrasco, M.J., Alishetty, S., Paige, M., Alameh, M.G., Weissman, D., 2021. Nanomaterial delivery systems for mRNA vaccines. Vaccines 9, 65. Available from: https://doi.org/10.3390/vaccines9010065.

Dolgin, E., 2021. The tangled history of mRNA vaccines. Nature 597, 318−324. Available from: https://doi.org/10.1038/d41586-021-02483-w.

Nanduri, S., Pilishvili, T., Derado, G., Soe, M.M., Dollard, P., et al., 2021. Effectiveness of Pfizer-BioNTech and moderna vaccines in preventing SARS-CoV-2 infection among nursing home residents before and during widespread circulation of the SARS-CoV-2 B.1.617.2 (delta) variant—National Healthcare Safety Network, March 1−August 1, 2021. MMWR Morb. Mortal. Wkly. Rep. 70, 1163. Available from: https://doi.org/10.15585/mmwr.mm7034e3.

Ndwandwe, D., Wiysonge, C.S., 2021. COVID-19 vaccines. Curr. Opin. Immunol. 71, 111. Available from: https://doi.org/10.1016/j.coi.2021.07.003.

8

Antiviral agents

After reading this chapter you should understand the following:

- Most antiviral therapies are highly virus-specific.
- Many different steps in virus replication cycles are targets of antiviral therapies.
- Monoclonal antibodies and immune sera can be used to block virus attachment and/or entry.
- Many approved antiviral drugs are nucleoside analogs that target viral nucleic acid synthesis.
- Detailed knowledge of a virus replication cycle is used for targeted drug design.

The first antibiotics were commercially available in the 1930s and penicillin was widely used by the mid-1940s to treat bacterial infections. Antibiotics dramatically reduce deaths from bacterial infection and are successful because they target metabolic processes specific to bacteria. In contrast, viruses use host cell machinery to replicate, making it difficult to inhibit their replication without significant damage to the host. Antiviral drugs often target only specific, or very small groups, of viruses. For example, the first antiviral drug was approved in 1963 was idoxuridine, a modified form of deoxyuridine. It is a topical drug that can be used to treat herpes virus keratitis. While there are over 100 known human viruses, only a handful are treated with approved antivirals. These include human immunodeficiency virus (HIV), human influenza viruses, some human herpesviruses, hepatitis B virus (HBV), hepatitis C virus (HCV), respiratory syncytial virus (RSV), and most recently SARS-CoV-2. The number of antiviral drugs and therapies used to treat animal diseases is even more limited. It is fortunate that many viral diseases of humans and animals can be prevented by vaccination. This chapter provides an introduction to antiviral agents with emphasis on their mechanism of action as related to the viral replication cycle. It should be noted that many compounds that are effective at inhibiting virus replication in cultured cells are ineffective, or toxic in animals. Discovery and development of drugs that have therapeutic value is a complex and lengthy process that is beyond the scope of this text.

Theoretically, any step of a virus replication cycle could be the target of an antiviral therapy (Fig. 8.1). Monoclonal antibodies (mABs) block the earliest steps in viral replication (attachment and entry) while many antiviral drugs target genome synthesis. Some of these drugs were initially developed to treat cancers but have found application as antiviral treatments. The discovery of interferons (IFNs), key players in innate immunity to viruses, was initially heralded as major break though in the treatment of viral infections, but in fact their use is quite limited. A major advance in the development of successful antiviral therapies is our ability to examine, with great precision, each step in a virus life cycle, down to the details of each protein–protein interaction. Information that might have once been thought arcane, and of interest only to virologists, has been successfully exploited to engineer powerful antiviral treatments. This is a testament to the power of basic research. Viral protease inhibitors, that block key steps in HIV and HCV

Viruses.
DOI: https://doi.org/10.1016/B978-0-323-90385-1.00024-8

99

© 2023 Elsevier Inc. All rights reserved.

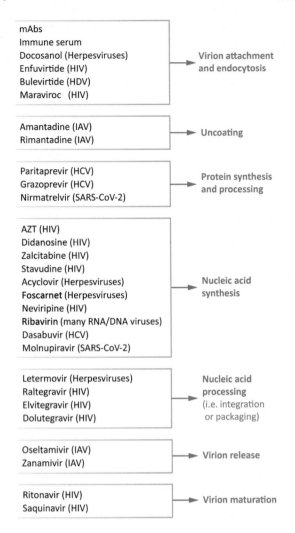

FIGURE 8.1 Steps in virus replication cycles targeted by approved antivirals. Any step in a virus replication cycle is a potential target for the development of therapeutics.

replication are examples of highly effective drugs developed through targeted design.

Inhibitors of attachment, entry, or uncoating

The first step in a virus replication cycle is attachment to, and entry into, specific host cells. The power of viral vaccines is their ability to marshal our natural defenses to block these key steps in replication. As a way to quickly mimic these natural responses, immunoglobulins or immune sera have long been used for prevention of a few viral infections, including rabies virus and HBV, for which they are often administered prophylactically (after suspected exposure). This process is called "passive immunization." mAB treatments also provide passive immunity but through use of very specific and highly targeted antibodies. A mAB

targets a single epitope and is derived from a single antibody-producing cell. To generate mABs against a specific target, a mouse is first immunized (with the target antigen) and its spleen is harvested. Spleen cells are then fused with immortalized myeloma cells. The fused cells can be specifically amplified and screened for production of a desired antibody (Fig. 8.2). Techniques have been developed whereby thousands of cells can be readily screened. In the case of a viral target, mAbs might initially be screened for their ability to bind the targeted viral antigen (e.g., a surface glycoprotein). Further screening might select for mAbs with strong virus neutralization activity. Cells producing the selected mAbs are grown in large quantities and the antibodies purified. New technologies involve use of molecular cloning to produce recombinant mAbs in yeast. mAb treatments were among the first approved therapies to treat or prevent severe SARS-CoV-2 disease. One drawback of mAb therapies is that they may become ineffective against viruses that mutate rapidly and indeed, this was the case with the first mAb treatments produced for SARS-CoV-2. They became ineffective as variants arose. mAbs are administered by infusion which requires significant medical infrastructure. mAb therapy is one of the few approved treatments for RSV which results in thousands of hospitalized infants each year.

A few nonimmunoglobulin drugs targeting viral entry have been developed. These have been designed by identifying key receptors and modeling virus attachment. In theory, a drug to block attachment might bind either the virus or the host receptor and there are examples of both. Maraviroc is a small molecule antagonist of the chemokine receptor 5 protein (CCR5), expressed on the surface of CD4 + T-cells and macrophages. The CCR5 protein is one of the coreceptors for HIV and maraviroc prevents attachment of the HIV surface protein (SU) to CCR5. Bulevirtide is a 47-amino acid peptide that blocks attachment of hepatitis delta virus (HDV) and HBV to the sodium/bile acid cotransporter protein expressed on hepatocytes which serves as the receptor for both of these viruses. Bulevirtide was approved for medical use in the European Union in July 2020. In the case of enveloped viruses, fusion is another step that can be targeted. Enfuvirtide is a 36-amino acid peptide that is an HIV fusion inhibitor. Enfuvirtide is homologous to part of the HIV transmembrane protein (TM or gp41). By interacting with a site on HIV TM, the synthetic peptide prevents it from assuming the necessary configuration to mediate envelope fusion. Docosanol, also known as behenyl alcohol is a 22-carbon saturated fatty alcohol that is the active ingredient in topical medications used to treat cold sores caused by the human herpes simplex virus (HSV or HHV-1). Though its mechanism of action has not been proven conclusively, it is thought to inhibit herpesvirus fusion with target cells.

Amantadine and rimantadine are related compounds that inhibit influenza A virus (IAV) replication by blocking the ability of viral nucleocapsids to reach the cell nucleus. They do this by inhibiting the IAV M2 protein. Both drugs are small organic compounds [1-aminoadamantane and 1-(1-adamantyl) ethanamine, respectively] that bind M2 and prevent its activation. M2 is viral ion channel protein (sometimes called a viroporin) that is normally activated after virion endocytosis. Low pH in the endosome triggers M2 to assume its active state and active M2 transports hydrogen ions from the endosome through the virion envelope. When the interior of IAV becomes acidified, strong interactions between the matrix (M) protein and the viral nucleocapsids are disrupted. Upon fusion of the viral and endosomal membranes the nucleocapsids are free to traffic to the nucleus. In the presence of amantadine or rimantadine the pH of the interior of IAV remains near neutral pH and nucleocapsids are not released after fusion. Unfortunately a single amino acid substitution in M2 is all that is required to confer resistance to these drugs. Drug resistance among human IAVs is so prevalent that these drugs have practically no value as an antiviral therapy (although the drugs have other applications).

Inhibitors of viral nucleic acid synthesis

Nucleic acid synthesis is a valuable target for drug development and there are many drugs that target DNA or RNA synthesis, particularly for viruses that do not use host DNA polymerase for replication. There are two broad groups of inhibitory agents. The first includes the nucleoside/nucleotide analogs that are competitive inhibitors of polymerases. These drugs are incorporated into a growing nucleic acid chain and either stall or inhibit further addition of nucleotides. The second group are noncompetitive inhibitors that binds and inactivate polymerases.

Nucleoside/nucleotide analogs are the most common group of therapeutic antiviral drugs. In order to be useful a nucleoside analog must be transported into cells and subsequently phosphorylated as phosphorylated compounds are not able to efficiently cross the plasma membrane. A few nucleotide analogs have been developed. These are monophosphorylated but have been designed such at the negatively charged phosphates are shielded. Nucleoside/nucleotide analogs should also have higher affinity for the target viral polymerase than cell polymerases, to reduce toxicity. Over 30 nucleoside analogs have been approved over

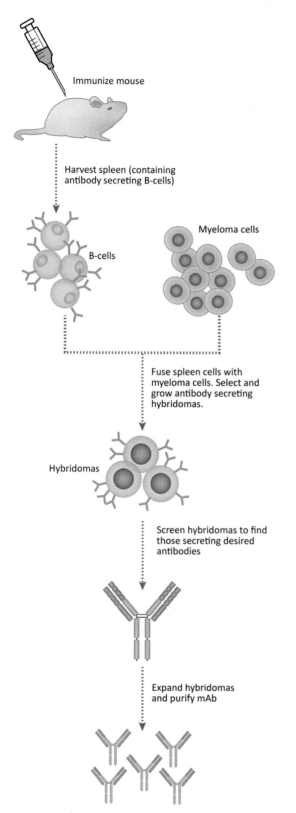

Immunize mouse

Harvest spleen (containing antibody secreting B-cells)

B-cells

Myeloma cells

Fuse spleen cells with myeloma cells. Select and grow antibody secreting hybridomas.

Hybridomas

Screen hybridomas to find those secreting desired antibodies

Expand hybridomas and purify mAb

FIGURE 8.2 Overview of the production of mAbs. The process begins with immunizing mice with an antigen of interest. Spleens are harvested as a source of antibody-producing plasma cells. These are fused to myeloma cells to make immortal cell hybrids and these are screened for production of desired antibodies. Selected cells are cultured for production of large quantities of specific mAbs.

the past 50 years. Many of these function as chain terminators. A chain terminator is a modified nucleotide that can be incorporated into a growing nucleic acid chain but which lacks a 3′-OH group so that polymerization cannot continue (Fig. 8.3). Chain-terminating nucleoside analogs are a major class of drugs used to treat HIV infection. These are frequently referred to as nucleoside reverse transcriptase inhibitors (NRTIs) and several are listed in Table 8.1. Zidovudine (azidothymidine, AZT) was the first drug approved to treat HIV infection. AZT is very effective at interfering with reverse transcription, but when used as a sole treatment, resistant HIV strains often arise in the patient. The early recognition of AZT resistance led to the development and use of additional NRTIs and multidrug treatment regimens to prevent the emergence of resistant viruses. These multidrug treatments are often called "highly active antiretroviral therapy" or HAART. HBV also replicates its genome using reverse transcriptase and there are approved NRTIs that are used for treatment of chronic HBV. A few drugs are approved for treatment of both HIV and HBV.

Acyclovir is a highly effective antiherpesvirus drug. It is a nucleoside analog and chain terminator discovered in 1971 and approved for use in 1981. Acyclovir

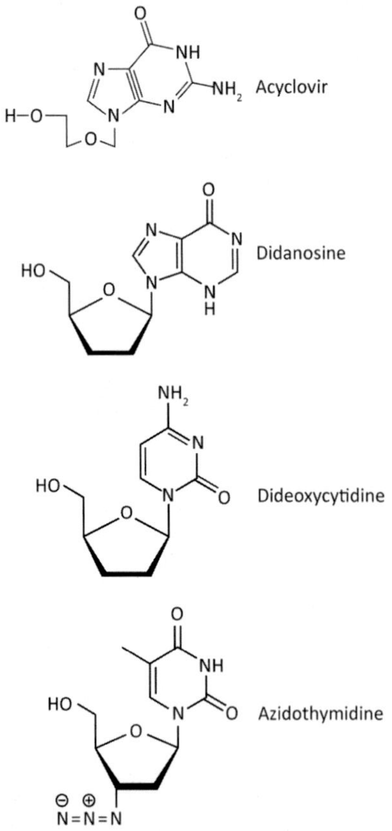

FIGURE 8.3 Chemical structures of a few chain-terminating nucleoside analogs.

is highly selective for herpesviruses and has little or no adverse effects on the host. It can be taken orally or used topically. The high specificity of acyclovir is due to the inability of the host cell to phosphorylate the drug. The drug is initially phosphorylated only by herpesviral thymidine kinase, after which cellular kinases add the second and third phosphates to generate a fully active compound (Fig. 8.4). In addition, the herpesviral DNA polymerase has higher affinity for acyclovir than do host polymerases. Several additional nucleoside analogs are also available for treatment of herpesvirus infection, including codofovir and ganciclovir.

Remdesivir is a broadly acting nucleotide analog originally developed for treatment of HCV. It has been investigated for treating Ebola virus and Marburg virus infections and was approved for use against SARS-CoV-2. Incorporation of remdesivir into a growing RNA chain does not immediately stop further synthesis, instead up to five additional nucleotides are added before termination of the growing chain. For this reason the drug can circumvent the proofreading exonuclease of SARS-CoV-2. The drug has been classified as a delayed chain terminator.

Ribavirin is a broad-spectrum guanosine nucleoside analog effective in vitro against a variety of DNA and RNA viruses. It has been clinically approved together with $IFN - \alpha$ as a treatment for chronic HCV. Ribavirin is also approved for treating infections caused by RSV, adenovirus, hantavirus, and some hemorrhagic fever viruses. Ribavirin has multiple mechanisms of action. It is a competitive inhibitor of inosine monophosphate dehydrogenase which is involved in the synthesis of inosine monophosphate and GTP. Thus ribavirin causes intracellular depletion of GTP, impacting both viral and cellular processes. The 5′-phosphorylated form of ribavirin can also directly inhibit the RNA-dependent-RNA polymerase (RdRp) of RNA viruses and can interfere with the formation of the 5′-cap structures of viral RNAs. It is possible that ribavirin enhances viral mutagenesis by substitution of RTP for GTP, because most RdRps lack proofreading abilities. However, ribavirin can cause serious side effects, including birth defects.

Molnupiravir is a nucleoside analog recently approved for treatment of SARS-CoV-2 (2021 in the United Kingdom and for emergency use in the United States). The mechanism of action of molnupiravir is quite interesting as it is a mutagen, causing G-to-A and C-to-U transitions. The highly mutated genomes produced in the presence of the drug are not infectious. The approval of molnupiravir for treatment of SARS-CoV-2 was quite important as chain-terminating nucleoside analogs are largely ineffective against coronaviruses due to an exonucleolytic proof-reading activity that can remove misincorporated nucleotides from the 3′-end of a nascent RNA chain.

TABLE 8.1 Examples of nucleoside and nucleotide analogs approved for treatment of viral infections.

Name	Drug type/mechanism of action	Virus(es) targeted
• Zidovudine (azidothymidine, AZT) • Didanosine (dideoxyinosine, ddI) • Zalcitabine (dideoxycytine, ddC) • Stavudine (d4T) • Abacavir • Emtricitabine (FTC)	Nucleoside analogs Chain terminators Target RT	HIV
• Lamivudine	Nucleoside analog/chain terminator. Targets RT	HIV, HBV
• Tenofovir	Nucleotide analog/chain terminator. Targets RT	HIV, HBV
• Adefovir • Entecavir	Nucleoside analogs Chain terminators Target RT	HBV
• Sofosbuvir	Nucleoside analog	HCV
• Acyclovir • Gancyclovir • Penciclovir • Cidofovir	Guanosine nucleoside analogs Chain terminators Target herpesviral DNA pol Phosphorylation requires herpesviral thymidine kinase	Herpesviruses (use and effectiveness vary for each drug)
• Ribavirin	Guanosine nucleoside analog. Multiple modes of action including: Competitive inhibition of IMP dehydrogenase thereby limiting intracellular concentrations of GTP. The 5'-triphosphate form can directly inhibit RdRp. Inhibits guanyl- and methyl-transferase activities to inhibit 5'-CAP formation. Could enhance mutagenesis.	Broad-spectrum antiviral activity. Active against both RNA and DNA viruses.
• Molnupiravir	Nucleoside analog/ acts as a mutagenizing agent that causes an "error catastrophe" during viral replication. Targets RdRp.	SARS-CoV-2
• Remdesivir	Nucleotide analog/chain terminator. Targets RdRp.	SARS-CoV-2

Another group of inhibitors of viral nucleic acid synthesis are the nonnucleoside inhibitors of viral polymerases (or other components of replicase complexes). Examples are listed in Table 8.2. There are several nonnucleoside RT inhibitors (NNRTIs) approved treating HIV infection. Other nonnucleoside inhibitors of genome replication target the HCV NS5A phosphoprotein, an essential component of the HCV replicase complex. Dasabuvir directly targets the HCV NS5B polymerase. Foscarnet (phosphonomethanoic acid) is primarily used to treat herpesviral infections. It is classified as a pyrophosphate analog DNA polymerase inhibitor.

A few drugs target viral genomes at steps after nucleic acid synthesis. For example, herpesviral genomes are synthesized as concatemers that must be cleaved into unit length genomes by the viral terminase complex during packaging. Letermovir is a drug with activity against the human cytomegalovirus (CMV) terminase complex. This drug was discovered by high-throughput screening. HIV DNA genomes must be integrated into host DNA as a required step in the retrovirus life cycle. Examples of approved drugs that target the HIV integrase protein include raltegravir, elvitegravir, and dolutegravir.

Protease inhibitors

Plus-strand RNA viruses (i.e., picornaviruses, flaviviruses, coronaviruses, togaviruses) and retroviruses produce some of their viral proteins as polyprotein precursors that must be cleaved by viral proteases. The plus-strand RNA viruses produce their nonstructural proteins, such as RdRp as polyproteins that include protease domains. Targeting these proteases inhibits the process of RNA synthesis very early on in the replication cycle as an active polymerase is never produced. Protease inhibitors have been approved for treatment of HCV (paritaprevir, grazoprevir) and SARS-CoV-2 (nirmatrelvir).

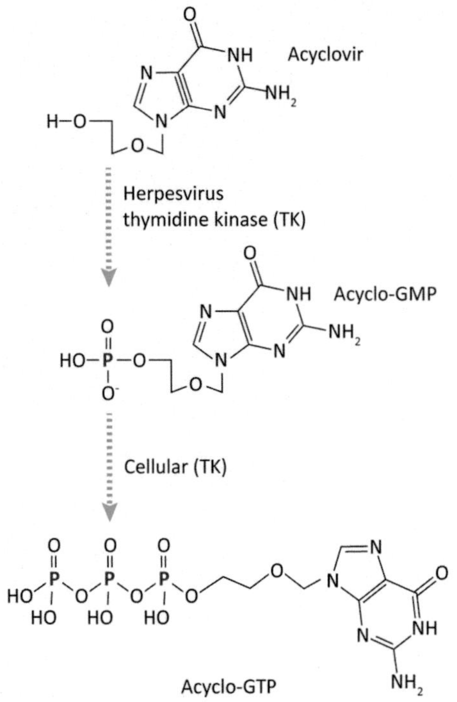

Acyclovir

Herpesvirus
thymidine kinase (TK)

Acyclo-GMP

Cellular (TK)

Acyclo-GTP

FIGURE 8.4 The specificity of acyclovir is explained by the need for phosphorylation by a herpesviral thymidine kinase. The drug is inactive in uninfected cells.

TABLE 8.2 Examples of nonnucleoside inhibitors of viral polymerases approved for treatment of viral infections.

Name	Virus(es) targeted
• Foscarnet	Herpesviruses
• Neviripine • Delavirdine • Efavirenz • Etravirine • Rilpivirine	HIV
• Daclatasvir • Ledipasvir • Ombitasvir • Elbasvir	HCV (through inhibition of the NS5A phosphoprotein)
• Dasabuvir	HCV (through inhibition of the NS5B polymerase)

Inhibitors of HIV proteases were among the first drugs developed by rational design, based upon detailed knowledge of protease structure and substrate binding properties. Protease inhibitors including ritonavir and saquinavir are synthetic, uncleavable, molecules that bind to the HIV protease active site. Retroviral proteases cleave viral polyproteins after assembly of immature capsid structures. Thus HIV protease inhibitors are able to inactivate virions produced by cells containing integrated copies of retroviral genomes, where inhibitors of reverse transcription

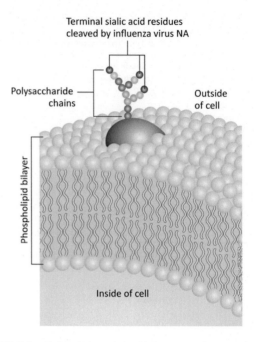

FIGURE 8.5 Sialic acid moieties (derivatives of neuraminic acid) are found at the ends of the polysaccharide chains that decorate glycoproteins. Sialic acid residues serve as attachment sites of influenza virus HA but later in the replication cycle must be removed from cell and viral proteins by influenza virus NA to facilitate release of newly formed virions.

are ineffective. For treatment of HIV, protease inhibitors are a powerful complement to RT inhibitors.

Neuraminidase inhibitors

Neuraminidase (NA) inhibitors specifically target influenza viruses. Influenza A and B viruses (family *Orthomyxoviridae*) have two spike glycoproteins on their surface. Hemagluttinin (HA) is the attachment and fusion protein while NA is a receptor destroying enzyme required for effective release of virions from cells. HA binds to sialic acid residues that are found at the terminal positions of the oligosaccharides that decorate glycoproteins and glycolipids (Fig. 8.5). Sialic acids are derivatives of the 9-carbon sugar neuraminic acid. Attachment of virions to sialic acid moieties is followed by endocytosis and fusion to initiate an influenza virus infection. However, at the end of the influenza virus life cycle, sialic acids must be cleaved from viral glycoproteins as well as from cellular proteins, to allow virions to spread efficiently to new cells. This is where the activity of NA is critical. In the absence of NA, or if its enzymatic activity is inhibited, virions form large clumps or rafts with greatly reduced infectivity, as virions attach to each other. The antiinfluenza drugs oseltamivir and zanamivir, are sialic acid analogs that bind to, and inhibit NA.

Interferons

IFNs are potent antiviral cytokines produced by cells in response to viral infections. When first discovered, IFNs were hailed as possible therapeutics to treat a broad range of viral infections. Type I IFNs are secreted by most cell types in response to viral infection. They act in an autocrine and paracrine fashion through binding to IFN receptors (IFNRs). Receptor binding activates signaling cascades that lead to the expression of a large set of IFN-stimulated genes, many of which have antiviral activities. Although viral infection stimulates IFN production, many viruses have evolved multiple pathways by which they downregulate the IFN response. Administration of exogenous IFN could theoretically overcome such blocks. In reality recombinant IFNs have very limited treatment value as antiviral agents due to the potential for severe side effects including fatigue, fever, influenza-like symptoms, depression, bone marrow suppression, and development of autoimmune illnesses. Pegylated IFN-α is currently approved for treatment of HBV. It has also been used to treat HCV in the past, but has been largely replaced by more effect drugs.

The future of antivirals

The ongoing SARS-CoV-2 pandemic has underscored the need to accelerate the pace of antiviral drug development. Development of antivirals is multifaceted, requiring an understanding of the virus replication cycle (including key interactions with host cell proteins), the ability to generate detailed models viral proteins, computer-assisted drug design, the ability to synthesize potential inhibitory compounds as well as safe assays and animal models to test drug efficacy. Computer modeling and simulation of molecular dynamics can greatly assist with assessing interactions of a ligand and receptor, as well as allowing an understanding of the fate of the complex over time. Other exciting developments include the design of broad platforms that can be rapidly modified to respond to new threats. One example is called proteolysis targeting chimera (PROTAC). PROTACs are heterobifunctional molecules that target viral proteins to the cell's ubiquitination−proteasome system. One end of a PROTAC binds to a viral target while the other end binds and recruits an E3 ubiquitin ligase leading to ubiquitination of the viral target. Another platform being investigated targets viral nucleic acids with a ribonuclease. These so-called ribonuclease targeting chimeras (RIBOTACs) bind RNAs and recruit specific RNases to the viral target.

In this chapter we have learned that:

- Most antiviral therapies target specific viruses.
- Many approval antiviral drugs target nucleic acid synthesis.
- Many antiviral drugs, for example, HIV protease and integrase inhibitors target very specific viral enzymes.
- Detailed knowledge of a virus replication cycle and virally encoded proteins are used to model specific antiviral agents.
- Although IFNs are a natural antiviral defense, recombinant IFN has limited utility in treating viral infections due to serious side effects.

9

Virus evolution and genetics

After reading this chapter, you should be able to answer the following questions:

- Did a single common ancestor give rise to all viruses?
- What are two discrete processes that drive virus evolution?
- How is the mutation rate of a virus described?
- How is the evolution rate of a virus described?
- How can endogenous viral elements reveal the age of viruses?
- What are the general mechanisms that generate novel viral genomes in nature?
- What are "nonessential genes" and are they always nonessential?
- What is "reverse genetics" and how is it used to probe the functions of viral genes?

This chapter presents, in very general terms, the origins of viruses and the ongoing process of virus evolution. An overview of the molecular genetics of some major groups of viruses is also presented.

Virus evolution

There is great diversity among viruses; they infect organisms from all Kingdoms of life and across all ecosystems. Animals, plants, single-celled eukaryotes, and prokaryotes are infected by a variety of viruses, both DNA and RNA. And some closely related viruses infect animals, plants, and bacteria. These facts suggest a very ancient origin for viruses.

There are three commonly proposed mechanisms for the origins of viruses:

Viruses descended from primitive precellular life forms. This theory posits that viruses originated and evolved along with the primitive self-replicating molecules that were destined to become cells. The agents we recognize as viruses today were originally self-replicating molecules in the precellular world. If this hypothesis is correct, it follows that cellular life forms were impacted by "viruses" from their earliest beginnings.

Viruses are "escaped" cellular genetic elements. This theory posits that viruses evolved *after* cells. Their origins were cell-associated genetic elements that acquired protein coats, allowing for efficient cell to cell transfer. The DNA genomes of some viruses do in fact resemble plasmids, and retroviruses are related to the large group of nonviral retroelements populating cellular genomes.

Retrograde evolution. The theory of retrograde evolution states that viruses were once complex intracellular parasites that lost the ability for all independent

© 2023 Elsevier Inc. All rights reserved.

metabolism; they retained only those genes required to manipulate the host cell and produce progeny virions.

Recent studies suggest virus evolution might be more complex than suggested by the list above. The International Committee on the Taxonomy of Viruses (ICTV) currently recognizes six realms of viruses, based on similarities of the member's genes and proteins. *The members of each realm are hypothesized to have arisen from a different common ancestor.* The virus realms currently recognized are: *Riboviria* (a group that includes viruses that use an RNA-directed polymerase to replicate), *Duplodnaviria* (a group that includes double-stranded DNA (dsDNA) viruses whose genomes encode a major capsid protein containing the HK97 fold), *Monodnaviria* (a group that includes single-stranded DNA viruses encoding a rolling-circle replication endonuclease of the HUH superfamily), *Varidnaviria* (a group that includes DNA viruses that encode a major capsid protein containing the vertical jelly-roll fold and forming pseudo-hexameric capsomers), *Adnaviria* (a group that includes filament-shaped viruses that infect archaea and share a shell structure), and *Ribozyviria* (a group that includes small RNA viruses with properties of ribozymes). It appears likely that the *Monodnaviria* arose via recombination between replication proteins of prokaryotic plasmids and capsid proteins of eukaryotic RNA viruses. These viruses would have had to evolve *after* the evolution of both bacteria and eukaryotic RNA viruses. Ribozymes are catalytic RNAs that may have been key to the evolution of prebiotic self-replicating systems and the *Ribozyviria* are so named because their genomes have self-cleaving domains that are critical to their replication. *Ribozyviria* genomes have enzymatic activity, thus they are called ribozymes. They also have a protein coding domain but replicate only in the presence of helper viruses, such as hepatitis B virus. Thus their evolutionary path must have been complex. Finally, large DNA viruses such as herpesviruses, poxviruses and African swine fever virus have almost certainly incorporated host cellular genes into their genomes, thus regardless of their origins, their evolution continued along with eukaryotic cells. Finally, there is good evidence that viruses have shaped the evolution of single-celled and multicellular organisms for billions of years. Animal genomes contain pieces of RNA and DNA genomes that have very ancient origins and there are numerous examples of gene flow from cells to viruses and vice versa. An example of animals coopting viral genes is evident in development of placental mammals. Proteins critical for implantation of embryos into the maternal endometrium, a process central to formation of the placenta, arose from the envelope proteins of endogenous retroviruses (ERVs).

While some virologists ponder the ancient origins of viruses, others examine the mechanisms that drive ongoing virus evolution. Examples of ongoing virus evolution include:

Cross-species jumps that allow viruses to find new hosts. Influenza viruses provide many examples of this type: A bird or swine influenza jumped into humans at the turn of the 20th century causing a major worldwide pandemic. Horse influenza virus moved into dogs in 2004 (Fig. 9.1). At least three human coronaviruses have jumped from animals into humans in recent years although only one, severe acute respiratory syndrome coronavirus (SARS-CoV)-2, became extremely well adapted to the human host. In 2002 SARS-CoV-1 jumped from bats into farm-raised Himalayan palm civets (*Paguma larvata*) and then into humans. Fortunately, SARS-CoV-1 was not easily transmitted among humans so overall cases remained only in the thousands before it was eliminated from the community. Middle Eastern respiratory syndrome coronavirus (MERS-CoV) appears to be a camel virus that occasionally infects humans. Like SARS-CoV-1, transmission between humans is poor, so outbreaks remain small. The exact origin of the virus, SARS-CoV-2, causing the current COVID-19 pandemic is unclear, but it is related to bat coronaviruses. Unfortunately, SARS-CoV-2 is extremely well adapted to replication in, and spread among, humans. It continues to evolve as it spreads rapidly across the globe. SARS-CoV-2 also infects animals such as dogs, cats, mink, and white-tailed deer; it remains to be seen if new animal coronaviruses will emerge. Among the DNA viruses, a dog-adapted version of a feline parvovirus emerged in the 1970s and rapidly spread worldwide through the domestic dog population. It is thought that jumps from one animal host to another are strong drivers of virus evolution and speciation.

Decreased virulence in a new host is another type of readily observed evolution. Viruses that are well adapted to their hosts often cause little or no disease. When a virus jumps to new host, the virus may cause

FIGURE 9.1   Virus moves to a new species: in 2004 equine influenza virus jumps to dogs.

severe disease, but coevolution of virus and host often restores more balance. An instructive example is provided by studies of rabbit myxoma virus, a member of the family *Poxviridae*. Rabbit myxoma virus causes localized skin infections in American rabbits but causes high mortality in European rabbit breeds. Rabbit myxoma virus was introduced into Australia in 1950 to control huge populations of imported European rabbits that were causing environmental havoc. While the virus initially produced high mortality and significantly reduced the rabbit population, the virus became much less virulent within 20 years. Thus rabbits and rabbit myxoma virus coexist in Australia to this day.

Emergence of drug-resistant viruses can occur rapidly, undermining the utility antiviral therapies. The mutation and replication rates of human immunodeficiency virus (HIV) are so high that successful long-term treatment requires use of drug cocktails, containing two to four active compounds. Amantadine and rimantadine are antivirals licensed for use against influenza A viruses in 1966 and 1993, respectively. These drugs were once widely used but rates of resistance to both drugs have been increasing globally since 2003. Resistance is now so prevalent that these once front line drugs are no longer recommended for treatment of influenza.

Immune escape mutants allow viruses to escape antiviral immunity on an individual or population level. HIV and influenza viruses are adept at escaping from host immune pressures. Point mutations in certain regions of influenza virus surface proteins are well tolerated by the virus and serve to thwart established immune responses. This is one reason that influenza viruses are closely monitored and vaccines are updated yearly. Immune escape is currently driving evolution of SARS-CoV-2.

Measuring virus evolution

Virus evolution is the outcome of two independent events (Fig. 9.2): the first event is genome mutation. The replication enzymes of RNA viruses have no proof reading functions so genome mutation is a common, and perhaps necessary event during their replication. The genomes of RNA viruses can also recombine during replication. DNA damage, mutations during replication and recombination all play roles in mutation of DNA viruses. Virus genomes can also capture genes from the host.

There are several ways to express mutation rates. A common method is to express them as *misincorporations per nucleotide synthesized*. Another way is to express them as *mutations/nucleotide/cell infection* as this accounts for different modes of viral genome replication. Accepted estimates range from 10^{-6} to 10^{-4} mutations/nucleotide/cell infection for RNA viruses and range from 10^{-8}

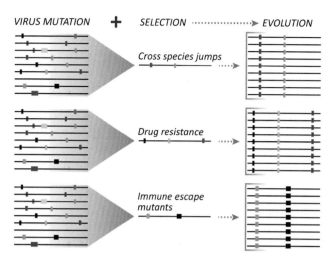

FIGURE 9.2 Evolution is the result of two independent processes: mutation and selection.

to 10^{-6} mutations/nucleotide/cell infection for DNA viruses. For some RNA viruses, these numbers translate into one or two mutations per genome replicated. Viral mutation rates can be difficult to measure, as genomes with lethal mutations are quickly eliminated from the population.

The second event that drives virus evolution is *selection* of mutants that are more fit for a particular environment. For example, beneficial mutations may allow viruses to expand their host-range, evade immune responses, or tolerate antiviral drugs. Thus virus evolution requires initial mutations followed by selective pressures. A measurement of virus evolution that includes the parameter of natural selection is *nucleotide substitutions per nucleotide site, per year*. This measure reflects mutations that become fixed in the virus population. If a virus is highly adapted to a particular host and environment, this number may be very low. In contrast, the rate may be much higher if a virus is in the process of adapting to a new host or environment. Different lineages of the same virus may evolve at different rates depending on the selective pressures they encounter.

Point mutation

General aspects of DNA mutation are not discussed herein, as descriptions are widely available in introductory biology and microbiology texts. In general, the rates of mutation of DNA virus genomes are slower than those of RNA viruses. It is generally assumed that some or all of the proof-reading mechanisms that are employed by host cells are available to DNA viruses (particularly those DNA viruses that use host cell DNA replication machinery). RNA viruses replicate their genomes using RNA-dependent RNA polymerases (RdRps) and in general do not employ proofreading (an exception seems to be coronaviruses). Studies of RdRps indicate rates of nucleotide misincorporations ranging from 10^{-6} to 10^{-4} mutations/nucleotide/cell infection. Detailed analyses of the poliovirus RdRp have identified single nucleotide substitutions that change rapid, error prone replication to a slower, more faithful replication phenotype. It is also clear RNA genome diversity impacts viral fitness. Each type of RNA virus appears to have an "optimal" amount of genetic diversity and shifting the level (either to higher or lower diversity) reduces population fitness. This finding has practical implications, providing a rationale for development of antiviral drugs that target RdRp fidelity.

Recombination

Homologous recombination between DNA virus genomes occurs by the same molecular mechanisms described for cellular DNA. In fact, DNA recombination is a method that researchers use to generate DNA viruses with desired mutations. Retroviruses insert a dsDNA genome into the host chromosome as a required step in their replication cycle. Recombination events between retroviral and host DNA, and among different retroviruses, is well documented. Recombination between endogenous and actively replicating retroviruses is so likely that is an important consideration when using retroviruses in gene therapy applications.

Viruses with RNA genomes also undergo genome "recombination" albeit by a different molecular mechanism. RNA virus recombination is achieved through a process called "copy choice." Copy choice involves template switching during genome replication (Fig. 9.3). The RNA RdRp physically dissociates from one genome and reassociates with another. Regions of genome complementarity probably serve to position the replicase complex on the new template. The frequency of copy choice recombination is dependent upon the mechanisms used by a particular virus family for genome replication and transcription. Copy choice recombination can also result in the insertion of fragments of cellular genes into the genomes of replicating viruses. A well-documented example of this is the development of highly pathogenic pestiviruses of cattle. Bovine viral diarrhea virus (BVDV) can cause persistent but asymptomatic infections of cattle. Persistent infection occurs when fetuses are infected and become immunologically tolerant. However persistently infected animals sometimes develop severe, lethal disease, caused by mutated BVDV. Often the mutated viruses have acquired fragments of cellular genes!

Reassortment

Reassortment is a mechanism available to viruses with segmented genomes (e.g., orthomyxoviruses, reoviruses, and bunyaviruses) allowing generation of novel viruses. Reassortment involves shuffling of complete genome segments. The orthomyxoviruses (influenza viruses) have segmented genomes (eight segments for influenza A viruses) and upon infection of a cell with two different viruses a variety of progeny can be released. A similar phenomenon has been well documented for rotaviruses. Influenza virus genome segments can also undergo copy choice recombination, but reassortment is a more common event.

Genotypes and phenotypes

The terms genotype and phenotype apply to viruses. The viral genotype is simply the sequence of its viral genome. But what kind of viral phenotypes can be measured? Quite an array as it turns out! Viral phenotypes include ability to: Productively infect different cell types (host range); replicate at elevated temperatures; form different types of plaques in cultured cells (large, small, clear, or cloudy); transform cells; replicate at an increased or decreased rate; cause disease; be neutralized by a particular antibody; replicate in the presence of various drugs; display increased or decreased mutation and/or recombination rates; be transmitted to new hosts. Other terms used to describe groups of viruses include variant, strain, clade, or lineage. These terms and their usage are discussed in Box 9.1.

Essential and nonessential genes

Compared to cells, most viruses have very small genomes. A smaller genome is an advantage when replicating as it takes few cell resources, and less time, to synthesize a small genome. When researchers began propagating viruses in cultured cells in the laboratory, they made a surprising finding: Not all viral genes appeared to be essential for replication! These so-called "nonessential" genes were initially identified because they were lost during passage in cultured cells and the mutated viruses actually grew more robustly than their wild-type parents. However, it was counterintuitive to some virologists to think that viruses, with their very ancient origins, maintained truly nonessential genes for millions of years. They hypothesized that while some genes are not essential *in cultured cells*, they must be advantageous under more exacting conditions, such as replication and spread in a natural population of susceptible hosts. And indeed, this appears to be the case. Over and over again, viral

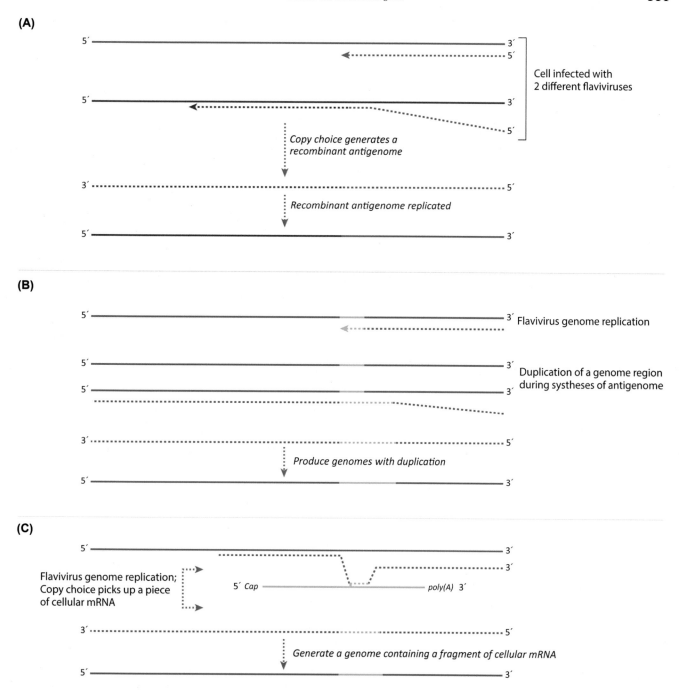

(A)

Cell infected with 2 different flaviviruses

Copy choice generates a recombinant antigenome

Recombinant antigenome replicated

(B)

Flavivirus genome replication

Duplication of a genome region during systheses of antigenome

Produce genomes with duplication

(C)

Flavivirus genome replication; Copy choice picks up a piece of cellular mRNA

5′ Cap ——— poly(A) 3′

Generate a genome containing a fragment of cellular mRNA

FIGURE 9.3 Copy choice recombination of RNA viruses.

genes identified as nonessential in cultured cells are required to maintain the virus in an animal population. For example, the herpesvirus thymidine kinase gene is nonessential for virus replication in cell culture but is essential for productive infection of natural hosts.

The loss of nonessential genes in cultured cells is virus evolution in action. Given an environment without physical barriers to infection, antiviral responses, or limited nutrients, viruses discard "useless" genes. The benefit is a smaller genome that it can be replicated at a faster rate, using fewer resources (Box 9.2). Thus viruses have a set of truly essential genes (i.e., genes for capsid proteins, polymerases, and replication origin-binding proteins), required under all conditions, and other genes that may only be required in a specific ecological niche. Finally, a gene may not be *absolutely* required in a particular niche, but its presence may confer a strong advantage. Thus not only is overall environment important, so too are the presence or absence of viral competitors!

BOX 9.1

Mutant, variant, strain, lineage: what do they mean?

The International Committee on Taxonomy of Viruses (ICTV) is the body that discusses and codifies viral taxonomy as laid out in a code, the International Code of Virus Classification and Nomenclature. The ICTV *classifies* viruses into taxa, which are manmade groups, with the lowest taxon being the virus species. The ICTV is not involved in naming individual viruses: those physical entities that can be grown, studied, sequenced, etc. This can be a confusing concept, but for the purposes of this discussion, the important take away is that the ICTV neither names individual viruses nor defines terms such as strain or lineage. We must look to general usage by groups of biologists or virologists, who may not always agree on usage, to get a sense of what these terms mean.

The term mutation does have a specific meaning in biology. A mutation refers to any change in a genome sequence. To define a mutation, we must have an accepted comparator, often called the wild-type sequence. We often use "mutant" to describe an organism that has a distinct phenotype. However two organisms, or viruses, may have different genetic sequences but may have virtually identical phenotypes. When considering viruses, mutant is a particularly poor word choice because there are cases (e.g., HIV) where it is actually uncommon to find identical genomes. In fact, research virologists often maintain viral "stocks" as cloned dsDNA to maintain genetic homogeneity. During

the COVID-19 pandemic the terms "mutant," "strain," and "variant" have been used describe distinct types of SARS-CoV-2. As discussed above, mutant is a particularly poor word choice.

In virology the term "variant" is often used to describe whole genome changes. Two viral genomes, with multiple nucleotide differences, are described as variants of one another (as neither is a "wild-type" sequence). In the case of SARS-CoV-2, most genomes are in fact variants of one another, so the term "variant of concern" was coined to describe a variant that might cause significant health problems. A variant of concern might be one that has multiple mutations in the spike protein, and whose prevalence is increasing among vaccinated individuals.

The terms clade and lineage are also used to describe viruses. Both are defined based on molecular phylogenies or "trees" developed from genome sequence data. Clades and lineages apply to groups of viruses that are clearly related, based on their positions in a "tree." As SARS-CoV-2 spreads worldwide, new clades and lineages are developing and this process will likely continue for some time. It can sometimes be helpful to name/designate clades or lineages, particularly from an epidemiological standpoint. In the case of SARS-CoV-2, the newsworthy, named "variants" are actually lineages that stand out in some way (i.e., higher transmission or virulence).

BOX 9.2

What are defective interfering particles?

Defective interfering particles (DIPs) can easily arise during virus culture. DIPs contain defective genomes, packaged into virions, that can only replicate in the presence of replication competent viruses (viruses with fully intact genomes). DIPs can arise in any type of virus, DNA or RNA. They were first identified as arising in cultures of RNA viruses. DIPs can become predominant in a stock of cultured virus as their genomes replicate more quickly than the complete viral genome, due to smaller size. In order to be replicated, the defective viral genomes (DVGs) must contain all of the cis-acting sequences required for genome replication and packaging. Therefore the study of

DIPs helped to define those important genetic elements, which are frequently found at the 5′ and 3′ ends of viral genomes. Infection of cells with large numbers of DIPs can reduce the titers of fully infectious virions. Thus in a research setting, generation of DIPs can be avoided by routinely plaque purifying viruses as a starting point to generate new viral stock. A culture is diluted and plated and a single plaque recovered, diluted, and plated again. The process can be repeated as necessary reduce DIPs. DVGs are not simply an artifact of laboratory culture however, they are found in nature and replicate in concert with helper viruses. They are often found in plants.

Coevolution of virus and host

One common question about viruses is their "age." Is HIV a few decades old, a few hundred years old, or many millions of years old? While we know that the virus we call HIV-1 *emerged* in the human population several decades ago, it is very closely related to viruses of nonhuman primates that are millions of years old. There are different methods for calculating rates of virus evolution (that will not be described here) and these can provide very different answers about the age of a virus. A relatively recent finding impacts our view of virus age: Sequencing of animal genomes has revealed that they contain small fragments of a wide variety of viral genomes. The insertion of viral genomes into host genomes is called *endogenization* and many of these events occurred tens or hundreds of millions of years ago. The insertion events can be timed by looking at closely related species of animals. If nearly identical virus insertions are present in two different species, it is assumed that the viral gene insertion occurred prior to the split of those two species. It is possible to recognize endogenized viruses because their genomes are *clearly related* to viruses widely circulating today. This means that viruses circulating today have very ancient origins. However, it should be noted that most endogenized viral elements (EVEs) are not capable of producing virions, as they are highly mutagenized.

Insertion of viral genomes (or fragments thereof) into host genomes is probably fairly common, and can be detected in cultured cells. However, insertion into germ line cells, and long-term maintenance in a population is a much rarer event. Many types of viruses have been endogenized, including a variety of DNA and RNA viruses (Table 9.1). EVEs derived from DNA viruses such as adenoviruses, polyomaviruses, and papillomaviruses probably entered genomes through nonhomologous recombination. EVEs derived from RNA viruses often contain poly-A sequences, suggesting that mRNAs served as the templates for integration, likely through interaction with host cellular retroelements. The EVEs of viruses other than retroviruses are usually just fragments. By far the most commonly endogenized viruses are the retroviruses, as these insert into the host genome as an obligate step in their replication. Endogenous retroviruses played important roles in evolution of the mammalian placenta and can also protect against infection with exogenous retroviruses (Box 9.3).

Genetic manipulation of viruses

The classical study of viral genetics (also called "forward" genetics) involved identifying a novel phenotype and then identifying the associated mutations. To accomplish this, virus stocks were subject to mutagenesis and large numbers of mutants individually analyzed. A variety of methods were used to sort out and classify mutants. The approach worked well for bacteriophages (bacterial viruses) and it was also applied to animal viruses when methods were developed to grow them using cultured cells. However, it can be challenging to design strategies for selecting a desired phenotype, and brute screening can take years.

Today, viral genetics depends largely on a general methodology called *reverse genetics*. The term reverse genetics is used to describe activities that involve *making specific mutations* in a viral genome and *then observing* the phenotype. Mutations can range from a single nucleotide substitution to partial or complete gene deletions. Sometimes the order or arrangement of genes is altered. The goal of these changes is to link gene function to altered phenotypes, thereby generating models that describe gene function. Most significant human/animal viral pathogens can be studied using reverse genetic approaches. Perhaps one of the most impressive (and controversial) feats in reverse genetics was the resurrection of the deadly 1918 pandemic strain of influenza virus, using genome sequences derived from decades-old, formalin-fixed lung autopsy materials and frozen tissues from victims buried in permafrost.

Reverse genetic analysis involves both *design* of a mutation and the ability to *incorporate* the mutation into a virus. Generating gene mutations uses a variety

TABLE 9.1 Nonretroviral EVEs present in mammalian and avian genomes.

Virus family	Host	Estimated age (MYAa)
Bornaviridae	Mammals	93
Filoviridae	Mammals	24
Parvoviridae	Mammals	99
Circoviridae	Mammals	5
Hepadnaviridae	Birds	8

*a*Million years ago.

BOX 9.3

Coevolution of virus and host

Animals have evolved a variety of strategies to control/limit/inhibit virus replication. In response, viruses have evolved mechanisms to "fight back." The ability to readily sequence both animal genomes and viruses is providing a detailed view into specific instances of evolution and coevolution of viruses and their hosts. The story we consider here is that of tetherin and its virally encoded antogonists.

Tetherin is a potent antiviral protein that inhibits a variety of enveloped viruses, including retroviruses. Tetherin expression is induced by interferon in response to viral infection. As the name implies, tetherin inhibits release of enveloped viruses from the cell surface (it "tethers" them to the cell). Tetherin appears to physically crosslink virions to the plasma membrane.

How then do viruses antagonize tetherin? The retroviruses simian immunodeficiency virus SIV*cpz* (from chimpanzees) and SIV*gor* (from gorillas) antagonize their hosts' tetherin with a protein called NEF. Each of these viruses encode a NEF protein that binds the tetherin protein of the host species. However, human tetherin has a deletion in the region that an SIV NEF would bind. SIVs replicate poorly in humans because human tetherin *is not* antagonized by SIV NEF. How then did HIV emerge in humans, as a virus able to replicate to high levels? HIV evolved a completely different tetherin antagonist! HIV encodes a protein called viral protein U (VPU) that counteracts tetherin by interacting at an entirely different site. HIV still encodes a NEF protein (it has a number of different functions that help HIV replicate) but it uses VPU to antagonize tetherin.

of methods available through modern molecular biology. (If anyone doubts the utility of "basic research" consider the powerful toolkit it has provided for the study of human and animal pathogens, genetic disorders, and cancer.) In the study of viruses, principles governing both the design of mutations and their incorporation into viruses benefit from having detailed knowledge of viral genome structure and replication strategy.

Growing viruses in cultured cells provides a straightforward method of obtaining large quantities of genomic material. This allowed detailed studies of viral genomes years before techniques were available to analyze cellular genomes. While purified virions are still an excellent source of viral genomes, the ready availability of reagents to perform DNA amplification by polymerase chain reaction (PCR) is used to rapidly generate large quantities of viral genes and genomes for downstream applications. When working with RNA viruses it is necessary to use the enzyme reverse transcriptase (RT) to generate a DNA copy of the RNA genome prior to the PCR reaction. Once a DNA fragment has been obtained it is incorporated into a plasmid (this process is commonly called gene cloning) (Fig. 9.4). Many different types of plasmids occur in nature and these have been engineered for diverse uses in biotechnology. Most plasmids have at least one gene for antibiotic resistance, allowing easy selection of bacteria that contain the desired plasmid. While the use of restriction enzymes, that cut DNA at specific sequences, once provided the backbone of genetic manipulation, novel methodologies allow virtually any DNA molecule to be rapidly inserted into a plasmid.

Plasmids have also been designed to specifically allow for production of proteins or RNA products. Some plasmids are capable of replication in both bacterial and eukaryotic cells of various types. Very large quantities of plasmids containing genes or viral genomes of interest can be obtained by growing a few liters of bacterial culture.

Reverse genetics of positive-strand RNA viruses

The genomes of positive sense RNA viruses serve as mRNA when they are released into cells. Translation begins in the absence of any preexisting viral protein. Thus in a research situation it is possible grow a positive strand RNA virus, purify the virions and strip away all proteins, leaving only purified nucleic acid. If the purified RNA is added back to cells (by injection, or by mixing with lipids that will fuse with the plasma membrane) infectious virions can be produced. No viral proteins need be added to initiate infection. Ribosomes assemble directly on the genome and synthesize viral proteins to begin the virus replication cycle. Alternatively, one can construct plasmids containing DNA copies of the RNA genome. The plasmids are designed to be transcribed, producing the viral genome in a cell, thus initiating the replication cycle. The basic steps for cloning a positive-strand RNA virus are illustrated in Fig. 9.5.

The first step in engineering the genome of a positive-strand RNA virus is to propagate the desired virus and extract genomic RNA. The enzyme RT is then used to copy the viral RNA, generating a DNA

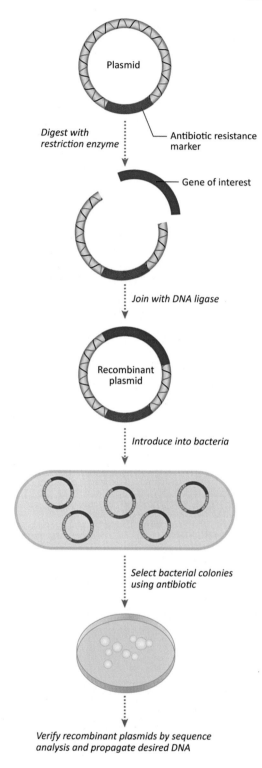

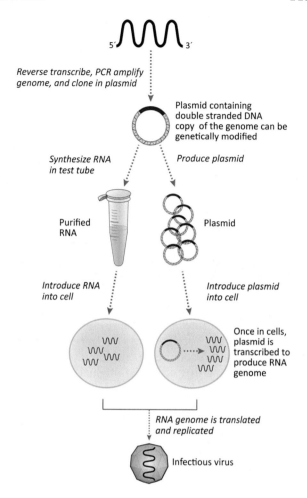

FIGURE 9.5 Overview of reverse genetics positive strand RNA viruses.

important to carefully analyze the recombinant plasmids, by sequence analysis, to select for clones containing exactly the desired sequences. At this point any desired mutations can be generated by manipulating the plasmid. In order to recover mutated virus the plasmid is introduced into eukaryotic cells where it can be transcribed to produce the viral genome. Alternatively, the plasmid can be used to produce the viral RNA in a test tube. Once the RNA is added to cells, the cultures are observed for any signs of virus production and spread. If virus is not recovered, the cell cultures can be analyzed to determine at which step replication was blocked, thereby getting information about the viral proteins that were mutated.

Reverse genetics of negative-strand RNA viruses

The *genomes* of negative sense, ambisense, and double-stranded RNA viruses are not infectious in the absence of viral proteins. In a natural infection with one of these types of viruses, the genome is closely associated with viral proteins, key among them an RdRp that

FIGURE 9.4 General scheme for cloning DNA fragments into bacterial plasmids for genetic manipulation and amplification.

molecule. This step is often combined with DNA amplification using PCR techniques, to generate dsDNA and greatly amplify the product. The PCR product is inserted into a plasmid suitable for generating mRNA. It is produces mRNA from the infecting genome after the virus gains entry into the cell. Therefore to accomplish reverse genetics of a negative-strand RNA virus one must introduce, or express within a cell, both a viral genome and a subset of viral proteins to initiate the replication cycle. The basic steps are illustrated in Fig. 9.6.

The initial steps are similar to those described for positive-strand RNA viruses: The desired virus is propagated, genomic RNA is extracted, and reverse transcription and PCR are used to generate DNA copies of the RNA genome. At this point the process becomes more complex as a set of plasmids must be constructed. One plasmid should contain a complete copy of the viral genome, cloned in a manner to allow transcription of the genomic viral RNA in a eukaryotic cell. Additional plasmids must also be constructed, these allowing for transcription of mRNAs for all viral proteins needed to replicate the viral genome. In the example in Fig. 9.6, our virus requires three proteins: the RdRp (called L in this case), a nucleocapsid protein (N), and a phosphoprotein (P). If these three proteins are present in a cell, along with genomic RNA, the replication cycle can commence and infectious virions can be generated. Note that while the original construction of plasmids is done in bacteria, the generation of infectious

virus requires that plasmids be introduced into eukaryotic cells permissive for the virus of interest.

Reverse genetics of retroviruses

The genomes of retroviruses are mRNA. However, retroviruses do not replicate in the manner of positive-strand RNA viruses, as their RNA genomes are not translated to initiate a productive infection. Instead, early in the infection process, retroviruses use the enzyme RT to synthesize of a DNA copy of their RNA genome. Therefore reverse genetics of retroviruses uses plasmids containing the DNA copies of retroviral genomes. A double-stranded retroviral genome, when introduced into a eukaryotic cell, is transcribed in the cell nucleus to produce the set of mRNAs needed to drive virion production. The basic steps are shown in Fig. 9.7.

The initial step of obtaining a retroviral genome is to obtain the DNA copy of the retroviral genome from infected cells. This is often done using a combination of conventional cloning strategies and PCR. The retroviral genome is inserted into a plasmid which is propagated in bacteria. At this point, any desired mutations can be readily generated.

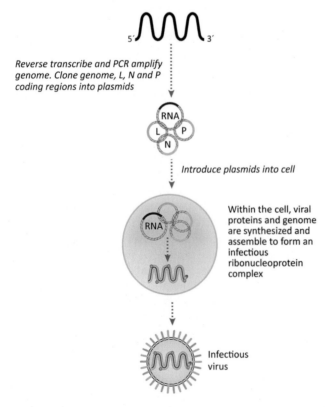

FIGURE 9.6 Overview of reverse genetics of negative strand RNA viruses. The L plasmid encodes RdRp. The N and P plasmids encode the nucleocapsid protein and a phosphoprotein required for assembly of a ribonucleoprotein complex that can be transcribed to produce viral mRNAs.

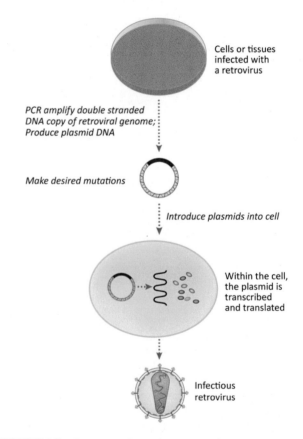

FIGURE 9.7 Overview of reverse genetics of retroviruses.

The design of the plasmid must allow for transcription of the retroviral genome when introduced into eukaryotic cells.

Reverse genetics of small DNA viruses

The genomes of small DNA viruses (e.g., paroviruses, polyomaviruses) are infectious and they can be synthesized, manipulated, and introduced back into cells as plasmids, with the resulting production of virions. The cloning processes may need to account for the specific strategies of genome replication used by a given type of DNA virus, but those details are beyond the scope of this book. The basic steps are shown in Fig. 9.8.

The first step is to obtain the DNA genome, usually from virus particles. If sufficient DNA is available it can be cloned directly into a plasmid. Often however PCR will be used to amplify the DNA prior to cloning. Once the viral DNA has been inserted into the plasmid vector, any desired mutations can be made. The viral genome may be removed from a plasmid and introduced into permissive cells as a linear molecule but in some cases the recombinant plasmid can be introduced directly into permissive cells.

Reverse genetics of large DNA viruses

A consideration in cloning large DNA viruses is the difficulty of handling large DNA molecules during routine manipulations. For example, very large DNA molecules are much more difficult to introduce into cells than small ones. Large DNA viruses such as herpesviruses (125–240 kbp genome size) are cloned and manipulated using special plasmids called bacterial artificial chromosomes (BACs). BACs are stably maintained in bacteria and can contain over 300 kbp of foreign DNA, enough to allow for cloning of entire herpesviral genomes. The process of cloning a herpesvirus genome involves inserting the BAC sequence (about 10 kbp) into a nonessential herpesviral gene, using homologous recombination. The BAC can be propagated and manipulated in bacteria before reintroduction into eukaryotic cells to generate infectious virus.

The dsDNA genomes of poxviruses range in size from 130 to 360 kbp. One method of introducing mutations into these viruses utilizes homologous recombination within an infected cell to exchange specific genome segments. The basic steps are illustrated in Fig. 9.9.

One or a few genes from the virus of interest are cloned into a plasmid. The genes may be obtained by

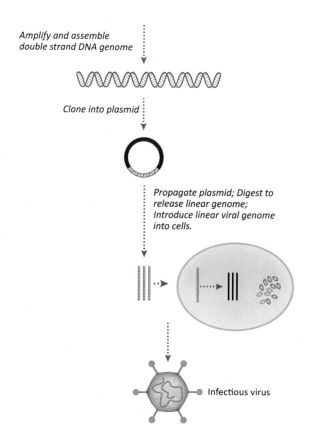

FIGURE 9.8 Overview of reverse genetics of small DNA viruses.

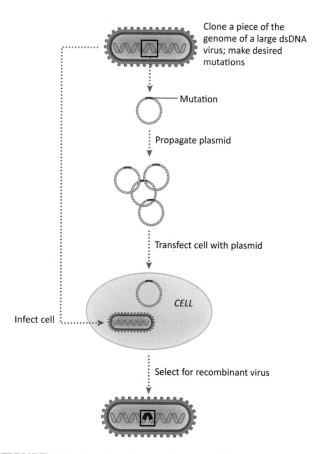

FIGURE 9.9 Overview of reverse genetics of large DNA viruses.

restriction enzyme digestion of purified poxviral DNA, or may be produced using PCR. Often a marker gene is added to the viral genes of interest. One commonly used marker is green fluorescent protein. It is small enough to be maintained in the viral genome and its presence is easily verified using a fluorescent microscope. Once the desired genes have been cloned, and any desired mutations have been made and verified, the plasmid is added to cells already infected with the parental or wild-type poxvirus. Within the infected cells, homologous recombination may occur, with the resulting insertion of the mutated genes into the genome of the wild-type viral genome. This is where the inclusion of a marker gene becomes critical because there must be a way to distinguish the rare recombinant virus from the parental virus. If green fluorescent protein is the marker, the cells can viewed and green plaques or foci can be selected.

In summary, methods have been developed that allowing genetic engineering or reverse genetics for many types of viruses. The methodology varies by virus: sometimes reverse genetics is rather straightforward, in other times, complex. In all cases, the ability to manipulate viral genomes has provided a mechanism to probe all facets of virus replication. Mutations as small as a single nucleotide substitution, to as large as changing the order of genes on a viral genome, have provided windows into the ways in which viruses are uniquely adapted to their hosts. As you read about the details of virus replication in the forthcoming chapters, keep in mind the types of genetic analyses and reverse genetic experiments that provided much of the basic information presented.

In this chapter we learned that:

- Two separate processes, mutation and selection are the drivers of virus evolution.
- Virus *mutation rates* are described in several ways, such as mutations per nucleotide synthesized, mutations per genome synthesized, or mutations/nucleotide/cell infection. For some RNA viruses, these numbers translate into one or two mutations per genome replicated. Estimates may try to account for the length of the virus replication cycle, the number of genomes produced, or the method of genome synthesis.
- The *evolution rate* of a virus is often described as nucleotide substitutions per nucleotide site, *per year*.
- Viruses have core sets of essential genes that are always required for replication. Examples include capsid proteins and polymerases. Other genes, sometimes called nonessential, are dispensable, and may be lost, under certain specific conditions (e.g., during replication in cultured cells).
- The process of reverse genetics is used to determine the function of viral genes. It involves cloning a viral genome, designing and making specific mutations, and reintroducing the genome into cells to obtain mutated virions whose phenotype can be determined.

10

Viral pathogenesis

After reading this chapter, you should be able to answer the following questions:

- What is disease?
- What is the meaning of the term "pathogenesis"?
- What is the meaning of the term "virulence"?
- Explain how modern genetic methods have improved development of animal models to study viral diseases.
- Provide an example of a disease process that benefits the viral pathogen.
- Provide an example of a disease process that does not benefit the pathogen.
- List some viruses associated with human cancers.
- What is the general, overall mechanism by which virus infection can lead to cancer?

Pathogens are disease-causing agents and pathogenesis (meaning "origins of disease") is the study of the disease *process*. In early days of infectious disease research pathologists examined organs and tissues (macroscopically and microscopically) to look for damage. Modern studies extend this approach to ask questions about the *underlying molecular events* that cause the observed damage. In other cases researchers seek to ferret out the causes of disease in the absence of obvious lesions or tissue damage.

Examples of the types of questions asked by those with an interest in viral pathogenesis include:

- What are the molecular differences between the attenuated version of a virus (used as a vaccine) and its highly virulent "parent"?

- What events turn an asymptomatic persistent infection into a lethal disease?
- How do some patients survive infections with highly lethal viruses? Conversely, what makes a usually innocuous virus lethal in a small percentage of infections?
- Is disease caused by direct damage by a virus or by immune responses (or a combination of both)?

When growing viruses in cultured cells, cytopathic effects can be attributed to virus replication. Cultured cells can be damaged in a variety of ways including virus inhibition of cell transcription and/or translation, changes in membrane permeability, alterations of cytoskeleton or trafficking pathways, cell–cell fusion, or induction of apoptosis. Within an animal, disease may arise from virally induced cell death but may also be caused immune by responses directed against infected cells. While necessary and essential, the immune system can damage the host in the process of clearing an infection. The effects of antiviral cytokines cause many of the symptoms of acute viral infections including lethargy and fever (Box 10.1).

Infectious disease always begins with infection, so we begin this chapter by looking at the infection process. Keep in mind that infection is not synonymous with disease. The molecular steps required for productive infection of a cell were discussed in earlier chapters; however, a cultured cell is not an intact organism so here we briefly consider the barriers that must be overcome to allow a virus to successfully

Viruses.
DOI: https://doi.org/10.1016/B978-0-323-90385-1.00020-0

119

© 2023 Elsevier Inc. All rights reserved.

BOX 10.1

Acute febrile illness

Many viruses cause "acute febrile illness" with symptoms including fever, lethargy, depression, prostration, and/or anorexia. These general symptoms may or may not be accompanied by more specific signs and symptoms such as diarrhea, rash, liver damage, or neurologic deficits.

Why do so many virus infections cause fever (particularly if replication is localized)? The answer is that the body's response to localized infection is often global, due to secretion of IFNs and other cytokines from infected cells. Many cytokines associated with a viral infection are pyrogens (fever-inducing molecules) that affect the hypothalamus. Cytokines that act as pyrogens include interleukin (IL)-1α, IL-1β, IL-6, IL-8, tumor necrosis factor-β, and IFNs-α, -β, and -γ.

---

reach target cells or tissues in the process of a natural infection.

Natural barriers to infection

Intact skin is a major barrier to viral infection. The outmost layers of the skin are made up of dead cells that cannot support virus replication. Viruses that directly infect the skin (e.g., papillomaviruses) require breaks or lesions that allow access to lower layers of dividing cells. Insect or animal bites provide another route past the skin barrier. Human activities provide additional avenues for infection. These include medical, dental, and veterinary procedures, cosmetic procedures (e.g., tattoos), and intravenous drug use. Mucosal surfaces of the eye, respiratory, gastrointestinal, and genitourinary tracts, although protected by a mucus layer and/or digestive enzymes, provide easier access to viral infection. Thus many viruses are transmitted by respiratory, fecal–oral, or sexual transmission.

Primary replication

Primary replication refers to the initial site of virus replication. This is often, though not always, an epithelial cell. Epithelial cell infection may remain localized in skin (e.g., warts), upper respiratory tract, or gastrointestinal tract. Viruses in the respiratory or digestive tract also encounter lymphoid tissues that may serve as primary sites of replication.

Microfold or M cells are specialized epithelial cells in the gastrointestinal tract (Fig. 10.1). They lack microvilli and are found primarily at sites of gut-associated lymphoid tissues, including Peyer's patches. While most epithelial cells provide a strong barrier to gut contents, the role of M cells is to sample gut contents. They do this by their high capacity for transcytosis. They take up material at their apical surfaces (luminal side of the gut) and release it basolaterally to immune cells. In a perfect system, the M cells alert immune cells that respond effectively. But some viruses target M cells to gain access to lymphoid tissues, thus facilitating replication and spread.

Movement to secondary replication sites

Some viral infections are very localized, with replication confined to the infection site. Examples include papillomaviruses and the poxvirus, molluscum contagiosum virus. However, many viruses spread from the initial site of infection to other organs. How do these viruses reach other sites in the body? There are three general mechanisms: blood, lymphatics, and nerve cells. The presence of virus in the blood is called viremia. Viruses in the blood gain access to many tissues and organs. Some viruses are "free" in the blood, present as extracellular virions in plasma. Other viruses in the "blood" are in fact, cell associated. Some viral infections produce a very short-term or transient viremia; other viruses are continuously found in the blood. Viruses that infect T-cell, B-cells, monocytes, or other circulating lymphocytes may be carried to numerous lymphoid tissues. A prime example is the human immunodeficiency virus (HIV), carried in an infectious form by dendritic cells to tissues rich in T-cell targets. Blood monocytes are also targets for HIV infection. These persistently infected cells transport virus out of the blood and into target organs.

Some viruses such as rabies, mammalian bornavirus, and some herpesviruses use neurons as their pathway through the body. Neurons can provide protected routes of travel and direct access into the central nervous system (CNS). Rabies virus enters neurons near the initial site of infection (e.g., a bite wound) and travels along neurons at rates of up to 50–100 mm/day using cell cytoskeletal elements. Rabies virus interacts with *dynein molecular motors* that move cargo from synapses to the cell body in a process called retrograde transport. *Kinesins* are motors that move cargo from the

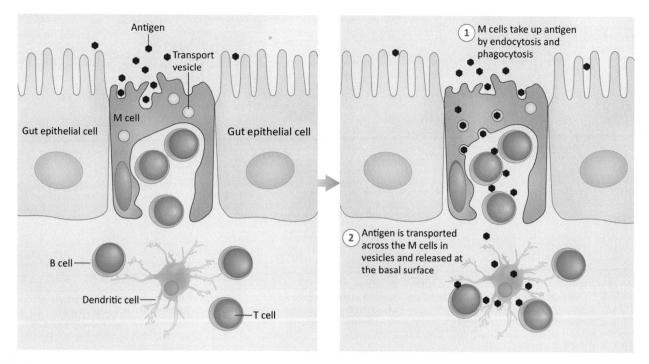

FIGURE 10.1 Microfold or M cells are specialized cells in the gastrointestinal tract. Their primary purpose is to sample the environment of the gut. Materials in the gut, including pathogens, cross M cells by transcytosis and are delivered to immune cells. Some viruses exploit M cells to gain entry into the host.

cell body to nerve endings, a process called antero-grade transport; rabies virus uses this system to move from the CNS back to salivary glands. While moving within a neuron, rabies virus is protected from immune surveillance. Herpesviruses both retrograde and antero-grade transport systems in neurons.

Some viruses use a combination of transit pathways Fig. 10.2. One example is measles virus which infects via the respiratory route. Virus released after primary infection causes a viremia and spreads to numerous lymphoid organs. Another example is the virus that causes chickenpox [human herpesvirus 3 (HHV3) or vericella-zoster virus]. HHV3 replicates in respiratory epithelial cells and is released into the blood. From the blood, HHV3 moves into the skin where it replicates to cause a rash. From the skin the virus moves into neu-rons to establish latency in the CNS. Decades later HHV3 may reactivate from latency, traveling back down the nerves to the skin to productively replicate once again.

Some viruses do not infect peripheral nerves but access the CNS by crossing the so-called blood–brain barrier (BBB). The brain is protected from most viruses by virtue of the BBB, a layer of tightly joined endothe-lial cells (Fig. 10.3). There are three major pathways across the BBB. One mechanism is via infected lym-phocytes. Other viruses cross the BBB when endothe-lial cells are damaged, for example, by excessive levels of cytokines. The third way to cross the BBB is by infecting the endothelial cells themselves. Viruses that

sometimes cross the BBB include poliovirus (PV), mea-sles virus, West Nile virus, equine encephalitis viruses, and HIV. While crossing the BBB has serious conse-quences for the infected host, none of the viruses listed here requires access to the CNS, as they all produc-tively infect nonneuronal tissues. In contrast, rabies virus must reach the CNS to replicate sufficiently to be transmitted to a new host.

Sometimes very closely related viruses differ in their ability to reach or invade certain target tissues or organs. Pathogens may reach sites that their avirulent relatives fail to reach. *Invasiveness* is the capacity of a virus to enter a tissue or organ. Viruses that can invade the CNS are more likely to cause neurologic disease than those that cannot. Some viruses cross the placenta to invade a fetus, while others cannot.

Genesis of disease

Disease is a disorder of structure or function of an organism. Disease can result from death, damage, or dysfunction of specific cell types. Damage may be directly due to virus infection (akin to cytopathic effects in cultured cells) or may be caused by immune responses to infection. Many diseases are probably a combination of both.

General types of diseases caused by viruses include:

- respiratory,

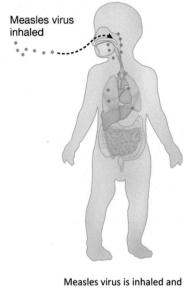

(A)

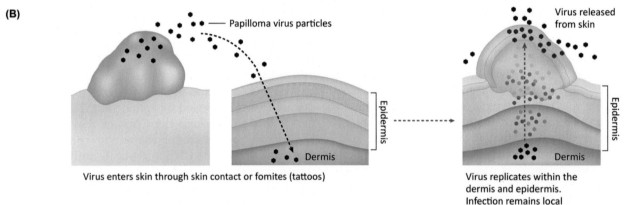

(B)

FIGURE 10.2   Generalized versus localized infection. Panel A: Measles virus gains access the body via the respiratory route. Virus released after primary infection causes a viremia and spreads to numerous lymphoid organs. Additional replication seeds the skin, causing a rash. However, virus is transmitted by release from respiratory epithelial cells. Panel B: Papillomaviruses cause warts, infecting skin and remaining highly localized. They are transmitted by direct contact with infected skin or fomites.

- intestinal,
- skin lesions/rashes,
- single organs (liver, kidney, heart, etc.),
- multiorgan/systemic,
- hemorrhagic,
- neurologic,
- immunosuppressive,
- benign and malignant tumors.

There are many different mechanisms by which viruses cause disease. Infection of the respiratory tract can lead to inflammation, generating the mucus typical of such infections. Inflammation may result when cells are directly killed by a virus and/or because of host responses to infection. Inflammation in the lung can result in pneumonia with reduced capacity to absorb oxygen. Damage to intestinal epithelial cells may inhibit absorption of nutrients and/or water into the gastrointestinal tract. The short-term effect can be diarrhea; a long-term effect may be weight-loss and wasting (Box 10.2). Direct and indirect damage to the immune system eventually leads to immunosuppression of the nontreated HIV patient. Direct or indirect damage to any organ (e.g., liver, kidney, brain, heart) may adversely impact normal function (Box 10.3).

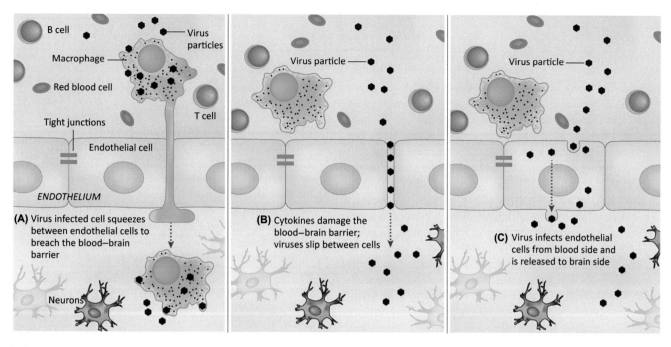

FIGURE 10.3 Endothelial cells lining blood vessels are key to maintaining the blood–brain barrier (BBB). However, some viruses cross the BBB, gaining access to the brain and spinal cord. Panel A: HIV infects T-lymphocytes and macrophages; these cells can carry HIV across the BBB. Panel B: Other viruses induce high levels of inflammatory molecules and cytokines that can damage the BBB. Panel C: Some viruses infect endothelial cells and can be released into the central nervous system.

BOX 10.2

Does disease benefit the virus?

There are many examples whereby a specific disease process benefits the virus. Often the benefit is transmission to a new host. For example, human noroviruses (family *Caliciviridae*) are intestinal pathogens and disease symptoms may include explosive vomiting and diarrhea. These highly unpleasant symptoms most certainly help disperse noroviruses to surfaces on which they remain infectious for extended periods of time. Respiratory viruses benefit from the coughing and sneezing that accompany those infections. A more extreme example is rabies virus. Infected dogs become aggressive, attacking people, or other animals without provocation. On the other extreme, rabies can render an animal lethargic and vulnerable to an unsuspecting predator (or to a

kindly human!). The virus is the only winner in this scenario.

However, it is also important to realize that while a disease may kill or debilitate a host, those outcomes do not always benefit the pathogen. For example, poliovirus (PV) infections most often result in asymptomatic replication in the gastrointestinal tract. PV replicates to high titer in the gastrointestinal tract and is shed into the environment and may go on to infect new hosts. On the other hand paralytic polio, a disabling disease, occurs only if PV invades the CNS. It is not at all obvious how this benefits the virus. In fact, the ability to infect the CNS has proven highly detrimental to PV, which is being pushed to extinction due to worldwide monitoring and vaccine programs!

Immunopathogenesis

The purpose of the immune system is to clear pathogens from the infected host (Fig. 10.4). Unfortunately this necessary activity can sometimes cause disease. Disease is a tolerable, even desirable, outcome if it results in clearance of a pathogen and lifetime protection. However in rare cases, the immune

system causes life-threatening damage (with or without virus clearance). For example, many viruses trigger the release of cytokines such as interferons (IFNs). IFNs trigger cell-signaling cascades that result in synthesis of antiviral proteins. However, IFNs also cause the fever, aches, and lethargy that accompany many viral infections. In extreme cases cytokines can disturb permeability of blood vessels resulting in fluid

BOX 10.3

Small mouse and small virus yield big results

Lymphocytic choriomeningitis virus or LCMV, a rodent virus, is a member of the family *Arenaviridae*. Arenaviruses are enveloped, ambisense RNA viruses that encode only four proteins. Most arenaviruses have rodent hosts but some infect humans as well. LCMV was first isolated in 1933 during an encephalitis outbreak in the United States. Shortly thereafter LCMV was linked to disease in mouse colonies. The natural host for LCMV is the house mouse.

Very early studies of LCMV disease in mice pointed directly to immune responses for causation, a novel finding for the time. In the LCMV system, infection of adult mice causes acute disease followed by virus clearance or death, depending on the route of inoculation. In contrast infection of mice in utero, or shortly after birth, results in a long-term persistent infection. Furthermore, immunosuppression of adult mice, by various means, turns a lethal infection into a persistent one.

Studies to investigate the cellular and molecular mechanisms of LCMV immunopathogenesis went hand in hand with advances in understanding adaptive immunity. LCMV had several advantages as a virus model: LCMV infection is noncytopathic and immune responses are the only cause of death of infected cells. The mouse model was highly tractable and inbred strains of mice provided a means of examining genetics of both host and virus. The small size of the virus made it relatively easy to link virus genotype and phenotype. Those advantages continued as reverse genetics systems for LCMV were developed and genetically engineered mice became a commonplace tool. The following model for LCMV immunopathogenesis is supported by an impressive number of studies:

- Virus replication is noncytopathic but in immunocompetent mice, virus-specific cytotoxic T-lymphocytes (CTLs) kill infected cells. This causes an acute disease episode, but clears the infection.
- Infection of mice, before complete development of their immune systems, results in tolerance (specifically lack of LCMV-specific CTL) and long-term persistent infection.
- During persistent infection, antibodies to LCMV are produced but they are not able to clear the infection. Instead, the formation and deposition of virus−antibody immune complexes causes glomerulonephritis and arteritis. Thus disease that eventually results from persistent infection is also immune-mediated.

Other important lessons learned from LCMV include:

- A variety of specific genetic factors, both host and viral, can alter the outcome of infection. Discrete and specific molecular interactions can explain those outcomes.
- Noncytopathic LCMV infection of differentiated cells can and does compromise cell function resulting in diseases such as a growth hormone deficiency syndrome.
- Persistent, noncytopathic LCVM infection measurably alters behavior and learning in mice.

---

imbalance, hemorrhage, and shock. Many aspects of viral hemorrhagic fever are caused by excessively high levels of cytokines. Thus overzealous inflammatory responses may cause tissue damage beyond that caused directly by the virus.

Virulence

Virulence is the *relative* ability of an infectious agent to cause disease. Thus virulent viruses have a greater propensity to cause disease in a greater proportion of infected hosts. Virulence determinants or factors are those genes and proteins that play key roles in disease development. Virulence determinants can range from surface/capsid proteins that determine cell/tissue tropism to small adapter proteins that alter cell-signaling cascades. Differences in virulence may even be determined by cis-acting viral sequences (i.e., gene promoter regions) that control the relative abundance of specific viral transcripts and their protein products. In the case of RNA viruses, the degree of fidelity of the RNA-dependent RNA polymerase can be a virulence determinant.

Virulence of a virus also depends upon genetics, age, and overall health of the host. And it would not be surprising to find that the microbiome (the mixture of commensal bacteria and viruses in the host) can have an impact on virulence. In experimental situations, the route of inoculation or dose administered can have a major effect on virulence. Thus the study of virulence determinants requires an understanding of the complex interactions between virus and host (and perhaps environment).

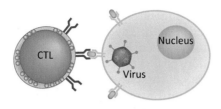

Cytotoxic T lymphocyte (CTL) binds
cognate antigen on virus infected cell;
antigen recognition activates CTL

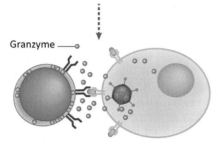

CTL releases granzymes that enter
target cell to activate caspaces

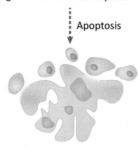

FIGURE 10.4 Killing of an infected cell after recognition by a cytotoxic T-cell.

Understanding the virulence determinants of a virus brings us full circle back to pathogenesis: The process by which a virus infects certain tissues, inhibits or induces various immune responses, or alters basic cellular processes. Thus the study of viral pathogenesis requires an understanding of virus molecular biology, host cell biology, and immune responses.

Why is it important to study pathogenesis? An obvious answer is that understanding pathogenesis is important for treatment. It allows the clinician to move past the empirical toward more directed treatments. It also provides a blueprint for understanding the likely course of an infection and may also provide a basis for administering supportive therapies that do not directly target the virus.

Approaches to study viral pathogenesis

Understanding the pathways from initial infection, to dissemination, to replication in target tissues is a complex process. Disease development may take days

to weeks, or years to decades. Studying natural infections can provide some insight into pathogenesis, but unraveling the molecular details often requires use of well-defined model systems.

Diseases caused by human viruses are often studied in animal models. Developing such models may include adaption (by repeated passage) of a human virus to growth in animals. Or human viruses may be used to inoculate animals by unnatural routes, or at very high doses. An alternative is to study the pathogenesis of an animal virus in its natural host to provide insight into pathogenesis of a related human virus. For example, the overall mechanisms by which bovine papillomaviruses cause warts in cows are instructive in understanding the development of human warts by human papillomavirus. Sometimes models must be developed to study animal viruses. Viruses that infect large animals, companion animals, or endangered species may be adapted to growth in smaller/more abundant laboratory animals (e.g., rodents). Developing a model system can be difficult, expensive, and time consuming. In the past it was largely an empirical process, but that has changed with the availability of modern methods to manipulate the genomes of animals and their viruses. Approaches to animal model development include:

- Adapting a virus to a new host by repeated passage.
- Genetic manipulation of a laboratory animal to support virus replication. For example, inserting the receptor for a human virus into the mouse genome.
- Identify specific genes that render a laboratory animal resistant to infection with a human virus, and remove them by genetic engineering.
- Use mice as hosts for *tissues or cells* from other organisms. For example, immunodeficient mice can be transplanted with human cells or tissues (tiny pieces!). Mice have been successfully "humanized" as models to examine both HIV and dengue virus pathogenesis.
- Genetically manipulate the virus to allow for replication in a model animal.
- Add one or more viral genes to an animal's genome to study their effects in the absence of virus replication. This allows one to assay for effects caused by a specific viral protein.

All model systems have their limits! Evaluating the data generated by use of a particular model requires an understanding of the strengths and limitations of the model. Indeed, comprehensive descriptions of viral pathogenesis often include results gained from several different model systems. Models for HIV pathogenesis include use of monkeys infected with simian immunodeficiency viruses (SIVs), chimpanzees infected with genetically engineered HIV/SIV chimeric viruses, humanized mice, and mice genetically engineered to express one or more HIV genes (in the absence of virus replication).

Virally induced cancers

In this chapter, we learned that:

- Disease is a disorder of structure or function in an animal. Colds are respiratory diseases that result from virus replication and/or inflammatory responses in the epithelial cells of the upper respiratory tract. Diarrhea is caused by fluid and electrolyte imbalance in the digestive tract. Many severe viral diseases result from death of infected cells in the CNS.
- Pathogens are infectious agents that cause disease and pathogenesis is the study of the disease process. Pathogenesis can be straightforward (rapid, virus-mediated destruction of a particular cell population) or may be a lengthy and complex process (development of immunodeficiency or a cancer many years after initial infection).
- Virulence is the relative ability of an infectious agent to cause disease. A highly virulent virus may cause debilitating disease while an avirulent virus is "without" virulence.
- The study of viral pathogenesis requires well-characterized animal models. Modern genetic methods have improved development of such models. Genes can be added to, or removed from model organisms. Viruses can be genetically altered as well.
- Disease may be beneficial to a virus, aiding in spread to new hosts, for example, via coughing, sneezing, or diarrhea. Alternatively disease may harm the host with no obvious benefit to the virus (paralytic polio). Finally, causing disease may be detrimental to a virus, if in response humans develop and aggressively utilize effective vaccines.
- Many viruses are associated with cancers. In most cases the viral infection is a first step in a long and complex pathway to cancer development.

Viruses associated with cancers include some retroviruses, papillomaviruses, and herpesviruses. Hepatitis B virus (HBV, a hepadnavirus) and hepatitis C virus (HCV, a flavivirus) are both associated with greatly increased risk of liver cancer in humans. Relatively recently a polyomavirus has been found to be highly associated with a rare but aggressive type of human skin cancer called Merkel cell carcinoma.

Cancers are comprised cells that have undergone a series of genetic changes. These so-called *transformed cells* fail to respond normally to the signals that govern cell growth within an organism. Transformed cells replicate out of control, eventually causing damage to organs and tissues. The genetic mutations that lead to cell transformation can be caused by ultraviolet light, radiation, or a variety of chemicals. Development of cancer is a complex, multistep process. Some retroviruses directly mutate cellular DNA (via insertion), leading to transformation. Many other cancer-associated viruses act by more indirect pathways: by a variety of means they activate growth-promoting genes and/or inactivate growth-suppressing genes leading to aberrant cell proliferation. Proliferating cells then accumulate additional mutations that lead to a transformed cell phenotype; as with other types of cancer-causing agents, this is probably a complex multistep process.

Animal retroviruses were some of the earliest viruses associated with cancers. Studies of oncogenic retroviruses date back to the early 1900s with descriptions of "filterable agents" capable of causing avian leucosis and avian sarcoma, two types of cancers of chickens. These filterable agents turned out to be tumor-causing retroviruses. Some oncogenic animal retroviruses mutate cellular DNA by a process called "insertional mutagenesis." In the process of replication, retroviruses stably insert their genomes into the host cell chromosome. In rare instances the insertion of retroviral DNA causes mutations that disrupt normal control of cell growth. For example, retroviral insertions can strongly activate growth-promoting genes that are normally under tight control. Uncontrolled expression of a growth-promoting gene leads to excess cell proliferation. As a result of cell proliferation, additional mutations may occur, enhancing the growth phenotype. A few oncogenic retroviruses (e.g., Rous sarcoma virus, RSV) cause transformation of many infected cells. RSV has actually captured a cellular protooncogene (protooncogenes are growth-promoting genes under normal conditions) and delivers it to each newly infected cell. The study of insertional mutagenesis by oncogenic animal retroviruses led to the discovery of many proteins involved in cell growth.

For decades after the discovery of oncogenic animal retroviruses, researchers looked for human retroviruses that caused cancer. None were found until the discovery of human T-cell leukemia virus (HTLV-1) by Robert Gallo in 1980. HTLV-1 is not a common virus but is endemic in certain parts of Japan, sub-Saharan Africa, the Caribbean, and South America. About 10% of infected individuals develop adult T-cell leukemia/lymphoma in the decades after infection. HTLV-1 does not cause insertional mutagenesis however. Instead HTLV-1 encodes a transcriptional activator protein that can interact with host proteins involved in the regulation of cell growth. "Excess" growth of infected cells is likely a first step in a longer process, whereby additional mutations develop, leading to T-lymphocyte transformation.

Papillomaviruses (PVs) are tumor-causing viruses. In most cases they cause only benign tumors (warts), that spontaneously resolve. A few PVs can cause malignant transformation, leading to invasive cervical cancers, anal cancers, penile cancer, or oral cancers. All PVs infect epithelial cells. They encode two small proteins, E6 and E7, that interfere with normal cell growth processes, leading to the proliferation of epithelial cells, a necessary event for virus replication. Epithelial cell proliferation leads to the development of a wart. Warts normally disappear as immune responses are activated, and infected cells are removed. The development of invasive, malignant PV-induced tumors is actually an *accident* of PV infection. It can occur when a small region of the PV genome, encoding E6 and E7, is inserted into the host genome. E6 and E7 proteins are continuously produced as a result of the insertion. In the normal PV replication cycle, E6 and E7 protein synthesis decreases as virions are made. The continuous production of E6 and E7 spurs cell division in an uncontrolled manner, which can lead to accumulation of additional mutations in the cell genome. Thus the PV infection is an initiating step in a multistep process that may produce a malignant cell. Cancers induced by PV infection do not actually produce the virus! The malignant cells retain only a small region of the PV genome. Thus cancer induction is of no value in spreading PVs. It should also be noted that the cancers caused by human PVs can be prevented by immunization. Preventing the initiating infection does indeed prevent cancer development.

Two herpesviruses, human herpesvirus-8 (HHV8) and Epstein Barr virus (EBV), also called human herpesvirus-4 (HHV-4), are associated with human cancers. EBV is a ubiquitous virus, best known as the cause of infectious mononucleosis. The vast majority of infected individuals are asymptomatic. EBV infects both B-lymphocytes and epithelial cells. B-cells are latently infected and seldom produce virus. Latency is probably driven by the development of immune responses to the virus. An association between EBV and B-cell tumors was suspected when it was observed that cultured primary B-lymphocytes, when infected with EBV, would spontaneously transform. EBV-associated tumors in humans include nasopharyngeal carcinoma, Burkitt's lymphoma, and some other B-cell lymphomas. The B-cells that make up a Burkitt's tumor are positive for latent EBV (they are not expressing virus particles) and express a subset of EBV genes that suppress apoptosis and activate cell division. But the Burkitt's tumor cells also show large chromosomal translocations that activate growth-promoting genes. The development of Burkitt's lymphoma and other EBV-associated malignancies may also involve immune system impairments. Thus the development of these cancers is associated with viral infection, but requires many additional steps.

HBV and HCV can both cause chronic liver infections. Both agents are associated with greatly increased risk of developing hepatocellular carcinoma. In both cases cancer development is complex, involving multiple viral proteins that interact with host cell proteins to alter cell growth pathways. In the case of HBV, which has a DNA genome, integration of the viral genome into the host cell chromosome seems to be an initiating step. Interestingly, HCV is an RNA virus. Although its genome is never integrated into host DNA, the virus expresses a variety of proteins and noncoding RNAs that alter normal cell proliferation processes.

In summary, most cancer-associated viruses do not cause cell transformation by a single-step mechanism. Instead, viral infection is a first step in a complex, multistep process that eventually leads to highly transformed cells. The infecting viruses, via a variety of viral proteins required for virus replication, subtly change cell growth properties, followed by additional accumulation of mutations. The exception to this scenario is the acutely oncogenic retrovirus of chickens, Rous sarcoma virus which has incorporated a copy of a mutated chicken protooncogene into its genome.

11

Introduction to RNA viruses

After studying this chapter, you should be able to:

- Define "positive-strand RNA virus."
- Describe "negative-strand RNA virus" and "ambisense RNA virus."
- Understand the activities of RNA-dependent RNA polymerase (RdRp).
- Describe three general methods for priming RNA synthesis.
- Describe some functions of the noncoding regions (NCRs) of an RNA virus genome.

Definition and basic properties of RNA viruses

RNA viruses use RNA as the genetic material and replicate their genomes using virally encoded RNA-dependent RNA polymerase (RdRp). The RNA genome is the template for synthesis of additional RNA strands. During replication of RNA viruses, there are at least three types of RNA that must be synthesized: the genome, a copy of the genome (copy genome), and mRNAs. Some RNA viruses also synthesize noncoding copies of mRNAs and some transcribe small noncoding RNAs that interfere with host functions. RdRp is the key player for all of these processes.

RdRps of RNA viruses probably arose from a common ancestor. In fact, the reverse transcriptases of retroviruses and hepadnaviruses probably arose from the same common ancestor. This ancient relationship can be elucidated from the highly conserved structures of these enzymes. To underscore this relationship, the ICTV has included all RNA viruses, retroviruses, and hepadnaviruses into the Realm *Riboviria*. The core domain of an RdRp is usually less than 500 amino acids (aa) and folds into the structure shown in Fig. 11.1. This is often described based on the structure of a "cupped" right hand. Fingertips and thumb are wrapped around the RNA which feeds through a trough analogous to the palm of the hand. The catalytic domain of the enzyme is associated with the core domain. It catalyzes RNA-template dependent formation of phosphodiester bonds between ribonucleotides, proceeding in the 5′ to 3′ direction, in the presence of divalent metal ions.

Many viral RdRps are much larger than 500 aa because they contain additional domains with catalytic activities. These often include methyltransferase or endonuclease domains. Methyltransferase domains serve to generate 5′ caps for viral mRNAs. The endonuclease domain of

© 2023 Elsevier Inc. All rights reserved.

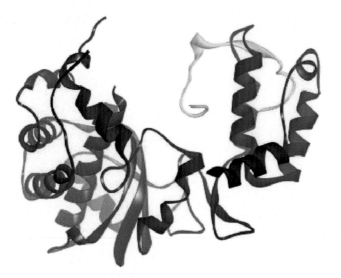

FIGURE 11.1 Ribbon diagram of poliovirus (family *Picornaviridae*) RdRp. The palm, finger and thumb subdomains are shown in purple, green and yellow, respectively (Tian et al., 2021). Source: *From Tian, L., Qiang, T., Liang, C., Ren, X., Jia, M., Zhang, J., Li, J., Wan, M., YuWen, X., Li, H., Cao, W., Liu, H., 2021. RNA-dependent RNA polymerase (RdRp) inhibitors: the current landscape and repurposing for the COVID-19 pandemic. Eur. J. Med. Chem. 213, 13201. https://doi.org/10.1016/j.ejmech.2021.113201. PMID: 33524687; PMCID: PMC7826122.*

TABLE 11.1　Selected families of positive-strand RNA viruses of animals.

Virus family	Virion morphology	Genome type
Arteriviridae	Enveloped, helical nucleocapsid	Unsegmented
Astroviridae	Nonenveloped, icosahedral	Unsegmented
Caliciviridae	Nonenveloped, icosahedral	Unsegmented
Coronaviridae	Enveloped, helical nucleocapsid	Unsegmented
Flaviviridae	Enveloped, icosahedral nucleocapsid	Unsegmented
Hepeviridae	Nonenveloped, icosahedral	Unsegmented
Picornaviridae	Nonenveloped, icosahedral	Unsegmented
Togaviridae	Enveloped, icosahedral nucleocapsid	Unsegmented

influenza virus RdRp cleaves host mRNAs to obtain 5′-methyl-capped primers required to prime viral mRNA synthesis. Helicase domains may be present, needed to unwind an RNA template.

RdRps usually require the help of other proteins, cellular or viral, to efficiently associated with, and transcribe or replicate viral genomes. The complex of proteins that is needed for efficient RNA genome synthesis is called the replicase complex. Some RNA viruses use two different RdRp complexes, one to replicate the genome and a second one to synthesize mRNA (the so-called transcription or mRNA synthesis complex). The RdRp in these complexes is the same, but the set of associated viral and/or host protein can differ. Some proteins in the replicase complex serve as a bridge between the RNA genome and the RdRp. The number of proteins in the replicase complex differs among virus families. There may also be a requirement for host cell proteins and/or cellular membrane complexes that act as scaffolds.

Positive-strand RNA viruses

RNA viruses can be subdivided into groups *based on the type of RNA that serves as the genome*. Positive or plus (+)-strand RNA viruses have genomes that are functional mRNAs (Table 11.1). Some viral genomes have 5′-cap structures and 3′-poly(A) sequences but others do not,

having evolved other strategies to coopt ribosomes. Upon penetration into the host cell, ribosomes assemble on the genome to synthesize viral proteins. During the replication cycle of positive-strand RNA viruses, among the first proteins to be synthesized is the RdRp. Thus the infecting genome has two functions: It *is* an mRNA and also *serves as the template* for synthesis of additional viral RNAs. A functional definition of a positive-strand virus is that purified or chemically synthesized genomes are infectious; no viral proteins need enter a cell along with the RNA genome (Box 11.1).

As shown in Fig. 11.2, after entry of the positive-strand virus genome into the cytoplasm, ribosomes assemble on the genome and translation commences. After synthesis of RdRp and other proteins required for genome replication, these copy the genome, from end to end to make a complementary RNA called the cRNA. The cRNA then serves as a template for synthesis of additional genome strands. Some positive-strand RNA viruses also use the cRNA as a template to make subgenomic mRNAs. The abundance of different classes of RNA can vary. There are usually many fewer cRNAs synthesized than genomic RNAs. This makes sense, as it is the genomic RNA that is packaged into virions.

Positive-strand RNA viruses often use large complexes of cellular membranes for genome replication. These viruses encode proteins that actively modify host cell membranes in order to construct viral replication scaffolds. Positive-strand RNA viruses of animals also use a common strategy to express RdRp. RdRp is a nonstructural protein, meaning that it is not found within the assembled virion. Instead it is translated directly from the infecting genome shortly after penetration. RdRp and other viral proteins needed for viral RNA synthesis are encoded as a *polyprotein* that is cleaved by virally encoded proteases. In the case of the picornaviruses and the flaviviruses, all viral proteins (structural and nonstructural) are synthesized as part

---

BOX 11.1

Key characteristics of positive-strand RNA viruses

- Purified genomes (or chemically synthesized genomes) are infectious if introduced into a permissive cell.
- The genome serves as an mRNA.
- The first synthetic event in the replication cycle is protein synthesis.
- Genome replication is cytoplasmic.

- Genomes of positive-strand RNA viruses fold into complex structures. These RNA structural elements have key roles in genome replication, transcription, and translation.
- Families include: *Picornaviridae*, *Flaviviridae*, *Togaviridae*, *Hepeviridae*, *Coronaviridae*, *Arteriviridae*, *Toroviridae*, among many others.

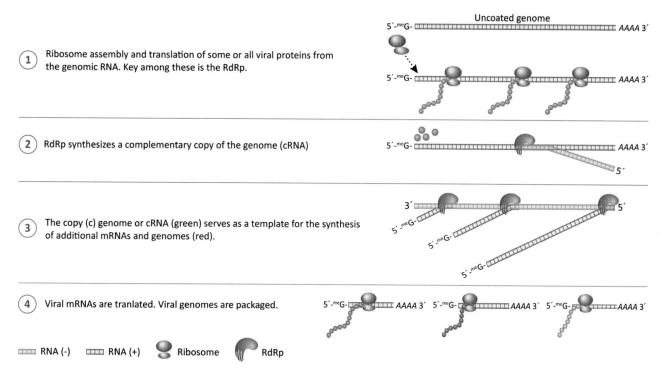

1. Ribosome assembly and translation of some or all viral proteins from the genomic RNA. Key among these is the RdRp.

2. RdRp synthesizes a complementary copy of the genome (cRNA)

3. The copy (c) genome or cRNA (green) serves as a template for the synthesis of additional mRNAs and genomes (red).

4. Viral mRNAs are tranlated. Viral genomes are packaged.

RNA (-)   RNA (+)   Ribosome   RdRp

FIGURE 11.2   General replication strategy of positive-strand RNA virus genomes. The genome of a positive-strand RNA virus is an mRNA that, upon entry into cells, is translated to produce proteins needed for transcription and genome replication (e.g., RdRp). After initial rounds of translation, the genome serves as the template for synthesis of cRNA. cRNA is the template for synthesis of additional genomes and subgenomic mRNAs.

of a single long polyprotein. Other positive-strand RNA viruses (i.e., togaviruses, coronaviruses, arteriviruses) synthesize an RdRp-containing polyprotein from genome-length mRNA, but use subgenomic mRNAs to encode structural and other proteins.

Negative-sense RNA, ambisense, and double-stranded RNA viruses

There are three additional groups of RNA viruses whose genomes are not mRNAs. They are the negative or minus (−)-strand RNA viruses, the closely related ambisense RNA viruses, and double-stranded RNA viruses (Table 11.2 and Box 11.2). For each of these groups of viruses, the first synthetic event after genome penetration is transcription. This is accomplished by viral proteins (including viral RdRp) *that enter cell with the genome.* Transcription/replication complexes usually contain between two and four proteins. They associate with the genome through interactions with RNA-binding nucleocapsid (N) or capsid proteins. Therefore naked (purified away from protein) genomic RNA is not infectious, cannot be translated, and will eventually be degraded in a cell. Before genome replication can begin, viral mRNAs must be transcribed and translated.

TABLE 11.2   Selected families of negative-strand and ambisense RNA viruses.

Virus family	Virion morphology	Genome type
Arenaviridae[a]	Enveloped, nucleocapsids form circular structures due to base pairing of RNA ends	Bisegmented, ambisense coding strategy
Bornaviridae[b]	Enveloped, helical nuclecapsid	Unsegmented
Filoviridae[b]	Enveloped, helical nuclecapsid	Unsegmented
Hantaviridae[a]	Enveloped, nucleocapsids forms circular structures due to base pairing of RNA ends	Trisegmented
Nairoviridae[a]	Enveloped, nucleocapsids form circular structures due to base pairing of RNA ends	Trisegmented, ambisense coding strategy
Orthomyxoviridae	Enveloped, helical nuclecapsid	Segmented (6–8 segments)
Peribunyaviridae[a]	Enveloped, nucleocapsids form circular structures due to base pairing of RNA ends	Trisegmented
Phenuiviridae[a]	Enveloped, nucleocapsids forms circular structures due to base pairing of RNA ends	Trisegmented, ambisense coding strategy
Pneumoviridae[b]	Enveloped, helical nuclecapsid	Unsegmented
Paramyxoviridae[b]	Enveloped, helical nuclecapsid	Unsegmented
Rhabdoviridae[b]	Enveloped, helical nuclecapsid	Unsegmented

[a]*Order bunyavirales.*
[b]*Order mononegavirales.*

Negative-sense RNA viruses

There are many negative-sense RNA viruses that are important pathogens of humans and animals. These include influenza viruses, rabies virus, Ebola virus, and measles virus among others. These viruses have linear, single-stranded RNA genomes. Some negative-sense RNA viruses encode all of their proteins on a single genome segment but others have genomes of two to eight segments. Influenza viruses package 7 or 8 separate RNA segments into each virion. When illustrating diagrams of the genomes of negative-sense RNA viruses it is not uncommon to draw them with the 3' end to the left (opposite to the usual convention of illustrating a strand of nucleic acid) in order to highlight the order in which mRNAs are synthesized.

Fig. 11.3 shows the major steps that occur when the genome of a negative-sense RNA virus accesses the cytoplasm. The first molecules synthesized are viral mRNAs that are transcribed from the infecting genome. This is accomplished by a transcription complex that enters the cell along with the genome. The viral mRNAs are translated, eventually resulting in a switch from mRNA synthesis to cRNA synthesis. The switch from transcription to genome replication requires synthesis viral N protein.

Because the virion of a negative-strand RNA virus contains RdRp, it is possible to synthesize viral mRNAs in a test tube (Fig. 11.4). If purified virions are gently lysed under appropriate buffer conditions, with the addition of NTPs, mRNAs will be transcribed in the test tube. However, genomic RNA will not be synthesized under these conditions because the mRNAs are not translated.

Ambisense RNA viruses

Negative-strand RNA viruses use their genome strand as a template for synthesis of all mRNAs. In

---

BOX 11.2

Negative and ambisense-strand RNA viruses

- Purified genomes (or chemically synthesized genomes) are NOT infectious.
- RdRp is packaged within the virion.
- Genomes serve as templates for transcription and the first synthetic event in the replication cycle is mRNA synthesis.
- Many replicate in the cytoplasm, a few replicate in the nucleus.

- Viral genomes are often tightly associated with a nucleocapsid (N) protein.
- Families of negative-strand RNA viruses include *Orthomyxoviridae, Paramyxoviridae, Rhabdoviridae, Bornaviridae,* and *Filoviridae.* The *Bunyaviridae* includes some members that use a negative-sense coding strategy and other that use an ambisense coding strategy.

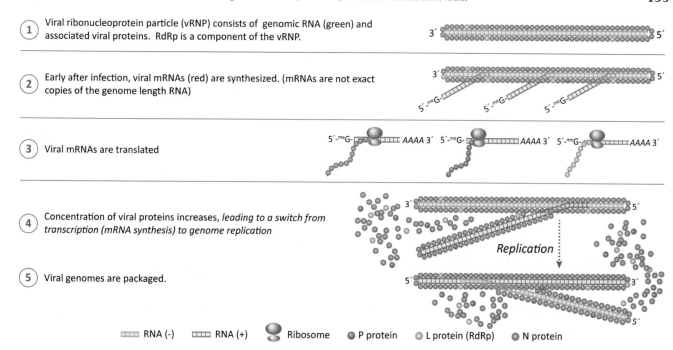

1 Viral ribonucleoprotein particle (vRNP) consists of genomic RNA (green) and associated viral proteins. RdRp is a component of the vRNP.

2 Early after infection, viral mRNAs (red) are synthesized. (mRNAs are not exact copies of the genome length RNA)

3 Viral mRNAs are translated

4 Concentration of viral proteins increases, *leading to a switch from transcription (mRNA synthesis) to genome replication*

5 Viral genomes are packaged.

RNA (-)   RNA (+)   Ribosome   P protein   L protein (RdRp)   N protein

FIGURE 11.3   General replication strategy of genomes of minus-strand RNA viruses. Genomes are associated with RNA-binding proteins and RdRp. Upon entry into the cell, the active *transcription complex* synthesizes mRNAs. This process can also occur in a test tube (see Fig. 11.4). Translation of mRNAs produces the proteins required for genome replication. Thus if protein synthesis is blocked in the infected cell, mRNAs continue to be synthesized but genome replication does not occur. Newly synthesized proteins provide the switch from transcription to genome replication.

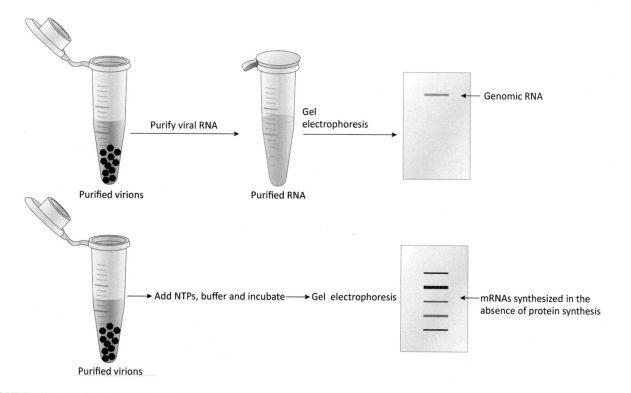

Purified virions   → Purify viral RNA →   Purified RNA   → Gel electrophoresis →   ← Genomic RNA

Purified virions   → Add NTPs, buffer and incubate → Gel electrophoresis →   ← mRNAs synthesized in the absence of protein synthesis

FIGURE 11.4   If RNA is extracted from purified virions of negative-strand RNA viruses such as measles virus (Family *Paramyxoviridae*) and analyzed by gel electrophoresis, a single RNA is seen (top panel). However, virions contain RdRp and if they are gently lysed, in a buffer solution containing NTPs, mRNAs are synthesized. The lower panel shows the products of mRNA synthesis as a set of subgenomic mRNAs of different sizes. Note that genome synthesis does not occur under these conditions.

TABLE 11.3 Selected families of double-stranded RNA viruses.[a]

Virus family	Virion morphology	Genome type
Birnaviridae	Nonenveloped, icosahedral	Bisegmented
Picobirnaviridae	Nonenveloped, icosahedral	Bisegmented
Reoviridae	Nonenveloped, icosahedral with multilayered capsids	Segmented (10–12)

[a]*Only those families with members that infect vertebrates are included in this list.*

contrast, viruses that use an ambisense coding strategy transcribe some mRNAs from the copy genome. Ambisense viruses that cause significant disease among humans and/or animals include arenaviruses (Lassa fever virus), phenuiviruses (Rift Valley fever virus), and nairoviruses (Crimean-Congo hemorrhagic fever virus). These viruses all have segmented genomes and at least one segment is ambisense.

dsRNA viruses

dsRNA viruses, package double-stranded RNAs. As dsRNA cannot be directly translated, it follows that these viruses produce mRNAs as a first step in initiating their replication cycle. The reoviruses are members of a large family, *Reoviridae*. Pathogens include the rotaviruses that cause childhood diarrhea. Reoviruses have segmented genomes, packaging 11–12 segments of dsRNA within unenveloped, double- or triple-layered capsids (Table 11.3). The genome segments are found within the innermost, $T = 1$ icosahedral shell. A very interesting feature of reovirus replication cycle is that their genome segments are transcribed from *within* the capsid. The mRNA products leave the capsid through pores at the vertices of the capsid. mRNAs are translated to produce viral proteins and are eventually assembled into capsid structures where they are converted into dsRNA.

Hepatitis delta virus uses host DNA-dependent RNA polymerase for replication

In contrast to the RNA viruses discussed above, hepatitis delta virus (HDV) uses a unique mechanism to replicate its RNA genome: host cell DNA-dependent RNA polymerase! Sometimes called a "satellite" virus, because it requires the help of another virus for transmission, the ICTV has recently dropped the term "satellite" from its descriptions. HDV is a member of the Realm *Ribozyvaria*, Family *Kolmioviridae*. HDV is a human virus but highly related viruses have been found in mice, fish, toads, and snakes. The name hepatitis delta derives from the fact that

it is associated with severe hepatitis in humans and is only found in association with hepatis B virus (HBV). HDV virions contain HBV surface protein. The HDV genome is only 1700 nt of ssRNA, but the circular genome is highly base paired and rigid in structure. The genome contains a single open reading frame that produces two proteins. One of these is required for genome replication.

HDV genomic RNA is considered negative sense as the genome is transcribed to produce an mRNA. Both genome replication and transcription occur in the nucleus. During replication both genomic and cRNAs are present in cells. Both RNAs are present mainly as monomers, but dimers, trimers, and higher oligomers are also present. It is hypothesized that host RNA polymerase II is the enzyme that replicates the HDV genome, using a rolling circle mechanism to produce contameric products that are cleaved by the RNA itself. A portion of the HDV RNA folds into a structure that cleaves the RNA strand to produce unit length genomes. The name ribozyvaria was derived from ribozyme, a term coined to describe RNAs with enzymatic activity. In the nucleus, the genome is transcribed to produce a capped polyadenylated mRNA that is transported to the cytoplasm and translated to produce the nucleocapsid protein (called delta antigen) that associates with genomic RNA.

Structural features of RNA virus genomes

The genomes of RNA viruses have some common general features. Obviously there are one or more open reading frames that encode the viral proteins. But there are also regions of RNA that do not code for protein. These noncoding regions (NCRs) or untranslated regions (UTRs) are often highly conserved within a virus family, indicating that they have important functions. The preferred usage can vary by virus family but the terms NCR and UTR are largely interchangeable. NCRs may have specific, critical nucleotide sequences but in some cases they are regions of the genome that fold into conserved structures, and *structure* may be more critical than a specific sequence.

All RNA genomes have NCRs at their 5′ and 3′ ends. NCRs vary in size, from very long (several hundred nt) in the case of picornaviral 5′ NCRs to just a short base-paired hairpin in the case of flavivirus 3′ NCRs. Complementary RNA sequences are sometimes found at the 5′ and 3′ ends of single-stranded RNA genomes, allowing them to circularize.

The 5′ and 3′ NCRs are required, and are often sufficient, to direct genome replication. Virally encoded proteins, such as the RdRp recognize specific sequences and/or structures at the ends of the genome. To simplify studies of the genome replication process, researchers often generate "minigenomes" that lack some virally encoded proteins. Of course a source of RdRp must be supplied. RdRp may be encoded in the minigenome or may be supplied *in trans* (by using a cell line stably expressing the viral RdRp, for example). The sequences required to direct RNA replication are often fairly simple and can be linked to virtually any RNA sequence to drive its replication.

Many RNA genomes also have promoters to direct synthesis of subgenomic mRNAs. These promoter sequences can be rather short but provide a means to direct the RdRp to internal sites on the genome. There may also be specific RNA sequences that signal polyadenylation. There are a variety of different strategies that RNA viruses use to regulate transcription and genome replication, but all involve RNA sequences found in the genome.

The RNA genomes of some viruses are highly structured and extensively base paired. An example is the *internal ribosome entry site* (IRES) at the 5′ end of the picornaviral genome. The IRES serves as a platform for ribosome assembly. Picornaviruses are among the RNA viruses that encode RNA-helicases to unwind highly structured regions of the genome, such as the IRES. Without the help of a helicase, the RdRp would not be able to disrupt the base-paired RNA in the IRES and genome replication would stall.

Priming viral RNA synthesis

While all RNA viruses use an RdRp for replication and transcription, there are a variety of strategies used for priming RNA synthesis, regulating RNA synthesis, and capping and polyadenylating mRNAs.

In the eukaryotic cell, RNA synthesis (from a DNA template) is primer independent. DNA sequences (promoters) direct RNA polymerases to the "correct" position on the DNA genome. Promoters in DNA sequences can be quite long and complex and promoter regions themselves are not transcribed. What mechanisms do viral RdRps use to prime viral RNA synthesis? It is particularly important, in the case of genome synthesis, that genetic information not be lost or modified; however, viral mRNAs are often capped and polyadenylated. Are the methods for priming viral mRNA synthesis the same or different from the methods of priming genome replication? The RNA viruses seem to have experimented widely. Various mechanisms by which RdRps initiate RNA synthesis are briefly described below.

- De novo or primer-independent transcription and replication employ the RdRp, the RNA template, an initiation NTP, and a second NTP. The initial NTP

can be considered the primer. The initial complex is not highly stable.
- De novo initiation is sometimes used to generate short RNA oligonucleotides that are subsequently used as primers in a mechanism known as prime and realign. The short internally initiated oligos slip back to extend from 5′ end of the template. Bunyaviruses and arenaviruses use this process. As a result the 5′ ends of the genome and copy genome contain nontemplated nucleotides.
- RNA structures such as hairpins can be used for priming in a process where the 3′ end of the template RNA loops back upon itself to serve as the primer. This process is "template priming" as part of the template (the genome) functions as the primer.
- Some RNA viruses (i.e., picornaviruses) use a protein to prime RNA replication. A tyrosine hydroxyl group on the primer protein (VPg) is linked to GTP to serve as the first nucleotide in the new strand. The protein interactions between VPg, RdRp, and the RNA template form a stable initiation complex.
- Some RNA viruses (e.g., influenza viruses) "steal" short (10−15 nt) capped oligonucleotides from cellular mRNAs to prime viral mRNA synthesis (a process called cap-snatching). The RdRp has an endonuclease domain that cleaves the host mRNA. This priming process cannot be used for genome synthesis and indeed, the genome segments of influenza viruses are primed by a de novo process.

Mechanisms to generate capped mRNAs

RNA viruses use different mechanisms to cap their mRNAs. In the case of the influenza viruses the fragment of cellular mRNA used for priming viral mRNA synthesis also provides the methyl-G cap. The RdRp proteins of many RNA viruses have methylase activities, thus the RdRp also synthesizes the cap. Some RNA viruses (e.g., piconaviruses) do not use capped mRNAs.

Mechanisms to generate polyadenylated mRNAs

RNA viruses also use a few different mechanisms to polyadenylate their mRNAs. For example, the picornaviruses use poly A tracts encoded in the genome. Among the negative-strand RNA viruses, those in the order *Mononegavirales* use a stuttering mechanism to synthesize long poly A tracts from short poly U tracts (Fig. 11.5). Other RNA viruses do not polyadenylated their mRNAs. The flaviviruses have a short hairpin at the 3' of their genome-length mRNA.

Mechanisms to regulate synthesis of genomes and transcripts

Even with fairly simple genomes, RNA viruses must, and do, regulate the amounts of genome, copy genome, and mRNAs that are synthesized during an infection. It would be "wasteful" if a positive-strand RNA virus had to make a new copy genome for synthesis of every genome. It is much more efficient to synthesize many genomes from each copy genome. *Having different structures at the 5' and 3' ends of the genome and copy RNA would be one way to accomplish this.* Togaviruses use slightly different versions of the replicase complex to synthesize genomes versus copy genomes, the different replicase activities depend on proteolytic cleavages. Internal promoters for mRNA synthesis can vary in sequence, controlling the relative affinity of the transcription complex for each mRNA.

A large group of unsegmented negative-strand RNA viruses (Order *Mononegavirales*) synthesize mRNAs sequentially, from the 3' end to the 5' end of the infecting genome. Those genes closest to the 3' end of the genome are more abundantly expressed, as the transcription complex always initiates synthesis at the 3' end of the genome.

RNA viruses and quasispecies

An important feature of RNA viruses is that many exist in nature as *quasispecies*. The term quasispecies is used to describe a group of closely related, but nonidentical genomes (Fig. 11.6). (Quasispecies are not unique to RNA viruses, some retroviruses, HIV, e.g., exist as quasispecies.) Poliovirus (PV) is a good example of a virus that forms a quasispecies. If one examines genome sequences from a mouse experimentally infected with PV, we find that the genomes are not identical, although they are all clearly related to one another. So, which one of these genomes is the "best" or the "fittest"? To the surprise of many virologists, it turns out that the *population* (quasispecies) may be

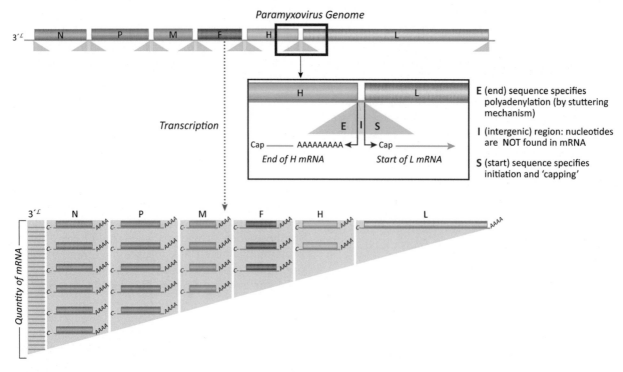

FIGURE 11.5 A strategy to regulate mRNA synthesis. This figure shows the organization of a paramyxovirus genome. Each protein-coding region is flanked by regulatory sequences that control capping and polyadenylation. The order of the genes on the genome regulates the relative quantities of mRNAs synthesized. The RdRp always initiates mRNA synthesis at the very 3' end of the genome. After polyadenylating an upstream mRNA, RdRp must remain associated with the genome to initiate synthesis of the downstream mRNA. If the RdRp dissociates from the genome, it cycles back to the 3' end to reinitiate transcription. Because RdRp *does* often dissociate from the genome during transcription, the downstream genes are produced in lower quantities. Note that the N protein (a structural protein required in large amounts) is positioned at the 3' end of the genome while the L protein (the RdRp), an enzyme is positioned at the 5' end of the genome.

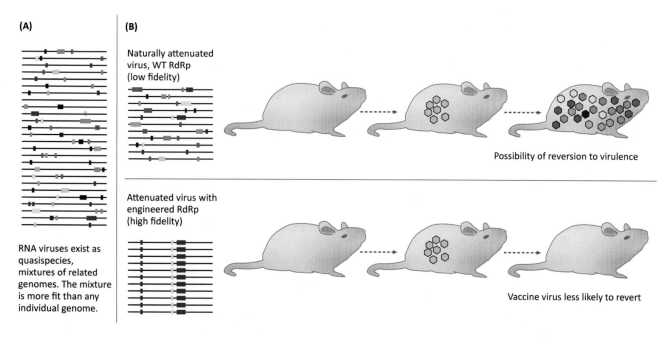

(A)

RNA viruses exist as quasispecies, mixtures of related genomes. The mixture is more fit than any individual genome.

(B)

Naturally attenuated virus, WT RdRp (low fidelity)

Possibility of reversion to virulence

Attenuated virus with engineered RdRp (high fidelity)

Vaccine virus less likely to revert

FIGURE 11.6 (A) Positive-strand RNA viruses exist as quasispecies, complex mixtures of related genomes. The mixture is more fit than any individual genome; fitness is maintained by generation of new variants in response to selective pressures. (B) Potential for safer vaccines. If the fidelity of RdRp is increased the population remains more homogeneous. Therefore an attenuated virus with a high-fidelity RdRp is more likely to remain attenuated.

BOX 11.3

RNA virus quasispecies and fitness

The RdRps of RNA viruses control RNA synthesis error rates. Molecular/biochemical studies have revealed that RdRp mutations can have a measurable impact on fidelity and this in turn measurably impacts virus fitness. A well-studied example is poliovirus (PV) a picornavirus. Vignuzzi et al. (2006) characterized a PV mutant with a high-fidelity polymerase. While the mutant replicated as well as the wild-type virus (produced equivalent numbers of virions), there was measurably less diversity in the population. The less diverse population performed poorly, when compared to the wild-type virus, when exposed to *adverse* growth conditions. (Adverse conditions included exposure to an antiviral drug in cultured cells and inoculation into mice.) The more diverse population was more adaptable and fit under these circumstances!

Based on this result, it might be tempting to assume that a PV mutant with a less faithful RdRp (generating a more highly mutated, thus diverse population) would have increased fitness compared with wild-type PV. However, this is not the case, as demonstrated in a study published by Korboukh et al. (2014). These researchers identified a PV with a single amino acid substitution in the RdRp that produced a "mutator phenotype." The progeny population was twofold to threefold more diverse than the wild-type population. This mutant population could not compete with the wild-type population and was driven to extinction in direct competition studies. Thus increasing the wild-type mutation rate was also deleterious. There were simply too many mutations in each genome produced; the progeny were not viable.

While these studies might seem esoteric, they actually provide practical and valuable insights. For example, they have led to proposals that drugs to increase RdRp error rates (as opposed to inhibiting these enzymes) will increase RNA virus mutation rates to a catastrophic level. On the other hand, attenuated vaccine strains might be engineered to have more faithful RdRps thereby decreasing their ability to revert to virulence (Fig. 11.6).

more fit than any individual genome. Or put another way, we cannot find any single genome in the population that replicates better than the group as a whole and in fact, most individual genomes replicate more poorly than the group. Why this occurs is not always clear, but an animal is a very complex ecosystem. Different members of the quasispecies may be better adapted to different niches in the animal.

How does a quasispecies form? PV genomes can be chemically synthesized and/or cloned to generate a single genome sequence. But as the cloned virus replicates, mutations accumulate. Mutations accumulate because the *fidelity* of PV RdRp is low. RdRps do not have proofreading activities (as do many DNA polymerases). If a mistake occurs, there are only two possibilities: RNA synthesis can stop, or RNA synthesis can continue beyond the mistake to generate a point mutation. The fidelity of RdRps has been studied and they generate transition mutations at a frequency of 10^{-3}–10^{-5} (one transition mutation for every 10^3–10^5 nt synthesized) and transversion mutations at a frequency of 10^{-6}–10^{-7} (one transversion mutation for every 10^6–10^7 nt synthesized). A rate of one mutation per 10^5 nucleotides synthesized ensures that during an infection, many progeny will contain a mutation. The significance of the quasispecies model for RNA viruses is that there is no single "fittest" or "best" genome sequence. *The population is fitter than the individual* (Box 11.3).

In this chapter we have learned:

- The definition of an RNA virus.
- The subgroups within the RNA virus group (positive strand, negative strand, ambisense, and double strand) and the differences in their replication strategies.
- The types of NCRs commonly found in the genomes of RNA viruses.
- The different mechanisms used to prime RNA synthesis.
- That some RNA viruses exist as quasispecies, mixed populations that are more fit than any individual member.

References

Korboukh, V.K., Lee, C.A., Acevedo, A., Vignuzzi, M., Xiao, Y., Arnold, J.J., et al., 2014. RNA virus population diversity, an optimum for maximal fitness and virulence. J. Biol. Chem. 289 (43), 29531–29544. Available from: https://doi.org/10.1074/jbc.M114.592303.

Tian, L., Qiang, T., Liang, C., Ren, X., Jia, M., Zhang, J., et al., 2021. RNA-dependent RNA polymerase (RdRp) inhibitors: the current landscape and repurposing for the COVID-19 pandemic. Eur. J. Med. Chem. 213, 13201. Available from: https://doi.org/10.1016/j.ejmech.2021.113201. PMID: 33524687; PMCID: PMC7826122.

Vignuzzi, M., Stone, J.K., Arnold, J.J., Cameron, C.E., Andino, R., 2006. Quasispecies diversity determines pathogenesis through cooperative interactions in a viral population. Nature 439 (7074), 344–348. Available from: https://doi.org/10.1038/nature04388.

12

Family *Picornaviridae*

After reading this chapter, you should be able to answer the following questions:

- What are the key features of all members of the family *Picornaviridae*?
- What strategies do picornaviruses use for protein expression?
- What is the primer for genome replication?
- What strategy is used by picornaviruses for ribosome assembly/translation initiation?
- At what site in the cell do picornaviruses replicate?
- What are some of the major human and animal diseases caused by picornaviruses?

Members of the family *Picornaviridae* are small (pico) RNA viruses. Family *Picornaviridae* is one of five named families in the Order *Picornavirales*, a large group of unenveloped positive-strand RNA viruses of plants, insects, and animals. In this chapter, picornavirus will be used to refer to family members. Virions are unenveloped with icosahedral capsids (Fig. 12.1 and Box 12.1). Capsids are assembled from 60 copies each, of 3 different capsid proteins.

Genome organization

Picornaviral genomes are a single molecule of positive-sense, single-strand RNA (Fig. 12.2). Genome sizes range from ~6700 to 10,100 nucleotides (nt). The 5′ terminus of all picornaviral genomes is covalently linked to a small protein called VPg (viral protein, genome linked). The picornaviral genome has a long untranslated region (UTR) at its 5′ end that ranges from 600 to 1200 nt in length. The 5′ UTR folds into a stem-loop structure that functions as an internal ribosome entry site (IRES). The IRES serves as a scaffold for ribosome assembly thus picornaviral mRNAs can be translated even though they lack a 5′-methyl-G cap. There is also a 40−330 nt UTR at the 3′ end of picornaviral genomes. The very 3′ end of picornaviral genomes is a stretch of polyadenine (polyA) that ranges from 35 to 100 nt. Another important structural region is found within the coding region of the PV genome: A 61 nt stem-loop structure required for genome replication and called the *cis-acting replication element*.

Between the 5′ and 3′ UTRs of all picornaviruses is a single open-reading frame (ORF) encoding a polyprotein of ~2300 amino acids. Structural (capsid) proteins are encoded at the 5′ end of the ORF while the RdRp (also called 3D^{Pol}) is encoded at the 3′ end of the ORF. In between are other nonstructural proteins required for replication, including proteases, helicases, and ATPases. The protein expression strategy of the picornaviruses is to generate all mature proteins from a single polyprotein precursor, by a series of stepwise proteolytic cleavages. The total number of protein products ranges from 11 to 15.

Viruses.
DOI: https://doi.org/10.1016/B978-0-323-90385-1.00018-2

© 2023 Elsevier Inc. All rights reserved.

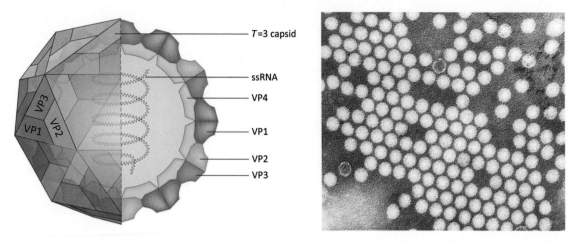

FIGURE 12.1   Structure of a virion (left) and negative stain electron micrograph of PV (right). Source: *From CDC/Dr. Fred Murphy, Sylvia Whitfield.*

BOX 12.1

General characteristics

Picornavirus genomes are a single strand of positive-sense RNA ranging in size from 6.7 to 10.1 kb. Genomes are covalently attached to a viral protein at their 5′ end as a result of protein-primed RNA synthesis. Genomes are polyadenylated at the 3′ end. A long UTR at the 5′ end of the genome folds into a complex stem-loop structure to form an IRES. Genome replication is cytoplasmic.

Genomes contain a single long ORF that encodes the full complement of viral proteins. Proteins are produced by proteolytic processing, by virally encoded proteases, of a long precursor. Virions are unenveloped, ∼30 nm diameter. Capsids are built from 180 capsid proteins (60 copies each of VP1, VP2, VP3). A fourth structural protein, VP4, is located inside of the capsid.

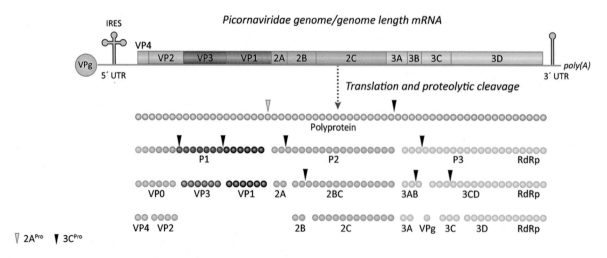

FIGURE 12.2   Schematic diagram of a picornaviral genome showing a single long ORF flanked by 5′ and 3′ UTRs. The 5′ end of the genomic RNA is covalently linked to a small viral protein, VPg. A highly base-paired structure in the 5′ UTR, called the internal ribosome entry site or IRES, is the site of ribosome assembly.

Virion structure

Virions are unenveloped with icosahedral symmetry, about 30 nm in diameter. Capsids are assembled from 180 capsid proteins: 60 copies of each of 3 viral capsid proteins, viral proteins (VP)1, VP2, and VP3 assemble to form the outer shell. A fourth structural protein, VP4 is inside the capsid. The structural proteins are produced as part of a long precursor polyprotein. VP2 is actually part of a precursor (VP0) that undergoes a maturation cleavage after capsid formation to generate VP2 and VP4.

Picornavirus replication (poliovirus model)

For many centuries paralytic poliomyelitis, although endemic worldwide, was a relatively rare disease. However, this changed in the early 20th century with major epidemics that began in Europe. By 1910 epidemics of paralytic polio were occurring throughout the developed world, including the United States, and the infectious agent, poliovirus (PV) had been isolated. By the 1940s and 1950s, polio was killing or paralyzing over half a million people per year. In the United States, president Franklin Roosevelt (a polio survivor) pushed for creation of the National Foundation for

Infantile Paralysis. Today the Foundation is known as the March of Dimes. The Foundation supported basic research on PV and supported development of vaccines. Fundamental aspects of picornaviral structure, genome organization, protein expression, and assembly were first described for PV. The replication cycle of PV, summarized here (Fig. 12.3), serves as a model for many other picornaviruses (Box 12.2).

Binding and penetration

PV virions bind to a cellular protein called the poliovirus receptor (PVR). The PVR is an integral membrane protein and member of the immunoglobulin superfamily of proteins. The PVR molecule has three immunoglobulin-like domains that extend from the cell surface. [The receptors for other picornaviruses include a wide variety of cell surface molecules including CD55 of the complement cascade, ICAM-1 (a cellular adhesion molecule), integrins, and heparin sulfate.] Binding of PV to the PVR triggers a rearrangement its capsid proteins. The capsid protein, VP1, is inserted into the plasma membrane (PM) to form a protein channel through which the RNA genome enters the cell. PV appears to enter cells at the PM although other picornaviruses are taken up by endocytosis (these viruses presumably use low pH to trigger capsid rearrangements for interaction with endosomal membranes).

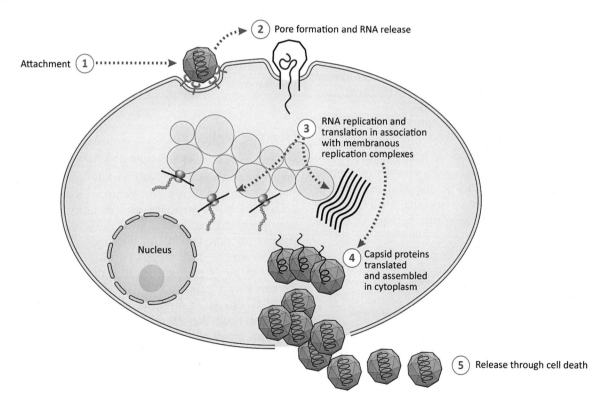

FIGURE 12.3 Overview of the picornavirus replication cycle showing replication in the cytoplasm. RNA synthesis is membrane-associated. Virions assemble in the cytoplasm. Replication is often cytopathic.

BOX 12.2
─────────────

Taxonomy

Picornaviruses are members of the realm *Ribovaria*, order *Picornavirales*. The family *Picornaviridae* contains over 60 genera. In this chapter we mention members of following genera:

Genus *Apthovirus*
Species include
  Bovine rhinitis A virus
  Bovine rhinitis B virus
  Equine rhinitis A virus
  Foot and mouth disease virus
Genus *Enterovirus*
Species include

  Enterovirus A—Enterovirus L
  Rhinovirus A—Rhinovirus C
Genus *Hepatovirus*
Species include
  Hepatovirus A—Hepatovirus I
Genus *Kobuvirus*
Species include:
  Aichivirus A—Aichivirus F
Genus *Parechovirus*
Species include
  Parechovirus A—Parechovirus F

Translation

The picornavirus genome is an mRNA and the full complement of viral proteins is synthesized from the poly(A)-containing genome length mRNA. VPg, a small peptide covalently linked to the 5′ end of the genome, is quickly removed from the infecting genome and the mRNA is translated. Thus PV mRNA differs from virion RNA only by the lack of VPg. Ribosomes rapidly and efficiently assemble on the IRES (Fig. 12.4). There are distinct types or classes of IRES and the mechanisms for ribosome assembly vary somewhat. (A few viruses, other than picornaviruses, also use an IRES to direct ribosome assembly and IRES-containing cellular genes have also been identified.)

The theoretical product of translation of the picornaviral genome is a very large polyprotein. However, examination of PV-infected cells reveals little to no complete polyprotein as it is rapidly cleaved into smaller products. In the case of PV, the first polyprotein cleavage is carried out *in cis* by the viral protease (2A^{pro}) to release the P1 polyprotein (Fig. 12.2). PV encodes a second protease (3C^{pro}) that performs additional cleavages.

Very soon after PV infection, translation of host proteins is inhibited. This happens because a PV protease (3C^{pro}) cleaves the cell "cap-binding protein" a critical component of the ribosome initiation complex. Inhibition of cap-dependent translation allows the virus exclusive access to the cell's translation machinery. In addition, inhibiting cellular protein synthesis interferes with the cell's ability to mount an antiviral defense. Picornaviruses vary in their ability to shut off host cell translation, and among those that do so, the mechanisms vary.

Genome replication

In order for the PV genome to be replicated, it must be free of ribosomes. Both virally encoded and host proteins are required for genome replication, which occurs in the cytoplasm and is associated with cell membranes. Picornaviral RdRp (3D^{pol}) uses a protein primer to initiate RNA synthesis. VPg is a short (22−24 amino acids) polypeptide found covalently linked to the 5′ ends of newly synthesized viral RNAs. To serve as a primer, VPg is first uridylated by 3D^{pol}. UMP is linked to VPg via the hydroxyl group on a tyrosine near the N-terminus of VPg. Additional U residues are then added. This generates the uridylated VPg that interacts with the polyA tail on the genome strand, to initiate synthesis of the "negative" or copy RNA (cRNA) strand (Fig. 12.5). The cRNA is not an mRNA and is not translated. It functions as the template for synthesis of additional genomes. Uridylated VPg primes the synthesis of all picornaviral RNAs,

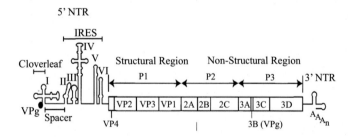

FIGURE 12.4   Internal ribosome entry site. Regions at the 5′ end of the picornavirus genome fold into complex base-paired structures that form sites for cap-independent ribosome assembly.

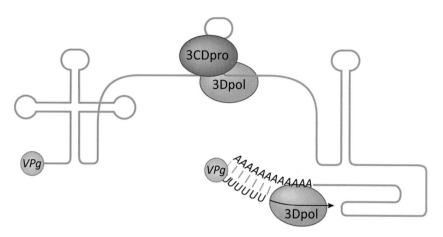

FIGURE 12.5   Minus-strand RNA synthesis. Schematic representation showing picornaviral RdRp ($3D^{Pol}$) using a protein primer to initiate RNA synthesis. VPg is uridylated by $3D^{Pol}$ (as part of protein complex, not shown) to produce polyuridylated VPg, which binds the 3' end of the genome strand to serve as a primer.

thus all copies of viral RNAs are covalently linked to a molecule of VPg.

The synthesis of viral RNAs heavily favors genome strands. The ratio of genome to cRNA is 30–70:1 in the infected cell. Thus each cRNA is the template for many genomes. The synthesis of genome strands is favored because formation of replication complexes is much more efficient on the 3' end of the cRNA than on the 3' end of the viral genomic RNA.

Assembly and maturation

The process of capsid assembly begins when local concentrations of proteins are sufficiently high. VP1, VP3, and VP0 (VP2 + VP4) form protomers and five protomers assemble to form a pentamer. It is likely that protomers form as cleavages between the proteins are made by the viral polymerase (Fig. 12.6). Twelve pentamers assemble to form an empty capsid. RNA is then inserted into the empty capsid. PV virions are released upon cell death. After release VP0 is cleaved to produce VP2 + VP4. VP4 is located inside the capsid, associated with RNA. Now the virion is ready to infect a new cell.

Diseases caused by picornaviruses

Genus enterovirus

To date there are over 80 named genera in the family *Picornaviridae*. Of importance to human health are several members of the genus *Enterovirus*. Primary sites of replication of members of the genus are epithelial cells and lymphoid tissues of upper respiratory tract (rhinoviruses) and gastrointestinal tracts. In the gastrointestinal tract, enterovirus infections are very often asymptomatic. PV is a member of the genus

Enterovirus and is transmitted by the fecal–oral route, for the most part causing inapparent infections. However, PV can cause a viremia and may cross the blood–brain barrier resulting in the clinical disease we call paralytic poliomyelitis (Fig. 12.7). PV can also infect the spinal cord to cause aseptic meningitis. (See also Box 12.3, history and pathogenesis of PV.)

There are hundreds of enteroviruses in addition to PV and some of these also cause human disease. Hand, foot, and mouth disease is a generally mild disease of infants and children younger than 5 years old. Symptoms of hand, foot, and mouth disease [not to be confused with foot and mouth disease (FMDV) of cattle, sheep, pigs, and horses caused by a different picornavirus] include fever, mouth sores, and a skin rash. Large outbreaks are uncommon in the United States; however, in some areas of Asia large outbreaks are common.

In rare cases, enteroviruses other than PV reach and replicate in organs such as the CNS, heart, vascular endothelium, and others. Infection of these organs can result in diseases ranging from paralysis to acute aseptic meningitis syndromes to pericarditis to nonspecific febrile illness. Enteroviruses account for approximately 90% of cases of viral meningitis, for which an etiological agent was identified.

Despite their name, enteroviruses do not cause epidemics of diarrheal disease. Within the genus *Enterovirus* are three species of rhinovirus, which are generally transmitted by aerosols, replicate in the upper respiratory tract, and cause the common cold (mild upper respiratory tract disease). The rhinoviruses are not acid stable, prefer to replicate at 33°C, and there are hundreds of serotypes. The large number of serotypes is the reason that there is no vaccine for the common cold. (There are also viruses other than picornaviruses that cause the "common cold.")

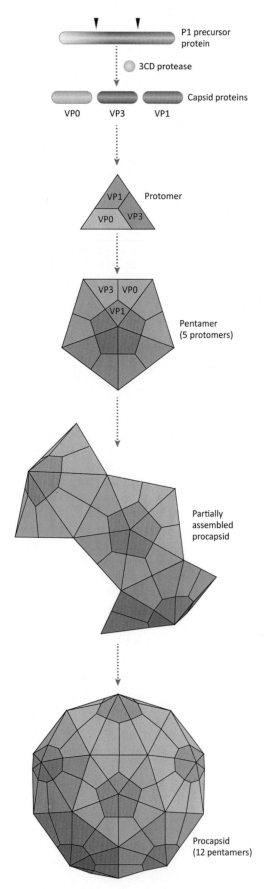

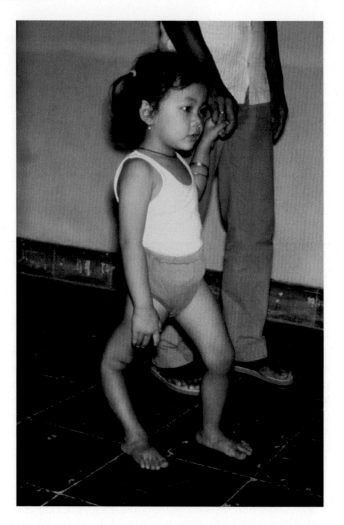

FIGURE 12.7   Child with a leg deformity resulting from paralytic polio. (CDC/PHIL image 5578, content provider CDC.)

Genus hepatovirus

Hepatitis A is a liver disease that results from infection with the hepatitis A virus (HAV). Disease can range from a mild illness lasting a few weeks to a severe illness lasting several months. Most infected individual recover but a small number of infected individuals die from fulminant hepatitis. Hepatitis A is contracted when a person ingests food or water contaminated with fecal matter containing HAV. HAV is resistant to high temperatures and low pH. HAV infects humans and other primates. A safe and effective vaccine is available and currently all children in the United States are routinely vaccinated, as are travelers to regions where HAV is prevalent. Highest risk of

FIGURE 12.6   Assembly of a picornaviral capsid. Protomers contain one copy each of VP1, VP0, and VP3. Protomers assemble into pentamers and 12 pentamers assemble to form a capsid.

BOX 12.3

Poliovirus history, vaccines, and progress toward eradication

For many centuries, paralytic polio, although endemic worldwide, was relatively rare. However, this changed in the early 20th century with major epidemics that began in Europe. By 1910 epidemics of paralytic polio were occurring throughout the developed world (including the United States) and the infectious agent, PV had been isolated. By the 1940s and 1950s, polio was killing or paralyzing over half a million people per year.

Vaccines were first introduced in 1935 but failed because it was not known that there were three distinct serotypes of PV. Not until 1948–49 it was understood that vaccine preparations must protect against all three serotypes. Early studies of PV were hampered by the belief that it replicated only in neurons; this assumption made perfect sense at the time, as the disease was neurological. An important breakthrough occurred in 1948 when Drs. John F. Enders, Frederick C. Robbins, and Thomas H. Weller, working at the Children's hospital in Boston, demonstrated that PV replicates in other cell types. Their work had not been focused primarily on PV culture but more generally on improving methods for virus culture. They were encouraged to try new culture techniques for PV with the knowledge that stool samples contained huge amounts of virus, indicating a replication site other than neurons. They were able to successfully culture PV in roller bottle cultures of embryonic skin and muscle cells and further showed that PV grown in this manner could be readily quantified and serotyped, procedures that previously required animal inoculation. Thus their studies greatly facilitated vaccine development. For their discovery, Drs. Enders, Weller, and Robbins were awarded the Nobel Prize in Physiology or Medicine in 1954.

How/why did a rare paralytic disease of children and infants suddenly explode into a worldwide epidemic? Did the virus change, or was it something else? Evidence points to improved sanitation as a major cause of the PV epidemic. Historically, infants were exposed to PV very early in life but they were also protected from neurologic disease by maternal antibody. Most adults were immune to PV through repeated exposure as children. However, as clean drinking water became the standard in the developed world, infants were no longer exposed to PV. By the early 20th century millions of children and adults had never been exposed to the virus thus were not immune. The existence of a large susceptible population set the stage for notable outbreaks of paralytic disease.

How does an enteric virus cause paralytic disease? PV is usually transmitted by the fecal–oral route,

replicating in epithelial cells in the gastrointestinal tract and lymphoid tissue such as the Peyer's patches. When virus is limited to the gastrointestinal tract, there is no paralytic disease even though large amounts of virus are produced and excreted. However, PV can move out of lymphoid tissues to produce a viremia. Rarely (0.5%–2% of infections) the virus gains access to the brain or spinal cord. Within the CNS, PV has a tropism for motor neurons and their destruction results in muscle weakness and/or paralysis (Figs. 12.7 and 12.8).

Two different types of vaccines were developed to protect against poliomyelitis. A comparison of these vaccines is quite instructive. Dr. Jonas Salk led the effort to create a killed vaccine. In brief, he developed methods

FIGURE 12.8 Emerson respirator, also known as an iron lung, which was used by polio patients whose ability to breath was paralyzed due to this crippling viral disease. The chamber is used to create a negative pressure around the thoracic cavity, thereby, causing air to rush into the lungs. (CDC/PHIL image 6529, content providers CDC/GHO/ Mary Hilpertshauser).

BOX 12.3 *(cont'd)*

to culture all three PV serotypes, purified the virions, and inactivated them (with formalin) to produce an injectable vaccine. The inactivated, injectable vaccine induces production of neutralizing antibodies to PV capsid proteins. Most of the antibody response is found as circulating IgG, which protects against CNS disease quite efficiently. Salk's vaccine did not protect against infection however, orally ingested PV could replicate in the gastrointestinal tract.

Dr. Albert Sabin led the effort to create an oral vaccine containing live-attenuated (weakened) viruses. The attenuated viruses replicate in the gastrointestinal tract, stimulating development of immune responses that protect against infection with wild-type viruses. The weakened viruses in the oral vaccine are unable to reach the CNS.

Both vaccines have relative advantages and disadvantages:

Salk's vaccine (inactivated PV or IPV) must be injected. As IPV cannot replicate, it is very safe. Vaccination does not lead to infection and the vaccine itself cannot be transmitted. Most people get vaccinated when they are children. Current recommendations call for four doses of IPV.

Sabin's vaccine is given orally (today it is called oral poliovirus vaccine or OPV), thus is easier and cheaper to administer. However, OPV contains live virus and should not be given to immunocompromised individuals. OPV replicates in the gastrointestinal tract thus can be transmitted to contacts of vaccinated individuals. In the mid-20th century this was considered an advantage by some, as it expanded the number of protected individuals in the population. However, replication meant that the virus could (and did) change. The so-called "revertant" viruses would (rarely) cause paralytic disease in vaccinated individuals or their contacts.

The recommended use of killed and attenuated PV vaccines has changed over the years. The Salk vaccine was the first PV vaccine licensed in the United States. However early on, a poorly prepared batch of vaccine (containing replication competent virus) resulted in many cases of vaccine-associated paralytic disease, diminishing initial enthusiasm for the Salk vaccine. While the live-attenuated Sabin vaccine did (rarely) cause vaccine-associated cases of paralytic polio, for many years its advantages were thought to outweigh the risk. However, as the incidence of vaccination increased and cases of natural disease plunged, risks of the attenuated vaccine became unacceptable and OPV was banned in the United States by the year 2000.

OPV continues to be used in parts of the world where health care infrastructure does not allow for use of IPV. Use of OPV is a key part of the World Health Organization's drive to eradicate PV. Indeed, use of OPV has reduced cases of paralytic poliomyelitis by 99%. The original OPV formula included three attenuated strains, PV1, PV2, and PV3. Recently it has been suggested that OPV be reformulated to eliminate PV2. All cases of PV2-associated paralytic polio since 1999 have been vaccine-associated and PV2 was declared eradicated in 2015. Thus there is measurable risk, but no known benefit, to including PV2 in current OPV formulations.

The goal of worldwide elimination of polio remains elusive, primarily due to political instability, conflict, hard-to-reach populations, and poor health care infrastructure. Sporadic cases continue to be reported in central Africa, Afghanistan, Pakistan, China, and Indonesia. The complete elimination of polio will require finding political and social solutions to reach as yet unprotected populations. Even then the PV story will not be completely closed, as there will be need for continued discussions about the eventual destruction of vaccine strains and laboratory stocks.

---

infection is among persons without safe drinking water and living under conditions of poor sanitation and hygiene.

HAV is stable in the environment, as might be expected of a virus that is transmitted by the fecal–oral route. Experiments have shown that HAV can remain infectious for weeks or months in water or soil sediments. HAV is concentrated from seawater by bivalves (clams, oysters, etc.) so they are a common source of infection. Persons with any type of immune disorder (old age is enough) are at high risk for contracting HAV from uncooked muscles, clams, or oysters. HAV is also highly resistant to drying, thus infectious virus can also contaminate fruits and

vegetables, such as lettuce, sprouts, strawberries, or green onions. HAV is also resistant to inactivation by detergents and solvents.

Genus kobuvirus

Aichiviruses were first defined after an oyster-related outbreak of gastroenteritis in the Aichi Prefecture in Japan. They have since been identified worldwide and can be recovered from sewage and polluted waters. There is high seropositivity among adults, but relatively few outbreaks of severe disease have been documented, suggesting many subclinical infections. In addition to humans,

Aichiviruses have been recovered from cattle and pigs with gastroenteritis.

Genus parechovirus

Parechovirus-A (PeV-A) was first identified in 1956 from children with diarrhea but can also cause severe central nervous system disease, most often in infants. PeV-A is found worldwide and seropositivity increases rapidly in children ages 2–5 years. Seropositivity among adults is very high. Transmission is presumed to be fecal–oral as virus can be detected in both stool and nasopharyngeal samples. There are several distinct genotypes. Genotype 1 is associated with acute gastroenteritis, upper respiratory tract symptoms, fever and rash in young children. In infants, geneotypes 1 and 3 have been associated with severe diseases including acute flaccid paralysis, meningitis, and encephalitis. Long-term neurodevelopmental sequelae have been associated with some genotype 3 infections. Neonatal sepsis, also known as systemic inflammatory response syndrome, is the most common severe clinical manifestation of genotype 3 infection.

Genus apthovirus

FMDV is a serious disease of cattle, sheep, pigs, and horses. The virus is found worldwide (except in the United States and parts of Europe). While mortality is low, morbidity is very high. FMDV is a "vesicular" disease, causing painful fluid-filled vesicles in the mouths and on the hooves of cattle and swine. FMDV is spread by contact, respiratory, and fecal–oral routes. It is *very* easily transmitted, moving over long distances (miles) in the air. The virus is also easily transmitted via fomites (cars, trucks, shoes) thus poses a high risk to food animal biosecurity. FMDV is also very stable in the environment and infectious virus has even been detected in smoked meat products.

FMDV is endemic in many parts of the world and vaccines are available. However, there are multiple serotypes and it is not feasible to vaccinate against them all. Countries that are free of FMDV take great precautions to retain their "FMDV-Free" status. Such countries, including the United States, exclude importation of meat from countries where the virus is endemic, or where vaccination is practiced. Rare outbreaks in FMDV-free countries disrupt normal agricultural activities causing huge economic losses and destruction of large numbers of animals.

An informative example is the FMDV outbreak that occurred in England in 2001. The first cases were reported in February 2001. The likely source of the outbreak was uncooked waste food fed to pigs. While only ~2000 cases of disease were confirmed, the outbreak was contained by slaughter and disposal of over 10 *million* sheep and cattle. This amounted to the slaughter of 80,000–90,000 animals per week. Veterinarians from the United States traveled to Great Britain to help in the control effort. (Because of the high infectivity and stability of the virus, once veterinarians had visited an infected farm, they could not visit another holding until they, their vehicles, and instruments were thoroughly decontaminated.) During the outbreak, all movement of livestock was halted. Also, with the intention of controlling the spread of the disease, public rights of way across farmland were closed, severely disrupting tourism in rural England. Economic losses due to curtailing tourism may have been greater than the agricultural losses.

In this chapter we have learned that:

- Members of the family *Picornaviridae* are small, unenveloped, positive-strand RNA viruses with icosahedral capsids.
- Picornavirus replication is cytoplasmic. Genome replication requires host cell membrane complexes.
- Genomes are covalently linked to a viral protein (VPg) that serves as the primer for RNA synthesis.
- The genome length RNA serves the single mRNA for expression of all viral proteins.
- Translation initiation is dependent on a stem-loop structure called an IRES found near the 5′ end of the genome.
- The family *Picornaviridae* is large and contains pathogens of humans and animals. The genus *Enterovirus* includes isolates that replicate in upper respiratory tract and viruses that replicate in the gastrointestinal tract.

13

Family Caliciviridae

After reading this chapter, you should be able to answer the following questions:

- What features are shared by members of the family *Caliciviridae*?
- What are the characteristic features of genome replication and translation initiation?
- At what site in the cell do caliciviruses replicate?
- In what ways are caliciviruses similar/different from picornaviruses?
- What are the major human diseases caused by caliciviruses?
- Why are human caliciviruses difficult to study?

Caliciviruses are small, unenveloped positive-strand RNA viruses with a particle size of 27—40 nm in diameter (Fig. 13.1 and Box 13.1). The name calicivirus is derived from the Latin word "calyx" meaning cup or goblet, as many strains have visible cup-shaped depressions on the capsid surface. Caliciviruses have been isolated from humans, cattle, pigs, cats, rabbits, chickens, reptiles, dolphins, and amphibians. Neither their replication nor pathogenesis have been well studied because many, including human noroviruses, are difficult to grow in cultured cells. Feline calicivirus (FCV) and murine noroviruses are often used as surrogates for human noroviruses. Novel caliciviruses continue to be isolated, at least in part, in an effort to find better model systems. An enteric calicivirus recently isolated from rhesus monkeys (Tulane virus) grows in cultured cells and may prove to be a useful model for human noroviruses.

There are currently 11 genera in the family *Caliciviridae* (Box 13.2). The genera *Norovirus* and *Sapovirus* contain human isolates that cause enteric disease. The genus *Vesivirus* includes FCV a common upper respiratory tract pathogen of cats. *Lagosviruses* and *Neboviruses* infect rabbits and bovids, respectively. Rabbit hemorrhagic disease virus (RHDV) emerged in China in 1984 and has since spread worldwide.

Genome organization

Calicivirus genomes are unsegmented positive-sense RNA ranging in size from 7.4 to 8.3 kb. Genome replication is cytoplasmic. Caliciviruses prime RNA synthesis with a protein primer (VPg). VPg is also required for translation. Genomes contain two or three open reading frames (ORFs) and are polyadenylated (Fig. 13.2). Nonstructural (NS) proteins, including the RNA-dependent RNA polymerase (RdRp) are synthesized from genome-length mRNA to produce a polyprotein processed by viral proteases. The major capsid protein of noroviruses and vesiviruses is expressed from ORF2 using subgenomic mRNA. The gene for the major capsid protein of lagosviruses and sapoviruses is found in the same reading frame as the NS proteins, and could be produced by cleavage of a

Viruses.
DOI: https://doi.org/10.1016/B978-0-323-90385-1.00007-8

© 2023 Elsevier Inc. All rights reserved.

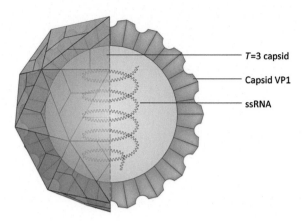
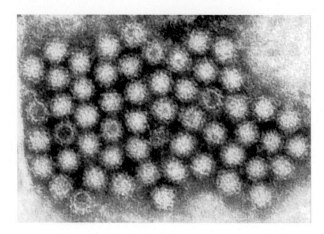

T=3 capsid

Capsid VP1

ssRNA

FIGURE 13.1   Virion structure (left). Negative stain electron micrograph of human norovirus (right). Content providers: CDC/Charles D. Humphrey.

BOX 13.1

General characteristics

Caliciviral genomes are unsegmented positive-sense RNA ranging in size from 7.4 to 8.3 kb. Genome replication is cytoplasmic. Caliciviruses prime RNA synthesis with a protein primer (VPg). Genomes contain two or three ORFs and are polyadenylated. Nonstructural (NS) proteins are synthesized from genome-length mRNA generating a polyprotein which is processed by viral proteases. The capsid protein is expressed from a subgenomic mRNA. VPg is required for translation. Virions are unenveloped with icosahedral symmetry, ~27–40 nm diameter. Capsids are built from 90 dimers of the capsid (C) protein. Human noroviruses are a leading cause of acute gastroenteritis in humans.

BOX 13.2

Taxonomy

Realm *Riboviria*
Order *Picornavirales*
Family *Caliciviridae*
The family contains 11 genera. Members of seven genera infect mammals, members of two genera infect birds, and members of two genera infect fish. Genera considered in this chapter include:

Genus *Lagovirus*
Rabbit hemorrhagic disease virus

Genus *Nebovirus*
Newbury-1 virus (bovine origin)
Genus *Norovirus*
Norwalk and Norwalk-like viruses
Genus *Sapovirus*
Sapporo virus
Genus *Vesivirus*
Feline calicivirus, vesicular exanthema of swine virus

---

long polyprotein (Fig. 13.3). However, transcription of a subgenomic mRNA is also used to produce these capsid proteins. ORF3 encodes the minor capsid protein VP2; it is produced from the subgenomic mRNA by a ribosomal termination–reinitiation mechanism.

The 5′ and 3′ nontranslated regions of calicivirus genomes are very short; however, recent evidence suggests that highly base-paired regions of secondary structure form using sequences that extend into coding regions.

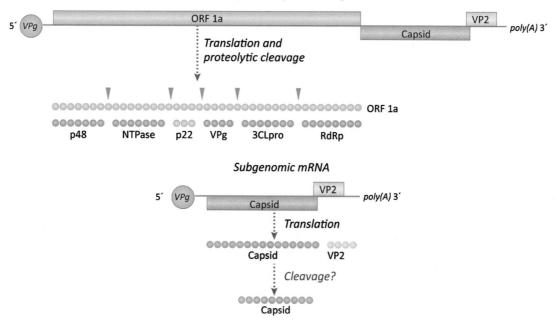

FIGURE 13.2 Example of calicivirus genome organization and protein expression strategy.

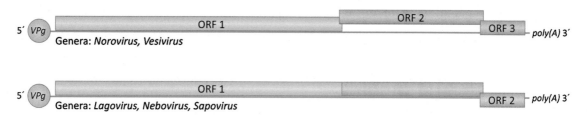

FIGURE 13.3 Genome comparison. The genome organization of the sapoviruses, lagoviruses, and neboviruses differ, as the capsid protein (VP1) is continuous with that of the NS proteins and can be released by proteolytic cleavage. All caliciviruses encode a second, minor structural protein (VP2) from an ORF near the 3′ end of the genome. Murine noroviruses contain four ORFs.

Virion structure

Calicivirus capsids assemble from 90 dimers of VP1 and a few molecules of VP2 to produce virions of 27–40 nm. Recombinant VP1 (e.g., expressed in baculovirus) can efficiently self-assemble to form virus-like particles (VLP). The process does not require RNA or the minor capsid protein VP2.

General replication cycle

Many medically important caliciviruses are difficult to grow in cultured cells, in particular the human noroviruses. As new caliciviruses are discovered, new details will likely emerge. Below is a general description of calicivirus replication.

Attachment/penetration

Noroviruses interact with histo blood group antigens, carbohydrates on the cell surface. Other caliciviruses interact with protein receptors. Different modes of penetration into the cell have been documented. Some caliciviruses penetrate at the plasma membrane while others penetrate after endocytosis.

Genome replication

As with other positive-strand RNA viruses, translation of the infecting genome immediately follows penetration, producing a long precursor polyprotein that is proteolytically processed to produce NS proteins that include a protease and the RdRp. Following translation of NS proteins, RNA synthesis can begin. RNA synthesis takes place on host cell membranes and is primed by VPg such that caliciviral RNAs have VPg covalently linked to the 5′ end. A genome-length copy RNA serves as template for synthesis of both genomes and subgenomic mRNAs.

Translation

The calicivirus VPg protein (13−15 kDa) is much larger than the ∼22 amino acid VPg of picornaviruses and it seems to play a larger role in replication. In addition to its role in genome synthesis, calicivirus VPg is essential for translation of viral RNA. Removal of VPg from caliciviral RNA decreases infectivity as well as the ability of the RNA to support translation in cell-free systems. Conversely, addition of recombinant VPg to cell-free extracts inhibits both cap- and IRES-dependent translation suggesting that caliciviral VPg competes for initiation factors, effectively removing them from translation systems. VPg proteins from various caliciviruses have been shown to interact with eukaryotic initiation factors (eIF3, eIF4E, eIF4G), further supporting a role in translation.

Caliciviruses use a variety of protein expression strategies. In the case of vesiviruses and noroviruses, genome-length mRNA is used to express an ORF1 polyprotein that is proteolytically processed to produce the NS proteins NS1−NS7 (including NTPase, VPg, protease, and RdRp). A subgenomic mRNA is produced and contains one or two ORFs. The major capsid protein (VP1) is encoded by ORF2. The genome organization of the sapoviruses, lagoviruses, and neboviruses is a bit different as the capsid protein (VP1) is continuous with that of the NS proteins. However, VP1 is also produced from a subgenomic mRNA. All caliciviruses encode a second, minor structural protein (VP2) from an ORF near the 3′ end of the genome. Murine noroviruses contain four ORFs. Two different strategies are used to produce the minor capsid protein VP2 from the bicistronic mRNA. Noroviruses use leaky scanning for expression of VP2. In this process, the ribosome bypasses the initial start codon on the mRNA and instead protein synthesis initiates at a downstream start codon. In the case of FCV (vesivirus) and RHDV (lagosvirus), VP2 is expressed through RNA termination−reinitiation or stop−start, a process in which the ribosome completes synthesis of a protein then reinitiates synthesis of a second protein, without disassembly from the mRNA.

Assembly/release

Calicivirus capsids assemble from 90 dimers of VP1 and a few copies of VP2. If the VP1 gene is cloned and expressed in cells, the recombinant VP1 protein assembles to form a VLP. This experimental result indicates that neither VP2 nor viral RNA is absolutely required for formation of capsids. Calicivirus-infected cells undergo lysis to release progeny.

Diseases

Acute gastrointestinal disease

In the United States, the Centers for Disease Control and Prevention (CDC) estimates that about half of all outbreaks of food-related illnesses are caused by noroviruses. Noroviruses are very contagious and are quite stable on contaminated surfaces; therefore the majority of norovirus infections are not a *direct* result of ingesting contaminated food or water, but rather result from contact with contaminated surfaces. The CDC estimates that each year in the United States noroviruses cause 19−21 million cases of acute gastroenteritis. Common symptoms are stomach pain, nausea, vomiting, and diarrhea lasting from 24 to 48 hours. Virus can be shed in feces for several days. While most norovirus infections are self-limiting, serious illness may occur in young children and the elderly. Unfortunately, norovirus infections are not uncommon in health care and long-term care settings. Several large norovirus outbreaks have been recorded on cruise ships (cruise ship disease). The CDC reports that over 90% of diarrheal disease outbreaks on cruise ships are caused by noroviruses. It is difficult to control these outbreaks due to crowded conditions and shared dining areas. Also, noroviruses are very resistant to many common disinfectants so may survive routine cleaning.

Sapoviruses are also associated with acute gastroenteritis of humans. Although all age groups are affected, disease burden is highest in children under the age of 5 years. Common symptoms include vomiting and diarrhea; symptoms usually resolve within a week. Similar to noroviruses, transmission is through food and water.

Feline calicivirus

FCV causes variable and sometimes complex disease syndromes in cats. FCVs are genetically diverse, and different strains can vary widely in their virulence. FCV can be isolated from about half of the house cats presenting with upper respiratory tract infections. In the case of FCV infection, ulceration of the mouth is a common finding. A few strains of FCV cause more severe systemic (multiorgan) disease. FCV is transmitted via saliva, feces, urine, and respiratory secretions and is very stable in the environment. It can be transmitted through the air, orally, and by fomites. Some cats become persistently infected, serving as an ongoing source of infection. Within persistently infected cats, it has been documented that mutant viruses, with changes in the capsid protein, arise. This allows the virus to escape from immune responses and may also lead to the generation of strains with increased virulence. Vaccines are available for FCV, but due to genetic variation, protection is not complete. FCV is one of the few caliciviruses that grows well in cultured cells.

Rabbit calicivirus

RHDV, genus *Lagosvirus*, emerged in China in 1984. It is a highly infectious and often fatal disease that affects adult wild and domestic rabbits of the species *Oryctolagus cuniculus*. This species includes European or common rabbits, including all breeds of domesticated rabbits. Other species of rabbits, such as wild rabbits found in the United States, are resistant. RHDV is considered a major rabbit pathogen in some countries where it threatens rabbits raised for food and kept as pets. In some areas of Europe, the virus has reduced wild rabbit populations to levels that threaten endangered predators such as the Iberian lynx. In stark contrast, RHDV has been introduced into Australia and New Zealand as a means of controlling overwhelming populations of nonnative European rabbits. The virus is transmitted by a variety of mechanisms; either by direct contact between rabbits or indirectly through contaminated fomites. Blood feeding insects may also be vectors.

Like other caliciviruses, RHDV is very stable in the environment. The disease is characterized by acute necrotizing hepatitis and disseminated intravascular coagulation. Disease progression can be rapid, sometimes resulting in sudden death in the absence of clinical symptoms.

In this chapter we have learned that:

- Members of the family *Caliciviridae* are small, unenveloped, positive-strand RNA viruses.
- Calicivirus replication is cytoplasmic. Genome replication requires host cell membrane complexes and is primed by a virally encoded protein, VPg.
- In addition to its role as a primer for RNA synthesis, VPg is required for translation of genome length and subgenomic mRNAs.
- Among family members there is some variability in protein expression strategies but genome-length RNA is translated to produce a large polyprotein that is processed by viral proteases. A subgenomic mRNA encodes one or two capsid proteins.
- Human caliciviruses (noroviruses and sapoviruses) cannot be cultured in the laboratory so surrogates are used to probe virus replication and environmental properties of caliciviruses.
- Norwalk virus and Norwalk-like viruses (noroviruses) cause acute gastroenteritis in humans. These agents are transmitted by the fecal—oral route. Virions are stable in the environment and are highly infectious. Outbreaks of cruise ships have involved hundreds of passengers.

14

Family Hepeviridae

After studying this chapter, you should be able to:

- Describe the shared characteristics of members of the family *Hepeviridae*.
- Describe the general strategies used by hepeviruses for transcription, translation, and genome replication.
- List modes of hepatitis E virus (HEV) transmission.

Members of the family *Hepeviridae* are small, unenveloped RNA viruses (Fig. 14.1 and Box 14.1). The family name was derived from the name of the clinical disease "hepatitis E," caused by the first identified member of the family. Hepatitis E, the most common cause of human hepatitis worldwide, is usually acute and self-limiting. The notable exception is an approximately 20% case fatality rate among pregnant women in their third trimester. In resource poor countries HEV circulates among humans and the most common genogroups are 1 and 2. Endemic areas include Asia, Africa, Southern Europe, and Mexico. In endemic regions, large waterborne outbreaks with thousands of people affected have been recorded. In more developed countries, HEV is more sporadic and infection is zoonotic, associated with consumption of raw or undercooked meats (particularly pork), raw shellfish, and contamination of drinking or irrigation water with animal manure. The host range (pigs, chickens, rabbits, rats, mongoose, deer, and possibly cattle and sheep) of HEV is expanding as more animals are being surveyed (Box 14.2).

Phylogenetic analysis of HEV has revealed that the hepeviruses arose via an ancient recombination event between a lepidopteran virus (family *Alphatetraviridae*), which donated the nonstructural (NS) proteins and a chicken astrovirus (family *Astroviridae*), which donated a capsid protein.

Genome organization

Hepevirus genomes are unsegmented, 5′ methyl G capped, polyadenylated positive-strand RNA. Genome size ranges from 6.4 to 7.2 kb. The genome contains three open-reading frames (ORFs) flanked by short 5′ and 3′ untranslated regions (Fig. 14.2). The ORF1 polyprotein is translated from genome-length RNA to encode the NS proteins. The RNA-dependent RNA polymerase (RdRp) is located at the carboxyl terminus of the polyprotein. Other functional domains identified include: methyltransferase, protease, and helicase. It is not known if the NS polyprotein is cleaved or functions as an uncleaved protein with multiple active domains. ORF2 encodes the capsid protein. ORF3 encodes a phosphoprotein with roles in virion assembly, release, and pathogenesis.

Virion morphology

Two types of HEV virions are produced: naked and quasienveloped. Naked virions have icosahedral symmetry with a diameter of ~32–34 nm. Detailed structures of the capsids of infectious virions are not yet available, however, when expressed alone, capsid proteins assemble into virus-like particles (VLPs) with T = 1 or T = 3 icosahedral symmetry.

© 2023 Elsevier Inc. All rights reserved.

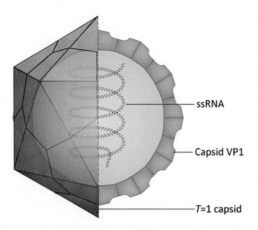
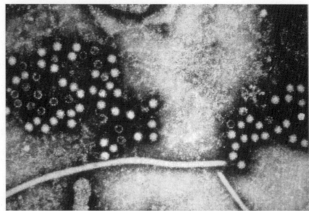

FIGURE 14.1   Capsid structure (left). Negative strain, electron micrograph of hepatitis E virus (HEV) (right). Source: *CDC. Public Health Image Library Image #5605.*

BOX 14.1

Description

Members of the family *Hepeviridae* are positive-strand RNA viruses. Genomes (6.4–7.2 kb) have 5′ methyl G caps and 3′ poly (A) tails. Genomes contain three ORFS that are flanked by short 5′ and 3′ noncoding regions. There are two mRNAs (genome length and a subgenomic mRNA). Genome-length mRNA is translated to produce proteins required for RNA replication. The subgenomic mRNA contains two overlapping ORFs used to produce the capsid protein and a phosphoprotein. Naked and quasienveloped particles are produced. Capsids are icosahedral with a size of ∼32–34 nm.

BOX 14.2

Taxonomy

Family *Hepeviridae*.
Genus *Orthohepevirus* (four species).
Species *Orthohepevirus A* contains multiple isolates and genogroups including HEV. Infect a wide range of mammals including humans, swine, deer, sheep, and camels.

Species *Orthohepevirus B* (multiple members, infect birds).
Species *Orthohepevirus C* (small mammals).
Species *Orthohepevirus D* (bats).
Genus *Piscihepevirus* (fish, one species).

General replication cycle of HEV

HEV is a hepatotropic virus and useful cell culture systems were lacking until recently. As such systems have been developed, the details of replication have become clearer. HEV replicates in the cytosol and its RNA genome is translated to produce a polyprotein (∼1693 amino acids) that contains an RdRp domain. Other identified domains include methylatransferase, helicase, and cysteine protease. Replication of the genome produces a copy RNA that serves as the template for synthesis of genome-length and subgenomic mRNAs.

The single subgenomic mRNA is bicistronic, translated to produce the ∼660 aa capsid protein (from ORF2) and a NS phosphoprotein (from ORF3). When the capsid gene is cloned, and the protein expressed in cultured cells, the resulting products include a 74 kDa protein (unglycosylated) and an 88 kDa *glycosylated*

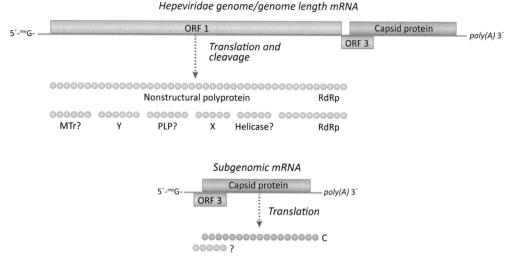

FIGURE 14.2   Hepevirus genome organization.

protein. The capsid protein contains an N-terminal signal sequence that directs the protein into the ER where it acquires N-linked glycosylation. Mutation of the N-linked glycosylation sites prevents formation of infectious virions suggesting that glycosylation is important, though its exact role is unclear.

Currently evidence favors the production of two distinct types of HEV virion. One is unenveloped and the other is called quasienveloped. The two populations of virions can be distinguished by ultracentrifugation on sucrose density gradients. Unenveloped virions, which are abundant in feces, have a buoyant density of 1.27–1.28 g/mL. These virions are readily neutralized by sera from recovered patients. In contrast, HEV particles from serum have a lower buoyant density of 1.15–1.16, suggesting that the particles contain lipid. They are also resistant to neutralization by anti-HEV antibodies. The lipid-containing particles seem to bud from intracellular membranes, and genetic studies have shown a requirement for ORF3 protein to generate these particles. Mutation of ORF3 results in the production of only naked particles. The term quasienveloped was coined to described the lipid-containing HEV particles (and similar particles identified for hepatitis A virus), because the lipid coat contains no virally encoded envelope proteins. Even though ORF3 protein is required for production of quasienveloped HEV virions, it has not been detected on their surface.

Disease

There are multiple genotypes of HEV (HEV-1 to HEV-8) and their distributions differs (Fig. 14.3). HEV-1

and -2 are strictly human viruses that cause large waterborne outbreaks via contaminated drinking water in endemic regions of South and Southeast Asia, Africa, and Mexico. HEV-infected mothers can transmit the virus to the fetus and transmission can occur via blood transfusion. HEV-3 and -4 infect humans and a variety of animals. They cause sporadic cases in humans and transmission routes include consumption of contaminated shellfish, raw or undercooked meat, contact with infected animals, environmental contamination by animal manure run-off, and via transfusion of blood products. HEV infection is very common among swine (based on serosurveys) but does not appear to cause any disease. In the United States HEV is most often associated with consumption of uncooked or undercooked pork or deer meat.

The World Health Organization (WHO) estimates that there are 20 million HEV infections leading to 3 million cases of disease yearly. In 2015 WHO estimated deaths caused by HEV infection at about 44,000. Symptomatic infection appears to be most common in young adults aged 15–40 years. Studies from endemic areas have shown excess mortality in pregnant women (ranging from 20% to 30% in the third trimester). High mortality among pregnant women seems to be restricted to genotypes 1 and 2. The reason for increased mortality is unclear and continues to be debated. While most cases of HEV are acute, chronic hepatitis may develop in high-risk groups such as immunocompromised individuals. A vaccine to prevent HEV infection has been developed and is licensed in China, but is not available elsewhere.

In this chapter we learned that:

- Members of the family *Hepeviridae* are positive-strand RNA viruses.

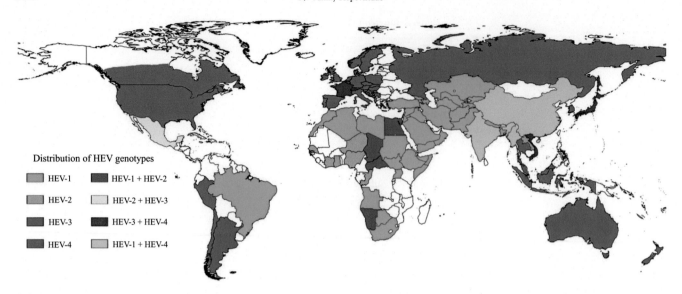

FIGURE 14.3   World-wide distribution of hepatitis E virus (HEV) by genotype (Pallerla et al., 2020). Source: *From Pallerla, S.R., Harms, D., Johne, R., Todt, D., Steinmann, E., Schemmerer, M., et al., 2020. Hepatitis E virus infection: circulation, molecular epidemiology, and impact on global health. Pathogens 9, 856. https://doi.org/10.3390/pathogens9100856.*

- The HEV genome is capped and polyadenylated and contains three ORFs. The NS proteins are translated from genome-length mRNA. A subgenomic mRNA is used to produce the capsid protein.
- Unenveloped virions are shed in feces, but quasienveloped particles are present in serum. The enveloped particles resist antibody neutralization.
- In developed countries transmission is largely zoonotic and sporadic. In resource poor countries, waterborne epidemics may affect thousands.
- Disease is usually acute and self-limiting but severe disease can occur in pregnant women.

- Many aspects of HEV replication and pathogenesis are unknown, due to lack of tractable experimental systems.

References

Berto, A., Grierson, S., Hakze-van der Honing, R., Martelli, F., Johne, R., Reetz, J., et al., 2013. Hepatitis E virus in pork liver sausage, France. Emerg. Infect. Dis. 19, 264. Available from: https://doi.org/10.3201/eid1902.121255.

Pallerla, S.R., Harms, D., Johne, R., Todt, D., Steinmann, E., Schemmerer, M., et al., 2020. Hepatitis E virus infection: circulation, molecular epidemiology, and impact on global health. Pathogens 9, 856. Available from: https://doi.org/10.3390/pathogens9100856.

15

Family Astroviridae

After reading this chapter, you should be able to discuss the following:

- What are the major structural and replicative features of astroviruses?
- What disease is most frequently associated with astrovirus infection?
- How are human astroviruses (HAstVs) most often transmitted?
- How is the astrovirus capsid protein processed to produce the proteins found in the mature virion?

Astroviruses are unsegmented, positive-sense RNA viruses with genomes ranging from 6.2 to 7.7 kb. They have a viral genome-linked protein (VpG) at the 5′ end and a poly (A) tail. The small, unenveloped viruses have spikes that project about 41 nm from the surface of the capsid, giving them a star-like appearance (Fig. 15.1 and Box 15.1).

Astroviruses (from the Greek, astron meaning star) were discovered in 1975, in association with an outbreak of diarrhea in humans. Since that time they have been isolated from many other mammals including pigs, cats, minks, dogs, rats, bats, calves, sheep, and deer as well as from marine mammals such as sea lions and dolphins. Astroviruses have also been isolated from birds; they can cause significant disease in turkeys, ducks, and chickens. There are two named genera (Box 15.2). Human AstVs (HAstV) cause gastroenteritis in children and adults. Symptoms last 3—4 days and include diarrhea, nausea, vomiting, fever, malaise, and abdominal pain. For the most part, disease is self-limiting. In rare cases among humans, astroviruses may be associated with neurologic disease. Several animal astroviruses are highly neurotropic.

Genome organization

Astrovirus genomes are unsegmented positive-strand RNA; genomes are linked to a small VpG protein. Genomes have three long overlapping reading frames (ORFs) that encode polyproteins (Fig. 15.2). There are short untranslated regions at the 5′ and 3′ ends of the genome which is polyadenylated. Similar to other positive-strand RNA viruses (e.g., togaviruses and coronaviruses), two ORFs, 1a and 1b are found in the 5′ half of the genome. The first two ORFs encode nonstructural proteins (NSPs) that include proteases, membrane-associated proteins, an NTP-binding protein, and the RNA-dependent RNA polymerase (RdRp). Synthesis of the complete ORF1a, 1b polyprotein product likely requires a ribosomal frame-shift. The structural proteins are encoded from ORF2 and include the capsid protein and a spike protein, produced from a polyprotein precursor by proteolytic cleavage.

Virion morphology

Astrovirus particles have icosahedral symmetry. The capsid protein is encoded from ORF2, expressed from a subgenomic mRNA. The capsid precursor

© 2023 Elsevier Inc. All rights reserved.

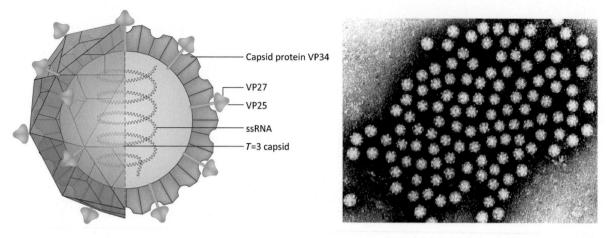

FIGURE 15.1   Virion structure (left) and electron micrograph (Content provider: Graham Beards at English Wikipedia).

BOX 15.1

General characteristics

Astrovirus genomes are single stranded, unsegmented positive-strand RNA ~6.8–7.9 kb. The genome is linked to VpG protein at the 5' end and has a poly (A) tail. Genomes have three overlapping ORFs that encode polyproteins. The first two ORFs produce nonstructural proteins (NSPs) that are cleaved by viral proteases. A third ORF encodes the capsid precursor.

The capsid precursor is cleaved by host proteases. Two mRNAs are present in infected cells (full-length and a single subgenomic mRNA). Replication is cytoplasmic.

Virions (~40–45 nm in diameter) are unenveloped and have spike-like projections at the vertices.

BOX 15.2

Taxonomy

Family *Astroviridae*
Genus *Avastrovirus* (three numbered species from birds)
Astrovirus 1 (turkey)
Astrovirus 2 (chicken)
Astrovirus 3 (duck)
Genus *Mamastrovirus*
Mamastrovirus 1, 6, 8, 9 (humans)

Mamastrovirus 2 (feline)
Mamastrovirus 3 (porcine)
Mamastroviruses 4, 11 (sea lion)
Mamastrovirus 5 (canine)
Mamastrovirus 7 (bottlenose dolphin)
Mamastrovirus 10 (mink)
Mamastroviruses 12, 14–19 (bat)
Mamastrovirus 13 (sheep).

---

undergoes multiple cleavages. The full-length precursor (VP90) is cleaved by cellular proteases (caspases) to generate the VP70 product. If capsase inhibitors are added to infected cells, release of virions is blocked. However, VP70-containing capsids are likely noninfectious until VP70 is further processed by trypsin-like proteases to generate mature virions containing three polypeptides (VP25, VP27, and VP34). Addition of trypsin to cultured cells produces infectious particles and similar enzymes are present in the intestine during a natural infection. The capsid core is formed by VP34 while VP25 and VP27 form the spikes on the virion surface. Binding sites for neutralizing antibodies map to VP25 and VP27.

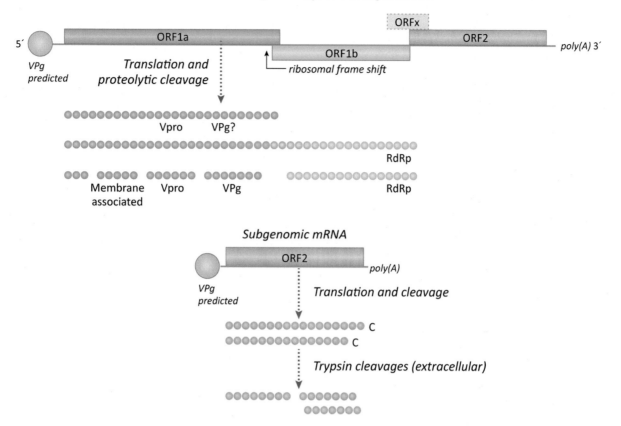

FIGURE 15.2   Astrovirus genome organization.

General replication strategy

Due to the lack of robust cell culture systems, many details of astrovirus replication have not been confirmed. However, the overall replication cycle of astroviruses is predicted to be quite similar to that of other positive-strand RNA viruses. Evidence suggests uptake of virions by endocytosis into clathrin-coated pits followed by genome release from acidified endosomes. Replication is cytoplasmic and there is evidence for the requirement for VpG in genome replication (pretreatment of genomes with proteases prior to their introduction into permissive cells greatly reduces infectivity). Replication is usually nonlytic.

Diseases caused by astroviruses

HAstVs are thought to be the second or third most common cause of viral diarrhea in young children. They have also been isolated from sporadic outbreaks of acute gastroenteritis in adults. A few studies have associated astroviruses with chronic diarrhea in immunocompromised children and adults. HAstVs are found worldwide.

The main mode of human astrovirus transmission is by contaminated food (including bivalve mollusks) and water, although direct person-to-person transmission has also been documented.

There are multiple serotypes of human astrovirus and the main target cells are enterocytes (epithelial cells of the intestinal tract). Astrovirus infection does not notably alter intestinal architecture and does not induce inflammation. It has been proposed that pathogenesis may be caused by apoptotic death of infected epithelial cells. Symptomatic infections are most common in children younger than 2 years of age, and it is estimated that 5%—9% of cases of viral diarrhea in young children are caused by astroviruses. In the US population the presence of antiastrovirus antibodies is very high, indicating that most infections are asymptomatic or very mild. Outbreaks of astrovirus-associated diarrhea have been reported among elderly patients and military recruits. Food-borne outbreaks, affecting thousands of individuals, have occurred in Japan. In temperate climates astrovirus infection is highest during winter months while in tropical regions prevalence is highest during the rainy season.

Rarely, astroviruses have been isolated from organs other than the gastrointestinal tract. They have been isolated from a few children with CNS disease although disease causation has not been confirmed. However, there is a good example of CNS-associated astrovirus infection in an animal model. Shaking mink syndrome is a neurologic disorder of farmed minks. Outbreaks have occurred in Denmark, Sweden, and Finland. Examination of diseased mink revealed brain lesions (nonsuppurative encephalomyelitis) and experimental infection of brain homogenates into healthy mink recapitulated the disease, a result highly suggestive of an infectious agent. Attempts to culture an infectious agent were unsuccessful but the agent was finally identified using metagenomics. Nucleic acid sequences were obtained from brain material of diseased and healthy mink. Comparisons revealed an astrovirus genome associated only with diseased mink. The CNS-associated astrovirus shares about 80% nucleotide identity with an enteric mink astrovirus. Astrovirus-associated neurologic disease has also been documented in pigs, cattle, sheep, and alpaca.

In this chapter we learned that:

- Astroviruses are unenveloped, positive-strand RNA viruses. Their name derives from their star-shaped virions.
- Astroviruses were first identified in association with outbreaks of gastroenteritis.
- HAstVs are most often transmitted by the fecal oral route, through contaminated food and water.
- The astrovirus capsid protein is processed by host proteases. One cleavage is mediated by intracellular caspases and others by extracellular trypsin-like proteases.

16

Family *Flaviviridae*

After reading this chapter, you should be able to answer the following questions:

- What structural features are shared by all members of the family *Flaviviridae*?
- What are the characteristic features of genome replication and translation initiation?
- At what site in the cell do members of the family *Flaviviridae* replicate?
- What are some of the major human and animal diseases caused by members of the family *Flaviviridae*?
- What is the basis for heptotropism of Hepatitis C virus (HCV)?

The *Flaviviridae* family is a large family of unsegmented positive-strand RNA viruses (Box 16.1). Virions are enveloped but do not contain the prominent spikes often seen on enveloped virions (Fig. 16.1). The family is quite large and contains many important human pathogens including yellow fever virus (YFV), dengue fever virus (DFV), Zika virus (ZIKV), and HCV. There are four genera within the family (Box 16.2):

Genus *Flavivirus*: Most members of the genus *Flavivirus* are transmitted by insect vectors; thus they are also often referred to as *arboviruses* (arthropod-borne means "transmitted by insects"). There are close to 70 identified members in the genus *Flavivirus*, and several are significant human pathogens [YFV, DFV, ZIKV and encephalitis viruses such as West Nile virus (WNV)].

Genus *Hepacivirus*: The type strain of the genus *Hepacivirus* is HCV, a blood-borne hepatitis virus. HCV constitutes a large group of related viruses that can be subdivided based on genome sequence. Even within an infected individual, viral genomes are not identical. Instead "swarms" of genetically related viruses are found. Thus HCV exists as a quasispecies. It is likely that genetic variability contributes to the ability of HCV to cause chronic infection, persisting for many decades in the infected host. There are also members of the genus *Hepacivirus* that infect other mammals; they have been recovered from dogs and horses where they do not appear to cause major disease outbreaks, but may cause sporadic or subclinical hepatitis. Both animal and human hepaciviruses are difficult to culture, but because they are important pathogens, details of their replication cycles have been elucidated using a variety of model systems.

Genus *Pestivirus*: There are three pestiviruses of veterinary significance: bovine viral diarrhea virus (BVDV), border disease virus (BDV) of sheep and goats, and classical swine fever virus (CSFV). These three pestiviruses are routinely transmitted by contact, but also by vertical transmission to the fetus. Vertical transmission may result in persistently infected offspring that are viremic, antibody negative, and constantly shed virus. Thus these viruses can be difficult to eradicate from livestock herds (CSFV has been eradicated from the United States).

© 2023 Elsevier Inc. All rights reserved.

BOX 16.1

General characteristics

Flavivirus genomes are positive-strand RNA, unsegmented, and range in size from 9.5–12.5 kb. Some have 5′ cap structures and the 3′ ends of the nonpolyadenylated genomes form hairpins. The genome serves as the sole mRNA. Viral proteins (structural and nonstructural) are synthesized as part of a large polyprotein precursor. Translation is interesting as some of the polyprotein is targeted to ER while other regions remain cytosolic. Replication is cytoplasmic.

A single capsid (C) protein forms the capsid. Enveloped virions contain two or three membrane-associated glycoproteins. Virions have a smooth, golf ball-type appearance due to the arrangement of the surface glycoproteins that lie flat on the envelope.

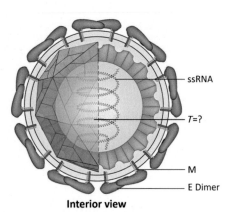
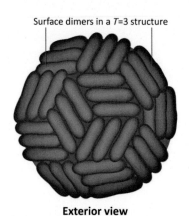
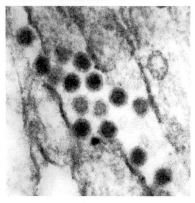

FIGURE 16.1 Virion structure in cross section (right). Surface view of the virion (middle). Negative stain of St. Louis encephalitis virus within a mosquito salivary gland tissue sample (left). CDC Public Health Image Library image ID# 10228. Content providers Dr. Fred Murphy and Sylvia Whitfield.

BOX 16.2

Taxonomy

Family *Flaviviridae*
Genus *Flavivirus*
Species: Over 50, most arthropod vectored. Viruses mentioned in this chapter include: dengue fever virus, yellow fever virus, Zika virus, and West Nile virus.
Genus *Hepacivirus*

Species *Hepacivirus A−N*. Hepatitis C virus is a member of the species *Hepacivirus C*.
Genus *Pestivirus*
Species: *Pestivirus A−K*. Viruses mentioned in this chapter include: bovine viral diarrhea virus (BVDV), border disease virus (BDV, infects sheep and goats), and classical swine fever virus (CSFV).

Genus *Pegivirus*: Pegiviruses cause persistent infections and have been detected in humans and animals. Relatively little is known about their role in disease.

Genome structure/organization

Members of the family *Flaviviridae* have positive sense, single-stranded RNA genomes. Genome sizes range from ~9000 to 12,500 nt (Box 16.1). Members of the genus *Flavivirus* have a methyl-G-cap protecting the 5′ end of their genomes. However, pestiviruses and hepacivirus genomes are not capped. The genome 3′ ends of all members of the family *Flaviviridae* have a hairpin structure that stabilizes the genome and is probably required for genome replication; genomes are not polyadenylated. Genomes contain 5′ and 3′ non-coding regions (NCRs) that precede and follow the single long open-reading frame (ORF) (Fig. 16.2). 5′ NCRs are ~300–400 nt. The 5′ NCR of HCV contains an internal ribosome entry site (IRES). Pestiviruses also have stem/loop structures in their 5′ NCR.

Between the 5′ and 3′ NCRs of all members of the family *Flaviviridae* is a single long ORF encoding a polyprotein. A single capsid (C) and two or three envelope proteins are encoded at the 5′-end of the ORF while several nonstructural proteins (NSPs) including the RNA-dependent RNA polymerase (RdRp, the NSP5B protein) are = encoded at the 3′ end of the ORF.

the capsid. Two or three glycoproteins (Fig. 16.3) are associated with the envelope. Flaviviruses have a unique morphology for enveloped viruses as the glycoproteins lie flat on the surface of the mature virion, giving it a smooth golf ball-like appearance (Fig. 16.1). The structure of the envelope glycoproteins is pH sensitive. They form spikes at low pH and the spike form mediates membrane fusion.

Virion structure

Virions are enveloped with a diameter of about 30 nm. A single capsid protein (C) assembles to form

General overview of replication

After attachment (various receptors that differ by virus), virions are endocytosed and fusion between

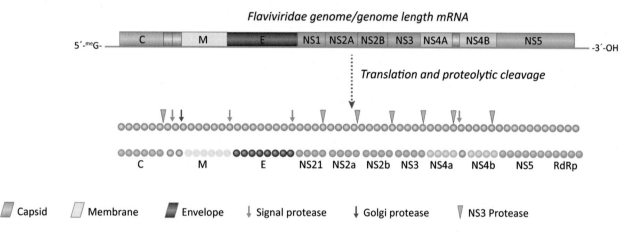

FIGURE 16.2   Genome and gene expression strategy (genus *Flavivirus*). The genome contains a single long ORF. No subgenomic mRNAs are expressed. NS5 is the RdRp.

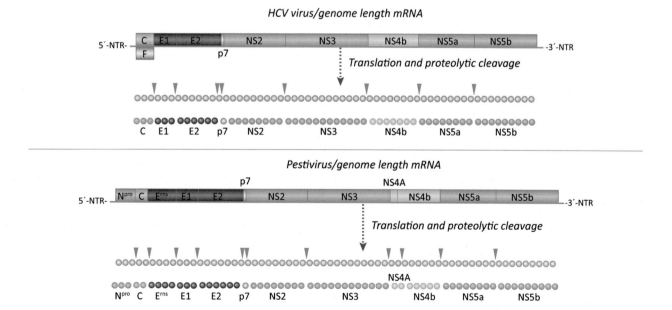

FIGURE 16.3   Comparison of hepacivirus and pestivirus genomes. Note that HCV encodes two envelope (E) proteins; pestiviruses encode three E proteins. Pestiviruses encode a protease (N^Pro) upstream of the structural proteins.

viral envelope and cell membrane occurs at low pH (Fig. 16.4). The low pH environment triggers viral envelope protein rearrangement so that the envelope (E) proteins change conformation from a "flat" to an unfolded (spike) structure, driving their insertion into the endosomal membrane. Membrane fusion releases the infecting genome, which is efficiently translated.

The process of translation initiation varies by genus. The hepaciviruses and pestiviruses have an IRES in the 5' NCR while the members of the genus *Flavivirus* use cap-directed translation. The protein expression strategy of the members of the family *Flaviviridae* is to generate the majority of their proteins from a single polyprotein precursor, by a series of stepwise proteolytic cleavages. This is very similar to the protein expression strategy of the picornaviruses. Some products of the polyprotein remain cytosolic (e.g., the capsid protein and several NSPs) while others (e.g., the envelope proteins) cross the ER membrane during synthesis. Cytosolic regions of the precursor polyprotein are cleaved by virally encoded proteases while cleavages in the ER are accomplished with host cell proteases. Viral proteins residing in the ER lumen traffic through the ER and Golgi where they are glycosylated.

After translation the infecting genome is replicated. Genome replication occurs on cell membrane scaffolds and requires the viral NSPs. Viral genomes associate with capsid proteins in the cytoplasm. Capsid proteins also interact with portions of the viral glycoproteins displayed on the cytoplasmic face of the ER. Viral assembly occurs by budding of the capsid into the lumen of the ER. Virions are released when the ER vesicles fuse with the plasma membrane, releasing their cargo outside the cell, a process called *exocytosis*. It is interesting to note that immature virions have spike-like E proteins that change conformation to lie flat on the surface of the mature virion at neutral pH. This shows that pH controls the conformation of the E proteins, and that E protein conformation is reversible.

Protein products

The overall genome organization of members of the family *Flaviviridae* is conserved such that the structural

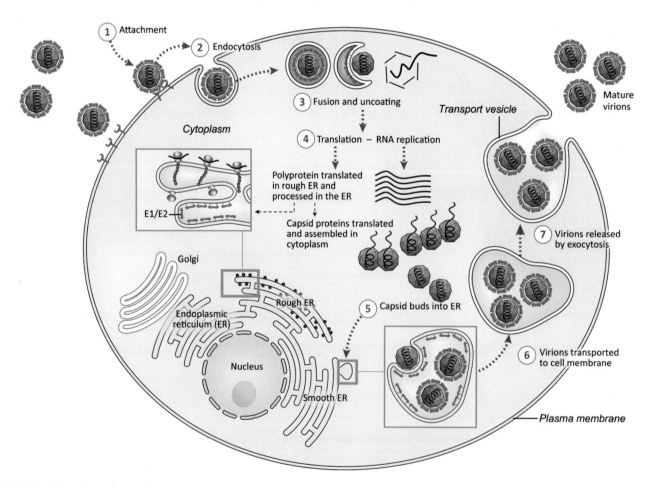

FIGURE 16.4   Flavivirus replication cycle.

proteins (capsid and envelope) are encoded at the 5′ end of the single long ORF. The RdRp (NSP5B) is encoded at the very 3′ end of the ORF. However, the pestiviruses are notably distinct in that they encode a protease domain (N^{pro}) upstream of C. N^{pro} cleaves the polyprotein to generate the correct N-terminus of the C protein (Fig. 16.3).

All members of the family *Flaviviridae* encode a single capsid (C) protein. In addition, there are two or three envelope glycoproteins. Members of the genus *Flavivirus* encode prM and E, hepaciviruses encode E1 and E2 while pestiviruses encode Erns, E1, and E2. Erns is an unusual glycoprotein that is found in both membrane anchored and secreted forms. Erns is a ribonuclease (RNase) with a preference for single-stranded RNA.

The NSPs of members of the family *Flaviviridae* include the RdRp (NS5B), proteases, helicase, and others involved in genome replication and subversion of immune responses. For our purposes, the hepacivirus NSPs provide a well-studied example (Table 16.1).

Diseases

Hepatitis C

About 3% of the world's population is infected with HCV. While blood screening has virtually eliminated blood transfusion-associated HCV infection, virus continues to be transmitted by other routes such as illegal injection drug use. Transmission is aided by the silent nature of many HCV infections. HCV has always been, and continues to be, a challenging virus to study as most isolates are impossible to culture and there is no inexpensive small animal model (Box 16.3).

The HCV pandemic is relatively recent, with worldwide increases in HCV infections paralleling the increase in manufacture and use of syringes for medical purposes (and illegal injection drug use). In some cases medical injections resulted in large numbers of infected individuals because needles and syringes were reused without proper sterilization. Such was the case in Egypt. In contrast, in the United States an epidemic of illegal injection drug use from the 1950s to the 1980s is believed to have caused many HCV infections. It has been estimated that there were probably less than 500,000 persons with chronic HCV infection in the United States in the early 1950s; by the mid-1990s an estimated 5 million infections had occurred. In the United States today, serosurveys reveal a high rate of HCV infection among people born between 1945 and 1964. Rates of new infections have decreased since 1980s, largely due to development of blood-screening tests (Box 16.4).

Humans are the only known reservoir of HCV. Acute HCV infection is usually asymptomatic with only a few patients presenting with typical symptoms of acute viral hepatitis. An inapparent acute infection becomes an inapparent chronic infection in ~70% of cases. However, chronic infection is accompanied by an increased likelihood of developing serious medical problems such as fatty deposits in the liver, liver disease, and hepatocellular carcinoma. The large numbers of chronically infected individuals have encouraged development of novel antiviral drugs to combat HCV infection.

Currently the US CDC recommends that one-time testing of all adults for HCV and testing for pregnant women during each pregnancy. High risk individuals (i.e., persons who use injectable drugs) should be tested regularly. One impetus of these testing recommendations is the availability of antiviral drugs that is highly effective against HCV. Currently licensed drugs include compounds that target the zinc binding metalloprotein NS5A (ledipasvir, velpatasvir, pibrentasvir), the RdRp NS5B (sofosbuvir), and the protease NS3/4 A (glecaprevir). These compounds are used in various combinations and treatment options are dictated by factors that include the treatment history of the patient,

TABLE 16.1   NSPs of hepatitis C virus (HCV) and their functions.

NSP	Function
P7	A 63 amino acid peptide that oligomerizes in membranes to form a functional viroporin, a cation-selective ion channel that allows leakage of protons from acidic vesicular compartments. Required for virion production
NS2	A cysteine protease
NS3	The N-terminal domain is a serine protease and the C-terminal domain is an RNA helicase-NTPase
NS4A	A cofactor for the NS3 serine protease
NS4B	Recruits and rearranges ER membranes forming a membranous structure that serves as the site of viral genome replication
NS5A	A zinc-binding metalloprotein that binds the viral RNA and various host factors
NS5B	The RdRp

BOX 16.3

Discovery of Hepatitis C virus

Michael Houghton, Qui-Lim Choo, George Kuo, and Daniel W. Bradley codiscovered Hepatitis C virus in 1989 (Choo et al., 1989). Dr. Houghton reviewed the details of this discovery in a 2009 manuscript (Houghton, 2009). Dr. Houghton's paper is an accessible and well-written manuscript that may be of interest to anyone who wishes to truly understand how many scientific "breakthroughs" are achieved. For the less adventuresome, a brief summary of this important discovery is presented here.

Hepatitis is liver disease and can be caused by several different viruses. By the 1970s it was understood that there were many cases of posttransfusion hepatitis that were not caused by either the hepatitis A virus or the hepatitis B virus. The so-called non-A, non-B hepatitis (NANBH) agent appeared to be a virus that could not be replicated in cultured cells. The NANBH agent was transmissible to chimpanzees, a characteristic that eventually led to the solution of this puzzle of a virus.

As described in a breakthrough study published in the journal *Science* in 1989, Dr. Houghton's group cloned random pieces of nucleic acid present in plasma of an experimentally infected chimpanzee. They started with a highly infectious plasma sample and centrifuged it at very high speed to recover any/all virus particles or nucleic acids that might be present. They then used the enzyme reverse transcriptase to generate copy DNA (cDNA) from any nucleic acid present in the sample. These short pieces of cDNA were cloned into a bacteriophage that had been engineered to produce protein products from the cloned cDNA. (A bacteriophage is a virus that infects bacteria.) The recombinant bacteriophages were grown on lawns of susceptible *Escherichia*

coli cells to produce millions of plaques. The proteins and nucleic acids from these plaques were transferred to membranes, followed by incubation with immune sera from infected chimpanzees. The researchers were looking for plaques that would bind antibodies present in sera from infected chimpanzees but not present in sera from uninfected controls. The researchers screened over a million plaques before identifying a single clone (called 5-1-1). This particular bacteriophage encoded a small peptide that bound to antibodies present *only* in chimps infected with the NANBH agent.

The gene fragment was sequenced and was recognized to be a piece of a virus. This first piece of viral genome was then used to identify additional overlapping pieces using nucleic acid hybridization technology. In this manner the researchers pieced together almost the entire genome of the agent we know today as HCV. It was soon possible to inject the genome into chimpanzees and recover infectious HCV to verify their findings. Obviously the inability to culture the virus and the need to use chimpanzees as experimental animals made for a very challenging system. However, the high toll of HCV (millions of blood-borne infections) and the need to develop a blood-screening test provided ample incentive to study this intractable virus.

It is highly unlikely that the studies described above would be performed today. Instead nucleic acid sequencing techniques would be used to generate and analyze all nucleic acids (host, viral, and bacterial) present in any sample and these would be compared to all known sequences in databases. Nucleic acids present in experimentally infected and uninfected animals would be compared, with hopes of finding unique viral sequences in infected animals.

coinfection with other viruses, and genotype of the infecting virus. One of the first treatments for HCV infection was a combination of recombinant interferon alpha and ribavirin. However, that treatment is not well tolerated, is only partially successful, and is not recommended when other treatments are available.

Diseases caused by members of the genus *Flavivirus*

There are many human and animal pathogens within the genus *Flavivirus*. Most are transmitted by insects. YFV, DFV, WNV, and ZIKV virus are transmitted by mosquitos. YFV and DFV are major human pathogens

while WNV is primarily an avian virus that only occasionally infects mammals. ZIKV caused a large-scale outbreak in 2015 in Brazil. YFV, DFV, and ZIKV are transmitted between human and mosquito hosts in a "simple transmission cycle." YFV infects both human and nonhuman primates (Fig. 16.5) while DFV is strictly a human pathogen (Fig. 16.6). The mosquito hosts of YFV, DFV, and ZIKV are members of the genus *Aedes*. These viruses actively replicate in their mosquito hosts, a requirement for their eventual retransmission to humans.

Yellow fever virus

YFV has two transmission cycles (Fig. 16.5), the so-called urban transmission cycle and a sylvatic (forest)

BOX 16.4

HCV: why the liver?

HCV is very highly tropic for the human liver, replicating only in hepatocytes. Most isolates of HCV will not even replicate in cultured hepatocyte cell lines. What is so special about the liver?

First entry of HCV requires several proteins, some of which are preferentially expressed on hepatocytes. Adaptation of hepatocytes to growth in culture does not preserve liver architecture or the expression of the full complement of necessary receptor proteins. Once HCV gains entry into a cell, its replication also requires the presence of a micro-RNA (miRNA) that is only expressed in liver cells. miRNAs are small ($\sim$22 nt) noncoding RNAs that are important regulators of protein synthesis. Their regulatory activity is based on their ability to base-pair with complementary sequences in

mRNAs. The miRNA required by HCV is called miR-122. miR-122 is a liver-specific miRNA. In the adult human liver miR-122 constitutes $\sim$70% of total miRNA; it is clearly very important for normal functioning of liver cells.

In the case of HCV replication, miR-122 base pairs to two complementary sites at the 5′ end of the viral genome. A protein called Argonaute-2 binds to miR-122 generating a double-stranded RNA/protein complex. Experiments have shown that this complex *protects* the 5′ end of the viral genome from degradation by a ubiquitous host cell *exo*-RNase that normally functions to degrade decapped miRNAs. Recall that the HCV genome is not capped. Thus in the absence of miR-122 HCV RNAs are not stable.

transmission cycle. Nonhuman primate hosts maintain YFV in the sylvatic cycle, maintaining a virus reservoir that can move back into urban environments. YFV can cause severe illness and death; up to 50% of severely affected persons will die without treatment. Worldwide, the numbers of YF cases are $\sim$200,000 per year with about $\sim$30,000 deaths. Vaccination is the single most important preventive measure against yellow fever. There is a single serotype of YFV and the vaccine is safe, affordable, effective, and appears to provide long-term protection. The YFV vaccine is a live-attenuated strain called 17D. The same vaccine has been used since the 1940s. At least 18 countries require proof of YFV vaccination as a prerequisite for entry. Other countries require vaccination depending on the country of origin of the traveler.

Dengue fever virus

DFV is transmitted between humans and mosquitos in a simple transmission cycle. Before World War II, DFV was restricted to two small areas in Southeast Asia. However, during the war, cargo ships carried infected *Aedes* mosquitos throughout the world. Since the 1960s the number of countries reporting DFV and the numbers of cases of DFV have skyrocketed, from less than 1000 cases reported from 1955 to 1959 to close to a million reported cases between 2000 and 2007. World Health Organization data show an eightfold increase in reported cases of dengue fever from 2000 to 2019. Modeling puts the number of infections at $\sim$300 million per year, of which $\sim$100 million are clinically significant.

Humans are the only mammalian hosts for DFV. Mosquito control in urban areas, including use of screening on houses and eliminating mosquito habitat (standing water), is currently the only way of slowing the spread of DFV. Primary infection with DFV is usually a self-limited febrile illness. A notable symptom is severe bone pain, thus the disease is also called "breakbone fever." More severe illness, dengue hemorrhagic fever and dengue hemorrhagic shock syndrome, are less common but have high mortality.

There are four serotypes of DFV and infection with one does not protect against infection with the others. In fact, infection with a second serotype increases risk of complications such as dengue hemorrhagic fever and dengue hemorrhagic shock syndrome. Why does a second infection with DFV increase the risk of severe disease? As shown in Fig. 16.6, the immune system responds to the primary infection by producing a mixture of neutralizing and binding antibodies. The neutralizing antibodies are serotype-specific, while binding antibodies are not. In fact, binding antibodies can increase infection of macrophages. Recall that macrophages are professional phagocytic cells that display Fc receptors on their surface. Fc is the constant fragment of the IgG molecule (the tail of the Y-shaped IgG molecule). The Fc portion of IgG binds to the Fc receptors on macrophages to facilitate antigen uptake. However, in the case of infection with a second DFV serotype, nonneutralizing or poorly neutralizing antibodies bind to virions and antibody bound virions can more easily infect macrophages. As macrophages are a source of proinflammatory cytokines, it is postulated that

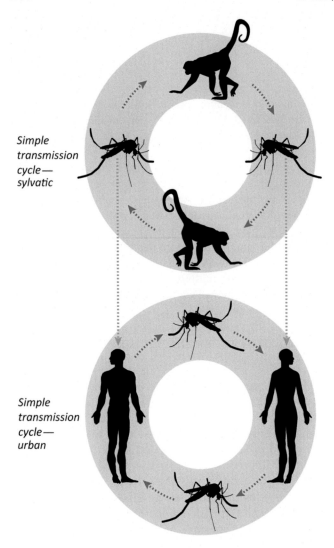

Simple transmission cycle—sylvatic

Simple transmission cycle—urban

FIGURE 16.5 Yellow fever virus (YFV) transmission cycles. Human and nonhuman primates are the only vertebrate hosts for YFV. Vaccination can break human to human transmission cycles but the sylvatic (forest) cycle remains intact allowing for possible reintroduction into humans.

increased levels of infection and death lead to dengue hemorrhagic fever and dengue shock syndrome.

Vaccine development for DFV has been slow because of the need to simultaneously generate protective responses against four distinct serotypes, the propensity for disease enhancement and the lack of an animal model for vaccine testing. There is currently one licensed vaccine for dengue fever virus: a tetravalent, recombinant vaccine produced by Sanofi Pasteur, sold under the brand name Dengvaxia. The vaccine was developed by creating chimeric flaviviruses containing the DFV preM and E protein coding regions on the backbone of the attenuated YFV vaccine strain, 17D. However, there are limitations to the use of Dengvaxia. It is licensed only for use in *seropositive* individuals. Data suggest that seronegative individuals are not fully protected by the vaccine and risk vaccine enhanced disease in the event of a natural infection. It is possible that DFV NSPs, which are not included in the vaccine, are required for full immunity. Currently this is only a hypothesis, based on the observation that vaccination of seropositive persons results in strong protection against all four serotypes. Additional vaccines are in development.

Zika virus

ZIKV infects humans and nonhuman primates and is spread by *Aedes* sp. mosquitos. The most common symptoms of ZIKV disease are fever, rash, joint pain, conjunctivitis, malaise, and headache. The illness is usually mild with symptoms lasting for several days to a week. Many infections are probably asymptomatic. Unfortunately, ZIKV infection during pregnancy can cause a serious birth defect called microcephaly. The head of a microcephalic baby is smaller than normal; the brain is often small and not properly developed. There is only one known serotype of ZIKV and once a person has been infected, they are probably protected from future infections.

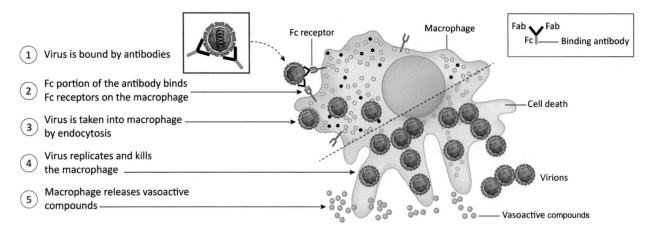

① Virus is bound by antibodies

② Fc portion of the antibody binds Fc receptors on the macrophage

③ Virus is taken into macrophage by endocytosis

④ Virus replicates and kills the macrophage

⑤ Macrophage releases vasoactive compounds

Fc receptor

Macrophage

Fab Fab
Fc ——— Binding antibody

Cell death

Virions

Vasoactive compounds

FIGURE 16.6 A model for immune enhancement of dengue fever virus (DFV) infection. In this model, nonneutralizing or poorly neutralizing antibodies bind to virions and enhance infection of macrophages.

ZIKV was first discovered in Uganda 1947 in monkeys and was detected in humans in 1952. Prior to 2015, cases were reported in tropical Africa, Southeast Asia, and the Pacific Islands but because the symptoms of ZIKV disease are similar to those of many other diseases, many cases may not have been recognized. The first large outbreak of disease caused by ZIKV was reported from the Island of Yap (Federated States of Micronesia) in 2007. ZIKV infection was associated with only mild illness prior to a large French Polynesian outbreak in 2013 and 2014, where severe neurologic disease was reported. In 2015 the first confirmed human case of ZIKV infection occurred in Brazil. By early 2016, the WHO declared ZIKV a worldwide public health emergency. A driving factor in declaring ZIKV a public health emergency was the association between ZIKV infection and microcephaly.

ZIKV is primarily transmitted to people through the bite of an infected *Aedes* mosquito. However, sexual transmission of ZIKV has been documented, and given the link to birth defects it is imperative that we learn more about this mode of transmission. Currently there is no vaccine for ZIKV. Prevention includes protection against mosquito bites. In areas where ZIKV is prevalent, pregnant women should be tested for possible exposure, should practice safe sex (use of condoms), and avoid mosquito bites. People who have been traveling to ZIKV endemic areas should be aware that they may be infected, even if they have not experienced symptoms. Men should adopt safe sex practices and couples planning a pregnancy should seek out medical advice.

West Nile virus

The natural hosts for WNV are birds (largely Passeriformes). Crows, robins, and blue jays are highly susceptible to lethal infection. To date, scientists have identified more than 138 bird species that can be infected, and more than 43 mosquito species that can transmit WNV. WNV was first reported in the United States in 1999. It was identified as the cause of bird deaths at the Bronx zoo in New York City. Within 5 years the virus spread to the West Coast of the United States, north to Canada, and southward to the Caribbean Islands and Latin America. WNV continues to adversely affect North American bird populations. In December 2013 WNV killed more than two dozens of bald eagles in Utah. The deaths of ~20,000 water birds around the Great Salt Lake in November 2013 was also attributed to WNV.

Humans are not natural hosts for WNV (Fig. 16.7) and approximately 80% of WNV infections in humans are subclinical. The most common clinical manifestation is West Nile fever. Symptoms include fever, headaches, fatigue, muscle pain or aches, malaise, nausea, anorexia, vomiting, myalgia, and rash. More serious

Complex transmission cycle

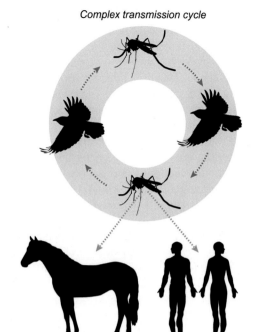

FIGURE 16.7 West Nile virus (WNV) transmission. The normal vertebrate hosts for WNV are birds. However, mosquitos may transmit virus to other mammals such as horses and humans which are dead end hosts.

are WNV infections that involve the central nervous system (CNS). These are rare, occurring in only ~1% of infected individuals. CNS complications include encephalitis, meningitis, meningoencephalitis, and WNV poliomyelitis (flaccid paralysis due to spinal cord inflammation). Elderly, the very young or immunosuppressed persons are most susceptible. Currently, no human vaccine against WNV infection is available. Rates of WNV infection can be reduced by adequate mosquito control, personal protective measures such as mosquito repellent, window screens, long sleeves, and pants. Vaccines approved for use in horses are available and should be used prior to mosquito season.

Genus *Pestivirus*

Three pestiviruses cause major diseases of livestock: BVDV (cattle), BDV (sheep), and CSFV (swine). These viruses sometimes cause inapparent, persistent, lifelong infections making it difficult to eliminate them from the environment.

BVDV and the results of persistent infection will be briefly described here. BVDV affects cattle and the clinical signs of BVDV are highly variable, ranging from few to no signs, to severe disease that quickly kills the animal. BVDV is transmitted by contact but can also be transmitted vertically from an infected cow to her calf. Congenital infections may result in resorption of the conceptus, abortion, stillbirth, or live-birth depending on the time of infection. Infected fetuses that survive may be *immunotolerant* (they do not make antiviral antibodies) and are persistently infected for life. These animals are the major source of virus for transmission.

A few persistently infected animals develop a lethal disease, called mucosal disease. Mucosal disease is characterized by fever, leukopenia, bloody diarrhea, loss of appetite, dehydration, erosive lesions of the nares and mouth, and death within a few days of onset of symptoms.

What has changed? What causes mucosal disease? Mucoal disease is caused by *mutated viruses that emerge in the persistently infected animal*. These viruses replicate much more robustly (in both cultured cells and in the infected animal) causing cell death. They are referred to as cytopathic (cp) BVDV. A variety of mutations can change BVDV to a cytopathic type. The mutations are often gene insertions and the inserted genes are of host origin (Fig. 16.8). The common denominator appears to be an increase in the rate of cleavage of the precursor polyprotein, from slow to very rapid. To summarize, BVDV is often of low virulence and can cause persistent infection of immunotolerant calves. This helps to maintain the virus within a herd. Rarely, recombinant viruses arise in persistently infected animals that replicate robustly and cause cytopathic infection. The mutant viruses cause extensive tissue damage leading to rapid death. Therefore these mutant viruses are not maintained long term in a herd. The change from non-cytopathic to cytopathic replication may be as simple as an increase in the relative rate of protein cleavage.

Diseases caused by members of the genus *Pegivirus*

Pegiviruses have been detected in a large number of mammalian species. Within the flavivirus family, they are most closely related to the hepaciviruses. There are two human pegiviruses, human hepegivirus (HPgV, also called hepatitis G or GB virus) and human pegivirus 2 (HPgV-2). Both viruses are transmitted by sexual or parenteral infection and there is no correlation with disease. It is estimated that from 1% to 5% of healthy blood donors in the United States have HPgV viremia. Blood is not screened for HPgV as it is not considered a threat. In fact, some studies have suggested some benefit to HPgV infection, due to the possible beneficial immunomodulatory effects (reducing immune activation) in patients with other viral diseases such as HIV infection, hepatitis B, and Ebola virus disease.

In this chapter we have learned that:

- All members of the family *Flaviviridae* are enveloped, positive-strand RNA viruses and their replication is cytoplasmic.
- Genomes have a single long ORF and individual proteins are generated by proteolytic cleavage.

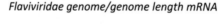

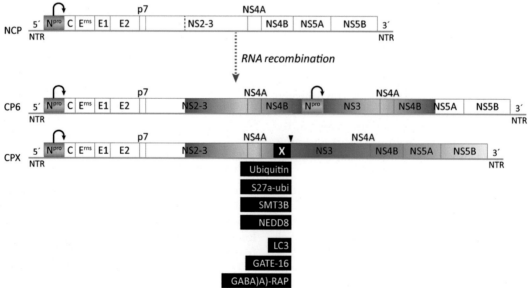

FIGURE 16.8   Examples of genome insertions that increase replication of the pestivirus bovine viral diarrhea virus (BVDV) to generate the cytopathic biotype. BVDV is generally noncytopathic and causes persistent infections. During the lifetime of a persistently infected bovid, the virus may accumulate gene duplications or insertions of host genes to generate a cytopathic biotype with high replicative ability. Increased replication is linked to enhance proteolysis of the precursor polyprotein.

- Some members of the family *Flaviviridae* have capped genomes and express proteins using cap-dependent translation initiation. Others have uncapped genomes and initiate protein synthesis in an IRES-dependent manner.
- HCV (genus *Hepacivirus*) is a significant human pathogen and major cause of blood-borne hepatitis. Most infections are asymptomatic and chronic. HCV infection is a major cause of liver damage, liver failure, and hepatocellular carcinoma.
- Members of the genus *Flavivirus* include human and animal pathogens that are transmitted by insects. They replicate in both animal and insect hosts.

- Members of the genus *Pestivirus* include important pathogens of food animals. They are spread by contact and may cause persistent infections.

References

Choo, Q.L., Kuo, G., Weiner, A.J., Overby, L.R., Bradley, D.W., Houghton, M., 1989. Isolation of a cDNA clone derived from a blood-borne non-A, non-B viral hepatitis genome. Science 244, 359. Available from: https://doi.org/10.1126/science.2523562. PMID: 2523562.

Houghton, M., 2009. The long and winding road leading to the identification of the hepatitis C virus. J. Hepatol. 51, 939. Available from: https://doi.org/10.1016/j.jhep.2009.08.004.

17

Families *Togaviridae* and *Matonaviridae*

After studying this chapter, you should be able to:

- Describe the characteristics of members of the family *Togaviridae*.
- Describe the general strategies used by togaviruses for transcription, translation, and genome replication.
- List some human and animal diseases caused by togaviruses.
- Describe transmission of togaviruses.
- Describe the characteristics of rubella virus.
- Explain the importance of vaccination to protect pregnant women from rubella virus infection.

Before 2019, the family *Togaviridae* included two genera, *Alphavirus* and *Rubivirus*. While the genus *Alphavirus* contained many members, there was only a single rubivirus, rubella virus the etiologic agent of German measles. While similar in genome organization there are important differences in virion structure and transmission modes between alphaviruses and rubella virus, thus in 2019 the family *Matonaviridae* was created, with the sole member being rubella virus. Since 2019, additional members of this family have

been identified. Both virus families are reviewed in this chapter.

Family *Togaviridae*

Togaviruses are enveloped viruses with unsegmented positive-strand RNA genomes (Figs. 17.1 and 17.2). The family contains one genus, *Alphavirus* with over 30 species (Box 17.1). The alphaviruses are arboviruses, transmitted by insects to vertebrate hosts including mammals and birds. Included among alphaviruses are pathogens such as Eastern equine encephalitis virus (EEEV), Western equine encephalitis virus (WEEV), Venezuelan equine encephalitis virus (VEEV), Sindbis virus (SINV), Ross River virus (RRV), Semliki Forest virus (SFV), and chikungunya virus (CHIKV). CHIKV is rapidly emerging as a worldwide human pathogen. The ability to infect mosquitoes is key to the natural transmission cycle of these togaviruses. Mosquitoes are infected by feeding on a viremic vertebrate host and the virus replicates within the salivary glands. Only after

Viruses.
DOI: https://doi.org/10.1016/B978-0-323-90385-1.00006-6

© 2023 Elsevier Inc. All rights reserved.

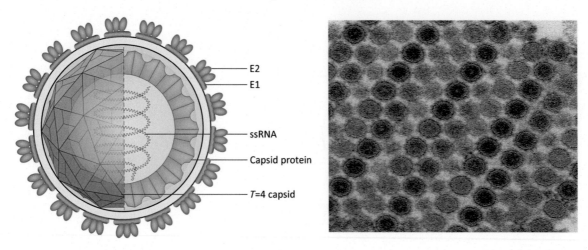

FIGURE 17.1   Virion structure (left) and transmission electron micrograph showing Chikungunya virions (right). Source: *CDC Public Health Image Library, Image #17369.*

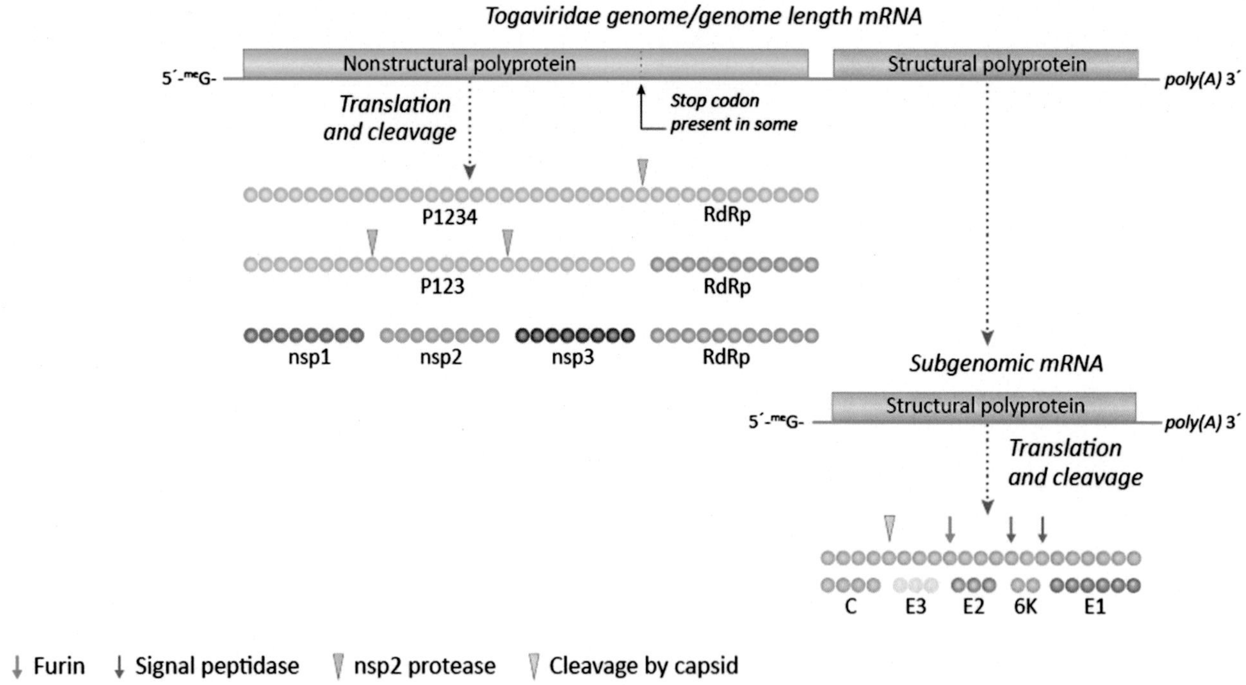

FIGURE 17.2   Togavirus genome organization and protein expression strategy.

BOX 17.1

Taxonomy

Family *Togaviridae*
Genus *Alphavirus*
    Thirty recognized species cluster into two groups. Old World alphaviruses include: Ross River virus, Semliki virus, Sindbis virus, and Chikungunya virus. The New World alphaviruses include EEEV and VEEV. Western equine encephalitis is a recombinant between a New World and an Old World alphavirus.
Family *Matonaviridae*
Genus *Rubivirus*
One species: *Rubella virus*

virus has replicated in the insect can it be transmitted to a new vertebrate host.

Genome structure

Togavirus genomes are a single piece of positive-strand RNA, 10–12 kb in length. Togavirus genomes have a methylguanosine cap at the 5′ end and a poly (A) stretch at the 3′ end. There are short 5′ and 3′ untranslated regions. Nonstructural proteins (nsps) are encoded at the 5′ end of the genome. Nsps are synthesized as a polyprotein that is cleaved by viral proteases. The RNA-dependent RNA polymerase (RdRp) is nsp4. Genes for the three structural proteins, C, E1, and E2 (and sometimes an E3 protein), are found at the 3′ end of the genome. They are synthesized as a polyprotein precursor from a subgenomic mRNA (Fig. 17.2). A short untranslated region upstream of the open-reading frame (ORF) encoding the structural polyprotein serves as a promoter for synthesis of the subgenomic mRNA (Box 17.2).

Virion structure

Togavirus capsids are assembled from 240 copies of a single capsid (C) protein to form a $T = 4$ icosahedron. The capsid is surrounded by a lipid bilayer that contains protein spikes assembled from the envelope glycoproteins, E1 and E2. E1 and E2 associate to form heterodimers and each spike is a trimer of heterodimers. The envelope glycoproteins associate with capsid proteins resulting in a very regular structure for the envelope (Fig. 17.1). Virions are ~ 70 nm in diameter.

Togavirus replication

Togaviruses replicate in both vertebrate and insect cells. In vertebrate cell lines alphaviruses replicate very rapidly, with the release of progeny virus typically within 4–6 hours after infection. Infection causes extensive cytopathic effects (CPEs) in vertebrate cells. In contrast, replication in insect cells is noncytopathic, producing a persistent infection.

Attachment and penetration

The receptors for togaviruses are most likely ubiquitous (common) cell surface molecules allowing the viruses to attach to both insect and vertebrate cells. Virions are endocytosed and fusion between the viral envelope and the endosomal membrane is triggered in the low pH environment.

Translation

After penetration and uncoating, genome-length RNA is translated to produce a polyprotein that is cleaved by virally encoded proteases to produce the nsps (including RdRp) required for genome replication. Most togavirus genomes have a single stop codon between nsp3 and nsp4. Therefore ribosomes must read-through or suppress the stop codon in order to synthesize the long polyprotein that contains nsp4.

Togavirus structural proteins (C, E1, and E2) are translated from a subgenomic mRNA that is synthesized during genome replication. The capsid protein, C, is cytosolic and is cleaved from the polyprotein precursor by a viral protease. E1 and E2 are transported across the endoplasmic reticulum (ER) membrane during synthesis. The polyprotein is cleaved by resident ER proteases. In vertebrate cells, E proteins are glycosylated and transported to the plasma membrane. The cleavages generating C, E1, and E2 also generate two small protein products with a shared amino terminus. They are called 6K (for 6 kDa MW) and transframe (TF). TF is formed by ribosomal frameshifting. Both proteins have a transmembrane region and they are both found, in low quantities, in purified virions. Mutations in 6K are associated with greatly decreased virion production and/or deformed virions. A description of the "discovery" of TF is provided in Box 17.3.

BOX 17.2

General characteristics

Flavivirus genomes are a single strand of positive-sense RNA ranging from 10 to 12 kb in length. Genomes have 5′-methyl-G caps and have poly(A) tails at the 3′ end. Genomes contain two ORFs. A polyprotein encoding the nsps is found at the 5′ end of the genome. The structural proteins (capsid and envelope glycoproteins) are synthesized from a polyprotein translated from a subgenomic mRNA.

Enveloped virions (~ 70 nm diameter) have $T = 4$ icosahedral symmetry as the $T = 4$ capsid is closely associated with 240 copies of the E1 and E2 proteins.

BOX 17.3

Protein discovery through bioinformatics (discovery of a new alphavirus protein)

SFV is one of the most extensively studied alphaviruses because it is relatively benign and grows quite well in cultured cells. It is a common model for studies of alphavirus replication and structure. The 3′ ORF of SFV encodes a precursor polyprotein that is cleaved to generate structural proteins C, E3, E2, 6K, and E1. The 6K protein contains a transmembrane domain and is present at low amounts in purified virions. The 6K protein is proposed to be a viroporin, possibly involved in virus budding.

Researchers studying 6K routinely noted an 8K protein along with it. It was thought to be a modified form of 6K protein, although this was never confirmed. The 8K product was finally identified in 2008, through use of bioinformatics tools, biochemistry, and genetics (Firth et al., 2008). Researchers analyzing viral genomes for potential ribosomal frameshifting sites found a previously unrecognized site in the genomes of most alphaviruses. The conserved sequence was located within the 6K coding region and had all of the structural hallmarks of a ribosomal frameshifting site; its use predicted an 8K protein. Using SFV as their model virus, the

investigators biochemically analyzed the 8K protein (using mass spectroscopy with peptide sequencing as well as immunoprecipitation) and demonstrated it was the product of a ribosomal frameshift in 6K. They named the new protein TF. Finally the investigators mutated SFV by introducing a stop codon just downstream of the predicted frameshifting site. They were able to grow the mutant virus, but the 8K product was missing! Since the 2008 publication, other groups have identified TF protein in virions of other alphaviruses, for example, Sindbis and Chikungunya.

The take-home message of this story is not that there is another alphavirus protein product to memorize, it is the following: After Firth, Chung, Fleeton, and Atkins developed and used a *bioinformatics approach* to identify a *potential* new frameshifting site, they did additional work to *confirm* the presence of the protein using a well-characterized model alphavirus. They used *two different biochemical methods* to characterize the protein and then followed up with a *genetic method* for additional confirmation. Their work set the stage for others to confirm the findings using pathogens of global importance.

Genome replication

Genome replication is cytoplasmic. Early after infection, copies of the genome (cRNA) are synthesized. Full-length cRNA serves as the template for synthesis of additional genomes as well as subgenomic mRNAs. In cases where the process has been studied in detail, it appears that an uncleaved nsp precursor (called P123) plus a molecule of cleaved nsP4 make up the complex that synthesizes the negative-strand cRNA. However, P123 is eventually cleaved to produce nsP1 + nsP2 + nsP3. When this occurs, the complex comprised of nsP1 + nsP2 + nsP3 + nsP4 synthesizes genomes and subgenomic mRNA (Fig. 17.3). In summary, the uncleaved precursor P123 binds to the 3′ end of the genome strand but the nsP1 + nsP2 + nsP3 complex interacts with the 3′ end of the cRNA to efficiently synthesize new positive-sense RNAs. This is a clever mechanism to control the relative amounts of cRNA (just a little) and genome RNA (a lot) without encoding different sets of replicase proteins. The details of this process were determined by generating viruses with mutations at P123 to eliminate cleavage sites. The RNA products produced by the mutated viruses were then analyzed. Togavirus nsps and activities are summarized in Table 17.1.

Assembly and release

In vertebrate cells, capsid proteins and genomes assemble in the cytoplasm (Fig. 17.4). Capsids associate with E proteins at the plasma membrane where budding occurs. In the case of animal cells infection is usually cytopathic; however, in insect cells the infection can be noncytopathic. In this case the virions probably bud into intracytoplasmic vesicles and are released by exocytosis. Large amounts of virus are made in animal cells, resulting in the high titer viremia that facilitates ingestion of the virus by blood-feeding insects.

Transmission

Most togaviruses are spread by mosquitoes and can be transmitted from the female to her eggs. In urban areas, the transmission cycle of CHIKV involves humans and mosquitos. In contrast, the transmission cycles of EEEV and WEEV are more complex. These viruses are usually transmitted between mosquitos and birds. Only infrequently do the viruses "spill over" into humans and horses (and other mammals).

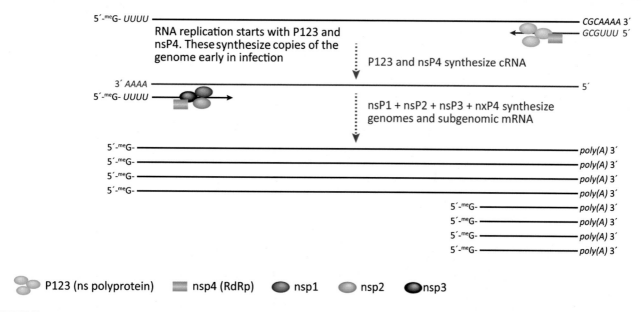

FIGURE 17.3   Togavirus model for regulation of synthesis of cRNA and genomic RNA. Proteolytic processing of the nonstructure polyprotein precursor controls synthesis of cRNA (unprocessed P123 plus nsP4) versus synthesis of genomic RNA and mRNA (nsP1, nsP2, nsP3, and nsP4).

TABLE 17.1   Togavirus nonstructural proteins (nsps).

Nonstructural protein	Function
nsP1	Enzymatic activities include: Methyltransferase and guanylyltransferase. These serve to cap togavirus mRNAs. An amphipathic helix and palmitoylation anchor the nsP1 to the cellular membranes.
nsP2	Enzymatic activities include: Nucleotide triphosphatase (NTPase) helicase, RNA triphosphatase, and protease. May act as a transcription factor to recruit the RNA synthetic complex to the subgenomic promoter. Involved in the shutoff of host macromolecular synthesis.
nsP3	Phosphoprotein important for minus-strand (cRNA) synthesis. Exact function not known. Binds to cellular amphiphysin-1 and -2, indicating a role in induction of membrane curvature.
nsP4	Enzymatic activities include: RdRp. Promoter binding and recognition.

Togaviruses associated with rash and arthritis

Chikungunya virus

In the Kimakonde language of Mozambique Africa chikungunya means "to walk bent over," thus is a very good name for this human virus that causes crippling joint pain with rapid onset. The virus is widespread; from 2004 to 2007, a large epidemic occurred in India and on islands in the Indian Ocean. From there the virus spread to Southeast Asia and Europe. In 2013 chikungunya virus was found on islands in the Caribbean. Up to a few hundred thousand cases occur worldwide, per year, according to the World Health Organization. Most cases of CHIKV infection identified in the United States

are acquired during travel to Caribbean, but there has been some limited local spread within the United States. The symptoms of CHIKV infection are similar to those of dengue virus, thus in areas where both viruses are common, these illnesses can be confused.

In urban areas CHIKV is transmitted by mosquitoes in a simple transmission cycle via *Aedes* sp. mosquitoes, including *A. albopictus* the Asian tiger mosquito (Fig. 17.5). In Africa and Asia, sylvan cycles also exist, with nonhuman primates presumed to be the major reservoir. Common symptoms of chikungunya virus infection are fever, joint pain, headache, muscle pain, joint swelling, or rash. While the disease is usually self-limiting and mortality is quite low, musculoskeletal symptoms are recurrent or persistent in approximately 40% of adults, lasting up to

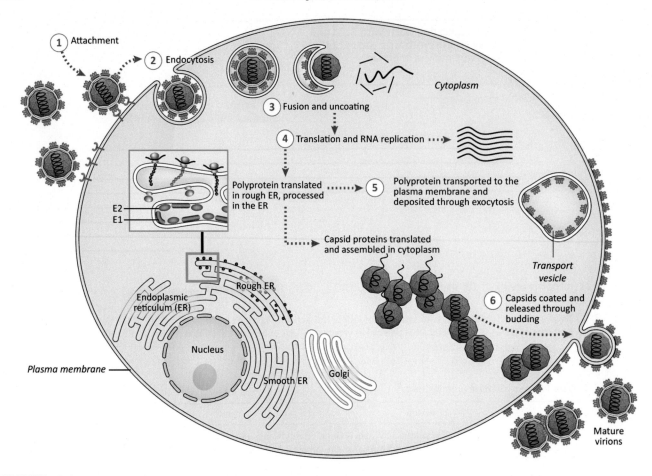

FIGURE 17.4   Overview of the togavirus replication cycle (vertebrate cell). Virion release from insect cells occurs by exocytosis.

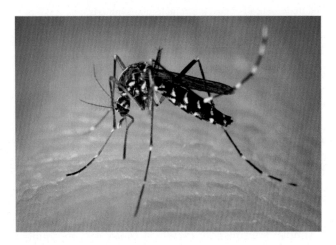

FIGURE 17.5   Asian tiger mosquito, *Aedes albopictus*, a vector for CHIKV and other arboviruses. Source: *CDC Public Health Image Library, Image #4487. Content providers CDC/ James Gathany.*

several months. This can result in significant disease burden in hard hit communities. There is no vaccine and the best way to avoid the disease is to avoid contact with mosquitos.

Other togaviruses of human concern

SINV was originally isolated from *Culex* mosquitoes in Egypt and is common in Africa. Small outbreaks have occurred in many places including Europe, Australia, India, and Africa. Symptoms include itching rash, arthritis, fever, and muscle pain. The normal host for SINV appears to be birds, with only occasional spillover into humans.

RRV is endemic in Australia. Many infections are inapparent but disease symptoms include swollen, painful joints, fever, and rash. RRV is the most common mosquito-borne pathogen in Australia, with approximately 5000 cases reported each year.

SFV is an African togavirus that was originally isolated from mosquitos. Serosurveys indicate that human infection is common, but disease is rare. The first documented case of human disease was associated with a fatal, laboratory-acquired infection (the individual probably had an underlying immunodeficiency). SFV has also been isolated from individuals with fever, persistent headache, myalgias, and arthralgias. Other viruses in this group include O'nyong'nyong virus (endemic in Africa) and Mayaro virus (endemic in South America).

Alphaviruses of veterinary importance

EEEV is a New World togavirus that is maintained in a bird to mosquito cycle in freshwater hardwood swamps. The mosquito, *Culiseta melanura*, feeds almost exclusively on avian hosts. Transmission to mammals requires mosquito species capable of creating a "bridge" between infected birds and uninfected mammals. Some *Aedes*, *Coquillettidia*, and *Culex* species serve as bridges to mammals. Humans and horses are considered "dead end" hosts for EEEV as the concentrations of virus in the bloodstream are not high enough for transmission back to mosquitos. Survival rates of infected horses are 70%–80% and a vaccine is available. Among humans there are on average, eight cases of EEEV-associated encephalitis reported yearly; the disease is fatal in about one-third of cases of encephalitis. However, many humans infected with EEEV remain asymptomatic.

WEEV is uncommon. WEEV is transmitted by *Culex* and *Culiseta* mosquitos. WEE is genetically very interesting as it is a *recombinant virus* generated from two other alphaviruses: One a Sindbis-like (Old Word) virus and the other an EEEV-like (New World) virus. WEEV is found primarily in the United States in states west of the Mississippi River (hence the name) and is also found in South America. The usual cycle of WEEV transmission is between birds and mosquitoes. Infection of horses or people is most often subclinical. The overall mortality of WEE in humans is low (approximately 4%) and is associated mostly with infection in the elderly.

VEEV. Some types of VEEV (the so-called *enzootic* subtypes) cycle among *mosquitoes and rodents* and they remain in localized areas. These sometimes "jump" to horses. Other strains of VEEV (the so-called *epizootic* subtypes) primarily infect horses as their mammalian hosts (*they cycle between horses and mosquitoes*). These strains can cause quite severe disease in horses. Infected horses may shed virus in nasal, eye, and mouth secretions, as well as in urine and milk. Therefore among horses, VEEV is not restricted to spread by mosquitos but can be transmitted by the respiratory route. Healthy adults (humans) often experience flu-like symptoms but the virus may cause a fatal encephalitis in very young, very old, or immunosuppressed persons.

Family *Matonaviridae*

Rubella virus

Until 2019 rubella virus was a member of the family *Togaviridae*, the sole member of the genus *Rubivirus*. In 2019 the International Committee on the Taxonomy of Viruses (ICTV) created a new family, *Matonaviridae*. The rationale for creating a new family was based, in part, on important differences in transmission and capsid structure. Rubella virus is not an arbovirus and is transmitted among human hosts by the respiratory route. The family name *Matonaviridae* is derived from George de Maton, a physician who first distinguished rubella from measles and scarlet fever. While most infections with rubella virus are mild and self-limiting, infection of pregnant women can lead to miscarriage or severe birth defects (Figs. 17.6 and 17.7).

Features and general replication cycle

In contrast to the highly ordered togavirus capsids, rubella viruses capsids do not have icosahedral

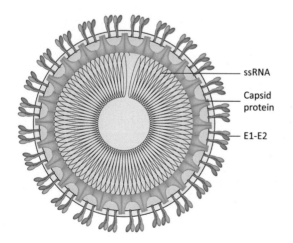
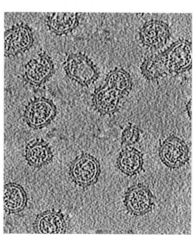
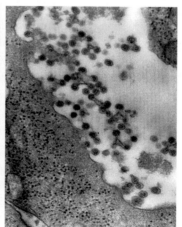

FIGURE 17.6 Left: Virion structure. Center: Section from a rubella virus tomogram showing the different morphologies of rubella virions. Right: Transmission electron micrograph revealing rubella virus virions budding from the host cell surface. *Source: (Center) From Mangala Prasad et al. (2017), (Right) From CDC/ PHIL, Image #10221, content providers CDC/Dr. Fred Murphy, Sylvia Whitfield.*

(Left diagram labels: ssRNA, Capsid protein, E1-E2)

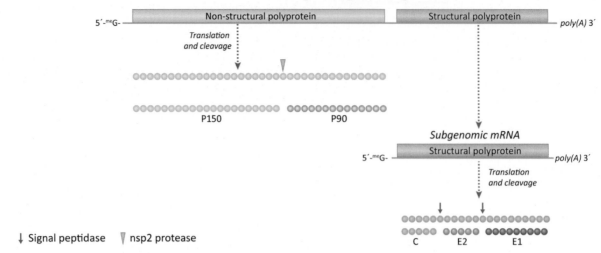

FIGURE 17.7   Rubella virus genome organization.

symmetry and the virions are heterogeneous, often cylindrical or tube-like with a diameter of ∼70 nm and up to 150 nm long. The E1−E2 spikes are arranged in parallel rows on the surface (Fig. 17.6). The rubella virus genome is unsegmented, positive strand RNA of ∼10 Kb with an overall organization similar to the togaviruses however the nonstructural polyprotein is not highly processed, producing only two products of 150 kDa and 90 kDa. The 90 kDa product is the RdRp (Fig. 17.8). The rubella virus genome is capped and polyadenylated. Nonstructural proteins are translated from genome length RNA and structural proteins are translated from a subgenomic mRNA. Structural proteins include a capsid protein and two envelope glycoproteins. The 33 kDa capsid protein is multifunctional, with roles in RNA replication and antagonizing cellular responses in addition to RNA binding and structural roles. E1 and E2 are membrane proteins that form heterodimers. E1 is glycosylated and serves at both the attachment and fusion protein. E2 is modified by N- and O-linked glycosylation as well as fatty acylation.

Disease

Rubella has an incubation period of 2−3 weeks and virus spreads primarily via the aerosol route. The virus is also found in urine, feces, and on the skin. Infections are acute and there is no prolonged shedding or carrier state. In children disease is mild and of short duration (1−3 days) and may go unnoticed, however a rash is common (Fig. 17.7). Persons infected as adults can have a more serious and prolonged disease course. Rubella virus does the most damage in pregnant women. If infection occurs

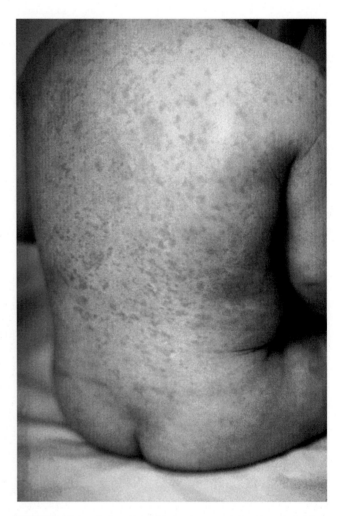

FIGURE 17.8   Torso of a child revealing a skin rash, due to a case of rubella, also referred to as German measles or 3-day measles. Source: *CDC/PHIL image 712. Content provider CDC.*

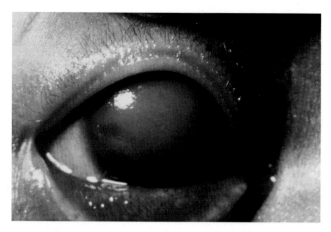

FIGURE 17.9   This image is a close view of the eye of a 3-year-old who exhibited symptoms of congenital glaucoma due to a case of congenital rubella. Source: *CDC/PHIL image 17618. Content providers CDC/Dr. Andre J. Lebrun.*

before the third trimester, the virus can severely damage the developing fetus. The earlier in the pregnancy a woman is infected, the more damage the virus causes as multiple organs in the fetus can be affected and develop abnormally. Miscarriages occur about 20% of the time, and children born with congenital rubella syndrome (CRS) face serious life-long health issues. Children born with CRS can shed virus for months, transmitting it to other infants and pregnant women. Infants born with CRS can present with deafness, eye abnormalities (Fig. 17.9), and congenital heart disease. Other problems may include low birth weight, microcephaly, and transient spleen, liver, or bone marrow problems. Children may also have developmental delays and intellectual disability. As of 2015 the World Health Organization estimates 110,000 babies are born with CRS every year. An effective rubella vaccine is available. In the United States it is usually is administered in a cocktail with measles and mumps vaccines (MMR vaccine). Rubella vaccination is very important for nonimmune women who may become pregnant. Widespread vaccination during the past decade has virtually eliminated rubella and CRS in many resource rich countries.

In this chapter we have learned that:

- Members of the family *Togaviridae* have enveloped virions with $T = 4$ icosahedral symmetry. Genomes are unsegmented positive-strand RNA (capped and polyadenylated).
- Togavirus nsps, including RdRp, are translated as a polyprotein from capped genome-length mRNA. Togavirus structural proteins include a capsid protein (C) and two or three envelope glycoproteins. Structural proteins are translated as a precursor polyprotein from a subgenomic mRNA.
- Most alphaviruses are transmitted by mosquitos. Chikungunya is an emerging human pathogen. EEEV and WEEV primarily cycle between avian and mosquito hosts but sometimes "spill over" into humans and horses. Some strains of VEEV cause quite severe disease in horses and can be transmitted by respiratory secretions.
- Rubella virus is a member of the family *Matonaviridae*.
- The genome organization of rubella virus is similar to that of the togaviruses but virion architecture is distinct.
- Rubella virus generally causes only mild disease in children but infection of nonimmune pregnant women can result in babies with serious birth defects. Rubella vaccination is very important for nonimmune women who may become pregnant.

Reference

Firth, A.E., Chung, B.Y., Fleeton, M.N., Atkins, J.F., 2008. Discovery of frameshifting in Alphavirus 6K resolves a 20-year enigma. Virol. J. 5, 108. Available from: https://doi.org/10.1186/1743-422X-5-108.

Mangala Prasad, V., Klose, T., Rossmann, M.G., 2017. Assembly, maturation and three-dimensional helical structure of the teratogenic rubella virus. PLoS Pathog. 13 (6), e1006377. Available from: https://doi.org/10.1371/journal.ppat.1006377.

18

Family *Coronaviridae*

After reading this chapter, you should be able to answer the following questions:

- What is the structure of a coronavirus virion?
- How is the coronavirus genome organized?
- Which coronavirus protein is responsible for attachment and fusion?
- By what mechanism are the nonstructural proteins involved in replication and transcription produced?
- By what mechanism are coronavirus mRNAs synthesized?
- Why do coronaviruses undergo high rates of RNA recombination?
- What human diseases are caused by coronaviruses?
- How does the virus that causes feline infectious peritonitis (FIP) virus differ from the virus that causes feline enteritis?

Members of the family *Coronaviridae* are large, enveloped, single-stranded RNA viruses. They are the largest known RNA viruses, with genomes ranging from 25 to 32 kb. Virions are spherical and range from 118 to 136 nm in diameter. Virions are notable for the large spike (S) glycoprotein that extends 15–20 nm from the virus envelope (Fig. 18.1 and Box 18.1). The family contains three subfamilies with the subfamily *Orthocoronavirinae* containing, by far, the largest

number viruses (Box 18.2). Orthocoronaviruses are widespread among mammals, often causing only mild respiratory or enteric infections but some cause a significant disease burden in human and animal hosts.

Over 60 coronaviruses (CoVs) have been isolated from bats (BtCoV) and most of these are in the genus *Betacoronavirus*. Bats serve as large (and highly mobile) CoV reservoirs; many bat species have their own unique BtCoV, suggesting a very long history of coevolution. Until 2002 CoVs were considered only minor pathogens of humans. However, an outbreak of severe acute respiratory syndrome (SARS) that began in 2002 was linked to infection with a new CoV (SARS-CoV-1). The outbreak increased interest in CoV replication, distribution, evolution, transmission, and pathogenesis. In 2014 a different coronavirus was isolated in connection with an outbreak of severe respiratory disease in the Middle East. This virus, called Middle East respiratory syndrome coronavirus (MERS-CoV), continues to cause sporadic cases of severe respiratory illness. In retrospect, SARS-CoV-1 and MERS-CoV caused minor outbreaks compared to the most significant pandemic in 100 years (ongoing as this chapter is being written) caused by SARS-CoV-2.

Several important animal diseases are also caused by CoVs. Infectious bronchitis virus (IBV) of chickens was the first CoV identified (in the 1930s). A pig

© 2023 Elsevier Inc. All rights reserved.

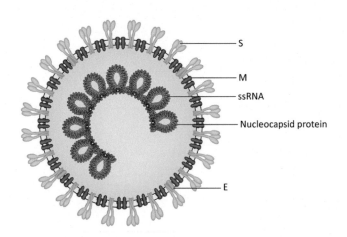

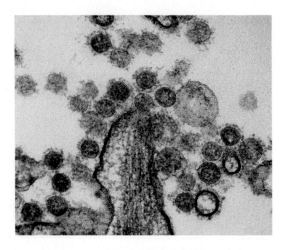

FIGURE 18.1   Virion structure (right). TEM image of MERS-CoV virions in culture (left). *Source: CDC Public Health Image Library #18107. Content provider National Institute of Allergy and Infectious Diseases.*

BOX 18.1

General characteristics

Genomes are unsegmented, single-stranded, positive sense RNA. Genomic RNA is capped and polyadenylated. Genome size ranges from 25 to 32 kb. Genomic RNA is used to translate two large polyproteins (REP1a and REP1b) that are autoproteolytically processed to produce a large set of nonstructural proteins (ns1−ns15) that participate in transcription and replication. The longer of the polyproteins is produced by ribosomal frameshift.

Enveloped virions are 118−140 nm. Within the envelope is a flexible nucleocapsid comprised of the genome and many copies of the nucleoprotein (N), a basic, RNA-binding protein. Envelopes contain a variable number of membrane proteins but all appear to have an integral membrane protein M and the heavily glycosylated spike (S) protein that extends 15−20 nm from the envelope.

Transcription and genome replication are cytoplasmic. Genome-length RNA serves as mRNA for a long polyprotein precursors. A set of 3′ coterminal mRNAs encode the structural proteins. Transciption is discontinuous, leading to high rates of template switching.

---

coronavirus (porcine epidemic diarrhea virus, PDDV) caused the deaths of millions of piglets in the United States in 2013−2014. Later in this chapter, we consider the unique pathogenesis of the feline CoV (FeCoV), the causative agent of feline infectious peritonitis (FIP), a deadly disease of domestic cats. A rodent coronavirus, mouse hepatitis virus (MHV), has served for many years as a useful model system for investigating CoV replication and pathogenesis. This chapter will cover the basics of coronavirus replication and will touch on animal and human coronaviruses other than SARS-CoV-2. More extensive coverage of SARS-CoV-2 replication and pathogenesis will be found in Chapter 19.

Coronavirus genome organization

The 25−32 kb positive-strand RNA genomes of CoVs contain 7−10 open reading frames (ORFs). Almost two-thirds of the genome encodes nonstructural proteins

(nsps) that are required for transcription and genome replication (Fig. 18.2). Among these is nsp12, the large 930 amino acid RNA-dependent RNA polymerase (RdRp). Nsp12 forms a multiprotein complex with other CoV nsps. The ORFs Rep1a and Rep1b, encoding CoV nsps, are located at the 5′ end of the genome and are translated upon release of the genome into the cytoplasm. CoV nsps are synthesized as long precursor polypeptides, cleaved by virally encoded proteases. The REP1a polyprotein is cleaved to produce 11 smaller products (nsp1−nsp11; listed in Table 18.1). The Rep1b ORF follows Rep1a in the genome, separated from it by a frame-shifting site. About 20%−25% of the time, ribosomes reaching the frame-shift site in Rep1a will slip into the minus-1 reading frame and a longer polyprotein, called REP1b is synthesized. REP1b codes for five additional nsps, nsps 12−16, among which are the RdRp (nsp12), a protein with helicase and phosphatase activities (nsp13), a protein with exonuclease and methyl transferase activities (nsp14), an endoclease (nsp15), and a second methyltransferase (nsp16). The

BOX 18.2

Taxonomy

The family *Coronaviridae* contains three subfamilies, including the subfamily *Orthocoronavirinae*, the subject of this chapter.

Subfamily *Orthocoronavirinae* contains four named genera.
Genus *Alphacoronavirus* includes 26 named species, among them are:
Human coronavirus 229E, Mink coronavirus 1, Porcine epidemic diarrhea virus, Human coronavirus NL63, and *Alphacoronavirus 1*. Viruses within species *Alphacoronavirus 1* include canine coronavirus type I, canine coronavirus type II, feline coronavirus type I, feline coronavirus type II, and transmissible gastroenteritis virus.
Genus: *Betacoronavirus* includes 14 named species, among them are:
Human coronavirus HKU1, Hedgehog coronavirus 1, Middle East respiratory syndrome-related coronavirus, severe acute respiratory syndrome-related coronavirus, and *Betacoronavirus 1*. Viruses within the species *Betacoronavirus 1* include bovine coronavirus, equine coronavirus, human coronavirus OC43, and porcine hemagglutinating encephalomyelitis virus.
Genus *Deltacoronavirus* contains seven named species.
Genus *Gammacoronavirus* contains five named species.

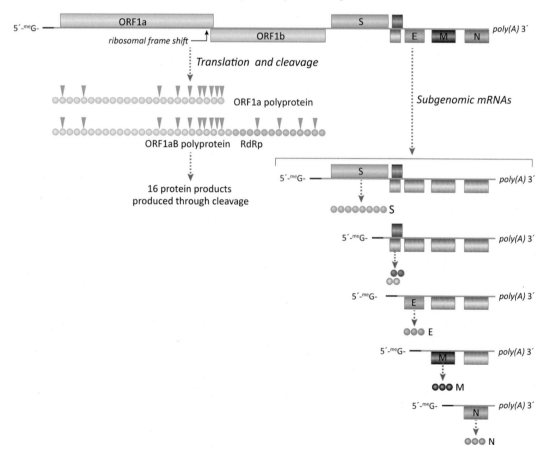

FIGURE 18.2 Coronavirus genome organization and gene expression. Nonstructural proteins are translated from genome length mRNA to produce large polyproteins which are processed by virally encoded proteases. Subgenomic mRNAs are used to produce structural and auxiliary proteins. Subgenomic mRNAs form a nested set and are produced by discontinuous transcription as illustrated in Fig. 18.4.

TABLE 18.1 Coronavirus nonstructural proteins (nsps).

Nonstructural protein	Function(s)
Nsp1	Interferon antagonist (not present in all coronaviruses) that mediates host mRNA degradation, translation inhibition, and cell cycle arrest. Notably, nsp1 does not degrade viral mRNAs.
Nsp2	Not known but associates with replicase transcriptase complex.
Nsp3	Papain-like protease domains and several other protein interaction domains. Interferon antagonist.
Nsp4	Transmembrane scaffold. Involved in membrane remodeling.
Nsp5	Main protease (Mpro) (also called 3C-like protease) required for polyprotein processing.
Nsp6	Transmembrane scaffold. Involved in membrane remodeling.
Nsp7	Forms a large complex with nsp8. Binds ssRNA.
Nsp8	Forms a large complex with nsp7. The complex may function as a processivity clamp for the RdRp. Noncanonical "secondary" RdRp with putative primase activity; forms hexadecameric supercomplex with nsp7.
Nsp9	Single-stranded RNA-binding protein that associates with the replicase transcriptase complex.
Nsp10	Dodecameric zinc finger protein, stimulates nsp16 methyltransferase activity.
Nsp11	Unknown
Nsp12	RdRp
Nsp13	RNA 5′ triphosphatase (cap synthesis), RNA helicase.
Nsp14	N7-methyltransferase (cap synthesis), Exo N 3′−5′ exonuclease provides a proofreading function for the coronavirus RdRp.
Nsp15	Nendo U endonuclease (cleaves single-stranded and double-stranded RNA downstream of uridylate residues, producing 2−3 cyclic phosphates). Downregulates host recognition of viral RNAs.
Nsp16	nsp16Ribose 2′-O-methyltransferase (cap synthesis).

guanine-N7-methyltransferase, nsp13, and the 2′-O-methyltransferase, nsp16, synthesize the CoV 5′-cap. In addition to the final cleavage products, long-lived intermediates may have activities different from those of the fully cleaved products.

Seven to ten additional ORFs lie downstream of the replicase-associated genes. The largest of these encodes the spike (S) protein. Other structural protein ORFs encode envelope (E), membrane (M), and nucleocapsid (N). The order of the structural proteins on the coronavirus genome (S, E, M, N) is well conserved. Dispersed among the structural protein genes are a variable number of additional small ORFs. These lie in-between, or overlap, the structural protein genes and they are not well conserved among even closely related CoVs. These are called accessory proteins and by convention they are named using the number of the shortest mRNA on which they are encoded (Box 18.3). Mutation studies show that at least some of the accessory proteins are not required for coronavirus replication in cell cultures. However, mutating the accessory proteins does have a profound effect on the ability of CoVs to replicate in their hosts and impact viral pathogenesis.

As expected for an RNA virus, the CoV genome contains cis-acting regulatory elements required for replication, transcription, and genome packaging. At the 5′ end of the genome, elements that participate in viral RNA synthesis extend past the 5′ untranslated region (UTR) and into the coding region of Rep1a. This region folds up into a set of seven stem-loops. Folding of the CoV 5′ RNA is dynamic, allowing for changing conformations that control critical aspects of RNA and protein synthesis. Functional analyses of cis-acting regulatory regions have shown that stability of RNA stem-loops can be critical for viral fitness.

At the 3′ end of the CoV genome, additional cis-acting elements are found in the 3′ UTR. The 3′ cis-acting elements are conserved (and are interchangeable) among betacoronaviruses. As with the 5′ UTR, this region is probably dynamic, with different conformations controlling different steps in RNA replication and transcription. Other important cis-acting sequences are those that direct ribosomal frame shifting near the end of the Rep1a gene. Finally, genome packaging signals have been localized to internal regions of the SARS-CoV-1 genome.

Virion structure

Coronavirus particles are enveloped with very prominent and distinctive spikes. Virions are spherical and range in size from 118 to 140 nm. Within the envelope is

---

BOX 18.3

Coronaviruses accessorize

Coronaviruses encode a heterogeneous group of proteins, many of which are not absolutely necessary for replication in cultured cells, collectively called accessory proteins. The numbers of accessory proteins encoded by CoVs differ among even closely related coronaviruses. The genes encoding accessory proteins are interspersed among the structural genes. They are numbered according to the mRNA from which they are produced. SARS-CoV-1 expresses eight accessory proteins (3a, 3b, 6, 7a, 7b, 8a, 8b, and 9b). They are all small peptides, ranging in size from about 45 to 180 amino acids. A few are found in purified virions and several are membrane associated. Several have been identified in infected cells from autopsy patients, good evidence that they are expressed during a natural infection. For the most part, their functions are not well characterized but a few activities have been determined: the SARS-CoV-1 3a protein forms an ion channel in the plasma membrane and may modulate virus releases. SARS-CoV-1 3b protein interferes with the interferon pathway, and protein 6 may be an antagonist of type I IFN. Both SARS-CoV-1 and MERS-CoV infections are characterized by a delay in type I IFN production and poor development of adaptive immune responses. It is likely that accessory proteins are involved in their pathogenesis.

---

a flexible nucleocapsid that consists of genomic RNA associated with the nucleoprotein (N). The S protein is the major glycoprotein that extends from the surface of the virion. Other membrane-associated proteins include M and E. M associates with both the viral envelope and E proteins and the nucleocapsid. E is a viroporin that forms ion channels in the envelope.

Coronavirus structural proteins

All CoVs encode four structural proteins: three membrane-associated proteins S, M, and E as well as a single nucleocapsid (N) protein. Some betacoronaviruses have an additional membrane protein with hemagglutinating and esterase activities, called HE. The order of the structural genes on the CoV genome is (HE) S, M, E, N (Fig. 18.2). The S forms a prominent projection or spike from the virus envelope and gives CoVs their characteristic appearance. Spikes are homotrimers of the heavily glycosylated S protein. S functions in both attachment and fusion. S contains a proteolytic cleave site and cleavage generates S1 (amino terminal) and S2 (carboxy terminl) proteins. S1 and S2 products remain noncovalently associated. Among the CoVs there are differences in the manner in which S is processed. In many betacoronaviruses S is partially or completely cleaved by host furin-like proteases in the ER, prior to assembly of new virions. The extent of proteolysis correlates with the number of basic residues at the S1/S2 cleavage site. A second cleavage, within S2 is also required to activate or prime S for fusion. This cleavage site is termed S2'. SARS-CoV-1 S is not cleaved by furin-like proteases and virions are released with uncleaved S proteins.

Cleavage of S is necessary for productive infection and is carried out by other proteases after attachment and/or endocytosis. Finally, some CoV S proteins are not cleaved, but the terms S1 and S2 are still used to refer to the corresponding N-terminal and C-terminal domains of the protein.

Comparisons of S1 sequences show that they diverge extensively and are not highly conserved, even within multiple isolates of a single type of coronavirus. We can reasonably speculate that sequence divergence is, at least in part, a result of host immune responses. In contrast to S1, the S2 product is highly conserved across the subfamily *Orthocoronavirinae*.

M, the membrane protein, is the most abundant protein in the virion. M contains three hydrophobic domains thus is tightly embedded with the viral envelope. M plays a major role in promoting membrane curvature and viral budding. It has a short extracellular domain that is modified by glycosylation. M also interacts with the N and E proteins. Expression of SARS-CoV-1 M (in the absence of other viral proteins) results in self-assembly and release of membrane-enveloped vesicles. Virus-like particles are released when M is coexpressed with either N or E.

The E protein is found in very small amounts in the virion (about 20 molecules per virion), although larger amounts of E are present in infected cells. Different CoVs have variable requirements for E during particle formation, ranging from required for particle formation to not essential. In fact, virus titers close to 1×10^6 pfu per mL have been reported for SARS-CoV-1 lacking the E protein. Studies show that E assembles in membranes to form ion channels, thus E is a viroporin. Viroporins influence the electrochemical balance in subcellular compartments.

The nucleocapsid (N) protein is the only protein found in the ribonucleoprotein particle. N forms homodimers and homooligomers and binds genomic RNA, packaging it into a long flexible nucleocapsid. In the infected cell, N localizes to the cytoplasm, and for some CoVs N is also found in the nucleolus. N interacts with other CoV structural proteins, thus has a role in assembly and budding. N also colocalizes with replicase—transcriptase components and is required for RNA synthesis. Other roles for N include modulating cell-cycle (promoting cell-cycle arrest) and inhibiting host cell translation.

An additional structural protein is found in most betacoronaviruses, including the human coronavirus HCoV-HKU1. The hemagglutinin-esterase (HE) is a 48 kDa glycoprotein that projects outwards 5—10 nm from the envelope. HE binds sialic acid units on glycoproteins and glycolipids. The esterase activity removes acetyl groups from O-acetylated sialic acid, thus may be a receptor-destroying enzyme.

Coronavirus replication cycle

The overall scheme of CoV replication is similar to that of other positive-strand RNA viruses (Fig. 18.3). Binding leads to endocytosis followed by fusion of the viral envelope with endosomal membranes to release the nucleocapsid into the cytoplasm. Translation of the genomic RNA produces the nps required to replicate genomes. The process for synthesis of coronaviral subgenomic mRNAs is unique and occurs by a discontinuous process.

Attachment

CoV S is the receptor-binding protein. The receptors for some CoVs have been identified. Three human CoVs, SARS-CoV-1 and -2, and HCoV-NL63 bind different regions of angiotensin converting enzyme 2 (ACE2). ACE2 is a cell-surface, zinc-binding carboxypeptidase important for regulation of cardiac function and blood pressure. ACE2 is expressed in epithelial cells of the lung and small intestine, as well as many other organs. MERS-CoV binds to dipeptidyl peptidase 4 (DPP4). Several other CoVs bind to aminopeptidase N. However not all CoVs bind to protein receptors. Bovine CV and HCoV-OC43 bind to sialic acid units found on glycoproteins and glycolipids.

Penetration

CoVs are enveloped, so penetration occurs as a consequence of a membrane fusion event. Just as cleavage of the S protein varies among CoVs, so do the sites and requirements of the fusion process. CoV S proteins contain a hydrophobic "fusion peptide" that becomes exposed upon a large-scale rearrangement of S. The location and details of S rearrangement are variable, but can be triggered by factors such as proteolytic cleavage of S and/or acid pH. Processing of SARS-CoV-1 has been well studied and will serve as our example. SARS-CoV-1 S is found as an uncleaved product on extracellular virions. There are two critical cleavage events that precede (and are required for) fusion. One cleavage is at the S1/S2 boundary and the other is a cleavage within S2 (S2′). Optimal cleavage of S to prime it for fusion appears to require transmembrane protease serine protease (TMPRSS2) which associates with ACE2 on the cell surface, as well as the cysteine protease cathepsin L in endosomes. Proteolytic processing and/or low pH cause largescale changes in S structure, revealing the hydrophobic fusion domain in S2.

Amplification

The first synthetic event in the CoV replication cycle is translation of the capped, polyadenylated viral genome by host cell ribosomes. REP1a and REP1b are the polyproteins that result from translation of genomic RNA; production and proteolytic processing of these polyproteins are necessary for virus replication to move forward. The longer REP1b polyprotein is generated by a ribosomal frame-shifting event. RdRp is part of the REP1b polyprotein. Some REP1a products (nsp3, nsp4, nsp6) have transmembrane domains and these serve to anchor replication—transcription complexes to cell membranes. Cell membranes are remodeled to form structures dedicated to viral RNA synthesis.

RNA synthesis

The CoV RdRp (nsp12) requires a primer, specifically a short RNA oligonucleotide to begin synthesis of RNA. It is thought that the required primer is synthesized by nsp8, which also has polymerase activity. Thus nsp8 gene product is thought to be a primase capable of synthesizing short oligonucleotides while nsp12 is the elongating polymerase. Other viral nsps in the replication—transcription complex are involved in cap synthesis. CoVs also encode two ribonucleases. The first, NendoU (nsp15) is a "Nidovirales endonuclease," so named because coronaviruses are in the order *Nidovirales* and most viruses in the order encode similar endonucleases. NendoU cleaves both single-stranded and double-stranded RNA, cutting downstream of uridylate residues. A primary function of

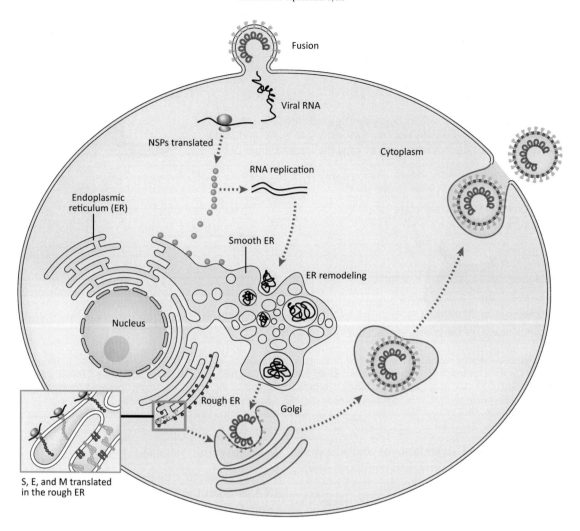

FIGURE 18.3 Overview of coronavirus replication cycle. Coronaviruses enter cells by fusion as the plasma membrane or within endosomes, releasing the nucleocapsid into the cytoplasm. Genome replication and transcription are cytoplasmic and take place on membranes. Genome length RNAs and the N protein assemble to form nucleocapsids that bud into vesicles, acquiring M, E, and S proteins. Virions are released by exocytosis.

NendoU seems to be evasion of host recognition of viral RNA by removal of 5′ poly(U) sequences from cRNAs.

The second coronavirus nuclease is ExoN (nsp14), a 3′–5′ exonuclease. CoVs with mutations in ExoN have enhanced mutation rates suggesting a role for ExoN in proofreading during RNA synthesis. For many years virologists were unsure how the large RNA genome of coronaviruses could be maintained with no proofreading. It appears that ExoN fulfills that proofreading function. The presence of ExoN has consequences for development of antiviral drugs to treat coronavirus infections. Many drugs that target nucleic acid synthesis are nucleoside analogs that are inserted into a growing nucleic acid chain, but lack a 3′-OH, thereby preventing addition of nucleotides. These so-called chain-terminators can be removed from CoV RNAs by ExoN, making these drugs less useful for treatment of CoV infections.

In addition to genome-length RNA, a set of subgenomic (sg) mRNAs is found in the infected cell. The sg mRNAs are used for the expression of structural and accessory proteins. All CoV mRNAs are capped and polyadenylated and they share a common 3′-end forming a so-called "nested set" of mRNAs (Fig. 18.4). A closer look at the sg mRNAs reveals that each contains an identical leader sequence of 70–100 nt at the 5′-end. The leader sequences found on all sg mRNAs are identical; however, this sequence is found only once in the genome, near the 5′-end. During mRNA synthesis leader sequences are fused to downstream sequences at identical, short, 8–9 nt motifs called the transcription regulating sequence (TRS). A TRS is found upstream of each ORF that encodes a structural protein. A TRS is also present just downstream of the leader sequence in the 5′ UTR. These findings provide clues to the unique strategy used for CoV mRNA synthesis: CoV sg mRNAs are

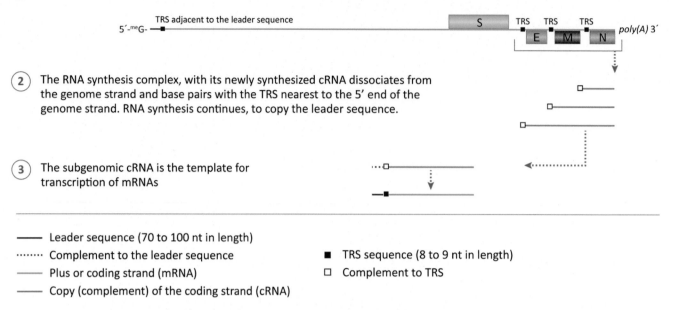

① The process begins with initiation of cRNA synthesis at the 3′ end of the genome RNA strand

② The RNA synthesis complex, with its newly synthesized cRNA dissociates from the genome strand and base pairs with the TRS nearest to the 5′ end of the genome strand. RNA synthesis continues, to copy the leader sequence.

③ The subgenomic cRNA is the template for transcription of mRNAs

—— Leader sequence (70 to 100 nt in length)

········ Complement to the leader sequence

—— Plus or coding strand (mRNA)

—— Copy (complement) of the coding strand (cRNA)

■ TRS sequence (8 to 9 nt in length)

☐ Complement to TRS

FIGURE 18.4 Synthesis of subgenomic (sg) mRNAs by discontinuous transcription. A set of sg mRNAs is found in the infected cell. The sg mRNAs are used for the expression of structural and accessory proteins. All sg mRNAs share a common 3′-end forming and all contain an identical leader sequence of 70–100 nt at the 5′-end. During synthesis of copy RNA the replicase complex "jumps" from a transcription regulating sequence (TRS) adjacent to an ORF for a structural gene to the TRS adjacent the leader sequence in the 5′ UTR.

generated by a process of discontinuous transcription illustrated (simplified) in Fig. 18.3. Discontinuous transcription very likely contributes to the high level of CoV genome recombination (Box 18.4). This type of RNA virus genome recombination is called a copy-choice mechanism (Fig. 18.5).

Assembly and release

Virion assembly takes place on membranes. Genomic RNA is bound by N protein and the ribonucleoprotein associates with M protein and buds into ER/Golgi membranes. M packs tightly into membranes and is thought to cause the membrane curvature that drives budding. S

BOX 18.4

Coronavirus recombination

Discontinuous transcription likely contributes to the readily observed examples of CoV genome recombination. If the replicase–transcriptase complex *must* dissociate/reassociate with template to generate subgenomic cRNAs, it follows that the same process sometimes occurs during synthesis of genome-length cRNA. Studies in the MHV experimental system show that genome recombination is relatively common event. Examination of feline and canine CoVs (CCoVs) provide real world examples of this phenomenon. FeCoVs are common in cats and usually cause mild enteric illness. On occasion enteric FeCoVs mutate within an infected cat and gain the ability to infect macrophages, an event

which is accompanied by production of severe systemic disease called FIP. However, among cats with FIP, some researchers found a few "novel" viruses. Sequencing these novel viruses revealed that they contained genes from both feline and CCoVs. As CCoV is a relatively common dog virus (and it is common to have both cats and dogs in a household), it was hypothesized that the novel FIP-associated virus arose in cats that became infected with *both* FeCoV and CCoV; in the coinfected cats, macrophage-infecting recombinants emerged, causing the clinical disease, FIP (Terada et al., 2014). The FIP-associated recombinant viruses do not appear to be transmissible from one cat to another.

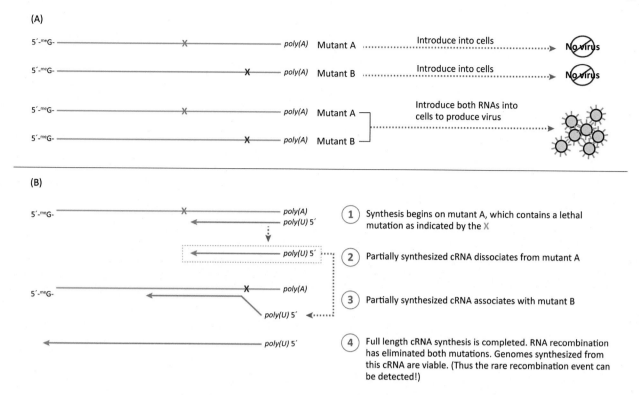

FIGURE 18.5 Coronavirus RNA recombination. (A) If individual mutant coronaviruses are used to infect cells there is no virus growth. However, if mutants are used together to infect cells, virus growth is restored suggesting that genome recombination has occurred. (B) This panel illustrates the mechanism of RNA copy choice "recombination."

and E are also membrane proteins and are acquired during the budding process. The ion channel activity of E is that of a viroporin; it alters cell secretory pathways to promote virus release. A function of E may be to increase the pH of the transport vesicles. Virus particles contained within membrane bound vesicles are released from cells by exocytosis.

Diseases caused by coronaviruses

The first coronaviruses isolated were from poultry with respiratory disease (infectious bronchitis) in the 1930s. IBV remains a worldwide problem, particularly in high-density commercial production facilities. Until 2002 HCoVs were associated only with mild respiratory tract disease, with estimates that they caused 15%—25% of all "common colds." That changed in 2002 when an HCoV was identified as the cause of an apparently new disease called SARS. The SARS outbreak was controlled, but in 2014 another novel CoV was isolated from patients hospitalized with severe respiratory disease in Saudi Arabia. As most infected patients lived in or had traveled to Middle East countries, the new disease was named MERS and the coronavirus responsible is called MERS-CoV. Most animal and human CoVs are transmitted by the fecal—oral route and their initial replication

site is in epithelial cells where virus production causes local respiratory symptoms or diarrhea. However, on occasion, CoVs cause severe to fatal disease. Some examples are presented here.

Severe acute respiratory syndrome

SARS-CoV emerged in the human population in China in 2002. Epidemiologic studies and genetic analysis indicated the virus most likely jumped from bats into farm-raised Himalayan palm civets (*Paguma larvata*) and then into humans, although other scenarios are possible. Once adapted to humans, transmission was by both respiratory and fecal routes. Total number of reported cases, worldwide, was approximately 8000 and the virus reached 26 countries. The case mortality was about 10% (774 recorded deaths). The most significant toll of the outbreak was financial, with huge economic losses recorded in Hong Kong (~5.9 billion $US). The WHO reported the last known case in July 2003.

Infection with SARS-CoV-1 results in a triphasic pattern of disease. The first phase is nonspecific with fever, cough, sore throat, and myalgia. Breathing difficulties (dyspnea) begin up 7—14 days after appearance of the first symptoms. The second phase of the disease includes shortness of breath, fever, onset of hypoxia, and often diarrhea. In the most serious cases, patients progress to a third phase with development of acute respiratory distress requiring hospitalization and mechanical ventilation. Three viral proteins have been implicated in SARS-CoV-1 pathogenesis. An accessory protein (encoded from orf3a) interferes with cell signaling pathways, another accessory protein (encoded from orf6) interferes with interferon signaling and the E glycoprotein is a strong inducer of proinflammatory cytokines. All of these proteins are dispensable for virus replication in cell cultures, but in mouse models of disease their deletion reduces disease.

Middle East respiratory syndrome

Cases of MERS are sporadic but occur primarily in Middle East countries. The virus is zoonotic, with transmission from camels to people the most likely route of human infections. However, infections often occur among people with no known contacts with camels. Viruses similar to MERS-CoV have been recovered from bats. Many questions about the epidemiology of MERS-CoV remain unanswered. Countries most affected include those in the Arabian Peninsula (Bahrain, Iran, Jordan, Kuwait, Lebanon, Oman, Qatar, Saudi Arabia, United Arab Emirates, and Yemen) and antibody positive camels are common in these countries. In humans, disease begins with coughing, fever, and breathing problems but may progress to pneumonia and kidney failure. The case fatality rate is greater than 30% but casual transmission from person to person is very rare. Most disease clusters occur among hospital workers caring for patients (if adequate infection control measures are not followed). Since 2012 cases have been reported in 27 countries and over 850 deaths have been recorded.

While MERS-CoV and SARS-CoV are both betacoronaviruses, they derive from different sources and have unique characteristics. SARS-CoV-1 has a tropism for ciliated respiratory epithelial cells and its receptor is ACE2. In contrast MERS-CoV has a tropism for nonciliated respiratory epithelial cells and its receptor is DPP4. Continued comparison of these two viruses will provide new insights into CoV transmission and pathogenesis. The propensity for CoV recombination, and their ubiquitous presence in bats, makes a case for close surveillance and study of these potential human pathogens.

Feline coronavirus and feline infectious peritonitis

FeCoV is the most common virus found in cat fecal samples and is spread through the fecal–oral route. Infection is usually subclinical or associated with a transient, mild diarrhea. Immunity to FeCoV is neither solid nor long lasting. As antibody levels decrease, cats may be reinfected and once again experience mild diarrhea. Most kittens infected with FeCoV clear the virus, but about 15% become chronic shedders. Chronic shedders are at highest risk for developing FIP, a systemic (multiorgan), lethal disease. FIP develops when enteric FeCoV mutates to become capable of infecting monocytes and macrophages. This results in systemic viral infection, as the virus is no longer confined to the digestive tract. The virus that infects monocytes and macrophages is called the FIP biotype.

What kinds of mutations change the cell tropism of enteric FeCoV and result in development of FIP? Mutation occurs independently within each cat, with every FIP biotype virus having unique genetic features. Specific viral genes (e.g., the orf3C protein) are important for the change in cell tropism. After changing cell tropism, the virus continues to mutate, becoming better adapted to peritoneal macrophages. This occurs despite preexisting host immune responses. Early signs of FIP are nonspecific and include anorexia, weight loss, inactivity, and dehydration. FIP can occur in any age cat, but cats less than 1 year of age or greater than 10 years of age are more susceptible. Constant virus replication in macrophages leads to B-cell activation and production of nonprotective (nonneutralizing) antibodies. In fact, immune complexes are damaging as they activate the complement system and lead to immune-mediated vasculitis. The classic lesions associated with the severest form of FIP are aggregates of macrophages, neutrophils, and lymphocytes that form in very small veins. These aggregates are called pyogranulomas and they are associated with development of edema and accumulation of large volumes of protein-rich fluids.

Porcine coronaviruses

There are six known coronaviruses that infect pigs. Four are alphacoronaviruses [transmissible gastroenteritis virus (TGEV), porcine epidemic diarrhea virus (PEDV), porcine respiratory coronavirus (PRCV), and swine acute diarrhea syndrome coronavirus (SADS-CoV)], one is a betacoronavirus [porcine hemagglutinating encephalomyelitis virus (PHEV)] and one is a deltacoronavirus [porcine deltacoronavirus (PDCoV)]. PEDV, SADS-CoV, PDCoV, and TGEV cause diarrhea in pigs and all present with similar symptoms that can include severe diarrhea and vomiting leading to dehydration and death of neonatal pigs. Their main route of transmission is fecal–oral. TEGV was first recognized in the United States in 1946. PEDV was endemic in Europe and Asia for about 30 years before first being detected in the United States in 2013 where it caused the deaths of millions of piglets.

PDCoV was first detected in Asia but its role in disease was not clear until it was isolated from sick pigs in the United States in 2014. SADS-CoV is a highly pathogenic CoV first detected in China in 2016—2017. The importance of pigs as a food resource underscores the importance of these viruses. In addition, pigs cannot be ruled out as a possible source of novel human CoVs. Therefore surveillance efforts to detect novel pig coronaviruses are of utmost importance. Unique recombinant pig coronaviruses (between TGEV and PEDV) have been detected in Italy and Germany as a result of such efforts.

In this chapter we have learned that:

- Members of the family *Coronaviridae* are large, positive-strand RNA viruses.
- They are enveloped, with a helical nucleocapsids.
- A long spike (S) protein form extends 16—21 nm from the surface of virions. S is the attachment and fusion protein.

- mRNAs are synthesized by a process of discontinuous transcription, a process that leads to high rates of RNA recombination.
- Most human CoVs cause mild respiratory or enteric disease. Notable exceptions are SARS-CoV-1, SARS-CoV-2, and MERS-CoV.
- Members of the family *Coronaviridae* are notable for the mechanism of transcription, which is discontinuous: RdRp moves from one location on the template RNA to a distant location. This process leads to high rates of RNA recombination during genome replication.

Reference

Terada, Y., et al., 2014. Emergence of pathogenic coronaviruses in cats by homologous recombination between feline and canine coronaviruses. PLoS One 9 (9), e106534. Available from: https://doi.org/10.1371/journal.pone.0106534.

19

SARS-CoV-2/COVID-19

After reading this chapter, you should be able to answer the following:

- When and where was the first outbreak of pneumonia linked to severe acute respiratory syndrome coronavirus 2 (SARS-CoV-2)?
- What aspects of disease and transmission allowed the virus to spread rapidly worldwide?
- How is the S protein activated or primed for binding and fusion?
- What is the host cell receptor for SARS-CoV-2 and what types of cells express it?
- Which SARS-CoV-2 protein is the target of most currently approved vaccines?
- What viral protein is targeted by nirmatrelvir?
- How do remdesivir and molnupiravir, two nucleoside analogs, differ in their mechanism of action?
- What are variants of concern (VOC) and why have variants emerged?

SARS-CoV-2 is a member of the family *Coronaviridae*, subfamily *Orthocoronavirinae*, genus *Betacoronavirus*. SARS-CoV-2 is the cause of the COVID-19 pandemic that originated in China, late in 2019, and is ongoing in 2022. The first genome sequence of SARS-CoV-2 was released on January 9, 2020, revealing its relationship with the coronavirus that caused a 2002 outbreak of SARS. An easily transmitted respiratory virus, SARS-CoV-2 spread worldwide in a matter of weeks. Spread was facilitated by the ability of the virus to replicate asymptomatically in some individuals, as the earliest detection efforts centered on patients with symptoms.

Coronaviruses are large plus-strand RNA viruses with a distinctive virion morphology. Enveloped virions are spherical and are decorated with long, prominent glycoprotein spikes (Fig. 19.1). Betacoronaviruses include: bat coronavirus (BtCoV), porcine hemagglutinating encephalomyelitis virus (PHEV), murine hepatitis virus (MHV), human coronavirus-4408 (HCoV-4408), human coronavirus-OC43 (HCoV-OC43), and human coronavirus-HKU1 (HCoV-HKU1), SARS-CoV-1, SARS-CoV-2, and Middle East respiratory syndrome coronavirus (MERS-CoV). A phylogeny of coronaviruses is shown in Fig. 19.2. SARS-CoV-2 is a close relative to CoVs from bats and pangolins in Asia.

Genome organization

The SARS-CoV-2 genome (Fig. 19.3) is ~29,700 nucleotides and its organization closely resembles that

Viruses.
DOI: https://doi.org/10.1016/B978-0-323-90385-1.00028-5

© 2023 Elsevier Inc. All rights reserved.

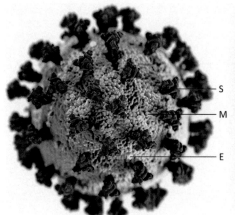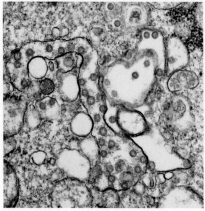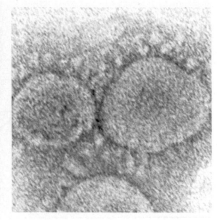

FIGURE 19.1   Left panel: Model and electron micrographs of SARS-CoV-2. Middle panel: Electron micrographic image of SARS-CoV-2 virions in cross-section. Virions are colorized blue. Virions contain cross-sections through the viral genome, seen as black dots. Right panel: Negative stain of SARS-CoV-2. Source: *(Left panel) CDC public health image library # 23312. Content providers CDC/Alissa Eckert, MSMI; Dan Higgins, MAMS. (Middle panel) CDC public health image library # 23591. Content providers CDC/Hannah A. Bullock; Azaibi Tamin. (Right panel) CDC public health image library #23641. Content providers CDC/Hannah A. Bullock and Azaibi Tamin.*

of other betacoronaviruses. It encodes 29 proteins including 16 nonstructural proteins (nsp1–nsp16), 4 structural proteins [spike (S), envelope (E), membrane (M), and nucleocapsid (N)] and is predicted to produce 8 accessory proteins (3a, 3b, 6, 7a, 8, 9b, 9c, and 10). There are long 5′ and 3′ untranslated regions (UTRs) that have important cis-acting regulatory functions. The majority of the genome encodes two large polyproteins, pp1a and pp1ab (alternatively called REP1a and REP1b polyproteins) that are proteolytically processed to generate the 16 nsps. The nsps have critical roles in genome replication, transcription, modulation of the cell environment, and counteracting host defenses such as interferon (IFN) signaling. The longer polyprotein, pp1ab, is produced by a ribosomal frameshift at a site just upstream of the open reading frame (ORF) 1a stop codon. A minus 1 frameshift occurs about 25% of the time, leading to production of the pp1ab. The shorter pp1a is proteolytically processed to produce nsps 1–11. The pp1ab product is cleaved to produce nsps 1–10 and 12–16. The cleavages are mediated by the virally encoded proteases nsp3 and nsp5, also known as the papain-like protease (PL^{pro}) and 3C-like protease (3CL^{pro} or main protease, M^{pro}), respectively.

Overall, SARS-CoV-2 shares 79% sequence identity with SARS-CoV-1 and 50% with MERS-CoV. Amino acid identity of the structural genes is 90% between SARS-CoV-1 and SARS-CoV-2. The S protein is the most divergent. Both viruses bind the receptor angiotensin converting enzyme 2 (ACE2) but their receptor binding domains (RBS) share only 73% amino acid similarity. The S1/S2 cleavage junction of SARS-CoV-2 has an insertion of four amino acids at the S1/S2 cleavage site making it a target for cleavage by furin-like proteases. It is likely that differences in S are a major factor in the differences in transmissibility and pathogenesis of the two viruses. SARS-CoV-2 binds ~10 times more tightly to ACE2 than does SARS-CoV-1. S1/S2 processing by furin-like proteases probably increases the host cell range of SARS-CoV-2.

Virion and structural proteins

The SARS-CoV-2 virion is similar to other coronaviruses; it is an enveloped virion, spherical in shape, with a diameter of 118–136 nm. There are three envelope-associated proteins, M, E, and S and a nucleocapsid protein (N) that binds to genomic RNA. The virion envelope derives from host cell endosomal membranes. M is the most abundant virion protein. It is a 25 kDa (222 amino acid) transmembrane protein that is the key mediator of budding. M contains three transmembrane segments, a short amino-terminal ectodomain, and a large carboxy-terminal endodomain. M interacts with both N and E to form scaffolds for virion assembly. M is glycosylated on its ectodomain. E is a 75-amino acid membrane protein that forms ion channels (thus E is a viroporin). High levels of E are expressed in infected cells and it is found in membranes of the endoplasmic reticulum Golgi intermediate compartment (ERGIC). E is a cation selective channel. Its ion channel forms by assembly of five transmembrane domains to form a helical bundle around a central pore. The E ion channel is permeable to Na$^+$, K$^+$, Ca^{2+}, and Mg^{2+}. As Ca^{2+} is an intracellular messenger and a trigger for the activation of the inflammasomes, E may be an important

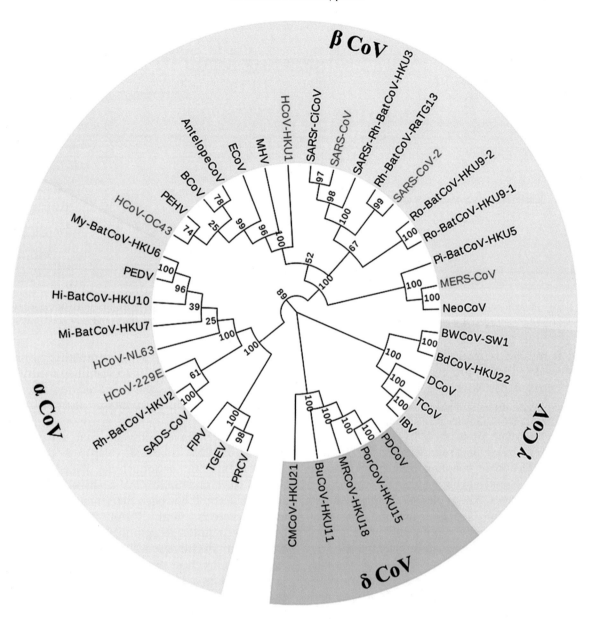

FIGURE 19.2 A phylogeny showing relationships among viruses in the subfamily *Orthocoronavirinae*. The phylogeny was based on nt sequences of RdRp, constructed by the neighborjoining method (Xiu et al., 2020). Source: *From Xiu, L., Binder, R.A., Alarja, N.A., Kochek, K., Coleman, K.K., Than, S.T., et al., 2020. A RT-PCR assay for the detection of coronaviruses from four genera. J. Clin. Virol. 128, 104391. https://doi.org/10.1016/j.jcv.2020.104391.*

determinant of pathogenesis. E also functions in mediating virion budding and release. Virions contain about 20 molecules of E.

S is a transmembrane glycoprotein of ~1273 amino acids. S assembles into trimers that form the visible spikes that extend 15−20 nm from the envelope. S monomers have 22 potential N-linked glycosylation sites and 2 potential O-linked glycosylation sites. Glycans have been identified as populating between 17 and 21 of the N-linked sites and both O-linked sites. S functions in attachment and fusion. The SARS-CoV-2 S precursor has a furin cleavage site at the S1/S2

boundary. A second cleavage site within S2 (S2′ site) is cleaved by host transmembrane proteases such at TMPRSS2. Both cleavages are required to fully activate S for attachment and fusion. N is the 419-amino acid RNA-binding protein. It associates with genomic RNA to form the nucleocapsid.

Nonstructural and accessory proteins

SARS-CoV-2 nsps are encoded by ORF1a and ORF1b to produce the polyproteins pp1a and pp1ab. A minus 1

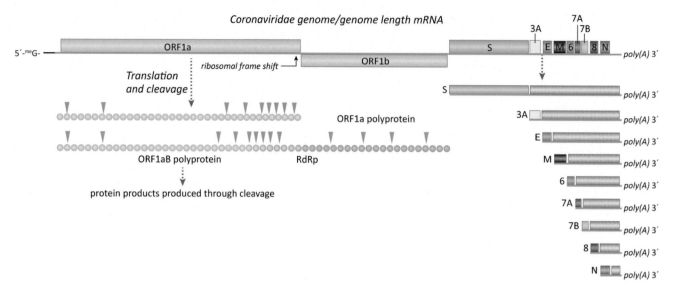

FIGURE 19.3   Illustration showing genome organization and gene expression strategy of SARS-CoV-2. Full-length mRNA is translated to produce polyprotein precursors for nonstructural proteins. Subgenomic mRNAs are used to produce structural and accessory proteins.

ribosomal frameshift is required for production of the longer polyprotein. Both pp1a and pp1ab are processed by virally encoded proteases to generate a total of 16 nsps. The cleavages are mediated by nsp3 (a papain-like protease, also referred to as PL^{pro}) and nsp5 (also referred to as the main protease, M^{pro} or 3C-like protease, 3CL^{pro}). Nsp3, nsp4, and nsp6 have transmembrane domains and function in coopting cellular membranes for viral replication and anchoring the viral replication complex to membranes. The replicase complex consists of nsp12, nsp7, and nsp8. Nsp12 contains the RdRp catalytic site but has little activity as a purified protein. Nsp7 is a ssRNA-binding protein that forms complexes with nsp8. These complexes may function as a "processivity clamp" that keeps the RdRp in contact with the RNA substrate to stimulate RNA synthesis. It has also been suggested that nsp8 may also have primase activity, synthesizing the short primers that are required by nsp12.

Nsp12 also possesses a recently described catalytic domain called nidovirus RdRp-associated nucleotidyltransferase (NiRAN) that can catalyze nucleotidyl monophosphorylation of RNA and protein substrates. The N-terminal region of nsp9, a small RNA-binding protein, binds the catalytic pocket of the NiRAN domain. Nucleotidyl monophosphorylation might be the first step in generating mRNA cap structures in association with the RNA triphosphatase activity of nsp13. An alternative proposal is that nsp9 itself may be GMPylated by the NiRAN to serve as a primer for RNA synthesis by nsp12. Proofreading activity is carried out by nsp14. Nsp14 contains two catalytic domains, an N-terminal exoribonuclease (ExoN) and a C-terminal N7-methyltransferase (N7-MTase) that functions in methylation of mRNA cap structures.

The ExoN activity of nsp14 is stimulated by binding to nsp10. Nsp10 also appears to stimulate methyltransferase activity of nsp16, also required for cap synthesis. Nsp13 has 5' triphopsphatase and RNA helicase activities.

Other nsps have activities that are not directly involved in RNA synthesis. Nsp15 is the nidovirales uridylate specific endonuclease, NendoU. Nsp15 of SARS-CoV-1 and SARS-CoV-2 share over 97% amino acid similarity. It has been determined that SARS-CoV-1 nsp15 protects viral RNAs from recognition by pattern recognition receptors by cleaving the 5'-polyuridines from negative-sense viral RNA. This dampens recognition by host sensors of viral infection. SARS-CoV-2 nsp1 is a 180 amino acid protein that also has endonuclease activity. It shares 84% amino acid identity with SARS-CoV-1 nsp1. SARS-CoV-2 nsp1 downregulates host cell translation by binding the 40S ribosomal subunit and cleaving host mRNAs near their 5' untranslated regions. SARS-CoV-2 nsp1 also binds to the 80S ribosome and obstructs mRNA binding. Nsp1 does not cleave viral specific mRNAs. It has been demonstrated that the leader sequence present at the 5' end of each coronaviral mRNA protects them against cleavage by nsp1. The exact mechanism by which this occurs is currently unknown.

In addition to structural proteins and nsps, a third group of coronviral proteins are broadly referred to as accessory proteins. The sequences of these proteins are not well conserved across the family and vary even among closely related viruses. Some have been determined to be nonessential for replication in cultured cells, but required for robust virus replication in natural hosts. Information about SARS-CoV-1 and

SARS-CoV-2 accessory proteins is beginning to emerge. The 3a protein (274 amino acids) has a transmembrane domain and forms an ion channel. Thus 3a is a viroporin. It appears to participate in regulating apoptosis and autophagosome activity of the cell. Accessory protein 6 is a potent IFN antagonist and SARS-CoV-2 accessory protein 8 regulates immune evasion by targeting major histocompatibility complex (MHC) class I for lysosomal degradation. In summary, SARS-CoV-1 produces 16 recognized nsps as well as additional accessory proteins involved in modulating cell physiology and antiviral responses and participating in viral mRNA synthesis. Any of these viral proteins are putative targets for development of antivirals; understanding their myriad roles in virus replication will inform ongoing drug development efforts.

Overview of replication

Binding and entry

The replication cycle of SARS-CoV-2 begins with binding of the S to the host cell receptor, ACE2. ACE2 is a plasma membrane-bound zinc metallopeptidase whose substrates are circulating vasoactive peptides. ACE2 removes a single amino acid from target peptides such as angiotensin, playing an essential role in the renin-angiotensin system which regulates cardiovascular homeostasis. ACE2 is expressed in the lungs, cardiovascular system, gut, kidneys, central nervous system, and even adipose tissue. In the cell membrane, ACE2 is closely associated with transmembrane protease/serine subfamily member 2 (TMPRSS2). Both ACE2 and TMPRSS2 are key players in entry of SARS-CoV-2.

The S protein has both receptor binding and fusion activities. S is produced in the infected cell as a precursor protein that requires cleavage at two sites in order to be primed to carry out receptor attachment and membrane fusion. The details of S processing have been the subject of numerous studies, in part to identify therapeutics that might inhibit the process. The cleavage site that produces S1 and S2 is a polybasic site (PRRAR) that is cleaved by furin-like proteases within infected cells. Cleavage at the S1/S2 boundary is required for conformational changes that allow efficient binding to the ACE2 receptor. The second cleavage is within S2 at a site referred to as the S2' cleavage site. This cleavage is mediated by transmembrane proteases such as TMPRSS2, after receptor engagement (Fig. 19.4). Some studies indicate that SARS-CoV-2 S can also be cleaved by cathepsins, lysosomal/endosomal cysteine proteases. The S2' cleavage activates S for fusion: major structural rearrangements allow insertion of the hydrophobic domain of S2 into host membranes thereby bringing host and viral membranes into close proximity. Membrane fusion allows release of the nucleocapsid into the cytoplasm.

Amplification

Production of viral proteins begins with translation of the viral genome (a capped and polyadenylated

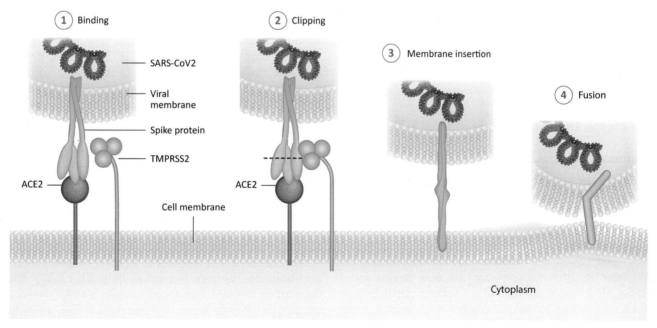

FIGURE 19.4 SARS-CoV-2 binds to ACE2 and TMPRSS2 on host cells. TMPRSS2 cleaves both S and ACE2. S cleavage leads to protein rearrangements exposing the hydrophobic domain on S2 that inserts into the host membrane.

mRNA) to produce the nsps required for RNA synthesis. Production of nsp1 shuts off host mRNA translation by binding to ribosomes; this allows the host translation apparatus to be directed to viral mRNA synthesis and dampens antiviral responses. Nsp1 does not inhibit the translation of viral mRNAs containing the viral 59 nt leader RNA sequence. This probably is achieved by the secondary structures assumed by the leader sequence and the manner in which they direct ribosome binding.

Nsp3 and nsp5 are viral proteases that cleave the precursor polyproteins pp1a and pp1ab to produce active products. Nsp5 is referred to as the coronavirus main protease, M^{pro} as it carries out the majority of the cleavages. Following production of nsps, viral replication organelles are formed by rearrangement of the ER. Synthesis of the full set of viral cRNAs and mRNAs occurs in association with membranes. Replication strongly favors synthesis of mRNAs.

Assembly and release

Synthesis of envelope-associated structural proteins S, M, and E takes place on rough ER. S is cleaved into S1/S2 products by furin-like proteases. S, M, and E are glycosylated. Synthesis of N is cytosolic. N binds to genome length mRNA to form nucleocapsids, and by association with M and E, directs budding into ER membranes. Studies show that M is densely packed into membranes and when expressed alone in cells drives the release of enveloped M-containing particles. This result shows that M is a major driver of membrane budding. When M is coexpressed with either N or E, virus-like particles are released. After budding into ER vesicles virions are release by exocytosis, a process whereby the vesicles fuse with the plasma membrane releasing their cargo.

Origins and pathogenesis of SARS-CoV-2

Origins of SARS-CoV-2

The origins of SARS-CoV-2 and the current pandemic (ongoing in 2022) began in December 2019, in the city of Wuhan, Hubei Province, China (Table 19.1). A cluster of 27 cases of unexplained pneumonia was reported to the World Health Organization on December 31. The earliest recognized cases in Wuhan date back earlier, to December 8, 2019. Given what we now know about SARS-CoV-2, the virus was likely widely circulating in Wuhan by early to mid-December and had already spread to distant countries. On January 9, 2020 China announced the discovery of a novel coronavirus as the cause of the pneumonia outbreak in Wuhan and

researchers released genome sequence data. By later in January human-to-human transmission was confirmed and on January 23 the city of Wuhan (home to more than 12 million people) was shut down and the government began constructing new medical facilities to care for the influx of patients.

In February, 2020 the International Committee on the Taxonomy of Viruses (ICTV) named the novel virus SARS-CoV-2, based on its high level of genetic similarity to SARS-CoV. The WHO adopted the name "coronavirus disease 2019" (COVID-19) as the name of the disease. Countries across the globe implemented a variety of strategies to stop the spread of SARS-CoV-2. Unfortunately, the high transmissibility of the virus by persons with asymptomatic infections thwarted all but the most severe measures. In the United States, persons entering from China were screened for symptoms of infection, such as fever. In reality this did little to prevent a US outbreak as the virus was most likely in the country before these measures were implemented, though transmission may have been slowed. Infected travelers from Europe (who were not being screened for the virus) brought the virus from there, resulting in a large outbreak centered in New York City. On March 11, 2020 the WHO officially declared COVID-19 a pandemic. Overall, the disruption of global and local economies during 2020 was massive: schools, businesses, and factories were closed; hospitals were overwhelmed; international shipping was disrupted; travelers were stranded far from home. Medical researchers worldwide dropped ongoing projects and refocused their efforts on developing diagnostic tests, vaccines, and antivirals. Local impacts changed as waves of the virus ebbed and flowed. By October, 2020 over 34,000 million cases, with over 1 million deaths worldwide were confirmed. As of April 30, 2022 the WHO reports over 500 million cumulative confirmed cases and over 6 million cumulative deaths. These numbers do not fully reflect the overall worldwide toll caused by disruption of basic services and health care infrastructure.

It is likely that the *exact* origins of SARS-CoV-2 will never be known with certainty. The first notable cluster of cases maps to Wuhan, China and the virus may have originated there. An early suspect for the origin of the virus was a large "wet market" in Wuhan where live animals were housed and sold. As coronaviruses infect a variety of animals, the wet market environment would allow for importation of a virus into the city, as well as spread among many species of animals, before jumping into humans. Samples from the wet market taken after the start of the outbreak failed to yield SARS-CoV-2-like viruses, but this does not rule out the wet market as the original source of the virus. A second scenario for the origin of SARS-CoV-2 points to the Wuhan Institute of Virology (WIV) where coronaviruses are one area of

TABLE 19.1   Timeline of developments in the SARS-CoV-2 pandemic.

Developments in the SARS-CoV-2 pandemic[a]

Dec 08, 2019	Onset of first recorded case of pneumonia of unknown origin in Wuhan, China
Dec 31, 2019	Report of 27 cases of pneumonia of unknown cause in Wuhan, China is released
Jan 09, 2020	Release of the genome sequence of a novel coronavirus associated with the pneumonia outbreak in Wuhan, China
Jan 2020	Evidence accumulates for person-to-person transmission as well as asymptomatic and presymptomatic transmission
Jan 23, 2020	City of Wuhan enters lockdown
Jan 21, 2020	US CDC announces the first US case of COVID-19 (from Washington State)
Jan 26, 2020	California, United States confirms its first case of COVID-19
Feb 11, 2020	ICTV names the novel coronavirus SARS-CoV-2. WHO names the disease COVID-19
Feb 2020	ACE2 identified as the cell entry receptor for SARS-CoV-2
Mar 01, 2020	New York State, United States confirms its first case of COVID-19
Mar 09, 2020	Italy institutes a national lockdown with closure of nonessential shops and businesses
Mar 11, 2020	WHO declares COVID-19 a pandemic
Mar 19, 2020	Governor of California, United States, issues state-wide stay-at-home orders
Mar 20, 2020	Governor of New York State, United States, issues state-wide stay-at-home orders
Mar 26, 2020	Neurological symptoms of COVID-19, including loss of taste and smell first described in Milan, Italy
Mar 23, 2020	United Kingdom announces first lockdown
Mar 31, 2020	SARS-CoV-2 shown to replicate in several species of domesticated and laboratory animals
April 2020	Over 100 vaccine candidates in development
Nov 05, 2020	WHO prequalifies use of dexamethasone in certain patients
Nov 21, 2020	US FDA issues an EUA for monoclonal antibodies (casirivimab and imdevimab) directed at the spike protein
Dec 11, 2020	US FDA authorizes Pfizer BioNTech mRNA vaccine for emergency use
Dec 18, 2020	US FDA authorizes Modern mRNA vaccine for emergency use
Jan 14, 2021	WHO approves Pfizer/BioNTech mRNA vaccine
Feb 02, 2021	US CDC requires masks for all travelers into, within, or out of the United States on public transport
Feb 03, 2021	WHO approves Moderna mRNA vaccine
2021	During 2021, eight vaccines are approved by the WHO
Dec 08, 2021	US FDA issues an EUA for monoclonal antibodies tixagevimab and cilgavimab, administered together.
Dec 22, 2021	US FDA approves emergency use of a combination of nirmatrelvir (coronavirus 3C-like protease inhibitor) and ritonavir (a booster of protease inhibitors) to treat COVID-19 in certain patients
Dec 23, 2021	US FDA approves emergency use of molnupiravir, a nucleoside analog, to treat COVID-19 in certain patients
Feb 09, 2021	US FDA approves emergency use of monoclonal antibodies bamlanivimab and etesevimab administered together to treat COVID-19 in certain patients
Feb 10, 2022	WHO prequalifies use of tocilizumab, a mAb with activity against the interleukin 6 receptor for some critically ill patients
Apr 08, 2022	US CDC no longer enforcing a mask mandate on public transportation.
Apr 22, 2022	WHO prequalifies use of nirmatrelvir and ritonavir in certain patients
Apr 25, 2022	WHO prequalifies use of remdesivir (a nucleoside analog, delayed chain terminator) in certain patients

[a]Red type indicates information pertinent to the SARS-CoV-2 virus, blue type indicates information pertinent to vaccine development and approval, green type indicates information pertinent to drug development and approval.

study. Scientists from the WIV conducted research on the origins of SARS-CoV-1 and hundreds CoVs from bats are stored there. Some of those viruses have sequences quite similar to SARS-CoV-1 and some were used in research directed at understanding how bat CoVs become adapted to other animals and humans. It is possible that a virus from the laboratory infected researchers and was then transmitted to individuals outside of the lab. The ability of SARS-CoV-2 to replicate in, and be transmitted among, inapparently infected persons makes it virtually impossible to point to a single point source for the virus. The important take home message is that coronaviruses can and do jump between species when given the opportunity.

Route of transmission and target cells

The main route of transmission of SARS-CoV-2 is airborne. Infection is acquired by exposure to microdroplets exhaled by infected individuals. Early in the COVID-19 pandemic, there was little detailed information available about transmission. It was unclear if transmission occurred via larger droplets and fomites (as is the case for influenza viruses) or fine aerosols (as is the case for the highly contagious measles and chickenpox viruses). Large respiratory droplets (often characterized as $5-20\,\mu m$) are released via coughing and sneezing. They remain in the air for a few minutes and rapidly fall out of the air column to contaminate surfaces near the infected individual. Transmission from a contaminated surface occurs when a person touches that surface and then touches their eyes, nose, or mouth. Transmission via aerosols involves particles less than $3\,\mu m$ that are expelled during normal breathing. Aerosols remain airborne for long periods of time, traveling many meters. SARS-CoV-2 can indeed be transmitted via aerosols. This has a huge impact on community spread and speaks to the need for wearing high quality masks and avoiding indoor spaces during an outbreak. Hand washing and cleaning surfaces are good practices, but will not prevent infection by the aerosol route. In addition to the respiratory tract, SARS-CoV-2 can replicate in intestinal epithelial cells and be shed in feces. SARS-CoV-2 RNA can be detected in sewage and this has been used as a method to detect the amount of virus in communities. It has not been determined if sewage contains *infectious* material and it is likely that the fecal–oral route is not a major factor of spread among humans, though it cannot be ruled out as a source of transmission to animals in the wild.

The main target cells for SARS-CoV-2 are bronchial epithelial cells and pneumocytes. Infection of these cells can cause acute lung damage. After establishing infection in the respiratory tract, virus spreads throughout the body as ACE2-positive cells are found in many tissues including cardiovascular system, gut, kidneys, central nervous system, and adipose tissue. The presence of a polybasic furin cleavage at the S1/S2 junction expands the range of cells capable of producing infectious virions beyond the respiratory tract.

Disease

COVID-19 is a disease with a wide spectrum of clinical manifestations ranging from asymptomatic to mild/moderate (common cold-type, fatigue, body aches, fever, pneumonia, and dyspnea) to severe (acute cardiac injury, multiorgan failure). Severe disease frequently requires hospitalization for provision of respiratory support. The inability of hospitals to care for large numbers of patients requiring ventilation was a factor in some governments calling for widespread lockdowns. The range of clinical outcomes is highly associated with risks factors including advanced age and chronic medical conditions such as obesity, cardiovascular disease, diabetes, or immunosuppression. The incubation period averages 5 days but ranges from 1 to 14 days. Healthy persons under 50 are the most likely to experience mild to moderate disease. Many of the symptoms of COVID-19 are similar to other respiratory infections such as influenza but some unusual symptoms include loss of taste and smell. COVID-19 patients who develop acute respiratory distress also have high rates of delirium and encephalopathy, pointing to central nervous system involvement.

COVID-19 is usually described as a respiratory disease but the virus often causes systemic infection comprising multiple organ systems. Complications of infection can be chronic with significant impacts on quality of life. Lingering consequences of SARS-CoV-2 infection are called "post-COVID19 syndrome" or "long COVID." The pathology of long COVID is not completely understood but probably results from direct effects of the virus as well as an aberrant inflammatory response. Long COVID generally refers to symptoms lasting longer than 2 months. These can be residual symptoms or organ damage remaining after acute infection or new symptoms that appear after an asymptomatic or mild infection. Reported symptoms include profound fatigue, muscle weakness, persistent breathlessness, headache, joint pain, chest pain, sleep difficulties, anxiety or depression, and memory or cognitive impairments. Most children infected with SARS-CoV-2 have asymptomatic or mild infections but very small numbers develop multisystem inflammatory syndrome 4−6 weeks after recovery from the initial infection. This is a hyperinflammatory condition whose symptoms include fever, abdominal pain, and diarrhea. Systemic vascular damage can involve coronary arteries and the brain. Patients have autoantibodies and elevation of specific inflammatory markers. The myriad possible outcomes of SARS-CoV-2 infection, from asymptomatic to a chronic lingering disease suggests multifactorial/complicated mechanisms of pathogenesis. Aberrant inflammatory responses and endothelial cell damage are two leading suspects. It remains to be seen how the availability of vaccines and antiviral therapies will impact the consequences of infection.

Vaccines

With the joint efforts of scientists from all over the world, more than 100 vaccines have been developed or are in development. The COVID19 Vaccine Tracker (https://covid19.trackvaccines.org) accessed on May 1, 2022, listed 195 vaccine candidates, 694 ongoing vaccine trials and 75 countries with vaccine trials. Of the vaccines in trials, 47 were in Phase 1 trials, 66 in Phase 2 trials, and 73 in Phase 3 trials. The number of vaccines approved for use in at least one country was 38. This is an amazing number, considering that viral sequences were first released in January 2020. Types of vaccines approved or in trials included subunit protein and peptide vaccines, virus-like particle recombinant vaccines, mRNA vaccines, DNA vaccines, inactivated (killed) SARS-CoV-2, replicating viral vectored and nonreplicating viral vectored vaccines. Most recombinant vaccines target the S protein. The approval of recombinant vaccines for human use, within a year of the identification of SARS-CoV-2, was a huge short-term achievement but was built upon decades of work preparing for the next pandemic virus. Outbreaks of Ebola hemorrhagic fever, SARS-CoV-1, and MERS-CoV in humans, as well as the spread of animal pathogens such as African swine fever virus, porcine reproductive and respiratory syndrome virus, and avian influenza among others, kept a needed focus on vaccine preparedness. The culmination of worldwide vaccine development efforts was approval of the first mRNA vaccines and the first DNA vaccine for use in humans. Traditional killed/inactivated whole viral vaccines have also been approved for use. Inactivated vaccines contain all of the structural proteins present in virions. These include S, E, M, and N. Some of the general strategies used to generate vaccines are illustrated in Fig. 19.5. In the coming months and years some vaccines will be discontinued and others approved. Vaccines approved to date have good records of safety and are moderately to highly effective at protecting vaccinated individuals from hospitalization and death. Changes will be made to reflect the emergence of new variants and use of new adjuvants to provide more durable protection will be investigated. New vaccine formulations may be targeted to specific populations. It is likely that vaccines will need to be reformulated to reflect the sequences of emerging variants. The use of vaccines that protect against severe disease, but not infection or transmission may actually drive the emergence of new variants. Vaccine development will remain a high priority for some time.

Treatments

Health care professionals on the front lines of the COVID-19 pandemic were faced with the enormous task of caring for critically ill patients with insufficient resources. Even the most basic personal protective equipment (N95 respirators, gowns, and gloves) was in short supply in the first months of the pandemic. Mechanical ventilators and oxygen supplies were often inadequate as well. There were no preexisting protocols or therapeutics specific to the treatment of COVID-19 patients. There were some missteps, but with surprising speed, clinical trials were established to vet proposed therapeutics and treatment modalities.

Currently there are a handful of approved drugs for treating COVID-19 patients. Some are monoclonal antibodies (mAb) that neutralize the SARS-CoV-2 S protein. It is likely that mAb therapeutics will need to be updated regularly to keep up with the emergence of SARS-CoV-2 variants. Antiviral agents that target virus replication include protease inhibitors and RNA replication inhibitors. Remdesivir is a nucleoside analog, categorized as a "delayed chain terminator." It is incorporated into a growing RNA chain and elongation is halted 3−5 nt downstream of the insertion. Because coronaviruses encode a proofreading enzyme (3′ to 5′-ExoN) that can remove a nt from a growing RNA chain, chain terminating ribonucleosides can be removed, but the delayed action of remdesivir appears to overcome this. Molnupiravir, is another nucleoside analog approved to treat COVID-19. It is not a chain terminator but instead is a mutagen that causes G-to-A and C-to-U transitions. The highly mutated genomes produced in the presence of the drug are not infectious. A protease inhibitor, nirmatrelvir, that targets the SARS-CoV-2 nsp5 protease (3CL^{pro} or M^{pro}) has also been approved. Dexamethasone, a systemic corticosteroid that dampens inflammatory responses, improves outcomes for some patients with severe disease. The WHO has prequalified the use of tocilizumab, a mAb with activity against the interleukin 6 receptor, for some critically ill patients.

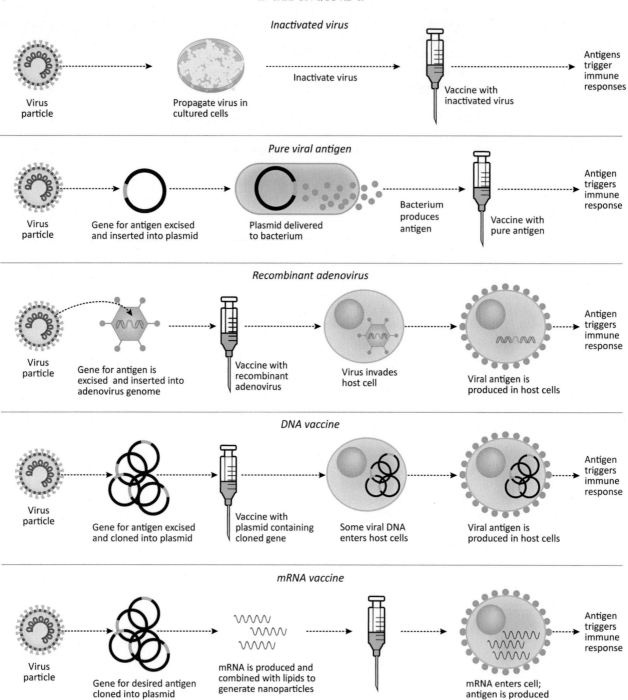

FIGURE 19.5 Various strategies used for development of SARS-CoV-2 vaccines.

This treatment also targets an overzealous immune response, rather than the virus itself.

Viral variation

RNA viruses have high mutation rates and selection can drive emergence of new viral lineages. This is well documented for influenza viruses and hepatitis C virus. During replication coronaviruses accumulate point mutations as well as small insertions and deletions, despite the presence of some proofreading by ExoN. The mechanism of discontinuous transcription that is required for synthesis of subgenomic mRNAs also leads to high rates of copy choice recombination. As illustrated in Fig. 19.6, high rates of mutation provide a pool upon which selection can act. Selection for a virus that replicated to high levels in a human host,

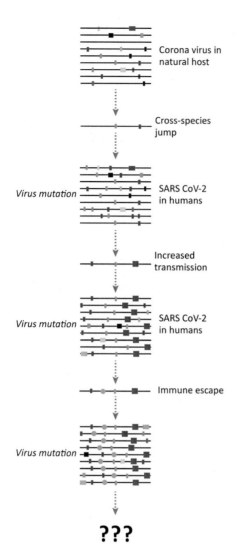

Corona virus in
natural host

Cross-species
jump

Virus mutation — SARS CoV-2
in humans

Increased
transmission

Virus mutation — SARS CoV-2
in humans

Immune escape

Virus mutation

???

FIGURE 19.6 A combination of mutation and selection drive emergence of SARS-CoV-2 variants.

and was readily transmitted, started this pandemic of historic proportions. As the virus spread among millions of individuals worldwide, new lineages or variants emerged. A challenge for biomedical researchers is to tease out which variants will have an impact on the course of the pandemic. As coronaviruses mutate readily, few isolates will have an identical sequence. Many tens of thousands of individual mutations have been documented in SARS-CoV-2 genomes. But most mutations are deleterious and will quickly be lost. Other have little impact on virus replication and transmission and again, will be lost. A few mutations, or combinations of mutations will become fixed and will spread through a population.

Researchers focusing on SARS-CoV-2 genome sequences pay particular attention to mutations in key viral proteins. For example, mutations in S impact receptor binding and fusion with versions of S that bind ACE2 more tightly being more readily transmitted. Three variants arose during 2020 that displayed increased transmissibility. Naming variants is complex, as phylogenetic analyses are required to delineate relationships among thousands of sequenced genomes. The WHO has attempted to provide a simplified nomenclature for the general public by naming variants of concern (VOC) using the Greek alphabet. The WHO has defined VOCs as being linked with increased transmissibility, pathogenicity, disease severity (increased hospitalization and death), immune escape, diagnostic failure, or vaccine (therapeutic) escape. The Alpha VOC emerged in the United Kingdom in 2020 and contained 10 point mutations and 3 deletions in S. It was more easily transmitted than the virus that first spread out of China to Europe (called B.1). However, the Alpha variant was still neutralized by mAbs based on B.1 virus and sera from vaccinated individuals had high neutralizing activity against the Alpha variant. The Beta VOC was first detected in South Africa and contained six amino acid substitutions and three deletions in S. Beta showed not only higher rates of transmission but appeared to be more lethal than B.1. The Beta variant also had significantly decreased susceptibility to mAb treatments based on B.1 and reduced neutralization by postvaccination sera. The Gamma VOC was first detected in Japan and Brazil while the Delta VOC was first detected in India in late 2020. The Delta variant was yet again more highly transmissible than previous lineages of the virus and by November 2022 had spread to over 179 countries becoming the predominant strains worldwide. The Omicron variant was first detected in South Africa and was spreading widely by late 2021. It has 30 mutations in the S protein and is highly transmissible. By January, the end of January 2022, it had been detected in 57 countries. It is quite clear that the large number of changes in S allows escape from neutralizing antibodies from patients either vaccinated or previously infected with other variants. As vaccines elicit both cell-mediated and antibody responses neutralizing antibody titers tell only part of the story; it appears that early vaccines still protect against hospitalization and death. But the spread of Omicron has prompted many countries to advise booster doses to previously vaccinated individuals, particularly those at high risk for severe disease. It is likely that Omicron will become the predominant SARS-CoV-2 variant and will likely infect much of the world's population, vaccinated or otherwise. The consequences of the Omicron wave remain to be determined.

This chapter has provided a brief summary of SARS-CoV-2 and the COVID-19 pandemic. We cannot predict the future course of this virus but there are likely additional COVID vaccines and boosters in our futures.

In this chapter we have learned that:

- SARS-CoV-2 shares high degree of sequence identity with SARS-CoV-1 and is a member of the family *Coronaviridae*, subfamily *Orthocoronavirinae*, genus *Betacoronavirus*.
- SARS-CoV-2 first emerged in Wuhan, China in late 2019.
- The WHO declared COVID-19 a pandemic on March 11, 2020.
- The receptor for SARS-CoV-2 is ACE2. ACE2-positive cells are found in many tissues including cardiovascular system, gut, kidneys, central nervous system, and adipose tissue.
- SARS-CoV-2 is transmitted via the respiratory route, in aerosols. Some variants are highly transmissible.
- The outcome of infection can range from inapparant to severe disease and death from multiorgan failure.
- Persons with the highest risk for severe disease are the elderly and persons with chronic diseases such as obesity, cardiovascular disease, diabetes, or immunosuppression.
- Long COVID-19 can develop after inapparant or severe infection. Symptoms may persist for weeks or months and include fatigue, muscle weakness, persistent breathlessness, headache, joint pain, chest pain, sleep difficulties, anxiety or depression, and memory or cognitive impairments.
- Antiviral drugs targeting SARS-CoV-2 protease and the replicase complex have been approved.
- Emerging variants may evade immune responses evoked by either vaccination or natural infection.
- Approved vaccines include mRNA, DNA, recombinant protein, viral vectored, and inactivated whole virus vaccines. Recombinant vaccines approved thus far target the S protein.

Reference

Xiu, L., Binder, R.A., Alarja, N.A., Kochek, K., Coleman, K.K., Than, S.T., et al., 2020. A RT-PCR assay for the detection of coronaviruses from four genera. J. Clin. Virol. 128, 104391. Available from: https://doi.org/10.1016/j.jcv.2020.104391.

20

Family *Arteriviridae*

After reading this chapter, you should be able to discuss the following:

- What are the major structural and replicative features of arteriviruses?
- By what mechanism are arterivirus mRNAs synthesized?
- What diseases are caused by arteriviruses?
- Equine arteritis virus (EAV) can cause a long-term persistent infection in stallions. What evidence supports a requirement for the hormone testosterone in this process?

The family *Arteriviridae* is one of eleven virus families in the order *Nidovirales*. Arteriviruses are enveloped viruses with unsegmented plus-strand RNA genomes (Fig. 20.1). Arterivirus genomes are considerably smaller than the coronaviruses but they share many characteristics with them, including overall genome organization and use of discontinuous transcription to synthesize subgenomic RNAs (Fig. 20.2). Arteriviruses lack the notable spike proteins of the coronaviruses. Members of the family include EAV, porcine reproductive and respiratory syndrome virus (PRRSV), lactate dehydrogenase elevating virus of mice, and simian hemorrhagic fever virus (SHFV). In the late 1987 PRRSV emerged as a serious pathogen of domesticated pigs in the United States; it is now found worldwide and some strains are quite virulent.

Genome organization

Arterivirus genomes are organized in the same overall manner as other virus families in the order *Nidovirales*. The 12.7–15.7 kb positive-strand RNA genome contains 10–13 open-reading frames (ORFs). Approximately three-fourths of the genome encodes the nonstructural proteins (nsp) required for transcription and genome replication. The nsps are encoded by two overlapping ORFs. ORF1a contains three to four protease domains and three transmembrane domains. The size of ORF1a is quite variable; for example, PRRSV ORF1a is ~800 bases longer than EAV ORF1a. ORF1b is much more conserved among the arteriviruses. It is expressed by a ribosomal frameshift and encodes the RdRp, a helicase, and the NendoU endoribonuclease. Arterivirus NendoU is related to the coronavirus protein of the same name and hydrolyzes single-stranded and double-stranded RNA. The exact role of NendoU in the arterivirus replication cycle remains unknown.

Structural proteins are encoded downstream of ORF1b. EAV and PRRSV encode eight proteins from short overlapping ORFs. The nucleoprotein (N) is encoded at the very 3′ end of the genome. The remaining structural proteins are all found associated with the lipid envelope. The SHFV genome encodes additional ORFs. These seem to have arisen by a gene duplication of ORFs 2a, 2b, 3, and 4.

Viruses.
DOI: https://doi.org/10.1016/B978-0-323-90385-1.00021-2

© 2023 Elsevier Inc. All rights reserved.

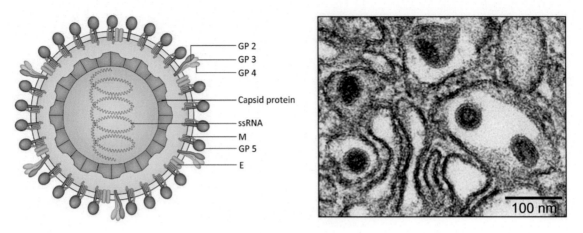

FIGURE 20.1   Virion structure (left) and electron micrograph (right). *Source: Wahl-Jensen et al. (2016).*

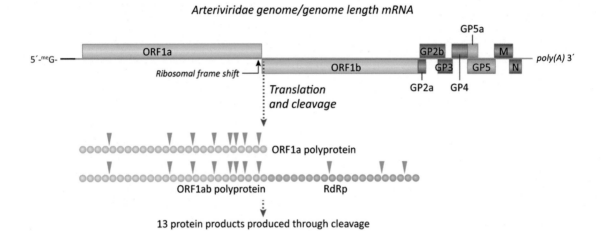

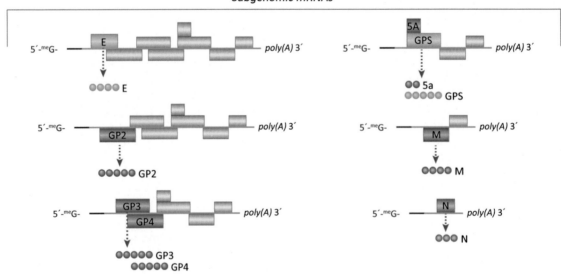

FIGURE 20.2   Genome organization and gene expression strategy of an arterivirus.

Virion structure

Arteriviruses are spherical, 50—60 nm in diameter with spherical nucleocapsid. Arteriviruses lack the notable spike proteins of the coronaviruses; in fact the envelope surface is very smooth. There are seven proteins associated with the envelope (GP2, GP3, GP4, M, E, 5a, and GP5). M, E, and 5a are nonglycosylated. GP2, GP3, and GP4 associate to form a heterotrimer. GP5 is a large glycoprotein with three membrane-spanning domains. All of the envelope proteins appear to be required for formation of infectious virions.

Replication cycle

The overall replication cycle of the arteriviruses is very similar to that of the coronaviruses. Entry occurs via endocytosis. The infecting genome is translated to produce the polyprotein 1a and polyprotein 1ab products; these are cleaved by viral proteases to generate the nsps. Replication is cytoplasmic and takes place in association with double membrane vesicles. Virions form by budding into membrane vesicles (endoplasmic reticulum/Golgi) and are released by exocytosis. Arterivirus genomes are considerably smaller than coronavirus genomes, but the overlapping ORFs encoding the nsps account for almost three-fourths of the total coding capacity (Box 20.1).

Structural proteins are expressed from a 3' coterminal set of subgenomic (sg) mRNAs. The model for mRNA synthesis is the same as previously described for coronaviruses. All sg mRNAs have the same leader sequence, produced by a process of discontinuous transcription. It is believed that discontinuous transcription occurs during synthesis of sg minus strands and these serve as templates for mRNAs.

Arterivirus nucleocapsids assemble in the cytoplasm and bud into the ER or Golgi compartments. Viruses accumulate in vesicles and are released by exocytosis at the plasma membrane. All arteriviruses replicate in macrophages in vitro and in vivo. They often establish persistent infections in their natural hosts. To date they have been isolated only from vertebrate hosts and there are no known human pathogens among the arteriviruses (Box 20.2).

Diseases caused by arteriviruses

Equine viral arteritis

EAV is a common infection of horses worldwide. In the United States, over 70% of horses are seropositive. Many infections are asymptomatic or cause mild upper respiratory tract disease; however, some infections are more severe. An infrequent outcome of infection is abortion in pregnant mares. Some infected stallions become persistent shedders of virus but can be cured of the infection by castration. This indicates a hormonal link (testosterone) to viral persistence although the molecular events associated with persistence have not been determined. During acute infection, virus is transmitted via aerosols and the primary sites of replication are in epithelial cells of the respiratory tract. The virus also infects macrophages and some lymphocytes thereby moving to regional lymph nodes. Infected cells then disseminate EAV throughout the horse. Virus is present in feces, urine, vaginal secretions, and semen.

BOX 20.1

Characteristics

Genomes are positive-strand RNA, 12.7—13.7 kb. Genomes have a 5' cap and a 3' poly(A) tail. Proteins required for RNA synthesis and polyprotein processing are encoded in two large overlapping ORFs, ORF1a and ORF1b, that are expressed from genome length RNA. ORF1a and ORF1b encode long polyproteins that are cleaved by viral proteases to generate a set of nonstructural proteins (nsps). ORF1b is expressed via ribosomal frameshifting. Nsps include an endonuclease (NendoU) common to all members of the order *Nidovirales*.

A leader sequence at the 5' end of the genome is found on all subgenomic mRNAs, supporting a model of discontinuous transcription similar to the coronaviruses. Structural and accessory proteins are encoded from a nested set of 3' coterminal mRNAs. Genome recombination by a copy choice mechanism is not uncommon. Virions are enveloped, 50—60 nm diameter with a spherical nucleocapsid. Structural proteins include an RNA-binding nucleoprotein (N) and seven envelope-associated proteins (M, E, 5a, GP2, GP3, GP4, and GP5). GP5 is the major envelope glycoprotein; it spans the envelope three times. None of the envelope-associated proteins have a long ectodomain, thus the surface of the virion is smooth in appearance.

BOX 20.2

Taxonomy

Order *Nidovirales*

Family *Arteriviridae*

The taxonomy of the family was updated in 2021 (Brinton et al., 2021). The family grouping contains 6 subfamilies, 13 genera, and 26 species. The genus and species names for the viruses discussed in this chapter are listed here.

Genus *Alphaarterivirus*

Species *Alphaarterivirus equid* (equine arteritis virus)

Genus *Deltaarterivirus*

Species *Deltaarterivirus hemfev* (simian hemorrhagic fever virus)

Genus *Betaarterivirus*

Species *Betaarterivirus suid 1* (porcine reproductive and respiratory syndrome virus 1)

Species *B. suid 2* (porcine reproductive and respiratory syndrome virus 2)

Porcine reproductive and respiratory syndrome

PRRSV is an economically important virus that made its appearance in the United States in 1987. The virus is found in both farm-raised and wild pigs. Once introduced into a naive herd, the virus most often spreads by direct contact but can also be spread by the aerosol route. The virus can be introduced into a herd via infected animals, semen, or contaminated fomites. Depending on the virulence of the infecting strain, health of the herd may be unaffected or animals may experience mild to severe disease.

PRRS has two distinct clinical presentations: reproductive failure and postweaning respiratory disease. Reproductive disease causes increased numbers of stillborn piglets, mummified fetuses, premature births, and weak piglets. Reproductive problems arise because PRRS can cross the placenta. If the virus crosses the placenta in the third trimester of gestation piglets may be viremic when born, and may transmit virus for 2–3 months. The respiratory form of the disease affects piglets, causing pneumonia and respiratory distress (Fig. 20.3). Pneumonia develops because the virus infects alveolar macrophage and the infection is cytopathic. PRRSV may also predispose pigs to secondary infections due to widespread destruction of macrophages. The emergence of PRRSV has resulted in the need for increased biosecurity on pig farms. To avoid introducing the virus into a herd, farmers use strict quarantine measures when introducing new animals. They must also purchase breeding stock and semen that is known to be virus free. Sanitation of transport vehicles and strict protocols of fomite and personnel movement between farms are also critical components of an effective program.

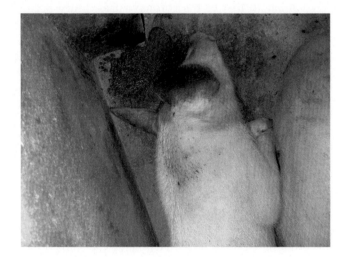

FIGURE 20.3 Pig with PRRS showing cyanosis of the ears. *Source: Dingar: https://commons.wikimedia.org/w/index.php?curid = 3308775.*

Simian hemorrhagic fever virus and related viruses

SHVF was first identified in 1964 in association with an outbreak of hemorrhagic fever that affected several species of captive Asian macaques. The disease was clinically similar to human hemorrhagic fevers and the mortality was quite high, approaching 100%. Macrophages are the primary target for SHFV and cytolytic infection is probably related to viral pathogenesis. It was long suspected that the natural hosts for SHFV were African monkeys. Recent studies have revealed that in fact many species of wild African monkeys are persistently infected with arteriviruses related to SHFV. However, there is a great deal of genetic diversity with viruses from different monkey species sharing only about 50% nucleotide sequence identity. In some areas, up to 40% of monkeys are persistently infected and virus titers in the blood are quite high. Thus it appears that many monkey arteriviruses are quite well adapted to their natural hosts but have the potential to cause severe disease upon transmission to another species.

In this chapter we learned that:

- Arteriviruses are smaller than coronaviruses but their genome organization and overall replication strategy is very similar to the coronaviruses. They replicate in the cytoplasm and use a discontinuous mode of transcription to generate subgenomic RNAs.
- Arteriviruses lack a distinctive spike protein. The N protein associates with genomic RNA to form a nearly spherical nucleocapsid. Seven proteins are associated with the viral envelope.
- To date no arterivirus has been isolated from humans. EAV and PRRSV are economically important animal pathogens. SHFV and related monkey arteriviruses seem to cause little disease in their "natural hosts" but can cause fatal hemorrhagic fever on cross-species transmission.
- EAV can cause a long-term persistent infection in stallions. However, castrating stallions allow the infection to be cleared indicating a link to the hormone testosterone. The molecular mechanism by which testosterone facilities viral persistence is unknown.

References

Brinton, M.A., Gulyaeva, A.A., Balasuriya, U.B.R., Dunowska, M., Faaberg, K.S., Goldberg, T., et al., 2021. ICTV virus taxonomy profile: *Arteriviridae*. J. Gen. Virol. 102, 1632. Available from: https://doi.org/10.1099/jgv.0.001632.

Wahl-Jensen, Victoria, Johnson, Joshua C., Lauck, Michael, Weinfurter, Jason T., Moncla, Louise H., Weiler, Andrea M., et al., 2016. Divergent simian arteriviruses cause simian hemorrhagic fever of differing severities in macaques. mBio 7 (1), e02009−15. Available from: https://doi.org/10.1128/mBio.02009-15.26908578.

21

Family *Rhabdoviridae*

After studying this chapter, you should be able to provide answers to the following questions:

- What are the main characteristics of the members of the family *Rhabdoviridae*?
- What is the general replication scheme of the unsegmented, negative-stranded RNA viruses?
- How does rhabdovirus N protein participate in the switch from mRNA synthesis to genome replication?
- How is rabies virus (RABV) transmitted and how does it reach the central nervous system (CNS)?
- Why is postexposure rabies vaccination an important part of rabies prevention?
- What type of disease is caused by vesicular stomatitis virus (VSV) and why is it of concern to food animal producers?
- What is "pseudotyping" and why is the VSV glycoprotein often the envelope glycoprotein of choice?

Viruses in the family *Rhabdoviridae* have unsegmented, negative-strand RNA genomes. Enveloped virions are bullet-shaped or rod-shaped with helical nucleocapsids (Fig. 21.1). A single glycoprotein (G) decorates the outer surface of virions. There are well over 100 named rhabdoviruses (the family name derives from rhabdos, Greek

for rod), isolated from hosts as diverse as insects, vertebrate animals, and plants (Box 21.1).

The most notorious rhabdovirus is RABV, a member of the genus *Lyssavirus*. The name rabies is based on the Latin *rabidus* meaning raving, furious, or mad, a name very descriptive of animals with this almost invariably fatal neurologic infection. The name of genus to which RABV belongs (*Lyssavirus*) also reflects symptoms of the disease, as in Greek mythology Lyssa was the goddess of mad rage, fury, and crazed frenzy.

VSV is a rhabdovirus in the genus *Vesiculovirus*. Not nearly as well known as RABV, it is nonetheless an important pathogen of hoofed-stock. Its name derives from the blister-like vesicles that form on the mouths, hooves, and teats of infected animals. VSV can be introduced into a herd by insects but can subsequently spread by contact. VSV is economically important, in part, because it is clinically indistinguishable from foot-and-mouth disease virus (FMDV), a member of the family *Picornaviridae*), thus its appearance in a herd is disruptive until confirmatory testing can be done. VSV can also infect humans, causing a generalized febrile illness. VSV is the most thoroughly studied of the rhabdoviruses and serves as the model for their replication. Taxonomy of selected rhabdoviruses is presented in Box 21.2.

© 2023 Elsevier Inc. All rights reserved.

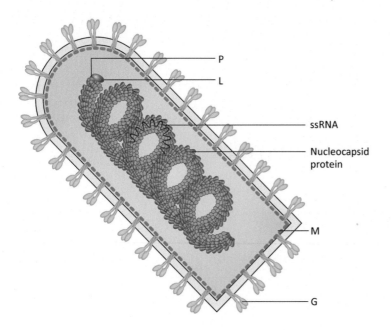
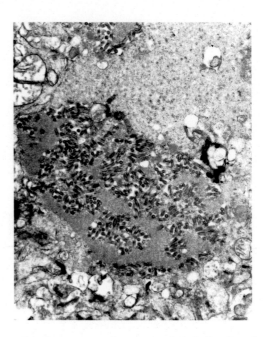

FIGURE 21.1   Virion structure (left) and electron microscopic (TEM) image (right) revealing the presence of numerous dark, bullet-shaped rabies virions within an infected tissue sample. Source: *From CDC/F. A. Murphy. Public Health Image Library ID# 1876.*

BOX 21.1

General characteristics

Rhabdovirus genomes are a single strand of negative-sense RNA, 11–15 kb. Genomes contain five open-reading frames. Short untranslated regions are found at the 3′ and 5′ ends of the genome and cis-acting regulatory sequences that instruct mRNA synthesis are present between each gene. Transcription initiates at a single site at the 3′ end of the genome. Genome replication is cytoplasmic and requires ongoing synthesis of N and P proteins.

Enveloped virions are bullet or rod shaped (~75 nm wide) with helical nucleocapsids. The nucleocapsid consists of the genome, bound by the nucleoprotein (N), a phosophoprotein (P), and the large (L) protein (the RdRp). The matrix (M) protein is associated with the inner side of the envelope. The glycosylated envelope protein (G) assembles into homotrimers to form the protein spikes extending from the outer surface of the envelope.

BOX 21.2

Taxonomy of selected rhabdoviruses

Order *Mononegavirales*
Family *Rhabdoviridae*
Genus *Ephemerovirus*
Species *Bovine fever ephemerovirus* (bovine ephemeral fever virus)
Genus *Lyssavirus*
Species *Rabies lyssavirus* (rabies virus)

Genus *Vesiculovirus*
Species *Indiana vesiculovirus* (vesicular stomatitis Indiana virus)
Species *New Jersey vesiculovirus* (vesicular stomatitis New Jersey)
Species *Chandipura vesiculovirus* (Chandipura virus)

Genome organization

Rhabdoviruses have unsegmented, negative-strand RNA genomes containing five open-reading frames. Rhabdovirus genomes range in size from 11 to 15 kb (Fig. 21.2). There are short ($\sim$50 nt) noncoding regions at the 3' and 5' ends of the genome. The genomes of rhabdoviruses (and other negative-sense RNA viruses) are often shown in diagrams with the 3' end to the *left*, opposite the usual convention of illustrating a strand of nucleic acid. Sequences at the 3' ends of genome serve as transcription promoters. There are also important regulatory sequences between each gene ($\sim$18 nt) that include a U-rich polyadenylation signal, a 2 nt untranscribed transcription termination signal followed by a transcription reinitiation signal. These sequences direct the transcription complex to cap and polyadenylate individual mRNAs.

Virus structure

Rhabdoviruses are enveloped particles with a rod or bullet (flat on one end) shape. Animal rhabdoviruses are 75–80 nm wide and $\sim$180 nm long. Virions contain a helical nucleocapsid. The major structural protein is the nucleocapsid protein, N, which interacts with the genome at a ratio of one N protein to nine bases of RNA. Nucleocapsids also contain several hundred copies of the phosphoprotein, P, and about 50 copies of the RdRp (L protein). A single glycoprotein, G, is associated with the envelope. G forms trimers that appear as spikes. G is uncleaved and serves as the attachment and fusion protein.

Overview of replication

Rhabdoviruses share a common replication strategy with other families in the order *Mononegavirales*. Key features of the rhabdovirus replication cycle summarized below are based, for the most part, on studies of VSV and RABV.

Attachment and penetration

Attachment and penetration (fusion) are mediated by the surface glycoprotein, G. G homotrimers form spikes on the surface of the virion. G is an uncleaved, membrane anchored protein. Attachment leads to endocytosis and within endosomes, low pH causes a conformational change to expose hydrophobic loops that participate in fusion. Fusion releases the ribonucleoprotein particle (RNP) into the cytosol, the site of transcription and genome replication. Due to its role in attachment and penetration, G is an important target of protective immune responses. Vaccines targeting RABV G are highly effective.

Transcription and translation

Synthesis of rhabdovirus mRNAs occurs after release of the RNP into the cytosol. The RNP is a helical structure formed by the N protein and genomic RNA. A few copies of the transcription complex (P and L proteins) are also associated with the RNP. Rhabdovirus transcription does not require ongoing protein synthesis and is a process that can take place in a test tube. Transcription generates a set of six RNAs, five of which are capped and polyadenylated

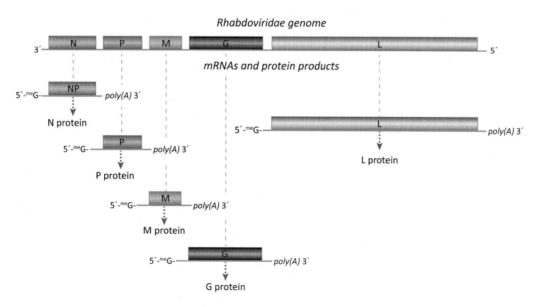

FIGURE 21.2 Rhabdovirus genome organization and gene expression strategy.

mRNA (Fig. 21.2). The sixth RNA is the short (~47 nt), uncapped, unpolyadenylated leader (*le*) RNA. Transcription of rhabdovirus RNA does not require a primer and proceeds sequentially from the 3' end of the infecting genome. Rhabdovirus L protein is the RdRp and also has cap synthesis and methyltransferase activities. Cis-acting RNA sequences that flank each gene provide signals for capping, polyadenylation, and transcription termination. Transcription "attenuation" is a general feature of nonsegmented, negative-strand RNA viruses. The term attenuation is used to describe the process by which the abundance of transcripts is regulated by their relative proximity to the 3' end of the genome. N mRNA is the most abundant transcript and L mRNA is the least abundant transcript. The process of attenuation occurs when L completes a transcript and then *dissociates* from the genome before starting transcription of the downstream gene. Studies with VSV have demonstrated that changing the locations of genes changes the relative abundance of mRNAs and reduces the overall efficiency of virus replication.

For the most part, each rhabdovirus mRNA encodes a single protein, thus five mRNAs are translated to produce N, P, M, G, and L. An exception is the P gene of vesiculoviruses, which encodes multiple proteins by use of alternative start codons and is sometimes called the P/C gene. An upstream start codon initiates translation of the P protein, whereas two downstream start codons initiate translation of an alternate reading frame that encodes two small basic proteins, C and C'. However, mutations that truncate C and C' (without affecting P) appear to have no detectable effect on virus replication in cell culture or pathogenesis in mice, so the functions of these products are not obvious.

N is the nucleoprotein, the major structural protein. N oligomerizes to form stable nucleocapsid-like structures in the absence of RNA, perhaps forming a scaffold for the RNA.

P is a phosphorylated protein present at small amounts in the virion. Phosphorylated P binds both N and L and is required for the RNA synthesis activities of L. Interactions between P and N are dynamic, and P is proposed to transiently bind and dissociate as the polymerase complex moves along the RNA template. In this model, the viral RNA always remains associated with N, even as it is transcribed.

M is a matrix protein associated with the inner side of the viral envelope. M serves as a bridge between the RNP and the virion envelope, interacting with both N and the cytosolic domain of G. The single glycoprotein, G, assembles into a trimer and is found in the plasma membrane of infected cells, where it serves to mark the location of virion assembly and budding. G serves both in attachment and fusion, but unlike many other envelope glycoproteins, G is not cleaved. VSV G has been extensively studied as a "model" integral membrane protein (Box 21.3). G is synthesized by ER-associated ribosomes and is transferred to the inside of the ER during synthesis. Translocation across the ER stops with the synthesis of the membrane anchor near the C-terminus of the protein. Only the final few amino acids of G remain on the cytoplasmic side of the ER membrane. G is initially glycosylated in the ER and associates with molecular chaperones that assist in folding. After release from chaperones, G monomers associate into trimers and are transported to the Golgi where further protein modifications take place (e.g., glycosylation is modified). G then traffics to the plasma membrane of infected cells where it is found in clusters that serve as site of virus budding.

BOX 21.3

Pseudotyping with VSV G

VSV infects most types of cells, suggesting that it uses a very common molecule for attachment. It has even been proposed that *nonspecific* electrostatic and hydrophobic interactions mediate attachment. Attachment is followed by endocytosis and membrane fusion at low pH. The ability of VSV to infect virtually all cell types has been harnessed as a means to get other, "fussier" viruses into a variety of cells by a process called pseudotyping. Pseudotyping is a general term describing the use of a "foreign" viral envelope glycoprotein to alter the tropism of a virus. Pseudotyping is used extensively in basic virology research as well as for clinical applications. In order for a viral glycoprotein to be useful in pseudotyping, it must be readily incorporated into the envelopes of heterologous viruses, should attach to most cell types, and should be able to efficiently mediate fusion. It is also helpful if the protein need not be cleaved by host cell proteases. VSV G meets all of these criteria. Of course if you wished to direct a virus to a very specific cell type, VSV G would not be the glycoprotein of choice.

At this point in the replication cycle of a rhabdovirus, the infected cell contains an RNP and newly synthesized viral mRNAs and proteins. However, in order for virus replication to proceed, mRNA synthesis must give way to genome synthesis.

The switch from mRNA synthesis to genome replication

Genome synthesis requires that instead of mRNAs, a *full-length* copy RNA (cRNA) must be synthesized. The cRNA is neither capped nor polyadenylated and serves at the template for synthesis of new genomes. It was recognized early on that translation is required to trigger the switch from transcription to genome replication. Thus purified nucleocapsids can support mRNA synthesis in a test tube but cannot support synthesis of cRNA or genomic RNA unless additional N protein is added to the mix. Thus N protein is the critical molecule enabling the switch from transcription to genome replication. The process requires that *soluble* N encapsidate the 5′ end of a newly initiated RNA strand. Short sequences present at the 3′ ends of both genomic RNA and cRNA serve as important cis-acting signals as they are the initial binding sites of N. What is less clear, in terms of the model, is just *how* the encapsidation of RNA signals the L polymerase to ignore signals for capping, polyadenylation, and termination of transcription.

Assembly/release

N binds genomic RNA at a ratio of one N protein to nine nucleotides to form the basic rhabdovirus RNP. P binds to both N and L, thereby bringing L to the RNA. RNPs traffic through the cytosol to the plasma membrane where they associate with M and G. Rhabdovirions bud through the plasma membrane. M plays a critical role in virion release as it contains a so-called *late domain*. Late domains are short amino acid motifs that interact with cellular proteins to allow release of budding virions.

Diseases

Rabies and rabies-related viruses

Rabies is a very ancient disease and human cases were clearly linked to the bite of a "rabid" dog. In the early 1800s it was formally shown that saliva from a rabid dog could transmit the disease. This provided the basis for studying the causative agent and by the mid-1800s Louis Pasteur and Emile Roux were searching for a cure. They passaged the rabies agent through rabbits and generated a vaccine from dried rabbit spinal cords. The vaccine worked to protect animals from rabies infection and in 1883 proof of effectiveness in humans was obtained when Pasteur used his vaccine to treat 9-year-old Joseph Meister. Joseph had been badly bitten by a rabid dog and it was almost certain he would develop rabies. Fortunately for Joseph (and Pasteur), the vaccine was successful. Although Pasteur's vaccine was not completely safe, it was used for decades as an effective postexposure treatment. Much safer and more effective vaccines are available today. In many parts of the world, routine vaccination of domestic animals significantly reduces the risk human of exposure to rabies virus. In the United States, RABV is found in bats, raccoons, skunks, and foxes, so it is important that dogs and cats receive regular vaccination for their own and their owner's protection. High-risk individuals, such as veterinarians, are routinely vaccinated. In addition, oral baits, containing live attenuated vaccines, are used to control rabies in wildlife. In the United States, the edible, oral rabies vaccination (ORV) baits contain RABORAL V-RG rabies vaccine. The vaccine is a vaccinia virus recombinant containing the rabies virus G protein and animals that eat the baits are protected from rabies infection in about 2 weeks. The baits are targeted to skunks, raccoons, and coyotes.

RABV is most often transmitted by bites but can also be contracted after scratches or any contact with saliva. RABV enters peripheral nerves and slowly travels up axons to reach the CNS (Fig. 21.3). After reaching the CNS, RABV replicates very rapidly. Brains of suspect animals are examined for the presence of virus (Figs. 21.4 and 21.5). Initial signs of rabies are nonspecific and include anorexia, lethargy, fever, and vomiting. These signs are followed by more obvious signs of CNS disease, including hyperexcitability and aggression (or sometimes extreme depression and lethargy).

After replicating in the CNS, virus moves back down the nerve fibers to locations such as the eye and salivary glands. Saliva is a source of abundant infectious virus. The neurologic changes brought about by RABV infection change animal behavior. Thus previously friendly animals become vicious, increasing the likelihood of transmission by bites. In contrast, wild animals that usually fear humans will become approachable or may even attack. After a bite, it may take days, weeks, or months for RABV to reach the CNS. Disease prevention in humans depends, in part, on postexposure vaccination. Postexposure vaccination is a critical part of postexposure treatment as it allows development of protective immune responses during the time that the RABV is "hidden" within neurons. Treatment often also includes administration of rabies immune globulin. In addition to routine vaccination of dogs and cats, baits containing recombinant rabies vaccines have successfully been used in the United States

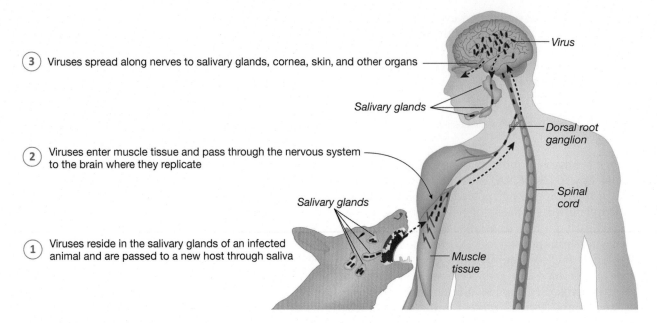

③ Viruses spread along nerves to salivary glands, cornea, skin, and other organs

② Viruses enter muscle tissue and pass through the nervous system to the brain where they replicate

① Viruses reside in the salivary glands of an infected animal and are passed to a new host through saliva

FIGURE 21.3   RABV transmission. RABV enters peripheral nerves and travels to the brain.

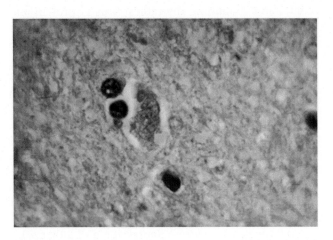

FIGURE 21.4   Hematoxylin-eosin (H&E) stain of brain tissue. Note the dark staining Negri bodies which are cellular inclusions associated with RABV replication. Source: *From CDC/Dr. Daniel P. Perl. Public Health Image Library ID# 14559.*

and Europe to reduce RABV in wild animals such as coyotes, foxes, and skunks.

The lethality of rabies infection requires that the public fully understand the risks of interacting with wildlife (including bats) as well as stray dogs and cats. Any/all animal bites should be reported in order that people and animals involved can be evaluated by health care professionals. If necessary, animals will be quarantined or tested (in the United States, laws vary by state). A good rule of thumb is to be aware of any animal that is exhibiting unusual behavior. That includes pets that become aggressive as well as wild animals that appear sick and approach, rather than flee, from humans.

There are very few cases of human rabies in the United States most years due to aggressive postexposure prophylaxis when exposure is suspected. The few cases that do occur usually involved contact with bats (thus *never* handle a bat without using proper infection control measures). Bats in the home are of particular concern as they may bite sleeping individuals. Thus it is advisable to contact a physician or public health authorities if a bat is found in a room where someone has been sleeping.

Within the genus *Lyssavirus* are several species; RABV is the type species. RABV is found worldwide and is the only species found in the United States. RABV has a wide host range that includes many mammals, including bats. However, there is also genetic variation within RABV and in the United States. Genome sequences are used to distinguish RABVs circulating primarily among coyotes, bats, raccoons, fox, and skunks. The ability to genetically differentiate among subgroups of RABV is important for epidemiologic studies; however, the subgroups are not strictly host specific and all produce the clinical disease we know as rabies. Thus humans or other animals (including livestock) infected with bat, coyote, skunk, or dog RABV can develop clinical rabies (Box 21.4).

Steps to take upon possible exposure to RABV are as follows: Thoroughly wash any bites (15—20 minutes) to remove infected saliva from the wound. Washing is very effective and can be done even when other medical care is not immediately available. Then seek medical care and/or contact public health officials to evaluate the situation. Additional postexposure treatment includes shots of immune globulin and vaccination.

(A)

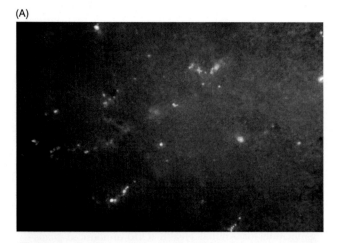

(B)

FIGURE 21.5   (A) Rabies viral antigen is present in a sample from a fox's brain. The viral antigen is revealed by staining with fluorescent antibody directed against the viral antigen. (B) Photograph of a rabid fox. The vast majority of rabies cases reported to the Centers for Disease Control and Prevention (CDC) each year occur in wild animals like raccoons, skunks, bats, and foxes. *Source: From (A) CDC/Dr. Hicklin. Public Health Image Library Image # 12644 and (B) CDC. Public Health Image Library ID# 2628.*

As mentioned above, clinical rabies is usually fatal. The story of Jeanna Giese serves both as a cautionary tale and a rare story of survival. In 2005 Jeanna was bitten while trying to "rescue" a bat. Unfortunately, her family did not understand the very real danger and failed to seek immediate medical attention. Later, when Jeanna developed signs of CNS disease, rabies was suspected. She was directed to the Children's Hospital of Wisconsin in Wauwatosa where she was treated by Dr. Rodney Willoughby Jr., a pediatric infectious disease specialist. He induced a coma to partially halt Jeanna's brain functions with the goal of protecting the brain from damage while giving the immune system time to defeat the virus. Antiviral drugs were also administered (however, it

is not clear that they had any beneficial effect). Jeanna's outcome was good and she was released from the hospital after 76 days. She suffered few permanent sequelae, attended college, and became a vocal advocate for rabies prevention education. However, it should be noted that the so-called Milwaukee treatment protocol is not a panacea, as survival rates of treated patients are less than 20% despite intensive hospital care. Therefore control of rabies in dogs and cats remains a high priority, as does education about the very real dangers of encounters with wild animals.

Chandipura virus

Chandipura virus (CHPV) is endemic to India and has been isolated from a variety of insects and animals, including humans. CHPV was first identified in 1965 after isolation from the blood of two patients. CHPV is associated with neurologic disease and those most at risk for developing encephalitis are children under 15 years of age. Vectors include mosquitoes and midges. While there is some disagreement about exact disease burden of CHPV in India, outbreaks continue to be reported as diagnostic methods improve. Case fatality rates range from 55% to 77% and death can occur within 24 hours of the onset of symptoms. CHPV neuropathogenesis is being actively studied using a suckling mouse model.

Vesicular stomatitis virus

VSV (genus *Vesiculovirus*) affects hoofed livestock in North and South America. Infections are generally mild but outbreaks can have significant economic impact. VSV infection causes formation of blister-like lesions of the mouth, tongue, teats, and hooves. VSV is one of several "vesicular" diseases of livestock that cannot be distinguished from FMDV without specific testing. Thus affected animals are quarantined and veterinary health officials put on alert until FMDV can be ruled out. Thus a VSV outbreak creates considerable angst among livestock producers.

VSV outbreaks are sporadic and are rare in the United States. Hosts of the virus include horses, cattle, swine, deer, and humans. VSV has been isolated from biting insects and this is how the virus is probably introduced into a herd. However, once an animal has been infected, transmission can continue as a result of contact. Human infections are generally associated with exposure to infected animals and in humans disease manifests as a general febrile illness.

BOX 21.4

Worldwide elimination of human rabies

Rabies is a disease that is feared worldwide. However, the state of rabies today is one of stark contrasts. While clinical rabies is rare in the United States and Europe, the World Health Organization (WHO) estimates the number of human deaths as high as 55,000 per year in Asia and Africa combined. In the United States, very rare cases of clinical rabies are most often associated with bats. In contrast, the vast majority of cases worldwide originate with dog bites and 40% of those bitten by suspect rabid dogs are children less than 15 years of age. It is difficult to image tens of thousands of cases of this dreaded and dreadful disease occurring annually, but the vast majority occur "out of site" in poor, rural populations. Cases are sporadic with no transmission among humans, thus no fear of the epidemic spread that so often is the catalyst for national or international attention. In the poor rural areas most affected by rabies, dogs may be largely feral and unvaccinated and postexposure treatment of human bite victims prohibitively expensive or completely unavailable. The zoonotic nature of RABV requires coordination and cooperation between human and animal health sectors to bring it under control. Vaccines for animals must be readily available and affordable enough to use routinely and consistently. Safe vaccines and protective immunoglobulin must also be available to bite victims. To address these needs, the WHO and the World Organization for Animal Health (OIE) announced plans in 2015 to work together to eliminate human rabies deaths by the year 2030. We have the medical and scientific expertise to drastically reduce cases of human rabies, if we acknowledge that the threat is ongoing and real in many parts of the world.

Other rhabdoviruses associated with disease

Other rhabdoviruses associated with disease include bovine ephemeral fever virus (BEFV) that affects cattle and buffalo in tropical and subtropical regions of the world. BEFV is transmitted by insects. Rhabdoviruses also cause fish diseases. These include spring viremia of carp virus, infectious hematopoietic necrosis virus, and viral hemorrhagic septicemia virus among others. These viruses cause significant mortality and morbidity in both wild and cultured fish.

In this chapter we learned that:

- Members of the family *Rhabdoviridae* are rod-shaped enveloped viruses with an unsegmented negative-sense RNA genome.
- As a family, rhabdoviruses have a wide host range, from plants to insects to animals. Some are transmitted by insects and they cause a variety of diseases.

- Their overall replication strategy is similar to other viruses in the order *Mononegavirales*: The L polymerase is packaged in the virion and begins transcribing mRNAs upon penetration of the RNP into the host cell. Synthesis of N protein triggers a switch from mRNA to genome synthesis.
- RABV is a dangerous pathogen of humans and animals and is found throughout most of the world. It is transmitted via infectious saliva (often as a result of a bite). In order to infect the recipient, RABV must gain access to neurons. Virus replicates very little until it completes its travel up neurons to reach the CNS where it replicates very efficiently in nerve bodies.
- The goal of postexposure RABV vaccination is to stimulate protective immune responses during the time it takes for the virus to reach the CNS.
- VSV causes a vesicular disease of livestock. VSV has a very broad cell tropism, thus its G protein is often used to pseudotype foreign viruses.

22

Family *Paramyxoviridae*

After studying this chapter, you should be able to answer the following questions:

- What are the main features of members of the family *Paramyxoviridae*?
- What is the general replication scheme of the unsegmented, negative-stranded RNA viruses?
- How does paramyxovirus N protein participate in the switch from mRNA synthesis to genome replication?
- Why do paramyxovirus-infected cells fuse with uninfected neighbors?
- Name two human diseases caused by a paramyxovirus.

The family *Paramyxoviridae* is a large group of enveloped, unsegmented, negative-strand RNA viruses (Fig. 22.1). The family is one of eleven within the order *Mononegavirales* (Box 22.1). Paramyxoviruses replicate in the cytoplasm of infected cells. Hosts include mammals, birds, reptiles, and fish. Paramyxoviruses are notable pathogens of humans and other animals. Some paramyxoviruses cause flu-like illnesses and together with the influenza viruses were initially referred to as myxoviruses (myxo meaning mucus or slime). When it was determined that distinct types of viruses caused flu-like illnesses, they were divided into two families: *Paramyxoviridae* and the *Orthomyxoviridae* (the family containing the "true"

influenza viruses). As we see later in this chapter, paramyxoviruses cause a variety of diseases in addition to respiratory disease.

Genome structure/organization

Paramyxovirus genomes range in size from 13 to 15 kb and encode 6–10 proteins from 6 to 8 capped and polyadenylated mRNAs (Fig. 22.2). The 3′ untranslated region (3′ UTR or 3′ leader) is short, ~50 nt, and the 5′ UTR (or 5′ trailer) is 50–161 nt. There is a single transcription start site, at the 3′ end of the genome and short cis-acting sequences flanking the open reading frames (ORFs) that signal polyadenylation and termination of the upstream mRNA, and initiation and capping of the downstream mRNA (Fig. 22.3).

Virion structure

Virions are enveloped with helical nucleocapsids. They are spherical to pleomorphic (sometime filamentous) and ~150–350 nm in diameter. When envelopes are gently removed from virion preparations, the helical nucleocapsids are evident (Fig. 22.1). All paramyxoviruses encode two surface glycoproteins that extend

© 2023 Elsevier Inc. All rights reserved.

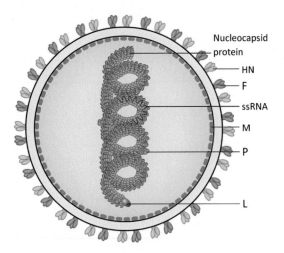
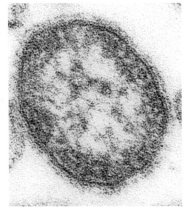
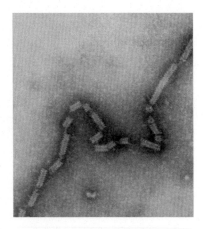

FIGURE 22.1   Virion structure (left). Transmission electron micrograph of measles virus (center). Measles virus nucleocapsids (right). Source: *(center) From CDC/Dr. F. A. Murphy. CDC Public Health Image Library Image #8758. (right) From CDC/Dr. F. A. Murphy. CDC Public Health Image Library Image #3366.*

BOX 22.1

Taxonomy

Order *Mononegavirales*

Family *Paramyxoviridae*

The family currently contains 4 subfamilies, 17 genera, and 78 species. Selected members (mentioned in this chapter) of three subfamilies are included in the list below:

Subfamily *Avulavirinae* (avian)

Genus *Orthoavulavirus* (avian paramyxoviruses including Newcastle disease virus)

Subfamily *Orthoparamyxovirinae*

Genus *Morbillivirus* (measles virus, rinderpest virus, peste des petits ruminants virus)

Genus *Respirovirus* (human parainfluenza virus 1 and 3, Sendai virus, bovine parainfluenza virus 3)

Genus *Henipavirus* (Hendra virus, Nipah virus)

Subfamily *Rubulavirinae*

Genus *Orthorubulavirus* (mumps virus, human parainfluenza viruses 2, 4, and 5)

8–12 nm from the envelope. Other structural proteins include the nucleocapsid (N) protein (an RNA-binding protein and the major protein of the ribonucleoprotein particle) and the matrix (M) protein that lines the inner layer of the envelope, forming a bridge between the nucleocapsid and the cytoplasmic tails of the two surface glycoproteins. A phosphoprotein (P) and the large (L) protein (RdRp) are also associated with the nucleocapsid.

Paramyxoviruses encode two surface glycoproteins: an *attachment protein* (variously called H, HN, or G) and a separate *fusion* protein (always called F). The attachment proteins of some paramyxoviruses bind to and agglutinate red blood cells, thus they are called "hemagglutinin" and are designated H. Some paramyxovirus attachment proteins also have neuraminidase activity and thus are called HN (hemagglutinin-neuraminidase). However, the attachment proteins of some paramyxoviruses are simply designated G to denote they are glycosylated, surface

proteins. All paramyxoviruses also encode an F protein that mediates fusion at the plasma membrane, at neutral pH. The presence of F in the plasma membrane causes neighboring cells to fuse, resulting in the formation of large multinucleate cells called syncytia. Syncytia form in both cultured cells and in infected tissues (Box 22.2).

Overview of replication

Attachment and penetration

The replication cycle of paramyxoviruses begins with interaction of the viral attachment protein (H, HN, or G) to cell receptors. Paramyxoviruses with H and HN attachment proteins bind to sialic acid residues. A general model for paramyxovirus entry is that attachment leads to important conformational changes in the F glycoprotein. The evidence for this is that

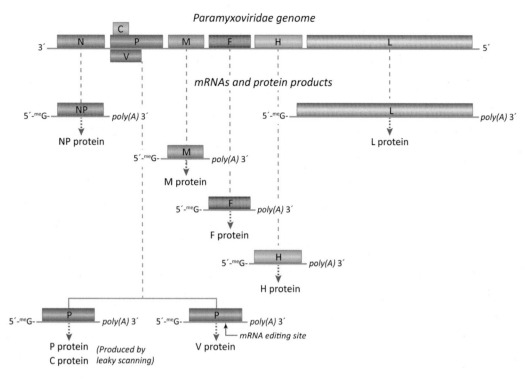

FIGURE 22.2   Genome organization and gene expression, paramyxovirus.

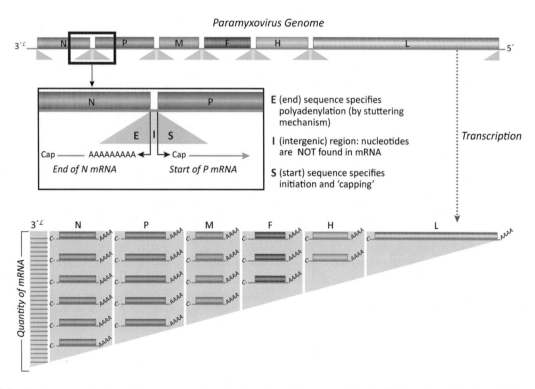

FIGURE 22.3   Genome organization of a paramyxovirus showing the relative amounts of mRNA produced by transcription and the organization of the intergenic region.

BOX 22.2

General characteristics

Paramyxoviruses have monopartite (unsegmented), negative-strand RNA genomes (~15−20 kb) transcribed to produce 6−8 mRNAs. Genome replication and transcription are cytoplasmic. Transcription (L protein is the RdRp) initiates at a single site at the 3′ end of the genome. As a completed transcript is released, the next transcript is initiated. This process is controlled by conserved sequences flanking each ORF. mRNAs are capped and polyadenylated by the L protein. Virions are enveloped with helical nucleocapsids. They are spherical to pleomorphic and 300−500 nm in diameter. Paramyxoviruses encode two surface glycoproteins. These are visualized as spikes extending 8−12 nm from the envelope. One serves as the attachment protein while the second is the fusion (F) protein.

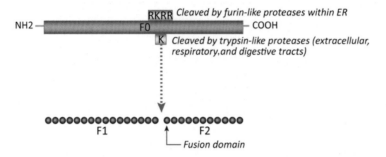

FIGURE 22.4 Cleavage of F0 produces F1 and F2 products.

fusion does not occur unless the virion has interacted with a receptor. After attachment, the hydrophobic fusion domain of F interacts with the cell plasma membrane to begin the fusion process. A key feature of the paramyxovirus replication cycle is that in order to be infectious, the F protein must be proteolytically cleaved to allow activation of the fusion domain (Fig. 22.4). Paramyxoviruses with uncleaved F proteins are not infectious. M is released from the nucleocapsid during entry.

Transcription

Transcription of paramyxovirus genomes begins when the nucleocapsid is released into the cytoplasm. The bulk of the nucleocapsid consists of the N, tightly bound to RNA at a ratio of 1 N protein to 6 nt of RNA. The complex of N protein and RNA is quite stable and serves as the substrate for transcription. The active viral transcription complex, also part of the nucleocapsid, consists of P (homo-tetramers) and L protein. There are about 50 copies of L in the virion. The set of mRNAs produced in the infected cell can be synthesized from nucleocapsids in a test tube. These include a short leader (*le*) RNA, neither capped nor polyadenylated, that is transcribed from the very 3′ end of the genome followed by 6−8 capped and polyadenylated mRNAs (Fig. 22.3). Most mRNAs encode a single protein, with the exception of the P mRNA from which *several* proteins can be produced. The abundance of each transcript is dependent on its position on the genome. Thus the *le* RNA is the most abundant, followed by the N mRNA. The process that regulates the relative abundance of mRNAs is sometimes called attenuation, reflecting the fact that genes positioned further along the genome (closer to the 5′ end) are produced at lower quantities (thus their transcription is attenuated).

Attenuation of transcription can be explained by the fact that there is a single promoter on the genomic RNA, present at the very 3′ end, and this is the only site where transcription can initiate. Following the short *le* RNA is a cis-acting sequence that directs cap synthesis for the N mRNA. At the 5′ end of the N gene are U-rich regulatory sequences that direct polyadenylation and termination. Following the termination signal for the N mRNA is a cis-regulatory sequence that directs initiation and capping of the P mRNA. After release of the N mRNA, successful synthesis of P mRNA requires that the transcription complex remain closely associated with the genome. If the transcription complex dissociates, it must relocate back to the very 3′ end of the genome, where it can interact with promoter sequences. Hence the position of ORFs on the genome regulates their relative mRNA abundance. It has been demonstrated experimentally (using reverse genetics systems) that altering the position of an ORF changes the relative abundance of mRNA and protein encoded by that ORF. This is a mechanism by which attenuated paramyxoviruses viruses can be generated for clinical applications.

Protein synthesis

N mRNA, the most abundant paramyxovirus mRNA, is cytosolic and N is synthesized as a soluble protein (sometimes called N0 to differentiate it from molecules of N tightly associated with RNA). N ranges in size from ~53 to 57 kDa. In the virion, N is found in the nucleocapsid, bound to genomic RNA. Each N protein binds six nucleotides. Expression of N protein (in the absence of other viral proteins) shows that N can self-associate to form nucleocapsid-like structures. However, N is not simply a scaffolding protein, it also has roles in RNA replication. Early after infection, newly synthesized N is soluble and in this form can interact with P.

P is synthesized in the cytosol, is ~400–600 amino acids, and is heavily phosphorylated. P is fairly abundant, present at ~300 copies in the virion. As noted above, P is a required part of the transcription/replication complex; RNA polymerization requires association of multimers of P with L. Several important domains of P have been mapped. The carboxyl terminal region has domains required for multimerization, interaction with L, and binding to N0. The importance of the carboxyl terminal region is demonstrated by the fact that, in some experimental systems, it can substitute for full-length P to support mRNA synthesis. The amino-terminal half of P is required for genome (and cRNA) synthesis, likely through interactions with soluble N0.

In addition to encoding the P protein, the P mRNA encodes additional proteins. Some of these are produced by *cotranscriptional mRNA editing*. The process is also called pseudotemplated transcription. It occurs when G residues are added to the P mRNA at a very specific location (at position approximately nt 400 of the P mRNA). The addition of G residues (usually one or two) shifts the reading frame. Thus edited mRNAs still code for the amino-terminal half of the P protein but lack the carboxyl terminal domains so critical for transcription (Box 22.3).

What are the functions of these additional proteins? Most paramyxoviruses encode a cysteine rich *V protein* of approximately 25–30 kDa (Table 22.1). V is produced by addition of a single G at the P mRNA editing site. Thus most of the V protein is identical to P, but editing produces a frame shift that replaces the carboxyl terminal domains of P with a short cysteine rich sequence. One role of V is a negative regulator of RNA synthesis (perhaps by binding to N0 and sequestering it from productive interactions with RNA). V also appears to be a suppressor of innate cell responses. Measles virus V interacts with several key cellular proteins involved in innate immunity to viruses, including MDA5 and LGP2.

The *W protein* is expressed from P mRNAs edited by insertion of two G residues. This results in a stop codon shortly after the editing site, thus W is essentially a truncated P protein. W may also regulate viral RNA synthesis. The *D protein* of bovine and human parainfluenza viruses is produced by addition of two Gs at the editing site. This new ORF extends for 131 amino acid residues beyond the editing site.

The editing scheme is a bit different for members of the genus *Rubulavirus*. Here the unedited mRNA encodes the cysteine rich V protein while it is P that is produced from an edited (+1 G) mRNA. Rubulaviruses

---

BOX 22.3

RNA editing or pseudotemplated transcription

Most eukaryotic nuclear mRNAs are modified by capping, splicing, and polyadenylation. A rare modification of transcripts, addition of nontemplated nucleotides, is a process often called RNA editing. The term nontemplated refers to nucleotides present in an mRNA that are not present in the genome sequence. Many paramyxoviruses edit their P mRNAs to produce multiple mRNAs from a single gene. Editing was first proposed for paramyxoviruses in the 1980s when cloned P mRNAs were found to have sequences that differed from the viral genome by insertion of 1–6 G residues. Further studies showed that the differences were consistent for a given

paramyxovirus and were only seen for the P mRNA. It is now clear that insertions occur at very specific sites and that the ratios of edited mRNAs to nonedited mRNAs are constant for a given paramyxovirus P gene. Thus insertions are driven and controlled by the *specific sequence of the viral RNA*. Because the process depends on cis-acting genome sequences, the process is also referred to as "pseudotemplated" transcription, a more specific term than "editing." Additional examples of pseudotemplated transcription have been found among the members of the order *Mononegavirales*: Ebola virus edits its glycoprotein gene.

TABLE 22.1   Products of paramyxovirus P genes.

Genus	Identified mRNA insertions and products[a]	ORFs expressed by use of alternative start codons[a]
Rubulavirus	+1 G or +4 G (W), +2 G (P) note that the unedited product is V	
Avulavirus	+1 G (V), +2 G (I)	
Respirovirus	+1 G (V), +2 G (W or D)	C, C′, Y1, Y2
Henipavirus	+1 G (V), +2 G (W)	C
Morbillivirus	+1 G (V), +2 G (W)	C

[a]*Vary by species.*

also produce W by insertion of one or four Gs during mRNA synthesis.

Some paramyxoviruses also encode *additional proteins* (C and C′) from P mRNA, by use of alternative start sites to access alternative reading frames. Sendai virus (genus *Respirovirus*), an extreme example, produces a nested set of four carboxyl coterminal proteins (C′, C, Y1, and Y2) that range in size from 175 to 215 amino acids. Translation of each of the C′, C, Y1, and Y2 ORFs is initiated at a different site, although translation is terminated at the same downstream stop codon. Morbilliviruses and henipaviruses express only one C protein and rubulaviruses do not express any C proteins. C proteins are basic polypeptides involved in regulation of RNA synthesis, counteracting host antiviral responses, and facilitating release of virions from infected cells.

There are two different mechanisms used to direct ribosomes to alternative start sites. Leaky scanning, whereby a ribosome fails to assemble on a start (AUG) codon (thus continues to scan the mRNA) is common. The sequences surrounding an AUG codon are important for defining a start site and if the AUG is not flanked by appropriate sequences the AUG will be bypassed by the ribosome. Sendai virus C and C′ are produced by leaky scanning. To produce C′ ribosomes must initiate translation at an ACG triplet at nt 81 of the P mRNA. This is not a common event as most ribosomes bypass this site and initiate translation of P protein at an AUG at nt 104 of the mRNA. However, some ribosomes also bypass the P initiation site and initiate the C protein by using an AUG codon at nt 114.

A different mechanism to direct ribosomes to alternative start codons is called ribosome shunting or stop-start. The mechanism differs from leaky scanning in that the ribosome completely bypasses some regions of the mRNA, effectively "jumping" over them to a downstream region. It requires two cis-acting sequences: an upstream donor and a downstream acceptor. The ribosome is shunted from the donor site directly to the acceptor site, effectively bypassing the intervening sequences. The acceptor site determines the translation initiation site, sometimes a non-AUG start codon. In the case of Sendai virus, translation of the Y1 and Y2 proteins occurs through a ribosome shunting.

What is the evidence for ribosome shunting? The process was initially proposed based on experiments that modified potential start codons in the Sendai virus P gene. In the leaky scanning scenario, changing the context of an upstream start codon should affect the amount of protein produced from downstream initiation codons. For example, if the Sendai virus C′ start codon was changed to an AUG, in a good context for ribosome initiation, we predict synthesis of more C′ protein and less of the P and C proteins. However, in the case of Y1 and Y2 proteins, modification of some upstream start codons had no effect on the levels of Y1 and Y2 produced, suggesting that ribosomes were not using a scanning mechanism to reach these translation start sites. Eventually, specific donor and acceptor sites were identified and it was confirmed that mutation of the intervening sequences has no effect on the levels of the downstream protein products. The functions of the various C and Y proteins have not been well defined. In the case of measles virus, C protein is not required for virus replication in some cultured cells but is required for replication in nonhuman primates. Evidence points to a role in antagonizing the interferon response.

The matrix (M) protein is a major structural protein. M proteins range in size from ~38.5 to 41.5 kDa. M proteins are highly basic (have a lot of positively charged amino acids) and are somewhat hydrophobic. M is associated with the inner face of membranes and can form two-dimensional arrays (sheets and tubes) under some conditions. M also associates with nucleocapsids by interactions with negatively charged surfaces. Thus M acts as a typical matrix protein, forming a bridge between envelope proteins in a membrane and the nucleocapsid. M proteins of several paramyxoviruses have so-called late domains, short amino acid motifs that interact with host cell sorting complexes to facilitate release of budded virions.

Translation of the two viral glycoproteins takes place on rough ER. Both are glycosylated on their transit through the ER and Golgi to the plasma membrane. The F protein is synthesized as a precursor (F_0) that must be cleaved into F_1 and F_2 fragments (Fig. 22.5). F_1 and F_2 remain associated via a disulfide bond. As noted previously, cleavage of F_0 is essential for virion infectivity. Some paramyxovirus F proteins are cleaved within the ER (by furin-like or cathepsin-like proteases); however, others are cleaved by extracellular, trypsin-like proteases (found in the respiratory and digestive tracts) to achieve particle maturation. The cellular location of F_0 cleavage plays a major role in paramyxovirus pathogenesis, as cleavage by extracellular proteases necessarily restricts productive virus replication to sites such the lungs or gastrointestinal tract. As might be expected, paramyxoviruses that cause system-wide infections (measles virus, canine distemper virus (CDV)) have F proteins that are cleaved in the ER, before they are even trafficked to the plasma membrane. In the case of Newcastle disease, virus of birds (an important poultry pathogen) virulence is clearly linked to F cleavage. Strains that contain F cleavage sites that facilitate cleavage in the ER are highly virulent, causing systemic disease and high mortality.

Translation of L mRNA is cytosolic and produces a protein of ~ 2200 aa (259 kDa). L is the catalytic portion of the transcriptase/replicase complex but it does not synthesize viral RNA in the absence of P. L is required for cap synthesis, polyadenylation of mRNAs, and mRNA editing. Among the paramyxovirus L proteins, the most conserved regions are in the central portion of the protein while the amino and carboxyl termini more variable.

Genome replication

Synthesis of N and P triggers a switch from transcription to genome synthesis. Genome synthesis requires synthesis of an end-to-end copy of the genome; the copy or cRNA is neither capped nor polyadenylated. Whereas mRNAs are not bound by the N protein, both cRNA and genomic RNA bind to N protein during their synthesis. In fact their synthesis requires that free N protein be available to encapsidate the RNA as it is synthesized. Each paramyxovirus N protein binds to exactly 6 nt of RNA. During efforts to generate efficient reverse genetic systems for paramyxoviruses, it was found that efficient virus replication requires that the genome length be an exact multiple of 6. This is called the paramyxovirus "rule of six."

Assembly, release, and maturation

Nucleocapsids assembly occurs in the cytoplasm. The nucleocapsid consists of genomic RNA, N, P, and L proteins. The nucleocapsid travels to the plasma membrane and interacts with M. Virions bud from the PM. Virions containing uncleaved F_0 are noninfectious unless F_0 is cleaved by extracellular, trypsin-like proteases. Recall that F cleavage is a necessary maturation step as it activates the hydrophobic fusion peptide at the N-terminus of F_2, a process critical for fusion to an uninfected cell. The cellular location of F cleavage is determined by the amino acid sequence at the cleavage site. Measles and CDVs have F proteins that are cleaved in the ER/Golgi by furin-like or cathepsin-like proteases. Thus these viruses have a wide tissue tropism in the infected animal. In contrast, human parainfluenza virus F_0 is cleaved by extracellular proteases found in the respiratory tract.

Diseases

Measles

Measles virus (a morbillivirus) is extremely contagious by the airborne route. It is estimated that 90% of susceptible persons coming in contact with an infected person will be infected. Measles (rubeola) most often is a disease of children and most deaths are among children less than 5 years old. Disease symptoms usually develop 10–12 days after exposure and persist for 7–10 days. Virus can be transmitted before symptoms appear. Measles starts with fever, runny nose, cough, red eyes, and sore throat, followed by a red, flat rash that usually starts on the face and spreads over the body. Small white spots (known as Koplik's spots) may form inside the mouth. Complications occur in

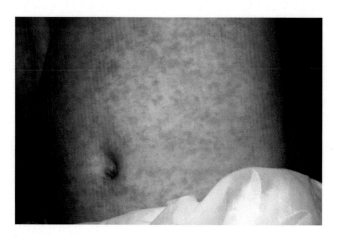

FIGURE 22.5 Torso of a child infected with measles virus, displaying the characteristic rash associated with this viral infection.

about 30% of infected persons and these include diarrhea, corneal ulceration and scarring, encephalitis, deafness, and pneumonia. A leading cause of deaths in children is the development of secondary infections that result from a transient but profound suppression of the immune system. A very rare complication of measles virus infection is subacute sclerosing panencephalitis (SSPE), a progressive neurological disorder of children and young adults. SSPE results from persistent infection of the CNS by *defective* measles virus that replicates very slowly. Cases of SSPE have essentially disappeared in countries that practice widespread immunization.

A measles virus vaccine was introduced in United States in 1963. It is a safe and effective vaccine. Numbers of measles cases in the United States were dramatically reduced to only thousands of cases per year by the 1980s. Worldwide, measles was estimated to have caused 2.6 million deaths per year during the 1980s. However, by 1990 estimated deaths were just over 600,000 and by 2013 estimated yearly deaths were less than 100,000. Why do we continue to see *any* deaths from measles virus? Despite a safe and effective vaccine, measles virus infects an estimated 20 million (unvaccinated) people per year. The risk of death is usually very low (0.2%), but may be up to 10% among malnourished children. Unfortunately, inadequate food and inadequate medical infrastructure often go hand in hand. Death rates due to measles virus infection are significantly higher (30%) among immunocompromised persons. Measles virus is thought to cause more *vaccine-preventable* deaths than any other virus.

Mumps

Mumps virus (genus *Rubulavirus*) is highly infectious and spreads rapidly among people living in close quarters. It is transmitted by respiratory droplets, direct contact, or contaminated objects. The incubation period is usually 14–18 days and patients shed virus a few days before the onset of symptoms that often include fever, muscle pain, headache, and lethargy. Initial symptoms are followed by painful swelling of the salivary glands (the parotid gland most often). The disease is usually self-limiting; however, infection of men can result in painful testicular swellings, which can cause decreased fertility and rarely sterility. Other complications include meningitis (in up to 10% of cases), ovarian inflammation (5% of adolescent and adult females), and acute pancreatic inflammation (4%). A very rare complication (0.005%) is profound hearing loss. The mumps vaccine was licensed in the United States in 1967 and since then cases have

dropped significantly. Between 2001 and 2008 the average cases per year was only 265. In the years 2006, 2016, and 2017 over 6000 cases were reported in the United States. This indicates the continued presence of mumps virus.

Respiratory tract diseases

Parainfluenza viruses most often cause upper respiratory tract diseases. Virus is restricted to respiratory epithelial cells (where proteases that cleave the F protein are expressed). Infants and young children are most often affected but natural immunity to parainfluenza viruses is not solid and reinfection can and does occur. Infections are usually self-limiting and mild (fever, runny nose, and cough) but croup and pneumonia can occur.

Canine distemper

CDV (genus *Morbillivirus*) causes a multisystem viral disease with high mortality. CDV affects the respiratory, gastrointestinal, and central nervous systems of young dogs. Canine distemper occurs worldwide and once was the leading cause of death in unvaccinated puppies. Development of a vaccine (in 1960) led to a drastic reduction in infections of dogs; however, the virus infects many other carnivores including raccoons, skunks, and foxes. In 1994 CDV wiped out a third of Africa's Serengeti lion population after an outbreak in feral dogs. CDV is now emerging as a pathogen that threatens wild tigers worldwide. Big cats in captivity have also fallen prey to CDV. While there are licensed (live attenuated) vaccines available for dogs and ferrets, they are *not* suitable for other mammals, such as big cats, in which they cause fatal disease.

Other animal paramyxoviruses of note include rinderpest virus, peste des petits ruminants virus, bovine parainfluenza virus, and avian paramyxoviruses. Rinderpest (also known a cattle plague) caused wide-spread outbreaks with high morbidity and mortality among domestic cattle until the virus was declared eradicated due to an aggressive vaccine initiative. Peste des petits ruminants or ovine rinderpest virus primarily affects sheep and goats. The disease involves oral necrosis, mucopurulent nasal and ocular discharges, cough, pneumonia, and diarrhea. Mortality can be as high as 100%. The virus is currently found in North, Central, West and East Africa, the Middle East, and South Asia. There are many avian paramyxoviruses but of particular concern is Newcastle disease virus, a variant that causes severe systemic disease in birds. The F protein cleavage site is a major virulence determinant; highly virulent forms of Newcastle disease virus have F proteins that are cleaved by proteases in the ER/Golgi.

Henipaviruses

Nipah virus is zoonotic and causes severe disease in swine and humans. Symptoms range from asymptomatic to acute respiratory syndrome with fatal encephalitis. The natural hosts of the virus are fruit bats. Nipah virus was first identified during an outbreak that took place in Malaysia in 1998. Other sporadic outbreaks have occurred since then in Singapore, Bangladesh, and India. Hendra virus was first identified in an outbreak that occurred in the suburb of Hendra, Australia, in 1994. The outbreak involved 21 horses and 2 human cases. Thirteen horses (case fatality of approximately 75%) and a trainer died. The natural host of Hendra virus is fruit bats.

In this chapter we have learned that:

- Family *Parmyxoviridae* is a member of the order *Mononegavirales*.

- Paramyxoviruses are enveloped viruses with a nonsegmented, negative-sense RNA genome. Replication is cytoplasmic.
- RdRp is packaged in the virion and can synthesize mRNAs in an in vitro reaction.
- During an infection, newly synthesized N triggers the switch between synthesis of mRNAs and genome-length RNAs.
- Paramyxoviruses encode two envelope glycoproteins, one serves as the attachment protein and the second (always called F) is the fusion protein. F cleavage is required for virion infectivity and the site of cleavage (intracellular vs extracellular) can greatly impact virulence. F protein is found in the plasma membrane of infected cells and can mediate fusion with neighboring uninfected cells to produce multinucleate cells called syncytia.
- Measles and mumps viruses are major human pathogens. There are protective vaccines available for both.

23

Family *Pneumoviridae*

After studying this chapter, you should be able to answer the following questions:

- What are the main characteristics of the members of the family *Pneumoviridae*?
- What is the general replication scheme of the unsegmented, negative-stranded RNA viruses?
- What diseases are caused by human respiratory syncytial virus? What age group is most adversely affected?

Members of the family *Pneumoviridae* are enveloped, unsegmented, negative-strand RNA viruses with helical nucleocapsids. Shapes range from spherical to long filaments (Fig. 23.1). Pneumoviruses share many similarities with paramyxoviruses and prior to 2016 were included in the family *Paramyxoviridae* (Box 23.1). Hosts include mammals and birds. Human respiratory syncytial virus (HRSV) is a notable human pathogen that can cause severe lower respiratory tract disease in infants.

Genome organization

The HRSV genomes is 15.2 kb and encodes 11 proteins (Fig. 23.2). Overall, the genome organization of the pneumoviruses is similar to the paramyxoviruses. Most genes are separated by short intergenic regions contain cis-acting sequences that direct capping and polyadenylation of transcripts. However, in contrast to the paramyxoviruses, the intergenic regions of the pneumoviruses are not highly conserved in either sequence or length and there is an overlap between the M2 and L genes of the human HRSV. The genome of human metapneumovirus (Fig. 23.3) lacks the NS1 and NS2 open reading frames found at the very 5′ end of the HRSV genome.

Virion structure

Pneumoviruses are enveloped, spherical to filamentous (up to 10 μm) in shape, with helical nucleocapsids. HRSV encodes two surface glycoproteins, G and F. G is the attachment protein while F mediates membrane fusion at neutral pH. The presence of F in the plasma membrane of infected cells causes neighboring cells to fuse, resulting in the formation of large multinucleate cells called syncytia to form in both cultured cells and in infected tissues, lungs in the case of HRSV. Other major structural proteins include a nucleocapsid (N) protein, matrix (M) protein, and a second matrix protein (M2−1). M forms a layer directly under the envelope and M2−1 forms a second layer beneath M. N binds to the viral genome (1 N protein per 7 nt) to generate a long, flexible, left-handed helix. Small amounts of a phosphoprotein (P) and the polymerase (L) are also found in the virion, in association with the nucleocapsid (Box 23.2).

Viruses.
DOI: https://doi.org/10.1016/B978-0-323-90385-1.00012-1

© 2023 Elsevier Inc. All rights reserved.

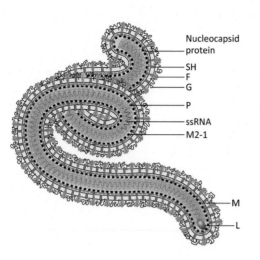
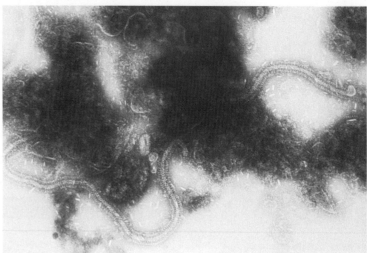

FIGURE 23.1   Virion structure (left), TEM of human respiratory syncytial virus (right). Source: *Centers for Disease Control and Prevention Public Health Image ID #276. Content providers CDC/Dr. Erskine Palmer.*

BOX 23.1

Taxonomy

Order *Mononegavirales*
Family *Pneumoviridae*
Genus *Orthopneumovirus* (Hosts are mammals, includes human respiratory syncytial virus and bovine respiratory syncytial virus)

Species *Human orthopneumovirus* (respiratory syncytial virus)
Genus *Metapneumovirus* (Hosts are mammals or birds; includes human metapneumoviruses)

Overview of replication

Attachment and penetration

The HRSV replication cycle begins with interaction of the viral attachment protein, G, to cell receptors. G has been shown to bind glycosaminoglycans (GAGs), which are unbranched disaccharide polymers linked to transmembrane proteins on the surface of many mammalian cell types. Other molecules proposed to be receptors for HRSV include intercellular adhesion molecule (ICAM)-1, heparin, annexin II, toll-like receptor (TLR) 4, the fractalkine receptor CX3CR1. In addition to G, the F protein also interacts with cell receptors. One piece of evidence that F serves as an attachment protein is the ability of HRSVs lacking G to replicate in cell culture systems. F has been shown to interact with nucleolin and blocking nucleolin expression reduces HRSV replication. Nucleolin is largely located in the nucleus and is abundant in dividing cells. However, some is found on the plasma membrane. Therapeutics targeting the F-nucleolin interaction are being tested in animal models. After attachment, F can induce fusion, at neutral pH, at the plasma membrane. In order to mediate fusion, the F precursor (F_0) must be cleaved and present on the surface of the virion as a trimeric spike of disulfide-linked F_1 and F_2 subunits. Fusion releases the nucleocapsid into the cytoplasm.

Transcription

The genome of HRSV (genus *Orthopneumovirus*) contains 10 genes while the genome of human metapneumovirus (HMPV, genus *Metapneumovirus*) contains 8 genes; 10 and 8 capped and polyadenylated mRNAs are produced, respectively. Most mRNAs have a single ORF but there are two exceptions: HMPV P mRNA has a second, short open-reading frame encoding a protein of unknown function. HRSV M2 mRNA encodes both the M2−1 and M2−2 proteins from overlapping ORFs. mRNA synthesis follows the same general process described for paramyxoviruses and rhabdoviruses.

There is a single transcription start site at the 3′ end of the genome and transcription requires the L, P, and

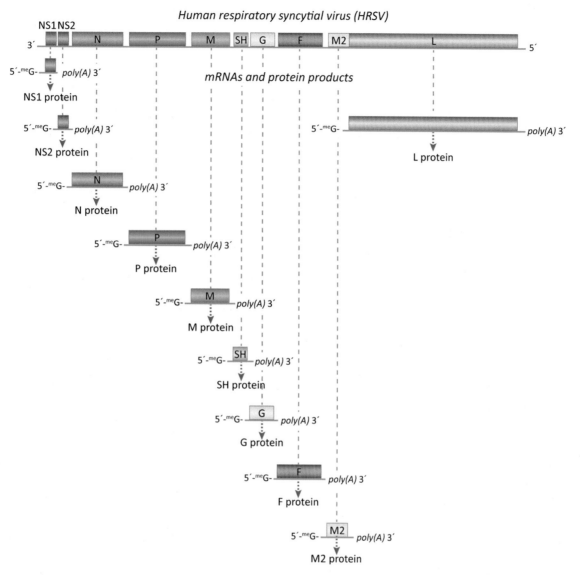

FIGURE 23.2   Genome organization and gene expression of human respiratory syncytial virus. The M2 mRNA is used to produce two proteins, M2−1 and M2−2. M2 and L are overlapping ORFs.

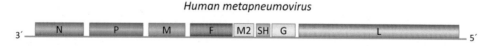

FIGURE 23.3   Genome organization of human metapneumovirus.

M2−1 proteins. L caps each mRNA as it is initiated and polyadenylates using a stuttering mechanism. Accumulation of M2−1 appears to mediate the switch from transcription to replication.

HRSV proteins

N mRNA, the most abundant pneumovirus protein, is cytosolic and N is synthesized as a soluble protein. N is an RNA binding protein and each N binds 7 nt of

RNA. The phosphoprotein (P) is a polymerase cofactor and L contains the catalytic domains for RNA polymerization and mRNA capping. N, P, and L are the major nucleocapsid-associated proteins. M and M2−1 are structural matrix proteins. M2−1 is also a transcription processivity factor. M2−2 is found in the infected cell and serves to shift RNA synthesis from transcription to genome replication.

G and F are glycoproteins, translated on rough ER. G is highly glycosylated with both N- and O-linked carbohydrates. G serves at a virion attachment protein,

---

BOX 23.2

General characteristics

Pneumoviruses have monopartite (unsegmented), negative-strand RNA genomes ($\sim 13.2-15.3$ kb) encoding 9−11 proteins from 8 to 10 mRNAs. Genome replication and transcription are cytoplasmic. Transcription (L protein is the RdRp) initiates at a single site at the 3′ end of the genome. As a completed transcript is released, the next transcript is initiated. This process is controlled by sequences flanking each ORF. mRNAs are capped and polyadenylated by the L protein. Virions are enveloped with helical nucleocapsids. Filamentous forms usually predominate and most often are 70−190 nm in diameter and up to 2 μm in length. Two layers of matrix proteins (M and M2−1) are under the envelope. Envelopes contain two glycosylated transmembrane viral glycoproteins, G and F. G is the attachment protein, F has a fusion domain but also plays a role in attachment.

---

but HRSV G is also found as a secreted form. The sequence of G is variable and during annual outbreaks, different strains cocirculate.

F is synthesized as a glycosylated precursor (F_0) that is cleaved by cellular furin-like proteases to produce disulfide-linked F_1 and F_2 subunits. Cleavage of HRSV F actually occurs in two locations, 27 amino acids apart, releasing a p27 peptide. (F proteins of members of the genus *Metapneumovirus* contain a single cleavage site.)

HRSV also encodes a small hydrophobic (SH) protein of 64−65 amino acids. A small amount of SH is incorporated into the viral envelope and is a putative viroporin. SH is not absolutely needed for HRSV replication in cell cultures.

Genes for two NS proteins are found at the very 3′ end of the genome (upstream of the N gene), and thus are the first genes transcribed after penetration. NS1 and NS2 mRNAs are abundant in infected cells and encode proteins of 139 and 124 amino acids, respectively. Viruses with deletions of NS1 and NS2 can replicate in cultured cells but not as well as wild-type viruses. NS1/NS2 deletion mutants are very attenuated in animals. NS1 and NS2 are multifunctional proteins with important roles in virulence. For example, they antagonize antiviral responses, such as apoptosis and expression of type I interferons.

Genome replication, assembly, and release

Genome synthesis requires synthesis of an end-to-end copy of the genome; this copy or cRNA is neither capped nor polyadenylated. M2−1 is required for the switch from transcription to genome replication. RNA synthesis and formation of nucleocapsids occurs in cytoplasmic viral inclusion bodies. F and G, processed in the ER/Golgi, are transported to the plasma membrane. Nucleocapsids associated with the M and M2−1 proteins are enveloped by budding at the cell surface plasma membrane at sites containing virus envelope proteins.

However, HRSV buds inefficiently from cultured cells, with most progeny virions remaining cell-associated. This finding led investigators to look at other possible mechanisms of virus spread. Cell−cell fusion to create syncytia, mediated by F is one possible mechanism. Another is use of filopodia, induced by HRSV infection, upon which virions could move to nearby cells.

Diseases

Human respiratory syncytial virus disease

HRSV infection of infants places a huge burden on healthcare systems and is the most common cause of acute respiratory tract infections. HRSV is the main viral cause of hospitalizations among infants and young children worldwide. It has been estimated that more than 3 million severe pediatric cases occur annually worldwide. Most children in the United States will be infected by HRSV by 2 years of age. It appears that infants hospitalized for HRSV infection have an increased risk of developing chronic wheeze and asthma. Immunity to HRSV is not solid and repeat infections can occur. In older children and adults, HRSV usually causes mild upper respiratory tract disease; however, severe disease can occur among the elderly or immunocompromised. Airway damage by HRSV also increases risk of secondary bacterial infection. HRSV is a respiratory virus transmitted via droplets. It can also be spread via fomites or direct contact (kissing). HRSV initially replicates in epithelial cells lining the nasopharynx and upper respiratory tract. Virus may then spread to bronchioles or alveoli of the lower respiratory tract.

HRSV outbreaks are seasonal. In the Northern Hemisphere, they tend to occur along with influenza seasons, between October and May, with a peak in January or February.

There are limited therapeutics available for HRSV and no approved vaccines, despite years of research. A humanized monoclonal antibody targeting the F protein is licensed for prophylaxis in at risk infants. Ribivirin is approved for treatment in children but its administration (inhaled) is difficult.

As of 2021, no licensed vaccine against RSV exists although many are now in development. For decades, RSV vaccine development languished after a failed vaccine was launched in the 1960s. The vaccine was formalin inactivated whole virus and it actually exacerbated disease in some children. A large number of previously unexposed, vaccinated children were hospitalized with severe lung inflammation and two died. The vaccine was quickly withdrawn. Current vaccine candidates have been developed using detailed structural knowledge of target proteins. The lack of a vaccine has driven detailed studies of HRSV replication, with the purpose of finding new avenues of intervention.

Human metapneumovirus disease

HMPV is a respiratory pathogen that causes a range of illnesses (from asymptomatic infection to severe bronchiolitis). It was first identified from children in the Netherlands in 2001. It is thought that HMPV may have caused 10% viral respiratory infections in which previously characterized viruses could not be detected.

Reports from all over the world now suggest that HMPV virus is ubiquitous and a common cause of severe childhood acute respiratory disease. Peak age of hospitalization from severe HMPV is between 6 and 12 months. Although this viral pathogen was first described in children, subsequent reports have highlighted the importance of HMPV as a cause of respiratory illness in adults, particularly among the immunocompromised. HMPV is found worldwide and outbreaks seem to be seasonal, January to March in the northern hemisphere and June to July in the southern hemisphere. HMPV infection does not induce solid, long-lasting immunity.

In this chapter we have learned that:

- The family *Pneumoviridae* is a member of the order *Mononegavirales*.
- Pneumoviruses are enveloped and have nonsegmented, negative-sense RNA genomes. Replication is cytoplasmic.
- Pneumovirus F protein is fusogenic at neutral pH. F can mediate fusion between uninfected and infected cells to produce multinucleate cells called syncytia.
- HRSV is a leading cause of infant hospitalizations; a safe and effective vaccine remains elusive.
- HMPV is an important respiratory pathogen first identified in 2001. It occurs worldwide and outbreaks are seasonal.
- Neither HRSV nor HMPV induce strong, long lasting immunity.

24

Family *Filoviridae*

- After reading this chapter, you should be able to answer the following questions:
- What are the general characteristics of members of the family *Filoviridae*?
- What is viral hemorrhagic fever (VHF)?
- What are the natural hosts for filoviruses?
- How are filoviruses transmitted?
- What special equipment and precautions are taken when working with filoviruses such as Ebola and Marburg viruses?

Members of the family *Filoviridae* are enveloped viruses with unsegmented, negative sense RNA genomes (Box 24.1). Ebola virus (EBOV) and Marburg virus (MARV) cause severe, often lethal, hemorrhagic fever in humans. During the years 2013–2016, an outbreak in West Africa was not contained and affected several countries. In time, an international response supplied medical personnel, equipment, medicines, and facilities to control the epidemic: But not before over 28,000 infections, resulting in over 11,325 deaths decimated the health care systems of affected communities.

Filoviruses are one of eleven families in the order *Mononegavirales*. The family *Filoviridae* includes five genera including *Ebolavirus* (EBOV) and *Marburgvirus* (MARV) (Box 24.2).

Filoviruses are notable for their bizarre filamentous shapes with virions that are 800–1200 nm long with a diameter of 80 nm (Figs. 24.1 and 24.2). The name filovirus is derived from the Latin, *filum* meaning "thread."

Genome organization

The overall organization of filovirus genomes is similar to that of other nonsegmented negative-strand RNA viruses. An RNA-binding nucleocapsid protein (NP) is encoded at the 3′ end of the genome and the L protein, the RdRp, is encoded at the 5′ end of the genome (Fig. 24.3). However, filovirus genomes are longer (~19 kb) than those of other viruses in the order *Mononegavirales* and there are some additional differences. A set of capped and polyadenylated mRNAs is transcribed using cis-acting signals that regulate capping, polyadenylation, and transcript termination. However, unlike rhabdoviruses and paramyxoviruses, the regulatory (stop and start) signals of some genes *overlap*, suggesting that the polymerase must somehow backtrack on the genome after completing the upstream transcript. Filoviruses also have longer noncoding regions at the 3′ and 5′ ends of the genome than are found on rhabdoviruses and paramyxoviruses.

Viruses.
DOI: https://doi.org/10.1016/B978-0-323-90385-1.00017-0

© 2023 Elsevier Inc. All rights reserved.

BOX 24.1

Taxonomy

Family *Filoviridae*
Genus *Ebolavirus* (hosts include primates and bats)
Species *Bombali ebolavirus*
Species *Bundibugyo ebolavirus*
Species *Reston ebolavirus*
Species *Sudan ebolavirus*
Species *Tai Forest ebolavirus*
Species *Zaire ebolavirus*
Genus *Marburgvirus*

Species *Marburg Marburgvirus* (bats)
Genus *Cuevavirus* (bats)
Species *Lloviu cuevavirus*
Genus *Dianlovirus*
Species *Mengla dianlovirus*
Genus *Striavirus*
Species *Xilang striavirus* (fish)
Genus *Thamnovirus*
Species *Huangjiao thamnovirus* (fish)

BOX 24.2

General characteristics

Members of the family *Filoviridae* are long, filamentous, enveloped viruses with unsegmented negative-strand RNA genomes of about 19 kb. Filovirus genomes are organized in a manner similar to other members of the order *Mononegavirales* with seven genes encoding seven to eight proteins. Transcription is also similar, with the production of a set of nonoverlapping capped and polyadenylated mRNAs. L protein contains the catalytic site for RNA synthesis but is only active as a complex with NP and VP35. An active transcriptase complex synthesizes mRNAs upon release of the nucleocapsid into the cytoplasm. Genome replication and transcription are cytoplasmic.

Virions are long (800–1000 nm) with a diameter of 80 nm. They are flexible and assume characteristic shapes. Spikes on the surface of the virion consist of a glycoprotein (GP) heterodimer that is produced by cleavage of a precursor. The amino terminal portion of the precursor (GP1) contains the attachment domain. The carboxyl terminal half of the precursor (GP2) contains the fusion domain. Fusion takes place in the environment of an acidified endosome.

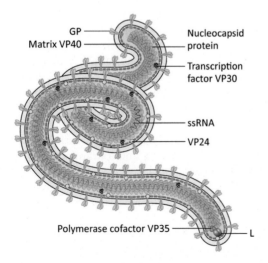

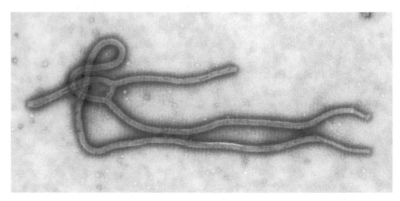

FIGURE 24.1   Virus structure (left). Transmission electron micrograph (right) of Ebola virus. Source (right panel): *CDC Public Health Image Library ID#1832. Content provider CDC/Cynthia Goldsmith.*

Virion structure

Filoviruses are enveloped and range in size from 800 to 1200 nm long. They have a very consistent diameter of ~80 nm. Their filamentous shapes are

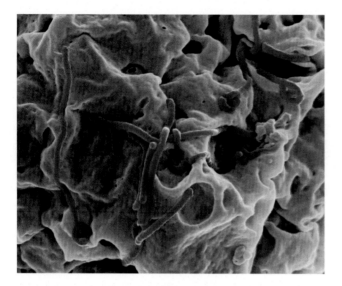

FIGURE 24.2 Digitally colorized scanning electron microscopic image depicting a number of filamentous Ebola virus particles (red) that had budded from the surface of a VERO cell (brown). Source: *CDC Public Health Image Library ID#17777. Content provider National Institute for Allergy and Infectious Diseases.*

quite distinct from those of other enveloped viruses of animals. Two nucleocapsid proteins (NP and VP30) bind to the genome, along with the proteins (L and VP35) that comprise the active RdRp to form the nucleocapsid. A single glycoprotein (GP) protrudes from the virion surface. The major matrix protein (VP40) is associated with the inner leaflet of the envelope. VP40 is the most abundant virion protein.

Overview of replication

As many filoviruses have the potential to cause severe disease, their study is restricted to high security, high containment laboratories, a very expensive undertaking (Fig. 24.4). Thus much of the work to decipher molecular aspects of their replication is done in "surrogate" systems that do not produce infectious virions.

Attachment/Penetration

The membrane associated glycoprotein contains both attachment and fusion domains. Infectious virions contain heterodimers of cleaved GP. GP1 is the amino terminal region and contains the attachment domain. GP2 is the carboxyl terminal region and contains a fusion domain. Ebolavirus GP mediates attachment to a wide variety of cells, including fibroblasts, hepatocytes, adrenal cortical

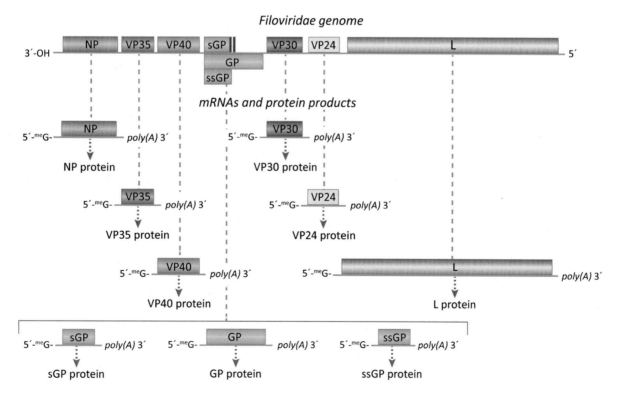

FIGURE 24.3 Genome organization and gene expression strategy of a filovirus.

"attachment" event whereby processed GP1 interacts with the Niemann Pick C1 (NPC1) protein, an endosomal protein. Cellular expression of NPC1 is required for productive infection and the interaction is believed to trigger a conformational change that allows insertion of the GP2 fusion domain into the endosomal membrane. Fusion releases the nucleocapsid into the cytosol.

Transcription, translation, genome replication

mRNAs are synthesized in the cytosol by the viral transcriptase/replicase complex that consists of VP35 and L. mRNAs are capped and polyadenylated and most are monocistronic (encode a single protein). Similar to the rhabdoviruses and paramyxoviruses, mRNA synthesis is controlled by cis-acting sequences flanking the open-reading frames. Unlike the rhabdoviruses and paramyxoviruses, these intergenic sequences are not highly conserved. Also, the biomechanics of transcriptase stopping and restarting is unclear as some genes contain overlapping transcription termination and initiation sites.

Most filovirus mRNAs encode a single protein (the exception is GP). The phosphorylated nucleoprotein, NP, is encoded at the 3' end of the genome and is the second most abundant structural protein. Ebolavirus NP oligomerizes into helical filaments, encapsidates the viral RNA genome, and serves as a scaffold for assembly of additional viral proteins to the viral nucleocapsid. As is the case for other families in the order *Mononegavirales*, soluble NP is thought to trigger the switch from transcription to replication.

Following the NP gene on the genome is VP35, a 35 kDa phosphoprotein. VP35 is part of the core ribonucleoprotein particle. VP35 binds to NP and regulates NP assembly. VP35 also interacts with L protein and serves as part of the polymerase complex. (Thus VP35 plays a role similar to the P proteins of rhabdoviruses and paramyxoviruses.) VP35 is also an interferon antagonist.

VP40 is the major matrix protein and the most abundant structural protein. VP40 is membrane associated and essential for virion assembly. VP40 has an affinity for membranes and is associated with the inner leaflet of the virion envelope. VP40 participates in budding, both initiating and driving the envelopment of the nucleocapsid at the plasma membrane.

The next gene on the filovirus genome encodes the major glycoprotein precursor, GP. Viruses in the genus *Ebolavirus* produce four proteins from GP mRNA. One version is the membrane-anchored virion spike protein consisting of heterodimers (GP_{12}) of furin-cleaved GP. GP_{12} mediates both attachment and fusion. GP_{12} is

FIGURE 24.4 Centers for Disease Control microbiologists in the process of suiting up in order to access the interior of the organization's Biosafety Level-4 (BSL-4) laboratory. The scientist on the left is attaching his supportive air hose, which provides a supply of filtered, breathable air. Source: *CDC Public Health Image Library ID# 10723. Content providers CDC/Dr. Scott Smith.*

cells, epithelial cells, macrophages, dendritic cells, and endothelial cells. Molecules identified as receptors include carbohydrate-binding proteins (C-type lectins, e.g., L-SIGN, DC-SIGN, and hMGL among others). One recently identified receptor is T-cell immunoglobulin and mucin domain-1 (TIM-1) protein found on the surface of epithelial cells. Studies to identify filovirus receptors often use pseudotyped viruses, for example, VSV (a rhabdovirus) expressing filovirus GP.

After attachment, virions enter cells by endocytosis. Some are taken up by macropinocytosis, but virions may also enter cells by clatherin or caveolae-mediated endocytosis. Regardless of the endocytic process used for uptake, acidification of the early endosome is required for membrane fusion. Within the acidified endosome, GP1 is first cleaved to a smaller form by the low pH-dependent cellular proteases cathepsin L and cathepsin B. This seems to trigger an additional highly modified: It is heavily N- and O-glycosylated, sialylated, acylated, and phosphorylated. GP_{12} is the longest GP product and a very surprising finding is that EBOV GP mRNA is edited to produce this product. The editing or pseudotemplated transcription adds one extra nucleotide to the mRNA. This occurs only about 20% of the time. (In contrast, MARV GP_{12} is produced from *unedited* GP mRNA.) GP_{12} contains attachment and fusion domains.

A second ebolavirus GP protein is a secreted form (sGP) that consists of uncleaved dimers of GP. This protein is synthesized from unedited GP mRNA and is produced in abundance. sGP shares the amino terminal 300 amino acids with membrane-anchored GP_{12}. However, sGP lacks the membrane anchor and is secreted from EBOV-infected cells. Its function is unknown but has been hypothesized to serve as an antibody "decoy." Perhaps the immune responses are directed to this version of the viral glycoprotein and fail to recognize the membrane-anchored version. It has been hypothesized that sGP may be a virulence factor of the highly lethal ebolaviruses. It must be noted, however, that in stark contrast to EBOV, MARV GP mRNA is not edited and no soluble forms of have been detected. As some MARVs are highly virulent, sGP cannot be the only factor contributing to filovirus virulence. A third ebolavirus glycoprotein is called the secondary secreted glycoprotein (ssGP), secreted as an N-glycosylated monomer. The fourth product, Δ-peptide, another secreted glycoprotein of unknown function is produced by cleavage of a short peptide from the carboxyl terminus of sGP.

Next on the filovirus genome is viral protein 30 (VP30), a 30 kDa phosphoprotein that binds genomic RNA. VP30 is a minor nucleoprotein. It interacts with L, forming a bridge from L to the nucleocapsid. It is followed on the genome by viral protein 24 (VP24), a 24 kDa phosphoprotein. VP24 is a minor matrix-associated protein. It is a hydrophobic protein that is found at the plasma membrane. The exact role of VP24 in filovirus replication is not yet known. The L protein, the RdRp, is encoded by the final third of the genome.

The general model of filovirus genome replication is consistent with other families in the order *Mononegavirales*. There appear to be some differences in the protein requirements for EBOV and MARV genome synthesis. In the case of EBOV, NP, VP35, VP30, and L are required to synthesize minigenomes in a recombinant model of replication while MARV genome replication requires only NP, VP35, and L.

Assembly

Filovirus virions assemble at the plasma membrane. GP is trafficked to the PM to serve as budding sites.

Matrix proteins (VP40 and VP24) are key players in assembly and budding. In some experimental systems, expression of VP40 alone results in production of virus-like particles (VLPs). However, VLP production is more efficient if GP and NP are added to the system. VP40 contains a late domain that promotes budding and release of budded vesicles, but it is not clear that it is absolutely required for virus production in cultured cells.

Disease

MARV was the first filovirus linked to human infection and disease. It was the cause of an outbreak that began in Germany with three cases of VHF among laboratory workers processing organs from African green monkeys. By the time it was contained, the outbreak totaled 31 cases with 7 fatalities. Marburg hemorrhagic fever was next documented in Africa in 1975, and there were a handful of additional reported cases in the 1980s. Larger outbreaks were recorded in 1999 (Democratic Republic of Congo, 154 cases) and 2005 (Angola, 252 cases, 227 deaths). The extremely high case fatality rate in 2005 was probably due to a large number of pediatric cases where infection may have been hospital acquired (parenteral infection).

Ebola hemorrhagic fever (EHF) was discovered with outbreaks in Democratic Republic of Congo (then called Zaire) and Sudan in 1976. The genetically distinct filoviruses associated with the outbreaks are Zaire ebolavirus (ZEBOV) and Sudan ebolavirus (SEBOV), respectively. In both outbreaks (318 cases in Democratic Republic of Congo, 284 cases in Sudan), high numbers of transmissions were caused by reuse of nonsterile needles and syringes at the treating hospitals. Sporadic outbreaks of EHF occurred during the next decades but most were limited in scale. That changed in 2013 with an outbreak in several West African countries. According to the World Health Organization, this outbreak (2013–2015) totaled 28,657 cases with 11,325 deaths (Box 24.3). As bad as these numbers are, they do not reflect the *overall impact* of an epidemic that closed borders to travel and trade, closed schools, and decimated health care infrastructure.

Transmission

EBOV infection begins with entry into the host through mucosal surfaces, breaks, or abrasions in the skin, or by parenteral introduction (needlestick). Most cases occur by direct contact with infected patients or cadavers, as large amounts of virus are present in blood and body fluids (perhaps as much as $10^7 - 10^8$

BOX 24.3

The West African Ebola epidemic

Ebola disease is an African disease. It was first described in 1976 when the disease broke out in South Sudan and the Democratic Republic of the Congo. These outbreaks, while highly lethal, were relatively small and were controlled quickly. The source of Ebola virus is unknown but it is generally believed that it is acquired from wild animals. Two of the leading candidates have been monkeys and bats. At the present time, bats are the most likely candidates since one Ebola outbreak occurred in miners in Angola and the mine was populated by bats. Other visitors to the same caves also contracted Ebola virus.

It was therefore a surprise when a case of Ebola was described, far away, in Guinea in West Africa in December 2013. The index case was a 1-year-old boy. The disease eventually killed his mother, sister, and grandmother. Other villagers spread the infection. They lived in the vicinity of a large bat colony. However, none of the bats from the village ever tested positive for the virus. It took a long time for the disease to be recognized as Ebola (there are many causes of infectious hemorrhagic fever in Africa) and was thus permitted to spread for several months unchecked. It was first reported as Ebola in March 2014 when almost 60 deaths had occurred. The disease spread into the capital of Guinea, Conakry, and by the end of May, the death toll had climbed to 180.

The disease spread into two neighboring countries, Sierra Leone and Liberia in April. By July there were 442 cases in Sierra Leone. As the epidemic grew, cases were reported in another neighboring country, Mali, and in more distant Nigeria. Eventually small numbers of cases were transmitted to Senegal, Spain, the United Kingdom, and the United States.

Ebola is a hemorrhagic disease. Affected individuals bleed from all their body orifices and this blood contains large amounts of virus. Virus is also abundant in vomit and diarrhea. As a result, individuals who come into contact with sick patients are at high risk. One especially important mode of spread is from dead bodies. In much of Africa, custom requires that the dead be washed before burial. As a result, the infection can be transmitted to family members who prepare the body. It has proven very difficult to prevent these burial practices and attempts to do so have met with resistance and hostility.

In order to manage and nurse Ebola cases, health care workers must wear full protective clothing to ensure that they do not come into contact with infected body fluids. This type of protective equipment is not readily available in much of Africa. Breaking the chain of transmission involves total isolation of the patient, a difficult feat when health care infrastructure is poor. Likewise many villages have essentially no access to health care, thus disease can spread in remote areas, without the authorities being aware of it. Monthly deaths peaked in Guinea and Liberia in August 2014 and in Sierra Leone in December. Subsequently, prevalence declined steadily to very low levels by late 2015. The World Health Organization declared the epidemic over in January 2016.

Like many viral diseases, there are few treatments for Ebola disease. During the aforementioned epidemic, some patients received blood transfusions from survivors in an attempt to passively immunize them by providing neutralizing antibodies. Subsequent analysis revealed no benefits of these treatments. Because Ebola hemorrhagic fever was, prior to this epidemic, a very uncommon sporadic disease, little effort had been made to develop a vaccine. As a result of the epidemic, this has changed. Many different experimental vaccines are under development including DNA vaccines, oral vaccines, recombinant viral vectors, and virus-like particles (VLPs). Unfortunately, due to the unpredictable nature of Ebola outbreaks, such a vaccine may be very difficult to test reliably in the field. Who should receive the vaccine—health care workers? Perhaps vaccine should be given to individuals in regions surrounding an outbreak in an attempt to prevent its spread. One feature of concern is the suspicion of Western medicine in parts of Africa. Local populations often suspect that the appearance of medical teams is the *cause* rather than a consequence of the disease.

However, it should also be noted that not all filoviruses cause disease in humans. Reston Ebolavirus (RBOV) was identified in an outbreak at an animal facility in Reston, Virginia (not far from Washington, DC) in 1989. The outbreak began among monkeys (macaques) imported from the Philippines. All of the macaques died from viral hemorrhagic fever and a filovirus was identified by electron microscopy examination of tissues from affected monkeys. The Centers for Disease Control and Prevention (CDC) was alerted and immediately dispatched an infection control team to the scene. The facility was closed and all remaining monkeys were euthanized. The prime concern for the infection control team was that personnel in contact with the monkeys might become ill. In the end, no people became ill although two animal handlers had antibodies to REBOV, a finding highly suggestive of infection. Apparently REBOV was deadly for monkeys but caused asymptomatic infections in humans. More recently, REBOV has been isolated from pigs in the Philippines, but again the virus was not linked to cases of human disease. This is a good example of the species specificity, particularly as regards pathogenesis, displayed by many viruses.

infectious particles per mL of serum). Those most at risk of contracting the disease are heath care workers or family members without access to appropriate infection control equipment. While virus is present in respiratory secretions, transmission still requires close contact with a symptomatic patient (or cadaver). There is no evidence that EBOV is transmitted by the "aerogenic" route in the manner of the highly transmissible measles or SARS-CoV-2 viruses. In fact, in the United States and Europe, transmissions have only been documented in a health care setting, never by causal contact. However, one surprise revealed in the 2013 outbreak is that filoviruses can be shed for pronged periods by recovered patients, and transmission by semen is possible.

Pathogenesis

VHF, an outcome of many human filovirus infections, is a fever and bleeding disorder. Leakage of blood from the vascular system and damage to clotting mechanisms may result in flushing of the skin, petechial rash, conjunctival bleeding, and hemorrhage of visceral organs. Edema results in swelling and is accompanied by low blood pressure and shock. Virus replication results in considerable direct organ damage as evidenced by necrotic lesions in the liver, spleen, kidneys, lungs, testes, and ovaries. Other symptoms of filovirus infection include malaise, muscle pain, headache, vomiting, and diarrhea (symptoms common to many viral infections). Mechanisms contributing to symptoms include liver damage, disseminated intravascular coagulation (DIC), and bone marrow dysfunction. In DIC, small blood clots form in blood vessels throughout the body, removing platelets, and reducing overall clotting ability. Damage to macrophages results in release of cytokines, chemokines, and other inflammatory mediators.

Ebolavirus and Marburg virus antagonize innate immune responses

It is clear that runaway virus replication is associated with severe disease. By what mechanisms are some filoviruses able to so completely subvert immune responses? The overall answer is that EBOV and MARV interfere with many proteins that are involved in development of a protective response. For example, there is severe suppression of interferon-stimulated genes. But other genes are affected as well, including many involved in immune regulation, coagulation, and apoptosis. The Type I IFN response is particularly important for establishing an antiviral state in host cells, to control virus replication during the period before development of adaptive or specific immune

responses. The Type I IFN response is also important for activating an adaptive response. Filovirus proteins that disrupt innate immune responses include VP35, VP40, and VP24. For example, VP35 suppresses production of IFN-α by binding and sequestering viral RNA. During infection with RNA viruses, it is the job of the cellular protein RIG-I to detect viral RNA by binding double-stranded RNA present during genome replication. But VP35, because it binds to viral dsRNA, blocks the activities of RIG-I. Other mechanisms by which filoviruses antagonize innate immune responses include:

- EBOV VP35 inhibits the phosphorylation of IRF-3/7 by the TBK-1/IKKε kinases.
- MARV VP40 inhibits STAT1/2 phosphorylation by inhibiting the JAK family kinases.
- EBOV VP24 inhibits nuclear translocation of activated STAT1 by karyopherin-α.

The end result of these protein interactions is to strongly inhibit the expression of Type 1 IFNs.

The virulence of filoviruses in humans is highly variable and multifactorial. Strain or species of virus is of primary importance. For example, ZEBOV is highly virulent while REBOV is on the very low end of the virulence spectrum for humans. Dose and route of infection also contribute to virulence and disease outcome. Higher fatality rates seem to be associated with parenteral infection, and a high-infecting dose is more likely to overwhelm immune responses. Host genetics, nutrition, overall patient health, and availability of supportive care all play important roles as well (Fig. 24.5).

FIGURE 24.5 The image taken in Uganda (2012) shows these responders from the Ugandan Red Cross in the process of donning their personal protective equipment before responding to a reported Ebola death in a village. Source: *CDC Public Health Image Library ID#:19916. Content provider CDC.*

Therapeutics and vaccines

There are currently three monoclonal antibody treatments approved by the US Food and Drug Administration (FDA). All target the EBOV glycoprotein. The FDA has also approved the Ebola vaccine rVSV-ZEBOV (tradename "Ervebo"). The vaccine is a recombinant virus consisting of vesicular stomatitis virus (VSV, a rhabdovirus) expressing the Zaire EBOV glycoprotein in place of the VSV glycoprotein. The rVSV-ZEBOV vaccine has been found to be safe and protective against only the species *Zaire ebolavirus*. The vaccine was used to vaccinate over 300,000 people in the Democratic Republic of Congo in a 2-year outbreak that was declared over in 2020.

In this chapter we have learned that:

- Filoviruses are filamentous-enveloped viruses with unsegmented negative-strand RNA genomes. Their overall replication strategy is similar to other viruses in the order *Mononegavirales*.

- Humans are not the natural hosts for filoviruses and many human infections result in development of severe disease. Viral hemorrhagic fever is caused by a number of factors including damage to vascular endothelium, dysregulation of clotting factors, and release of excessive amounts of proinflammatory molecules.

- Multifocal organ damage is common and is the direct result of filovirus replication.

- Filovirus virulence is associated with the ability to evade innate immune responses.

- The range of natural hosts of filoviruses has not been determined, but there is good evidence that it includes bats.

- Index cases of human filovirus infection are often associated with exposure to infected nonhuman primates (through hunting/butchering monkeys, for example) or bats. However, most human cases result from exposure to body fluids of symptomatic patients.

25

Family *Bornaviridae*

After reading this chapter, you should be able to answer the following questions:

- What are the defining characteristics of members of the family *Bornaviridae*?
- In what ways are bornaviruses different from other viruses in the order *Mononegavirales*?
- How would you describe the interactions of bornaviruses with infected cells?
- How were avian bornaviruses discovered?
- What diseases are caused by bornaviruses?

The bornaviruses are enveloped, monopartite, negative-strand RNA viruses. The family *Bornaviridae* is in the order *Mononegavirales* and shares many basic characteristics with other viruses in the order (Fig. 25.1). However, bornaviruses have several unique traits. They replicate in the nucleus, use cellular splicing machinery to produce some mRNAs, and are highly cell-associated (Box 25.1). They cause persistent infections with little or no cell pathology and are often highly associated with neurons in an infected host. Overall, bornaviruses are quite stealthy in their life styles and this is reflected in the history of the agent.

Borna disease was first described in horses in the 18th century in central Germany. Large outbreaks among cavalry horses in the 19th century, near the city of Borna, were the basis of the name. In the early decades of the 20th century, the infectious nature of the disease was confirmed using animal passage experiments. It would be another ∼50 years before persistently infected cell cultures were established for the virus. However, the study

of the agent remained difficult due to lack of any obvious cell pathology. While persistently infected cultures provided infectious material, they did not provide sufficient cell-free virus for convincing biochemical or structural studies. (In addition, the rarity of the disease and the limited geographic area of outbreaks did not move this agent to the top of most researcher's "must study" list.) Thus it was not until 1994 that the agent was cloned and sequenced. Genome sequence and organization revealed the similarities of Borna disease virus (BoDV-1) to other unsegmented, negative-strand RNA viruses (Fig. 25.2). BoDV-1 was the sole member of the family *Bornaviridae* until 2008 when avian bornaviruses were discovered (Box 25.2). Additional bornaviruses have been identified in snakes.

Genome organization

Bornaviruses have monopartite, negative-strand RNA genomes of ∼9 kb. They encode six proteins (N, P, X, M, G, and L) and the gene order is reminiscent of other members of the order *Mononegavirales*. However, the transcription strategy is unique in that splicing is used to generate some mRNAs. mRNA splicing is in line with the nuclear location of virus transcription.

As with other members of the order *Mononegavirales*, it is likely that there is a single promoter, near the 3′ end of the genome that drives initiation of transcription. Four transcription termination sites and three reinitiation sites have been mapped in the genome.

Viruses.
DOI: https://doi.org/10.1016/B978-0-323-90385-1.00037-6

© 2023 Elsevier Inc. All rights reserved.

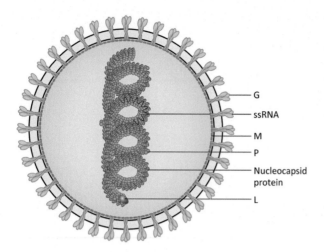
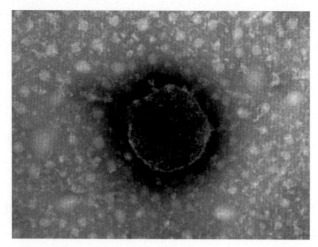

FIGURE 25.1　Virion structure (left). Virus-like particle (83 nm in diameter) from the eye fluid of an Eclectus parrot with confirmed PDD and ABV infection (right). The image was recorded with an FEI Morgagni 268 transmission electron microscope at a magnification setting of ×180,000. Source: *Courtesy Ross Payne.*

BOX 25.1

General characteristics

Bornaviruses have monopartite, negative-strand RNA genomes of ~9 kb. They encode six proteins (N, P, X, M, G, and L) and the gene order is similar to other members of the order *Mononegavirales*. Bornaviruses are unique among the nonsegmented negative-sense RNA viruses in that they replicate in the nucleus and splicing is used to generate some mRNAs.

Virions are enveloped, 90–130 nm in diameter but are poorly described. Bornaviruses are highly neurotropic and cause persistent infections. Many infections are likely inapparent, but fatal neurologic disease can develop.

Virion structure

In contrast to the precise, detailed structures that are available for many viruses, details of bornavirus ultrastructure remain somewhat fuzzy due to the low amounts of complete virions produced during an infection. Bornavirions are enveloped, ~100–130 nm in diameter and are assumed to have a helical nucleocapsid (Fig. 25.1). The genome encodes a single envelope glycoprotein.

Overview of replication

The bornavirus glycoprotein (G) is the likely candidate for receptor attachment and subsequent fusion. Candidate host receptors have not been determined but it is clear that they must be expressed on neurons. G is also the best candidate for mediating fusion through a hydrophobic domain present in the carboxyl terminal cleavage product, GP-C. Although the molecular details are not yet known, it is clear that the nucleocapsid (consisting of N, P, and L proteins and RNA) traffics to the nucleus.

N, P, and L are required for transcription. It is likely that L caps and polyadenylates viral mRNAs, while cellular splicing machinery also processes some transcripts. The most abundant transcript is the mRNA encoding the N protein. mRNAs are transported back to the cytoplasm where they are translated.

Bornaviral proteins are N, P, M, G and L (found in the virion), and X. N is the nucleoprotein and is found in two versions, p40 and p38, produced by use of alternative start codons. P is a phosphoprotein that serves as a cofactor for L. M is the matrix protein. G is the envelope glycoprotein and L is the catalytic subunit of the transcription/replication complex. In line with the nuclear location of RNA synthesis, N, P, and L proteins all have nuclear localization signals. The X protein is presumed to be nonstructural, with a role in RNA replication.

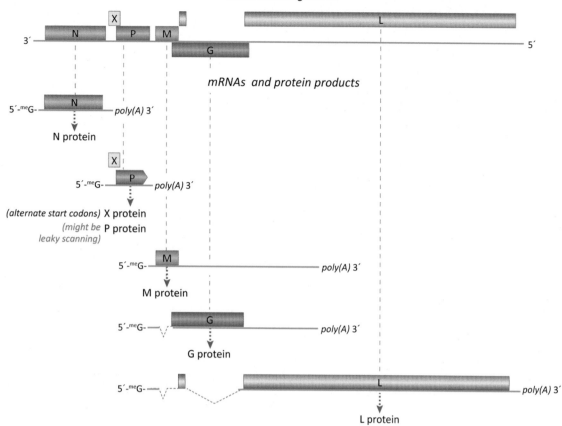

FIGURE 25.2   Bornavirus genome organization.

BOX 25.2

Taxonomy

Order *Mononegavirales*
Family *Bornaviridae*
Genus *Orthobornavirus*
Species:

Elapid 1 orthobornavirus
Mammalian 1 orthobornavirus (mammalian bornavirus 1, bicolored white toothed shrew)
Mammalian 2 orthobornavirus (variegated squirrel bornavirus)
Passeriform 1 orthobornavirus (passeriforme birds)

Passeriform 2 orthobornavirus (passeriforme birds)
Psittaciform 1 orthobornavirus (psittaciforme birds)
Psittaciform 2 orthobornavirus (psittaciforme birds)
Waterbird 1 orthobornavirus (anseriforme birds)
Genus *Cultervirus*
Species *Sharpbelly cultervirus* (fish)
Genus *Carbovirus*
Species *Queensland carbovirus* (snake)
Species *Southwest carbovirus* (snake)

The bornavirus glycoprotein, G, is translated on rough ER and posttranslationally modified by N-glycosylation to yield a 93–94 kDa primary product. A portion of G is cleaved to produce amino terminal (G-N) and carboxyl terminal (G-C) products. The virion may contain both cleaved and uncleaved glycoprotein products.

Genome replication

Genome replication is a nuclear event and requires N, P, and L. Cloning of genomes from persistently infected cells revealed truncations at ends of the genome strand, removing promoters for genome

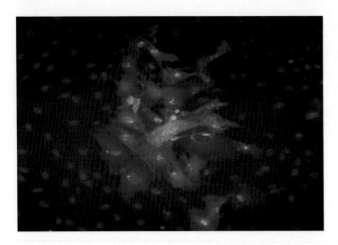

FIGURE 25.3 Fluorescent antibody labeling of duck embryo fibroblasts persistently infected an avian bornavirus. A focus of infected cells (green) sits amidst uninfected cells. The uninfected cells are visualized by DAPI staining (blue) of their nuclei.

replication and possibly transcription. It is speculated that this process leads to downregulation of replication/transcription and might play an important role in the persistent nature of bornavirus infections. Cleavage of the ends of full-length RNAs yields a 5′ monophosphate end, rather than the expected triphosphate, and this is a mechanism by which viral RNA can persist without activating innate immune responses. Recall antiviral IFN responses are strongly activated through retinoic acid-inducible gene-I (RIG-I) recognition of 5′ triphosphates on RNA.

Bornavirus nucleocapsids contain N and RNA, associated with P and L proteins. M is also part of the nucleocapsid. While the details are lacking, the process of nucleocapsid formation is presumed to be similar to other members of the order *Mononegavirales*. Nucleocapsid formation occurs in the nucleus and the RNP must be exported from the nucleus for the assembly of complete virions. In addition to transmission by extracellular virions, it is likely that nucleocapsids are the transmissible particle from cell to cell (Fig. 25.3). This is most clearly seen in cultured cells where BoDV nucleocapsids associate tightly with chromatin to segregate into daughter cells during cell division.

Diseases

Borna disease

Prototypical Borna disease is a rare neurologic disorder of horses and sheep with outbreaks most often seen in central Germany. While the virus can be experimentally transmitted between horses and sheep, this is clearly not the natural route of infection. The natural host for BoDV-1 has been identified as the bicolored white toothed shrew. The epidemiology of natural disease suggests transmission via infected urine or feces, though an olfactory route. Virus gains access to neurons and spreads by axonal transmission. BoDV-1 infection of neurons is noncytopathic, but persistent infection of an animal results in immunopathologic damage. Mammalian bornaviruses have also been recovered from other mammals with neurologic disease. In cats, infection with BoDV-1 is associated with staggering disease, a fatal neurologic condition. Recently, a handful of human cases of bornavirus encephalitis have been confirmed in the BoDV-1 endemic regions of Germany (Box 25.3).

The best-characterized animal model of BoDV-1 infection is the rat model, using specifically adapted strains of virus. In this model, the outcome of infection depends on the genetic background, age of infection, and immune status of the rat. Infection of adult, immunocompetent rats results in the development of encephalitis similar to that seen in horses and sheep. Early studies quickly demonstrated that disease is immune-mediated, with cytotoxic T-cells being the major players (through killing of infected cells). This is consistent with the pathology of natural infection, where microscopic examination of the brain reveals massive infiltration of lymphocytes. In contrast, experimental infection of neonatal rats (which are not yet immunocompetent) leads to persistent infection, in the absence of acute disease. However, infection is not without consequence as persistently infected rats are significantly smaller than mock-infected littermates and exhibit various behavioral, emotional, and cognitive impairments.

Birds and bornaviruses

In the early 1970s veterinarians began reporting cases of a mysterious and deadly disease of captive parrots. The birds were eating but were not able to digest their food and were wasting away. Some had severe neurological problems including incoordination, seizures, and blindness. The most obvious finding was *extreme* enlargement of a region of the birds' digestive tract called the proventriculus (akin to the esophagus in mammals). One of the common names for the disease is proventricular dilatation disease (PDD). The disease appeared to be caused by an infectious agent, but for over 30 years, the cause of the disease remained a mystery despite numerous efforts to identify a culprit. Therefore no diagnostic tests were available. As many large parrots are highly endangered in their natural habitats, a deadly disease among those in breeding programs was devastating. In 2008 the mystery was solved using molecular techniques. Virologists using genome-sequencing methodologies detected evidence of a bornavirus in the brains of

BOX 25.3

Human infection with bornaviruses

Beginning in 1985 and extending over a period of about 25 years, investigators published studies linking a number of different human psychiatric syndromes to BoDV-1 infection. Evidence of infection was provided through serologic studies and use of polymerase chain reaction. In fact it was early reports of an association between BoDV-1 and human psychiatric disease that encouraged interest in the exact nature of the Borna disease agent. Unfortunately, even as the molecular and biochemical nature of the agent became clearer, the association between BoDV-1 and human psychiatric disease remained murky, as many negative studies were published. While an association between BoDV-1 and human psychiatric disease cannot be completely ruled out, there is no strong evidence supporting one.

However, in 2015, a *novel* bornavirus was clearly linked to fatal, acute encephalitis in humans. The three victims were men (63, 62, and 72 years of age) from the state of Saxony-Anhalt, Germany. The cases occurred between 2011 and 2013. The clinical course for all three men was that of a progressive encephalitis that ended in death. The men were screened for a number of infectious agents associated with CNS disease, but results were negative. An intriguing epidemiologic finding was that all three victims were breeders of the variegated squirrel

(*Sciurus variegatoides*, a native of Central America) and had exchanged squirrels on multiple occasions. Some of these squirrels had also succumbed to neurologic disease but again, initial screenings for suspect infectious agents were negative. Eventually brain samples from the squirrels were examined by unbiased genome sequencing and sequences of a novel bornavirus were obtained from clinically affected animals. On revisiting stored human samples, a virus identical to the variegated squirrel virus was found. Variegated squirrel bornavirus is clearly distinct from BoDV-1 and was unknown before these studies. While the affected squirrels are native to Central America, it is not yet clear where, or how, the captive breeding stock was infected, as other mammalian bornaviruses are clearly endemic to Germany.

A handful of additional cases of human bornavirus encephalitis have since been found. Three cases were linked to organ transplantation and another two occur in previously healthy young persons. These cases of severe encephalitis with high mortality, while rare, have actually served to discredit the notion that BoDV-1 is associated with psychiatric disease. The encephalitis patients developed high titers of antibodies, and viral RNA was readily detected; such findings have never been documented among psychiatric patients.

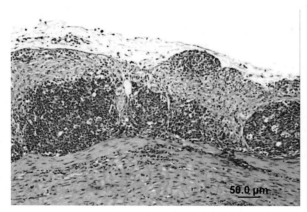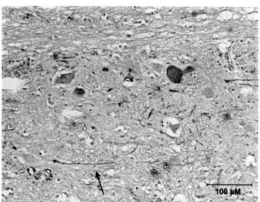

FIGURE 25.4   Microscopic lesions from a PDD bird, showing lymphocytic infiltration (dark purple cells) (left) and presence of bornavirus antigen (discrete areas of dark brown staining) in neurons (right).

affected birds. More importantly, the viral sequences were not found in healthy birds.

Now it was obvious why the PDD "agent" had eluded detection for so long: It was a bornavirus that caused persistent, noncytopathic, highly cell-associated infections in

cell cultures. Disease was primarily immune-mediated, with no evidence of virus particles in lesions. Once the initial identification of a bornavirus was made, immunological reagents developed for BoDV-1 proved useful for detecting the virus in cultured cells (Fig. 25.3) and for demonstrating that PDD birds have central nervous system (CNS) lesions containing abundant bornaviral proteins (Fig. 25.4). Most cases of PDD are seen in captive parrots but infection and (rarely) disease among wild waterfowl in the United States and Canada are also well documented. It should be noted that a high percentage of infected birds are completely asymptomatic. It is not yet known to what extent bornavirus affects wild bird populations.

In this chapter we learned that:

- Members of the family *Bornaviridae* have unsegmented, negative-sense RNA genomes that encode six proteins (N, P, X, M, G, and L). Virions are enveloped.
- Transcription of bornaviral mRNA and genomes is nuclear.
- Infections are noncytopathic and persistent, both in cultured cells and animals.
- Neonatal rats become persistently infected and immunotolerant after experimental infection. They display developmental, emotional, and cognitive impairments.
- In natural infections disease involves the CNS and is immune-mediated. BoDV-1 is highly neurotropic while avian bornaviruses seems to have a wider cell tropism.

26

Family *Orthomyxoviridae*

After reading this chapter, you should be able to discuss the following:

- What are the general characteristics of the orthomyxoviruses?
- What are the functions of influenza virus HA?
- What is the function of influenza virus neuraminidase (NA)?
- Describe genome variation and host range of influenza A viruses.
- What is antigenic drift? What are the consequences of influenza virus antigenic drift?
- What is antigenic shift? What are the consequences of influenza virus antigenic shift?
- Describe synthesis of influenza virus mRNAs. How does mRNA synthesis differ from genome synthesis?
- What is the function of the influenza A virus M2 protein?

- What classes of drugs are used to treat influenza virus infection? (Which viral proteins do they target?)

The family *Orthomyxoviridae* contains pathogens of humans and animals. Enveloped virions are spherical to filamentous, about 100 nm in diameter (Fig. 26.1) and genomes are segmented, negative-sense RNA (Box 26.1). Unlike most RNA viruses, replication is nuclear and some viral mRNAs are spliced. The family contains seven genera, at least three of which infect humans (Box 26.2). Influenza A virus is quite common and infects a wide variety of mammals and birds. It is a major human pathogen causing periodic epidemics and pandemics. Influenza A virus is often restricted to the respiratory or gastrointestinal tracts but the highly pathogenic avian influenza viruses (HPAIV) cause severe systemic infections of domestic poultry. Influenza A virus is genetically diverse, as a consequence of accumulating

Viruses.
DOI: https://doi.org/10.1016/B978-0-323-90385-1.00015-7

© 2023 Elsevier Inc. All rights reserved.

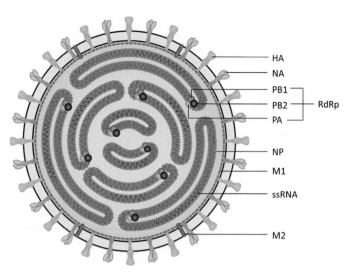
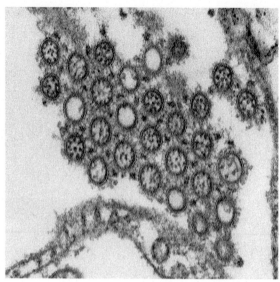

FIGURE 26.1   Virion structure (left). Transmission electron microscopic image of influenza virions (right). CDC Public Health Image Library Image ID#11746. Content providers CDC/Cynthia Goldsmith.

BOX 26.1

General characteristics

Influenza virus genomes are segmented, single-stranded, negative-sense RNA. Genomes contain six to eight segments that range in size from 2.3 to 0.9 kb. Segmented genomes lead to high rates of reassortment. Transcription and genome replication occur in the nucleus. mRNAs are primed by cap-snatching and are polyadenylated. Some viral mRNAs are spliced. Most RNA segments encode a single protein. Virions are enveloped with helical nucleocapsids. The envelopes of influenza A and influenza B viruses display two transmembrane glycoproteins, HA and NA, and also contains an integral membrane ion channel protein (M2). Virion shape is spherical to filamentous. The RNA-binding NP protein is the most abundant in the virion. A polymerase complex (PB1, PB2, and PA) is associated with each RNA segment.

BOX 26.2

Family *Orthomyxoviridae*

Genus: *Alphainfluenzavirus* (Circulate widely in humans, a variety of other mammals and birds)
Species *Influenza A virus*
Genus: *Betainfluenzavirus* (Circulate widely in humans)
Species *Influenza B virus*
Genus: *Deltainfluenzavirus* (Hosts include cattle and swine)
Species *Influenza D virus*
Genus *Gammainfluenzavirus* (Circulate widely in humans)

Species *Influenza C virus*
Genus: *Isavirus* (Identified from fish)
Species *Salmon isavirus*
Genus: *Quaranjavirus* (Infect arthropod and vertebrate hosts)
Species: *Johnston Atoll quaranjavirus*
Species *Quaranfil quaranjavirus*
Genus: *Thogotovirus* (Arboviruses transmitted by ticks; infect a wide range of mammals and birds)
Species *Dhori thogotovirus*
Species *Thogoto thogotovirus*

and tolerating point mutations, a wide host range, and the ability to reassort genome segments. In the early 20th century (1918–19), a particularly lethal strain of influenza A virus caused an estimated 40 million human deaths worldwide. Although a pandemic of that magnitude has not occurred since, epidemics and pandemics occur with regularity, assuring that influenza A viruses are carefully monitored and studied.

Genome structure/organization

Members of the family *Orthomyxoviridae* have segmented, single-stranded negative-sense RNA genomes (Fig. 26.2). Influenza viruses A and B as well as isavirus have eight genome segments, influenza viruses C and D have seven segments, and Quaranjaviruses and Thogotoviruses have six segments. Genome segments form helical nucleocapsids that contain N protein and an active polymerase complex. Most genome segments encode a single protein; however, there is some variation (some influenza A viruses encode up to 11 proteins). The three largest genome segments ($\sim$2300 nt each) encode proteins that form the replicase/transcriptase complex. The smallest segment ($\sim$900 nt) usually encodes one or two nonstructural proteins. The nucleoprotein (NP),

matrix (M1) protein, and glycoproteins are encoded by individual genome segments.

Virion structure

Members of the family *Orthomyxoviridae* are enveloped viruses containing 6–8 helical nucleocapsids. Virions are spherical to filamentous, about 100 nm in diameter with glycoprotein spikes that extend 10–14 nm from the outer surface of the envelope. Structural proteins include NP (forms the scaffold for RNA), M1 (a matrix protein that forms a bridges between nucleocapsids and envelope), an integral membrane protein (M2), and one or two envelope glycoproteins that appear as spikes extending from the surface of the virion. A molecule of polymerase complex is associated with each genome segment. In transmission electron micrographs, the individual structures formed by each genome segment can be clearly identified.

Members of the genera *Alphainfluenzavirus* and *Betainfluenzavirus* encode two glycoproteins. Hemagglutinin (HA) forms homotrimer spikes extending from the virion surface. HA serves as both the attachment protein and the fusion protein and is a major target of neutralizing antibodies. As suggested by the name, HA binds sialic acid

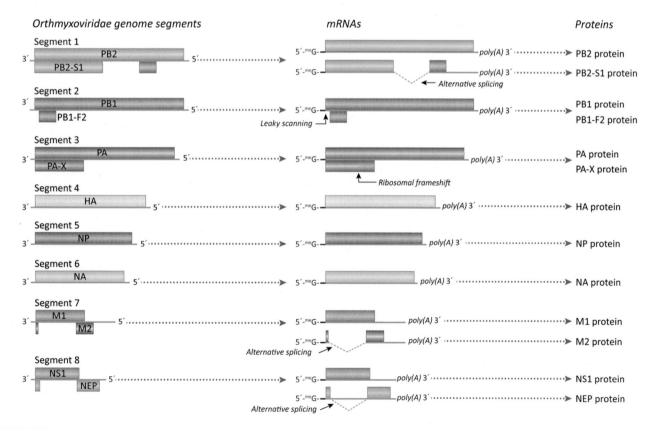

FIGURE 26.2   Example of orthomyxovirus genome organization and gene expression strategy.

residues on glycoproteins and agglutinates red blood cells. Neuraminidase (NA) spikes are homotetramers. NA is a *receptor-destroying* enzyme that cleaves sialic acid residues from cellular and viral glycoproteins. NA plays a key role in virion release from the infected cell. In contrast, members of the genus *Gammainfluenzavirus* encode a single major envelope glycoprotein called HA-esterase-fusion (HEF). HEF combines the functions of HA and NA. Thus influenza C viruses have only seven gene segments. The glycoprotein spikes of the isaviruses divide up their essential functions somewhat differently. Isaviruses encode two glycoproteins, HA acetylesterase and F. The HA acetylesterase binds sialic acid residues and hydrolyzes the acetyl group (thus has receptor binding and receptor-destroying activities). The second glycoprotein is the fusion (F) protein and as its name suggests is essential for membrane fusion. Thogotoviruses and Quaranjaviruses are arboviruses (transmitted by ticks). They encode a single envelope glycoprotein, GP that serves in attachment and fusion. Surprisingly Thogotovirus G shares 30% amino acid sequence identity with an envelope glycoprotein of baculoviruses, DNA viruses that infect a variety of insects.

Classification and nomenclature of orthomyxoviruses

There are currently seven recognized genera in the family *Orthomyxoviridae* (Box 26.2). Members of three genera, *Alphainfluenzavirus*, *Betainfluenzavirus*, and *Gammainfluenzavirus*, infect humans. Influenza A virus,

an alphainfluenzavirus, is a major human pathogen and infects other mammals and birds. In mammals disease is usually respiratory. Influenza B virus, a betainfluenzavirus, is primarily a human respiratory virus but has also been isolated from seals. Influenza B viruses do not infect birds and they usually cause milder human infections than influenza A viruses. Influenza C virus, a gammainfluenzavirus, also infects humans. In fact, most human adults have antibodies to influenza C virus suggesting that it is a widespread, common, and probably very mild upper respiratory tract virus.

The influenza A viruses are significant human and animal pathogens. They exhibit extreme genetic diversity and a wide host range. They tolerate (and benefit from) a high rate of point mutations and also undergo genome reassortment. A complex nomenclature system has been developed to track them (Box 26.3). Genome sequencing reveals that each genome segment has subtypes. For example, among influenza A viruses, there are 16 known HA subtypes (H1−H16) and 9 known NA subtypes (N1−N9). All NA and NA subtypes have been found among birds; the most common HA and NA subtypes that circulate in humans are H1, H2, and H3 and N1 and N2, respectively. Human HA and NA subtypes can be found in various combination (i.e., H1N1, H2N1, H3N2). However, important genetic and antigenic variation also exists within a subtype, thus all H1N1 viruses (even those that share all other segments) are not created equal. So names of influenza A viruses include information recording exactly when and where a particular virus was isolated (if that

---

BOX 26.3

Keeping a close watch

Given the history of repeated epidemics and pandemics, influenza virus surveillance is a global health imperative. The WHO sponsors a Global Influenza Program to collect and analyze influenza virus surveillance data from around the world. This allows the WHO to monitor global trends in influenza virus transmission and to support the selection of strains for vaccine production. In the United States, the Epidemiology and Prevention Branch in the Influenza Division at the CDC coordinates a robust influenza virus surveillance program. The CDC produces weekly reports (FluView and FluView Interactive), available at the CDC website (http://www.cdc.gov). The US surveillance system is a collaborative effort between CDC and state, local and territorial health departments, public health and clinical laboratories, and healthcare providers. The program includes documenting and reporting when and where influenza activity is

occurring, tracking influenza-related illness and deaths, identifying what viruses are circulating and perhaps most critically, detecting changes in circulating influenza viruses that may foreshadow an epidemic or pandemic. There are ∼350 laboratories in the United States that report (weekly) the total number of respiratory specimens tested and the number positive for influenza virus.

Cataloging and tracking influenza viruses generate a lot of information and it is imperative that each virus isolate be clearly named. In the case of human influenza viruses, names include the following critical information:

- Type A, B, or C,
- location the isolate was collected and the isolate number,
- year of isolation,
- HA and NA subtypes.

information is known). In the case of human influenza viruses, names include the following information: genus (designated as A, B or C), location the isolate was collected, isolate number, year of isolation, HA and NA subtypes. Thus "A/Texas/1/79 (H3N2)" denotes an influenza A virus isolated in Texas in 1979. It was the first isolate of the year and the HA and NA subtypes were H3 and N2. When an influenza virus is isolated from an animal (other than a human), the species is added to the name.

Influenza virus A proteins

This section will focus almost exclusively on influenza A viruses as we walk through a general replication cycle (Fig. 26.3). Much detailed information is available for each viral protein so we will begin the discussion by describing their major functions.

Hemagglutinin

HA is a transmembrane glycoprotein critical for both attachment and penetration (it has membrane fusion activity). HA spikes are assembled as homotrimers in the viral envelope. HA binds to sialic acid residues on glycoproteins. Species specificity of influenza A viruses depends on their preference for attaching to particular sialic acid linkages. The fusion activity of HA is dependent on proteolytic cleavage to generate a hydrophobic peptide with a free amino terminus

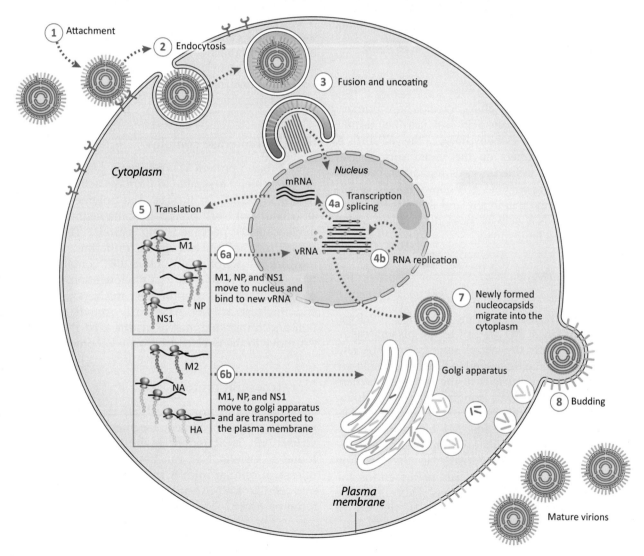

FIGURE 26.3 General replication cycle of influenza virus. Attachment is followed by endocytosis. After release from endosomes, the nucleocapsids move into the nucleus. Transcripts are produced in the nucleus. mRNA moves into the cytoplasm. Following translation some viral proteins must traffic back into the nucleus to support genome replication and nucleocapsid assembly. Nucleocapsids move back into the cytoplasm and traffic to the plasma membrane, the site of budding.

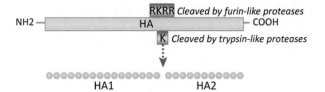

FIGURE 26.4   Schematic of influenza virus A HA cleavage site.

(Fig. 26.4). Many influenza virus HAs are cleaved by extracellular, trypsin-like proteases, restricting virus replication to sites (respiratory and gastrointestinal tract) where these enzymes are found. However, some influenza virus HAs are cleaved by furin-like proteases while moving through the ER, thus are fully processed before arrival at the plasma membrane. These viruses are not restricted to replication in the respiratory and/or gastrointestinal tracts. HPAIV, capable of causing systemic, multiorgan infections in chickens, have HA proteins that are cleaved by furin-like proteases.

Neuraminidase

NA is a transmembrane glycoprotein whose primary function is cleavage of sialic acid residues (neuraminic acid) from proteins and carbohydrates at the cell surface. NA forms homotetramers on the virion surface. NA functions late in the influenza virus replication cycle, as a receptor-destroying enzyme. NA allows release of virions from the plasma membrane and limits their aggregation. Inhibiting NA activity does not prevent virion formation but does limit spread.

Matrix protein (M1)

M1 is a major structural protein that serves as a bridge between nucleocapsids and the virion envelope. After synthesis, most M1 traffics to the plasma membrane where it binds the cytoplasmic tails of HA and NA. As virions are being assembled, nucleocapsids are transported from the nucleus to the cytoplasm where they interact with M1 (thus bringing them to budding sites). Key to M1 function is the pH dependence of its interaction with nucleocapsids. M1 binds nucleocapsids at neutral pH to facilitate virion assembly at the plasma membrane. However, early in the infection process (during entry), the low pH of the endosome disrupts M1 nucleocapsid interactions. This allows the nucleocapsids to move into the nucleus, the site of transcription and genome replication.

M2 ion channel protein

M2 is a 97-amino acid integral membrane protein that forms acid activated ion channels in the influenza

virus envelope. M2 is produced from the same genome segment that encodes for M1, but the M2 mRNA is spliced. M2 ion channel activity is required for virus infectivity and functions as a proton transporter. M2 becomes activated in acidic endosomes during viral entry and triggers viral uncoating by moving protons to the interior of the virus particle. Recall that low pH is required to facilitate release of the nucleocapsids from the M1 protein upon membrane fusion. The requirement of M2 is underscored by the ability of antiviral drugs such as amantadine and rimantadine to inhibit influenza virus replication. They do this by specifically blocking the M2 channel.

Nucleoprotein

NP is a basic protein that binds single-stranded viral RNA. After synthesis NP moves into the nucleus of the infected cell and genome segments wrap around NP to form the helical nucleocapsid. NP is a major structural protein and is also required for influenza virus genome replication. Thus genome replication requires ongoing synthesis of NP in the infected cell.

RNA polymerase complex

PB1 (basic protein 1), PB2 (basic protein 2), and PA (acidic protein) assemble to form the active polymerase complex, ~250 kDa in size. The polymerase complex associates with NP bound genome segments to form nucleocapsids packaged into virions. Thus early in infection, when nucleocapsids traffic into the nucleus, they are transcriptionally active. PB1 contains the catalytic domain active in chain elongation, PB2 binds methyl guanosine caps of host mRNAs, and PA has endonuclease activity. After early rounds of transcription and translation, newly synthesized PB1, PB2, and PA move to the nucleus to initiate genome replication.

Nonstructural protein 1

Nonstructural protein 1 (NS1) is encoded from the smallest genome segment (segment 8 of influenza A virus). NS1 is abundant in infected cells but is not found in virions. NS1 functions include: (1) inhibiting release of cellular mRNA from the nucleus; (2) inhibiting cellular mRNA splicing; (3) inhibiting the interferon pathway.

Nuclear export protein

Nuclear export protein (NEP) is nonstructural protein translated from a spliced mRNA produced from the same genome segment that codes for NS1. After synthesis NEP moves into the nucleus, the site of nucleocapsid assembly where it associates with nucleocapsids (and cellular proteins) to facilitate transport of RNPs from the nucleus.

Overview of replication

Attachment/penetration

HA spikes are critical for both attachment and fusion. HA binds sialic acid residues on cell surface glycoproteins and glycolipids. Given that sialic acid is a ubiquitous cell surface molecule, it is somewhat surprising that HA binding strongly influences host susceptibility. However, the linkage of the terminal sialic acid to the penultimate galactose in the carbohydrate chain specifies HA binding. Human influenza viruses preferentially bind to SAα2,6Gal while avian viruses preferentially bind to SAα2,3Gal. As we might guess, SAα2,6Gal is common in the upper respiratory tract of humans and SAα2,3Gal decorates carbohydrate chains in the avian respiratory and digestive tracts (Fig. 26.5). However, SAα2,3Gal is also found on cells in the lower respiratory tract of humans and this might partially explain why some HPAIVs are very poorly transmissible to humans, yet cause severe disease if an infection becomes established. It also has been demonstrated that a single amino acid substitution can suffice to change HA-binding specificity. Binding triggers endocytosis and within endosomes acidification results in rearrangement of HA to expose its fusion domain, thereby mediating fusion of virion and endosomal membranes.

For many years it was thought that the fusion process alone released nucleocapsids to traffic to the nucleus. However, we now understand that before fusion, the *inside* of the virion must be exposed to low pH. Acidification of the virion requires the ion channel protein, M2. M2 channel activity is itself activated by low pH in the endosome and then allows hydrogen ions into the interior of the virion (Fig. 26.6). Only after exposure to low pH are nucleocapsids released from their tight interactions with M1. Thus inactivation of the M2 channel blocks influenza virus replication and this is the mechanism of one class of antiinfluenza drugs (adamantanes).

Transcription

Nucleocapsids are actively trafficked to and imported into the nucleus. As with other negative-strand RNA viruses (i.e., order *Mononegavirales*), transcription and genome replication are discrete processes. In the case of the orthomyxoviruses, transcription depends on an *exogenous primer* that is derived from cellular mRNAs in the nucleus. The transcription complex binds to a newly synthesized, capped cellular mRNA and cleaves it ~10−13 nt downstream from the cap. This capped oligonucleotide remains associated with the transcription complex to serve as the primer for synthesis of an influenza virus mRNA. This process is called "cap-snatching" and the result is that the 5′ ends of influenza virus mRNAs are heterogeneous in sequence and are not exact copies of genome segments. The need for capped cellular RNAs explains the requirement for active host RNA polymerase II to support influenza virus replication. Prior to reaching the very 3′ end of the genome segment, the influenza virus mRNA is polyadenylated by the transcription complex.

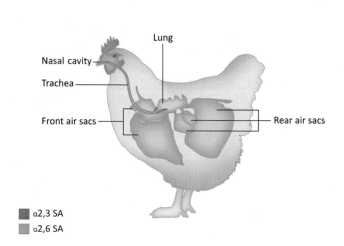

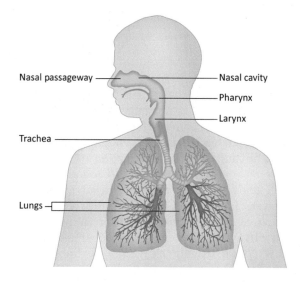

FIGURE 26.5 Avian flu transmissibility and virulence. Avian flu is poorly transmitted to humans as it does not bind A2,6 SA in the upper respiratory tract. However, if virus gets into the lung, A2,3 SA receptors support replication.

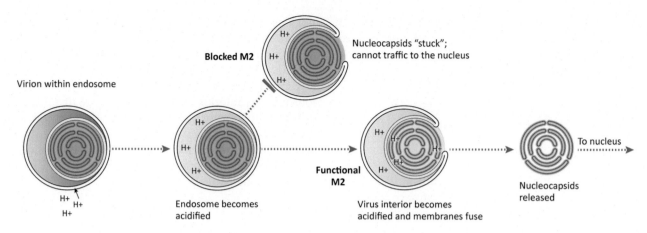

FIGURE 26.6   M2 ion channel function. The M2 protein is a membrane-associated, low pH activated ion channel. M2 is present in the envelope of the infecting virion. M2 is activated in the endosome allowing protons to move into the interior of the virion. At low pH, the matrix protein (M) "releases" nucleocapsids, allowing their transit to the nucleus.

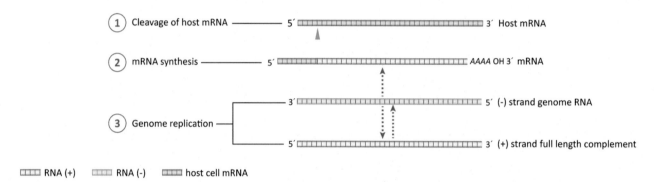

FIGURE 26.7   Comparison of processes to produce genomes and mRNAs. Genome replication produces an exact end-to-end copy of each genome segment. mRNAs are primed with a fragment of cellular mRNA. Transcripts terminate and are polyadenylated at sites upstream of the 5′ end of each genome segment.

Some mRNAs (notably products of the NS and M genome segments) are alternatively spliced. Viral mRNAs are exported from the nucleus.

Translation

Translation of influenza virus NP, M1, PA, PB1, and PB2 are cytosolic. The majority of the NA, PA, PB1, PB2 proteins, and some M1 protein, traffic back into the nucleus (recall that the genome of the infecting virus is in the nucleus). Translation of M2, HA, and NA are on rough ER. HA and NA are glycosylated in the ER and Goli and are transported to the plasma membrane. HA is produced as a precursor (HA0) that must be cleaved to produce fusion active, disulfide-linked HA1 and HA2. HA0 is sometimes cleaved in the ER but is usually cleaved by extracellular proteases. Most M1 protein traffics to the PM where it associates with the cytoplasmic tails of HA and NA, forming sites for virion assembly.

Genome replication

Before any new virions can be assembled, genomes must be replicated. Genome replication is a process very different from mRNA synthesis, although the same polymerase complex (PA, PB1, PB2) is used. The switch from transcription to genome replication is triggered by the presence of newly synthesized, unbound (to RNA) NP, PB1, PB2, and PA that have moved into the nucleus after translation. In the presence of soluble NP, replication initiates at the very 3′ end of each genome segment to produce an exact end-to-end copy (Fig. 26.7). The copy RNA strands are neither capped nor polyadenylated but they are bound to NP. Each cRNA serves as the template for synthesis of many genomes.

Assembly, budding, and maturation

As genomes are synthesized, they associate with NP and the polymerase complex to form nucleocapsids.

Also critical to the process is M1. Nuclear M1 may facilitate assembly of nucleocapsids and is required for their export from the nucleus. Nuclear export also requires the viral NEP and host cell proteins. NEP interacts with the cellular nuclear export machinery and with M1. Nucleocapsids traffic to the PM where they interact with M1 at that location. A major question in influenza virus biology is how seven or eight distinct genome segments are assembled into each virion. There is evidence of selective packaging that is regulated by RNA sequences unique to each genome segment. However, there is also evidence of random packaging, suggesting that the selection mechanism may be "sloppy." It is clear that if a cell is infected with two different influenza viruses, individual virions can contain segments from both parents (a process called reassortment, Fig. 26.8).

Virions bud from the PM and release requires NA. Recall that NA cleaves sialic acids from viral and host glycoproteins at the PM to release virions from the cell surface. In the absence of NA activity, large virion aggregates are seen at the cell surface (Fig. 26.9). These aggregates are less transmissible to new hosts. An important class of antiviral drugs inactivates NA.

For most influenza viruses, production of infectious virions requires an extracellular maturation event, cleavage of HA0 by extracellular trypsin-like proteases. These enzymes are found in the respiratory and

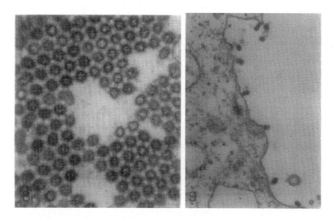

FIGURE 26.9 A packed array of influenza virions when neuraminidase is inactivated (left). Normally budding virions upon addition of exogenous neuraminidase (right) (Palese et al., 1974). Source: *From Palese, P., Tobita, K., Ueda, M., Compans, R., 1974. Characterization of temperature sensitive influenza virus mutants defective in neuraminidase. Virology 61, 397–410.*

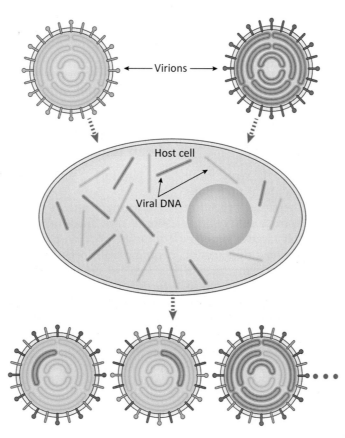

FIGURE 26.8 Reassortment of genome segments can occur in cells infected with more than one type of influenza virus. This is a powerful mechanism for generating genome diversity and leads to emergence of new strains.

digestive tracts, thus limiting the productive sites of virus replication. When growing influenza viruses in cultured cells, the virions must usually be exposed to trypsin to generate infectious particles. However, for some influenza viruses (e.g., HPAIV), HA0 is cleaved during its journey through the ER by furin-like proteases. The amino acid sequence at the cleavage site determines the type of protease required. HPAIVs have broad tissue tropism and can cause multiorgan, systemic infection in susceptible birds.

Antigenic drift and antigenic shift

Genetic variation enables RNA viruses to exploit variable and changing environmental conditions that include (among other things) genetics, immune status, and life style of potential hosts. Influenza A viruses exploit numerous hosts but key among them are aquatic birds. All known HA and NA subtypes have been found in aquatic birds, while a limited number are found in mammals.

Two types of mutations are key to the success of influenza viruses. The first are simple nucleotide substitutions (point mutations) that arise during genome replication. In the context of human influenza viruses, the point mutations of most consequence are those that allow escape from host immune responses. These are most often seen for HA and NA, as these surface glycoproteins are key targets of the adaptive immune response. The term *antigenic drift* is often used to describe point mutations in HA and NA (the term drift implying a slow, continuous process). Antigen drift in influenza A viruses is the process that mandates ever changing vaccines because a protective immune response against last year's flu virus does not provide complete protection against next year's flu virus. How can the HA and NA proteins tolerate point mutations given their important functions in the virus replication cycle? HA and NA both have unstructured regions (loops) of amino acids that are good targets for neutralizing antibodies. In contrast, functional domains of HA (attachment domain, cleavage domain, fusion domain) and NA (enzyme active site) are less accessible targets of the immune system.

Genome reassortment (*antigenic shift*) provides for larger, more sudden changes in influenza viruses. Given that there are eight genome segments, infection of a cell with two different influenza viruses could theoretically yield up to 254 recombinant viruses. In particular, new NA and HA combinations can generate a virus for which large segments of the human population have little preexisting immunity. The emergence of a "shifted" virus that is well-adapted to humans can cause a pandemic.

Disease

Human influenza

In humans, influenza (the flu) is a mild to severe respiratory tract disease whose common symptoms include rapid onset of fever, muscle aches, and fatigue. These may be accompanied by headache, runny nose, sore throat, or (occasionally) gastrointestinal tract symptoms. An indication of influenza virus infection (as opposed to a head cold) is the rapid onset of aches, pain, and fever. The most severe symptoms usually last only a few days, but fatigue and cough can last longer. In severe cases, the virus causes a primary viral pneumonia. Secondary bacterial infections are also common.

Transmission occurs via respiratory droplets. Often these contaminate our hands and are deposited on other objects. Thus a common pathway of infection is touching contaminated objects and then your nose or eyes. Frequent and thorough hand washing with soap and water limits the spread of influenza virus and should be practiced by those who are sick and those wishing to remain healthy.

Mammals other than humans catch influenza. For example, influenza A viruses infect pigs, dogs, and horses. Pigs are an important host as they can be infected by both mammalian and avian influenza (AI) viruses, providing a source of novel reassortants. Influenza virus infection can cause considerable economic loss, thus pork producers use biosecurity measures to keep their animals safe from infected humans and birds.

Avian influenza

Migratory aquatic birds are the natural hosts for influenza A viruses (mammalian influenza viruses likely originated in birds). When AI is in the news, stories are most often describing outbreaks among domestic poultry, particularly chickens and turkeys. In commercial flocks, mortality can be high and outbreak control is often accomplished by killing millions of birds.

AI is defined as low pathogenic (LP) or highly pathogenic (HP) for domestic poultry (Fig. 26.10). LP avian influenza viruses replicate primarily on mucosal surfaces of the respiratory and/or gastrointestinal tracts, and disease is generally mild, with low mortality. In contrast, HPAIV replicates both on mucosal surfaces and in internal organs. These systemic infections cause severe disease with mortality exceeding 75%. Historically outbreaks of HPAIV were sporadic, regional events. However, that changed with the emergence of the so-called Asian lineage of H5N1 HPAIV.

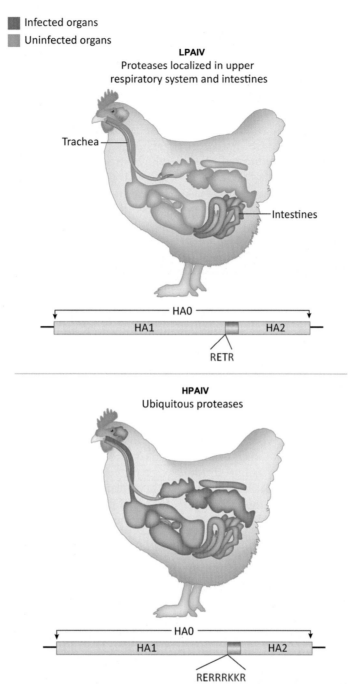

FIGURE 26.10   Tissue tropism of LPAIV and HPAIV. LPAIV are restricted to the respiratory and gastrointestinal tracts where extracellular trypsin-like proteases can cleave and activate HA. Some HPAIV have a much broader tissue tropism and can replicate in many organs as their HA protein is cleaved by intracellular furin-like proteases.

This virus emerged in southern China in 1996 and eventually spread across Asia to Europe and Africa. The first outbreak to gain international attention occurred in 1997 in the live bird markets of Hong Kong. In addition to avian cases, there were 18 documented human cases, with 6 deaths. The human cases resulted from direct contact with infected birds. The Hong Kong outbreak was controlled by the closure of live bird markets and slaughter of ~1.5 million birds. However, the virus continued to circulate in southern China, and in 2002 another outbreak in Hong Kong involved wild birds and birds in a zoological collection. Thus the highly pathogenic H5N1 virus was clearly not confined to domestic poultry.

By late 2003 HPAIV H5N1 had spread to South Korea. By 2004 genome sequencing revealed development of multiple sublineages of its HA protein. In the winter of 2006, H5N1 was associated with deaths of waterfowl in Europe, confirming fears that the virus would spread via migratory birds. Currently, H5N1 is known to be endemic in at least four countries (Vietnam, Egypt, Indonesia, and China). Sporadic human cases of H5N1 infection continue to be reported, and as of 2016 the World Health Organization (WHO) has tallied over 850 cases of human disease, with 450 deaths. All cases continue to be linked to direct contact with infected poultry, as the virus remains poorly transmissible among mammals. The H5N1 outbreak has led to increased surveillance for AIV among migratory birds and in domestic flocks. In the case of migratory birds, HPAIV is most frequently found associated with dead birds and is rarely isolated from healthy migrating birds. Thus there is still a lack of understanding about the basic ecology of HPAIV in wild aquatic birds.

Antiviral drugs

In the United States, four drugs are currently recommended for treatment of uncomplicated influenza. Oseltamivir phosphate (trade name Tamiflu), zanamivir (trade name Relenza), and peramivir (trade name Rapivab) are NA inhibitors. Recall that in the absence of NA, influenza viruses form large aggregates that remain attached to cells. The aggregates are less infectious and less easily transmitted to new cells. Baloxavir (trade name Xofluza) targets the influenza virus cap-dependent endonuclease activity of the PA subunit of the transcription complex, thereby inhibiting viral mRNA synthesis. Two other compounds, amantadine and rimantadine work by inhibiting the M2 ion channel. Though previously approved for treatment of influenza, they are no longer recommended due to high rates of resistance.

Vaccines

In the United States, both killed and live attenuated influenza virus vaccines are manufactured on a yearly cycle. The vaccines include three or four different virus strains. The ability to manufacture the "correct" vaccines depends on global tracking of human influenza viruses, with the hope of predicting which strains will arrive with the next flu season. The WHO coordinates a Global Influenza Surveillance Network that includes more than 100 National Influenza Centers based in the United States, Australia, Japan, and the United Kingdom. The centers work with academic laboratories, regulatory agencies, and manufacturers to ensure

that the appropriate vaccines are available each year. Circulating influenza viruses are tracked by the collection of swabs from sick patients. Collaborating centers analyze the viruses and compare their sequences to those from previous years. Surveillance data are reviewed and experts recommend the most appropriate viruses for inclusion in vaccines. However, this is always an educated guess! Countries consider the WHO suggestions but make the final decisions on which strains to include in vaccines. Once chosen, the selected viruses must be prepared for manufacture. This usually requires generating reassorted viruses that can be grown to high titers. Historically, eggs have been used for virus propagation but alternatives such as cell cultures have been established. Inactivated whole virus vaccines are administered via a shot, while live attenuated flu vaccines are administered as a nasal spray. Flu vaccines typically contain virus or viral antigens for three or four strains of virus.

In this chapter we learned the following:

- Orthomyxoviruses are enveloped viruses with segmented negative-sense RNA genomes. They replicate in the nucleus of infected cells, using newly transcribed capped host cell mRNAs as a source of primers for viral mRNA synthesis.
- Influenza A viruses are major human pathogens. They have two envelope glycoproteins, HA and NA. HA is the attachment and fusion protein. NA is required for virion release as it destroys the sialic acid residues to which HA binds.
- Aquatic birds are the major reservoirs for all influenza A viruses. Mammals are infected with smaller subsets of influenza A viruses. Genome reassortment (antigenic shift) leads to generation of new viruses, with altered host range and virulence determinants.
- The names of influenza A, B, and C viruses include the viral species, host name (if other than human), the date and location of isolation, and HA and NA types.
- HA and NA of influenza viruses are major targets of the immune response. However, a steady accumulation of point mutations generates new isolates that can escape (at least in part) from preexisting immune responses.
- To keep up with antigenic drift and shift, human vaccines are reformulated yearly.

Reference

Palese, P., Tobita, K., Ueda, M., Compans, R., 1974. Characterization of temperature sensitive influenza virus mutants defective in neuraminidase. Virology. 61, 397—410.

27

Family *Hantaviridae*

After studying this chapter, you should be able to:

- Describe the general characteristics of members of the family *Hantaviridae*.
- Explain modes of transmission of hantaviruses.
- List human diseases caused by hantaviruses.

In 2017 the International Committee of the Taxonomy of Viruses (ICTV) accepted a reorganized taxonomy for enveloped, negative-strand RNA viruses by creating a new order, *Bunyavirales* (Maes et al., 2019). The order currently contains 11 families that include viruses of plants, insects, and a wide variety of animals. The family *Bunyaviridae* was eliminated from this new taxonomy, and the viruses previously included therein were placed in the newly created families *Hantaviridae*, *Peribunyaviridae*, *Phenuiviridae*, and *Nairoviridae* among others. Molecular phylogenies provided a strong basis for this update. This chapter describes the hantaviruses, specifically those in the subfamily *Mammantavirinae* (Box 27.1). These are found worldwide, in close association with their rodent hosts. Many are exquisitely host specific and likely coevolved with hosts over millions of years. A few mammantaviruses infect humans, sometimes causing severe disease. Human infections occur via inhalation of dried, aerosolized, rodent droppings and sporadic cases of disease are most often associated with high rodent activity in or around dwellings.

Virion structure

Virions are spherical or pleomorphic, ranging in size from 120 to 160 nm in diameter. Genome segments have complementary sequences at their 5′ and 3′ ends that base pair, generating the circular nucleocapsids that can be seen in virions. Hantaviruses do not encode matrix proteins. The N protein associates with both the RNA and the cytoplasmic tails of the Gn and Gc glycoproteins. Gn and Gc dimerize, then form higher order tetramers in the envelope (Fig. 27.1).

Genome organization

Hantaviruses are enveloped viruses with single stranded, segmented, negative sense RNA genomes (Fig. 27.2 and Box 27.2). Genomes are comprised of three RNA segments, designated small (S), medium (M), and large (L), according to their relative sizes. The total genome size is ~11.8 kb and encodes four structural proteins and one nonstructural protein. The 5′ and 3′ ends of each RNA segment base pair to form panhandle structures causing the RNA segments to circularize. The S mRNA encodes two proteins, N and NSs (nonstructural protein encoded by the small RNA segment). A single mRNA is transcribed from the S genome segment; large quantities of N, an RNA-binding protein (~54 kDa), are produced while smaller amounts of NSs (90 amino acids) are produced by a +1

© 2023 Elsevier Inc. All rights reserved.

BOX 27.1

Taxonomy

Order *Bunyavirales*
Family *Hantaviridae*
Subfamily *Actantavirinae*. Three species, all infect fish.
Subfamily *Agantavirinae*. Wēnling hagfish virus (sole member).
Subfamily *Mammantavirinae*. Four genera and over 30 species. Infect rodents. Human pathogens

include Hantaan virus, named after the Hantaan river in Korea and Sin Nombre virus. The natural hosts are rodents but humans can become infected by inhaling dried urine.
Subfamily *Repantavirinae* Hǎinán oriental leaf-toed gecko virus (sole member).

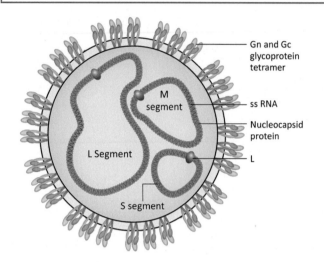
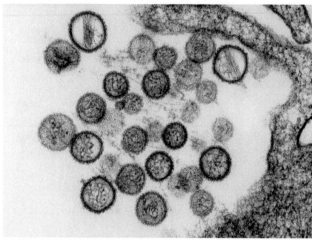

FIGURE 27.1　Virion morphology (left). Transmission electron microscopic image of a hantavirus known as the (Sin Nombre virus) (right). Source: *From CDC Public Health Image Library, Image ID #14340. Content providers CDC/Brian W.J. Mahy, PhD; Luanne H. Elliott, M.S.*

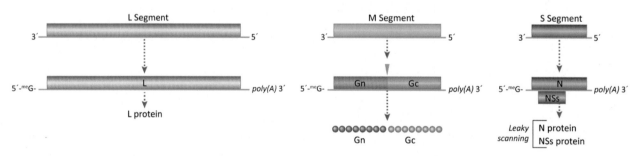

FIGURE 27.2　Genome and protein expression strategy of a hantavirus.

frameshift or by a leaky scanning mechanism. N oligomerizes and binds to both genome RNA and cRNA. N also interacts with L and the viral glycoproteins to facilitate virion assembly and morphogenesis. NSs also appears to be an inhibitor of interferon signaling. The M RNA segment encodes the envelope glycoprotein precursor that is processed in the endoplasmic reticulum to produce amino terminal (Gn) and carboxy terminal (Gc) glycoproteins. Gn and Gc are transmembrane proteins that associate to form tetramers in the envelope. Gn is predicted to have the receptor-binding domain while Gc is predicted to contain a fusion domain. The hantavirus L genome segment encodes the L protein, the RdRp. L is multifunctional with polymerase, helicase, and endonuclease domains. L protein catalyzes both replication and transcription of viral genomic RNA. Replication and transcription also require N. As might be expected with a segmented virus, there is evidence of genome reassortment among related strains of hantaviruses.

General replication strategy

Gn and Gc are the viral attachment proteins and fusion proteins, respectively. Hantaviruses enter cells by endocytosis and acidification of endocytic vesicles is followed by membrane fusion, releasing nucleocapsids into the cytoplasm. Hantaviruses replicate in the cytoplasm using a negative-strand RNA virus coding strategy with genomic RNAs serving as templates for synthesis of both mRNA and cRNA. Transcription and genome replication appear to occur in close association with membranes. Synthesis of hantavirus mRNA is primed using capped oligonucleotides (10−20 nt long) cleaved from host mRNAs present in the cytosol. This is accomplished via a cap-snatching endonuclease domain present at the very N-terminus of the L protein. In contrast to mRNA synthesis, genome synthesis requires production of uncapped full-length cRNAs from each genome segment. Similar to other negative-strand RNA viruses, there must be a molecular mechanism (yet to be defined) that triggers the switch from transcription to genome replication. As with other negative-strand RNA viruses, N protein is likely involved. Signals for the encapsidation of genomic RNAs are present in the noncoding 3′ and 5′ regions.

Diseases

Hantaviruses are spread among rodents through bites, or aerosolized urine, feces, or saliva. In rodents they frequently cause chronic infections, without disease. However, when hantaviruses occasionally infect humans, disease can be severe. Person-to-person transmission of Andes hantavirus (ANDV) has been reported, suggesting an adaptation to transmission among humans.

BOX 27.2

Characteristics

Hantaviruses are enveloped viruses with segmented, linear, single stranded, negative sense RNA genomes. Genomes have three segments, designated small (S) (~1.8 kb), medium (M) (~3.7 kb), and large (L) (~6.3 kb) for a total genome size of ~ 11.8 kb. RNA segments have complementary nucleotides at their 3′ and 5′ ends. Base-pairing of complementary nucleotides is predicted to form stable panhandle structures to form noncovalently closed circular RNAs. Virions range in size from 120 to 160 nm in diameter.

Hantaviruses encode a precursor envelope glycoprotein from the M segment that is processed in the endoplasmic reticulum to produce the mature glycoprotein heterodimers containing Gc and Gn. The polymerase (designated the L protein) is encoded by the largest genome segment and a nucleoprotein (N) is encoded by the smallest genome segment. N and L bind genomic RNA forming nucleocapsids within the envelope.

FIGURE 27.3   Deer mouse, *Peromyscus maniculatus*, among deteriorating sheets of fabric and bird feathers. Source: *From CDC Public Health Image Library, Image ID #13214. Content provider CDC.*

FIGURE 27.4   Dresser drawer in a home, showing evidence of rodent habitation (from home of a hantavirus patient). Source: *From CDC Public Health Image Library, Image ID #14340. Content provider CDC.*

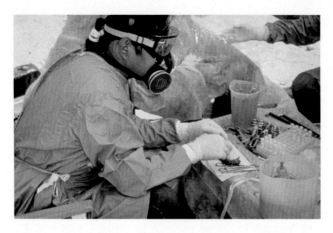

FIGURE 27.5 CDC scientist, wearing protective clothing while collecting specimens from trapped rodents. Source: *From CDC Public Health Image Library, Image ID #2. Content provider CDC/Cheryl Tryon.*

Hantavirus pulmonary syndrome

Members of the family *Hantaviridae* infect rodents worldwide. Hantaviruses are transmitted from rodent to rodent, usually without disease. But rarely, these viruses may be transmitted to humans. Human exposure results from contact with infected rodent urine, feces, and/or saliva (Figs. 27.3–27.5). Hantavirus pulmonary syndrome (HPS) was first described in the southwestern United States in 1993 (Box 27.3). The disease presents as a flu-like illness that *rapidly* progresses to severe respiratory disease. Severe disease may be caused by an overactive immune response. Other contributors to the pathogenesis of HPS are platelet dysfunction and breakdown of endothelial cell barriers. In the 1993 outbreak, the death rate was almost 50% among hospitalized patients. Sporadic cases of HPS have been associated with at least 10 different

---

BOX 27.3

Hantavirus pulmonary syndrome

The 1993 Four Corners hantavirus outbreak was the first time that human hantavirus infection was documented in the United States. Four Corners refers to the quadripoint where the boundaries of Colorado, Utah, Arizona, and New Mexico meet. This is a mostly rural, arid, rugged area of the United States. While only a handful of cases occurred, they were serious and clinically notable for their very rapid onset of severe shortness of breath and high mortality. The CDC was notified and sent a team to the region to investigate. The CDC initially identified five people with similar symptoms, all of whom had died. A total of 48 cases were identified in 1993. Most of the stricken were young Navajos living in New Mexico. Virologists at the CDC, using new molecular methods, were able to link the new disease with a previously unknown hantavirus. Hantaviruses had previously been associated with sporadic cases of human hemorrhagic fever in Asia and were known to infect rodents. Therefore investigators trapped many species of rodents in areas where patients lived and worked. Several types of rodents, including deer mice, were infected with hantaviruses. Deer mice were probably the link to human disease, as these rodents live in rural and semirural areas, sometimes inhabiting homes, outbuildings, barns, and woodpiles.

The reason for the notable outbreak in 1993 was probably environmental. There were unusually large numbers of deer mice due to an abundance of snow, rain, and vegetation. It was estimated that there were 10 times more deer mice in 1993 than in 1992. Once researchers had identified

the viral cause of the 1993 outbreak, they tested stored samples from patients who had previously died of unexplained adult respiratory distress syndrome and found many more hantavirus positive samples. The oldest was from 1959. Thus human infections had occurred in small numbers, for years.

Today we know that many species of rodent can be infected with hantavirus. Virus is excreted in urine and feces. Human infections occur mainly by inhaling infected dust from rodent infested buildings. The rodents that carry hantaviruses (e.g., the deer mouse and white-footed mouse) live in rural, as opposed to urban areas. Therefore infections, while rare, are more common in rural areas. Practices to avoid infection include airing out structures (i.e., cabins) that have been closed up for long periods of time, followed by cleaning with water containing household bleach or other disinfectant. It is very important to avoid generating airborne dust while cleaning.

In 2012, 10 confirmed cases (with three deaths) of hantavirus infection were identified among visitors to Yosemite National Park in California. Many of the infected had stayed in a relatively new set of "tent cabins" erected a few years earlier. Investigations revealed that the manner in which the structures were built provided very attractive nesting areas for rodents, thus they were subsequently torn down to prevent future outbreaks. However, as in the 1993 Four Corners outbreak, weather probably played a role, as rodent populations were high at the time of the outbreak.

hantaviruses, each associated with a different rodent species. The best prevention against hantavirus infection is adequate rodent control and limiting exposure to infected dust.

Hemorrhagic fever with renal syndrome

Hantaviruses are also endemic to Europe and Asia where they cause hemorrhagic fever with renal syndrome in humans. The clinical presentation and outcome varies, depending on the viral agent; case mortality rates vary from less than 1% to up to 15%. Factors contributing to disease are likely similar to those described for HPS. The clinical picture is similar to other viral hemorrhagic fevers, beginning with abrupt onset of high fever, chills, and body aches. Nausea and vomiting may occur as well. Days later, hemorrhages appear on mucosal surfaces and the skin. As the disease progresses, renal function may also become impaired.

In this chapter we have learned that:

- Hantaviruses are enveloped viruses with single-stranded, segmented RNA genomes (three segments).
- The overall replication strategy of hantaviruses is similar to other negative-sense RNA viruses.
- Hantaviruses are rodent viruses that cause little harm to their rodent hosts. However, human infections can result in severe diseases including hantavirus pulmonary syndrome (North America) or hemorrhagic fever with renal syndrome (Europe and Asia). The numbers of human cases each year are relatively small and most transmission to humans occurs via inhalation of infected rodent urine and feces.

References

Maes, P., Adkins, S., Alkhovsky, S.V., Avšič-Županc, T., Ballinger, M. J., Bente, D.A., et al., 2019. Taxonomy of the order Bunyavirales: second update 2018. Arch. Virol. 164, 927–941.

28

Order *Bunyavirales*: Families *Peribunyaviridae*, *Phenuiviridae*, and *Nairoviridae*

After studying this chapter, you should be able to:

- Describe the general characteristics of members of the families *Nairoviridae*, *Peribunyaviridae*, and *Phenuiviridae*.
- Describe the general features of genome organization shared by these viruses.
- List the major modes of transmission of these viruses.
- List some diseases caused by these viruses.

In 2017 the International Committee of the Taxonomy of Viruses (ICTV) accepted a reorganized taxonomy for enveloped, negative-strand RNA viruses by creating a new order, *Bunyavirales* (Maes et al., 2019). The order currently contains 11 families of viruses that infect plants, insects, and a wide variety of animals, a few of which are pathogens of animals and humans. The family *Bunyaviridae* was eliminated from the new taxonomy, and the viruses previously included therein were placed in newly created families (Box 28.1). Molecular phylogenies provided the strong basis for this update. This chapter contains general descriptions of the families *Peribunyaviridae*, *Phenuiviridae*, and *Nairoviridae*. Members of these families are enveloped viruses with segmented, negative sense or ambisense RNA genomes. Genomes have three segments, designated small (S), medium (M), and large (L) (Box 28.2). Total genome sizes range from 11 to 23 kb. The 5′ and 3′ ends of each genome segment base pair to form panhandle structures, causing the RNA segments to circularize (Fig. 28.1). Members of the three families encode a precursor envelope glycoprotein from the M segment, which is processed to produce the mature glycoprotein heterodimers containing Gc and Gn. Viruses in all three families encode a polymerase (designated the L protein) from the largest genome segment. The smallest genome segment encodes a nucleoprotein (N) that binds to genomic RNA. Differences among the three families include sizes of the RNA segments, as well as coding strategy and the total number of proteins produced.

Genome organization

Members of all three families have trisegmented genomes that are referred to by their size (S, M, and L). The overall genome organization is conserved with an RNA-binding N protein encoded by the S genome segment, envelope glycoproteins encoded by the M genome segment, and an RNA-dependent RNA polymerase (called the large (L) protein) encoded on the L genome segment. Members of the family *Peribunyaviridae* use a negative-strand RNA virus coding strategy (Fig. 28.2). The peribunyavirus S genome segment encodes two proteins, N

Viruses.
DOI: https://doi.org/10.1016/B978-0-323-90385-1.00002-9

© 2023 Elsevier Inc. All rights reserved.

BOX 28.1

Taxonomy

Order *Bunyavirales*
Families include:
Family *Peribunyaviridae*. Pathogens include *La Crosse virus*, *Oropouche virus*, *Schmallenberg virus*, and *Cache Valley Virus*.

Family *Phenuiviridae*. Pathogens include Rift Valley fever virus.

Family *Nairoviridae*. Most are tick-borne viruses. Pathogens include Crimean Congo hemorrhagic fever virus.
Family *Hantaviridae* (see Chapter 27).
Family *Arenaviridae* (see Chapter 29).

BOX 28.2

Shared characteristics

Members of the families *Peribunyaviridae*, *Phenuiviridae*, and *Nairoviridae* are enveloped viruses with segmented, linear, single stranded, negative sense, or ambisense RNA genomes. Genomes have three segments, designated small (S), medium (M), and large (L), and total genome sizes ranges from 11 to 23 kb. RNA segments have complementary nucleotides at their 3′ and 5′ ends. Base-pairing of complementary nucleotides is predicted to form stable panhandle structures to form noncovalently closed circular RNAs. Virions range in size from 75−115 nm in diameter.

Members of the three families encode a precursor envelope glycoprotein processed in the endoplasmic reticulum by host proteases. All encode a polymerase (designated the L protein) from the largest genome segment. The smallest genome segment encodes a nucleoprotein (N) that binds to genomic RNA and associates with the envelope proteins to drive virion assembly and budding. Differences among the three families include sizes of the RNA segments, details of their protein coding strategies, and the number of nonstructural proteins produced.

---

and NSs proteins. The NSs coding region overlaps N and may be produced by a leaky scanning mechanism. M encodes a precursor polyprotein that is cleaved in the ER by host proteases to generate three protein products: Gn, NSm, and Gc. The L RNA segment encodes the L polymerase.

The genome of Rift Valley fever virus (RVFV), a member of the family *Phenuiviridae*, is shown in Fig. 28.2. Phenuiviruses use an ambisense coding strategy to produce two proteins from the S genome segment. The N mRNA is transcribed from the genome strand while NSs mRNA is transcribed from the S segment cRNA (Fig. 28.2, middle panel). The M RNA encodes the NSm, Gn, and Gc proteins while the L polymerase is encoded by the L RNA segment.

Nairoviruses also use a negative sense coding strategy to produce protein products from its genome segments. N mRNA is transcribed from the S genomic RNA segment. (Some nairoviruses are reported to produce an NSs product as well.) The nairovirus M RNA segment encodes five proteins that include: A mucin like domain (MLD) protein, a 38 kDa glycoprotein (gp38), a nonstructural protein (NSm), and two envelope glycoproteins (Gc and Gn). These are produced by processing of a precursor polyprotein in the endoplasmic reticulum (ER). The nairovirus L genome segment, at ∼12,000 nt, encodes an L protein that is almost twice as large (∼459 kDa) as the L proteins of the peribunyaviruses and phenuiviruses. It contains, at its very N-terminus, an ovarian tumor (OUT)-like cysteine protease domain that suppresses the cellular immune responses by altering ubiquitination/deubiquitination of key cellular proteins.

Virion structure

Members of the families *Nairoviridae*, *Peribunyaviridae*, and *Phenuiviridae* are enveloped and package three nucleocapsid segments (Fig. 28.1). Envelopes are studded with Gn and Gc glycoproteins named according to their respective positions in the precursor polyproteins. By conventional electron microscopy, viruses in these three families display round or pleomorphic morphology. However, advances in imaging, specifically the use of cryo-electron tomography has recently revealed more detail about the morphology of these viruses, particularly the arrangements of the glycoproteins. In the case of the

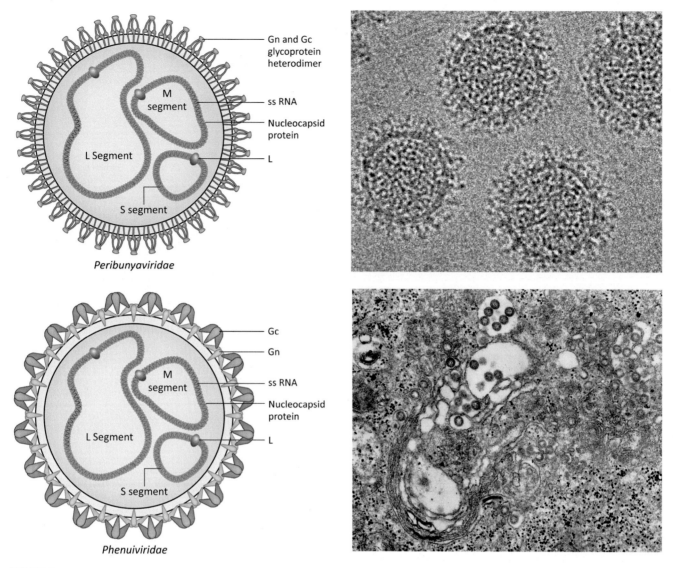

FIGURE 28.1 Top panel: Illustration and electron cryomicrograph of Bunyamwera orthobunyavirus, a member of the family *Peribunyaviridae* (Bowden et al., 2013). Bottom panel: Illustration and electron micrograph of Rift Valley Fever virus, a member of the family *Phenuiviridae*. Source: *(Top panel) From Bowden, T.A., Bitto, D., McLees, A., Yeromonahos, C., Elliott, R.M., Huiskonen, J.T., 2013. Orthobunyavirus ultrastructure and the curious tripodal glycoprotein spike. PLoS Pathog. 9, e1003374. (Bottom panel) From CDC Public Health Image Library ID #8242. Content providers: CDC/ F. A. Murphy; J. Dalrymple.*

peribunyaviruses, the glycoprotein arrangement appears to be a tripodal spike. Spikes appear to consist of Gn Gc heterodimers and three spikes associate to form a tripod. The base of each tripod appears to be in contact with neighboring tripods, completely covering the surface of the envelope. The cytoplasmic tails cross the envelope and contact the N protein within the virion.

General replication strategy

There are hundreds of named viruses in the families *Peribunyaviridae*, *Phenuiviridae*, and *Nairoviridae*. This chapter summarizes only general aspects of their

replication. Gc appears to be the primary attachment protein and Gn contains a fusion domain. Many of the peribunyaviruses, phenuiviruses, and nairoviruses enter cells by endocytosis; however, some fuse at the plasma membrane. The viruses that are the focus of this chapter all transcribe and replicate their genomes in the cytoplasm. Some, or all, of the genome segments are replicated using a negative sense RNA virus coding strategy but the S RNA segments of the phenuiviruses are ambisense. Their N mRNAs are transcribed from S genome RNA, but their NSs proteins are transcribed from cRNA. While it is theoretically possible that NSs mRNA is transcribed only after genome replication commences, this appears not to be the case. Studies with RVFV have

Peribunyaviridae genome, mRNAs and proteins

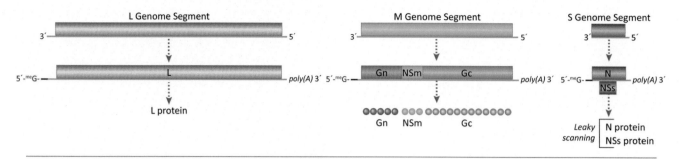

Phenuiviridae genome, mRNAs and proteins

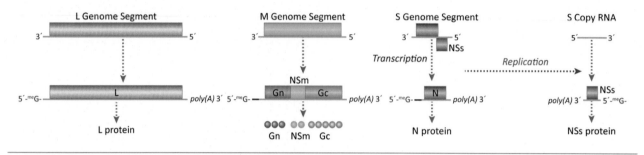

Nairoviridae genome, mRNAs and proteins

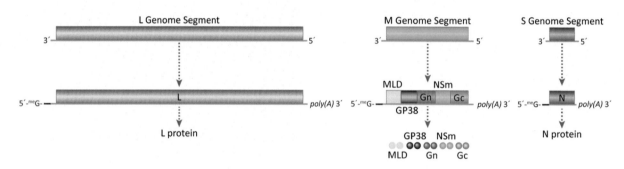

FIGURE 28.2   Genome and protein expression strategies.

demonstrated that cRNA from the S segment is packaged into virions. Thus some of these viruses package four RNAs. The advantage of packaging S cRNA is clear when one considers that the function of NSs is to antagonize interferon signaling. A delay in production of NSs would allow the host cell to respond quickly to the infecting virus.

Genome replication and mRNA synthesis are carried out by the L proteins, large RNA-dependent RNA polymerases (RdRp) that have polymerase, helicase, and endonuclease domains. Viral mRNAs contain capped oligonucleotides, (10–20 nt) obtained from host mRNAs, obtained by a process called cap-snatching. Thus the L proteins of the nairoviruses, peribunyaviruses, and phenuiviruses do not have capping activities per se but do have endonucleases that cleave the host mRNAs to prime viral mRNA synthesis.

Both transcription and genome replication appear to occur in association with membranes. Both processes probably require interaction of the complementary sequences found at the 5' and 3' ends of the genome-length RNAs. In contrast to mRNA synthesis, genome synthesis requires production of uncapped full-length cRNAs from each genome segment. Similar to other negative-strand RNA viruses, there must be a molecular mechanism (yet to be defined) that triggers the switch from transcription to genome replication.

N proteins are always encoded by the S genome segment and range in size from ~19 kDa (phenuiviruses) to ~54 kDa (nairoviruses). N proteins oligomerize and bind to both genome RNA and cRNA. N also interacts with L, Gn, and Gc proteins to facilitate virion assembly and morphogenesis. The S segments of peribunyaviruses encode a nonstructural protein called NSs (nonstructural protein from the S segment). There is little amino acid sequence homology among NSs proteins (even within a single genus), thus different activities and functions are likely. However, those NSs proteins that have been studied seem to have effects on host interferon responses. The flexibility of NSs sequence and function may be one of the factors that allow these viruses to exploit a wide host range.

Another genomic commonality among all the nairoviruses, peribunyaviruses, and phenuiviruses is the encoding of the envelope glycoproteins by the M genome segment. However, each family is different as regards the complement of proteins encoded by M. Nairovirus M RNA segments are the most complex: Five proteins that include a mucin-like domain (MLD) protein, a 38 kDa glycoprotein (gp38), a nonstructural protein (NSm), and two envelope glycoproteins (Gn and Gc) are encoded. NSm is located between Gn and Gc in the polyprotein precursor. Three proteins are encoded by peribunyavirus M RNA: Gn, NSm, and Gc. The M RNA segment of the phenuiviruses also encodes NSm, Gn, and Gc but note that the order of the coding regions differs from the peribunyaviruses. For all three families, the M protein precursor is processed by host proteases in the ER.

Signals for the encapsidation of vRNAs and cRNAs are present in their noncoding 3′ and 5′ regions. N proteins binds the genomic RNA segments to form the basis of the nucleocapsids. N also interacts with L, also part of the nucleocapsid. N also interacts with the cytosolic protein tails of Gn and Gc to drive virion assembly, thus playing the role of the matrix proteins of other negative-strand RNA viruses. Virion budding may be into Golgi membranes, followed by release by exocytosis, or budding may be directly from the plasma membrane. Finally, as viruses with segmented genomes, genetic reassortment can occur during RNA packaging. Reassorted viruses may be nonviable or they may emerge as novel pathogens.

Diseases

Members of the families *Peribunyaviridae*, *Phenuiviridae*, *and Nairoviridae* are transmitted by insects. The diseases described below include some significant pathogens of livestock that are occasionally transmitted to humans. It remains to be seen how these viruses will affect humans in the future, given changes in insect vector distribution caused by ongoing climate change.

Crimean Congo hemorrhagic fever

Crimean Congo hemorrhagic fever (CCHF) is caused by the nairovirus, Crimean Congo hemorrhagic

fever virus (CCHFV). CCHFV infects a wide variety of wild animals and livestock and usually causes little disease in these natural hosts. The virus is spread among animals by ticks of the genus *Hyalomma*. Distribution is quite widespread and includes parts of Africa, Central Asia, the Indian subcontinent, and Eastern Europe. CCHFV is zoonotic and human infections can cause severe disease. Symptoms are similar to other viral hemorrhagic fevers and include fever, muscle pains, headache, vomiting, diarrhea, and bleeding into the skin. Case fatality rates range from 10% to 40%. People that work with livestock (herders, farmers, veterinarians, slaughterhouse workers) are at highest risk for contracting the virus. Human infection can result from tick bites or directly from animals, through cuts or via mucus membranes (Fig. 28.3). Spread between humans, via body fluids, may also occur, particularly in hospital settings. The reasons for severe disease in humans are not known but may be due to exacerbated immune responses.

Oropouche fever

Oropouche fever is a widespread, acute, self-limiting febrile illness throughout Brazil and other South and Central American countries. Over 30 epidemics were reported in the period from 1960 to 2009. The disease is caused by Oropouche virus (OROV), a member of the family *Peribunyaviridae*, genus *Orthobunyavirus*. OROV is zoonotic and appears to be maintained in both sylvan (forest) and urban cycles. In the urban cycle, the vectors implicated in spread include *Culicoides* midges and *Culex* mosquitoes.

La Crosse encephalitis

La Crosse virus (LACV) is a member of the family *Peribunyaviridae*, genus *Orthobunyavirus*. According to the US Centers for Disease Control and Prevention (CDC), there are ∼80−100 reported cases of La Crosse encephalitis in the United States each year. The actual number of infections may be higher as most infected individuals do not exhibit symptoms. Children under the age of 16 are at highest risk for developing encephalitis. LACV is transmitted to humans by the mosquito *Aedes triseriatus*. This mosquito normally feeds on chipmunks and squirrels (Fig. 28.4). Infected female mosquitos transmit LACV vertically to offspring via their eggs. Studies have shown that infected females are more efficient at mating than uninfected females. Humans, although occasionally infected with LACV, are likely dead end hosts that do not play a role in maintenance of the virus in the environment.

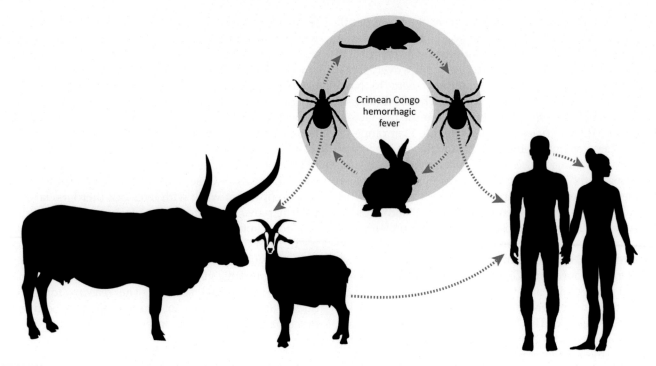

FIGURE 28.3   Transmission cycle of Crimean Congo hemorrhagic fever virus.

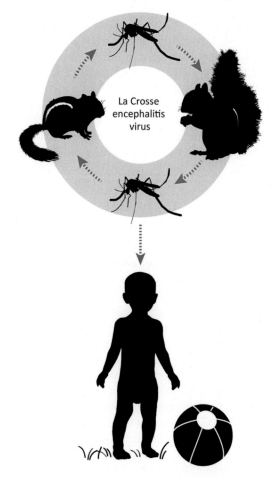

FIGURE 28.4   Transmission cycle of La Crosse virus in North America.

Schmallenberg and Cache Valley viruses

Schmallenberg virus (SBV) and Cache Valley virus (CCV) are orthobunyaviruses that infect livestock. Both viruses are transmitted by mosquitos and midges; their greatest impact is in ruminants, particularly sheep and goats. CVV causes a mild, acute infection in adult animals but causes decreased birth rates, abortions, stillbirths, and deformed offspring in affected herds. Among wild ruminants, deer in the United States have a high seroprevalence for CVV. CVV is a zoonotic virus, and there have been a handful of cases of human infection in the United States.

SBV was first identified in Central Europe in 2011 and quickly spread throughout the European continent. The disease was first noticed as fever, decreased milk production, and diarrhea in cattle. More serious consequences result from infection of naïve pregnant animals, including abortion, premature birth, stillbirth, or fetal malformation. Spread by *Culicoides* midges, SBV has reappeared every 2–3 years since it was initially discovered.

Rift Valley fever virus

Within the family *Phenuiviridae* RVFV is a pathogen of note. RVFV is endemic to Africa and the Arabian Peninsula; it is transmitted by mosquitoes, and common hosts are domesticated and wild ruminants. Sheep are the most susceptible host and the virus can cause abortion storms in herds; mortality is very high for new born lambs which present with severe necrosis of the liver. Humans are also susceptible to RVFV. In the year 2000, in Saudi Arabia, an outbreak involved over 500 individuals. Most infected persons present with a transient, febrile course of headache and muscle pains. However, some people develop hemorrhagic fever. In the 2000 outbreak, mortality was about 17%. High rates of abortion have also been associated with the infection. Many infected humans reported exposure to animal blood, but transmission from mosquitos to humans is also a likely route of infection.

In this chapter we have learned that:

- Peribunyaviruses, phenuiviruses, and nairoviruses have enveloped virions containing three segments of negative or ambisense RNA.
- An ambisense coding strategy means that mRNAs are produced from both genomic RNA and copy RNA.

- Members of the three virus families described in this chapter encode an N protein from the S genome segment, a glycoprotein precursor from the M genome segment, and a polymerase from the L genome segment.
- Crimean Congo hemorrhagic fever virus (CCHFV) is a tick-borne hemorrhagic fever virus. It causes little disease in animals but can cause severe disease in humans. Persons who work with infected livestock are at highest risk for acquiring CCHFV.

References

Bowden, T.A., Bitto, D., McLees, A., Yeromonahos, C., Elliott, R.M., Huiskonen, J.T., 2013. Orthobunyavirus ultrastructure and the curious tripodal glycoprotein spike. PLoS Pathog 9, e1003374. Available from: https://doi.org/10.1371/journal.ppat.1003374.

Maes, P., Adkins, S., Alkhovsky, S.V., Avšič-Županc, T., Ballinger, M.J., Bente, D.A., et al., 2019. Taxonomy of the order Bunyavirales: second update 2018. Arch. Virol. 164, 927–941 [PubMed: 30663021].

29

Family *Arenaviridae*

After reading this chapter, you should be able to discuss the following:

- What are the key features of all members of the family *Arenaviridae*?
- How is arenavirus genome replication similar to negative-strand RNA viruses?
- What is an ambisense genome segment?
- At what site in the cell do arenaviruses replicate?
- What are common mammalian hosts for arenaviruses?
- Under what conditions are people infected with arenaviruses?

Members of the family *Arenaviridae* have two or three negative or ambisense genome segments within enveloped virions (Box 29.1). Virions have a granular appearance that gave rise to the name arenavirus (from the Latin *arenosus*, sandy) (Fig. 29.1). Ribosome-like granules packaged within virions are responsible for the "sandy" appearance of the particles; the function of these structures remains unknown. There are four recognized genera within the family (Box 29.2). Genus *Mammarenavirus* contains viruses whose natural hosts are rodents. Human infections occur by contact with infected rodents or inhalation of infectious urine or feces. Many human infections are asymptomatic, but severe disease (hemorrhagic fever and/or neurological involvement) can occur.

Genome organization

Arenaviruses are segmented RNA viruses that replicate using an overall strategy common to negative-strand RNA viruses. Mammarenaviruses have two genome segments, large (L) and small (S). Both segments are ambisense and both contain two open reading frames (ORFs). However, the two genome segments (large (L) and small (S)) are ambisense: Each genome segment contains two ORFs, separated by an intergenic region (IGR) that serves as a transcription termination signal (Fig. 29.2). IGRs are predicted to form hairpin structures that signal the release of the polymerase from the substrate during transcription. The L genome segment encodes the L (RNA-dependent RNA polymerase, RdRp) and Z proteins. The S genome segment encodes the RNA-binding nucleoprotein (NP) and glycoprotein precursor (GPC). Arenavirus mRNAs are not polyadenylated but do have a hairpin structure at their 3′ end. Sequences at the 3′ end of the L and S segments are highly conserved; they are likely important cis-regulatory elements (i.e., promoter for polymerase entry). There is base complementarity between the 5′ and 3′ ends of each genome segment and they are predicted to form panhandle structures. Viral mRNAs are capped at the 5′ end followed by four to five nucleotides not found in the genome, a finding suggestive of a cap-snatching mechanism to prime mRNA synthesis.

© 2023 Elsevier Inc. All rights reserved.

BOX 29.1

Characteristics

Virions are enveloped with helical nucleocapsids. Virions are pleomorphic, ranging in size from 100 nm to more than 130 nm in diameter. Virions have a "sandy" or granular appearance. Penetration into the cell occurs after endocytosis and is low pH dependent. Transcription and genome replication are cytoplasmic. Arenaviruses have segmented RNA genomes. Mammarenaviruses have two genome segments. The large (L) segment is ~7.2 kb and the small (S) segment is ~3.5 kb. Arenavirus replication follows the pattern of negative-sense RNA viruses; however, both genome segments are ambisense. There are two open reading frames on each segment. mRNAs have four to five non-templated nucleotides and a cap structure at their 5′ ends, which are likely obtained from cellular mRNAs via cap-snatching mechanisms (whose details remain to be determined). mRNAs have a hairpin at the 3′ end. The mammarenavirus S genome segment encodes NP and the viral glycoprotein precursor (GPC). The L genome segment encodes the L (RdRp) and Z proteins.

BOX 29.2

Taxonomy

Order *Bunyavirales*
Family *Arenaviridae*
There are four recognized genera.
Genus *Antennavirus* (fish hosts)
Genus *Hartmanvirus* (snake hosts)
Genus *Mammarenavirus* (mammalian hosts; more than 24 species that cluster into two groups based on antigenic properties and genome sequences. The New World (or Tacaribe serocomplex) includes Junin virus, Machupo virus, and Tacaribe virus among others. The Old World viruses include LFV and LCMV).
Genus *Reptarenavirus* (snake hosts)

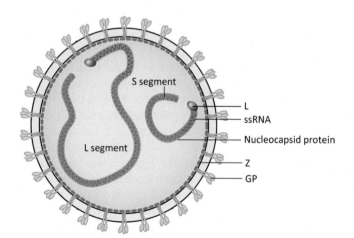
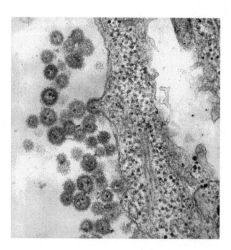

FIGURE 29.1   Virion morphology (left). Electron micrograph of the Machupo virus (right). Machupo virus is a member of the Arenavirus family, isolated in the Beni Province of Bolivia in 1963. Source: *From CDC/Dr. Fred Murphy; Sylvia Whitfield. Public Health Library Image #1869.*

Genome organization is not strictly conserved within the family. Mammarenaviruses are the only ones encoding a Z protein. Members of the genus *Antennavirus* have three genome segments: Their S segment encodes NS while a medium (M) segment encodes GPC.

Arenaviridae genome segments, vcRNAs, mRNAs and proteins

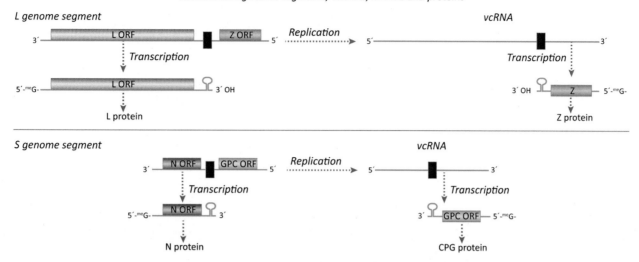

FIGURE 29.2   Genome and protein expression strategy.

Virion structure

Arenavirus virions are enveloped with helical nucleocapsids and are pleomorphic, ranging in size from 100 to 130 nm in diameter (Fig. 29.1). Within virions, genome segments are organized into circular structures. Nucleocapsids consist of genome segments bound to the NP protein. As with other negative-sense RNA viruses, the L protein (RdRp) is associated with nucleocapsids. Glycoprotein spikes extend from the virion surface. They have distinct stalks with round heads. Spikes are a complex of GP1 and GP2 (products of the GPC). Z protein is a zinc-binding protein that functions as a matrix protein. Detailed structures of mammarenaviruses reveal that the Z protein forms a layer under the cytoplasmic tails of the glycoproteins.

Overview of replication

Arenavirus attachment to and entry into the host cell is mediated by the glycoprotein spikes. The virion is endocytosed and membrane fusion is initiated in the acidified endosome.

Arenavirus transcription and genome replication are cytoplasmic. After penetration, L protein initiates transcription to produce NP and L mRNAs. The presence of short, nontemplated sequences at the 5' ends of arenavirus mRNAs suggests that L uses a cap-snatching mechanism to prime transcription but this has not been formally demonstrated. mRNA synthesis terminates at the noncoding IGR located between the ORFs (Fig. 29.2).

NP is the most abundant arenavirus protein. It has two distinct structural domains, N-terminal domain and a C-terminal domain connected by a flexible linker region. The N-terminal domain appears to bind RNA. The C-terminal domain of lymphocytic choriomeningitis virus (LCMV) counteracts the interferon response by preventing activation and nuclear translocation of interferon regulatory factor 3. L protein is the RdRp. It has characteristic protein motifs found in the L proteins of other negative-sense RNA viruses. L has an endonuclease activity that may have a function in cap-snatching, to prime mRNA synthesis.

As the concentration of NP in the cell increases, genome segments are replicated, to produce full-length copies (cRNA). cRNAs are now templates for transcription of additional mRNAs: GPC mRNA from the S segment and Z mRNA from the L segment. GPC is translated on rough endoplasmic reticulum (ER). Within the ER lumen, GPC is cleaved to produce a stable signal peptide (SSP) and is heavily glycosylated and cleaved to produce GP1 and GP2 products. SSP remains associated with GP1/GP2 and is virion associated. The GP1/GP2/SSP complex traffics to the plasma membrane. Z is a zinc-binding matrix protein that is not encoded by all arenaviruses. Mammarenavirus Z is myristoylated (myristic acid, a fatty acid, is linked to the protein near the N-terminus) appears to form a distinct layer under the lipid envelope. Assembled nucleocapsids (genome length RNA, NP, and L) associate with Z and CPG at the plasma membrane where progeny virions are released by budding. Budding is driven by the Z protein.

In cultured cells, many arenavirus infections are noncytolytic and long-term persistent infection is a common outcome. During persistent infections, few virions are released from cells; instead virus spreads when nucleocapsids partition into daughter cells during cell division.

Diseases caused by arenaviruses

Arenaviruses are commonly found in association with rodents. However, some can be transmitted to humans and cause disease. Humans become infected through contact with infected rodents or through inhalation of infectious rodent urine and feces. In humans, disease often begins with a gradual onset of fever and myalgia. More serious disease, such as viral hemorrhagic fever and/or neurologic impairment, may develop.

Lassa fever

Lassa fever virus (LFV) is endemic in parts of West Africa including Sierra Leone, Liberia, Guinea, and Nigeria. The natural hosts for LFV are rats of the genus *Mastomys*. Once infected, rats excrete virus for months and LFV is transmitted to humans via contact with materials contaminated with rodent urine or feces. *Mastomys* rats are numerous, breed frequently, and readily colonize homes. There are an estimated 100,000−300,000 LFV infections per year, with 5000 deaths. Severe disease affects organs such as the liver, spleen, and kidneys. Disease may include gastrointestinal involvement (abdominal pain, nausea and vomiting, diarrhea, or constipation). Other complications include bleeding, edema, pharyngitis, and conjunctivitis.

LFV can also be transmitted from person-to-person, particularly in health care settings when personal protective equipment is not available.

Lymphocytic choriomeningitis

LCMV infections in humans are rarely fatal, but serious disease (viral hemorrhagic fever or neurological complications) can occur. The natural host for LCMV is the common house mouse, *Mus musculus*. Other rodents, such as hamsters sold in the pet trade can be infected through contact with mice. LCMV infections of humans have been reported worldwide, wherever infected rodent hosts are found. As with other arenaviruses, contact with infected rodents, their urine or droppings is associated with most human infections. It is likely that most human infections with LCMV are asymptomatic or mild, seldom resulting in the severe neurologic disease for which the virus is named. Infection of pregnant women can cause fetal congenital defects. In the United States, infection through solid organ transplantation has been reported.

While LCMV is a rare disease of humans, this mouse virus has been a valuable model for dissecting diverse mechanisms of viral pathogenesis. LCMV infects mice (*Mus musculus*), a good laboratory model animal, with variable disease outcomes. Complex interactions between host and virus determine the outcome of LCMV infection. Acute disease can range from mild to lethal. Persistent infections can also develop; some with no pathology and some accompanied by immune-mediated disease. Some of the factors that determine LCMV outcome of infection include the virus strain, breed of mouse, route of inoculation, and age of mouse. Both direct effects of viral infection and the effects of immune responses are important in the outcome of infection. Many important concepts in viral immunology were first described for the LCMV system. It continues to be an important model for viral immunologists.

New World arenaviruses

Several New World (NW or Tacaribe complex arenaviruses) have been associated with human disease. These include Junin (Argentine hemorrhagic fever), Machupo (Bolivian hemorrhagic fever), and Guanarito (Venezuelan hemorrhagic fever). Other NW arenaviruses have been isolated from sporadic cases of hemorrhagic fever. Each of the aforementioned arenaviruses is associated with one or a few closely related species of rodents. NW arenaviruses continue to be isolated, with outbreaks depending on complex interactions humans, rodents, and the environment.

In this chapter we learned that:

- Arenaviruses are segmented RNA viruses.
- Virions are enveloped and ribosome-like granules are packaged into virions giving a "sandy" appearance.
- The overall replication strategy is similar to the negative-strand RNA viruses.
- Arenaviruses have ambisense genomes. Each genome segment is transcribed to produce an mRNA. However, after genome replication, mRNAs are also transcribed from full-length cRNAs.
- Many rodents are infected with arenaviruses and most of these infections are nonpathogenic. Humans can be infected by some arenaviruses (through contact with rodents).
- LFV, LCMV, Junin virus, and Machupo virus can cause severe disease in humans.

30

Family *Reoviridae*

After reading this chapter, you should be able to discuss the following:

- Describe the structure of reovirus genomes.
- Describe the structure of reovirus capsids. Do all reoviruses have the same capsid type?
- What is the "viroplasm" and why is it important in the reovirus replication cycle?
- Describe the unique features of reovirus transcription and genome replication.
- Discuss the transmission strategies of orthoreoviruses, rotaviruses, and orbiviruses.
- What are some important human and animal diseases caused by reoviruses?

Members of the family *Reoviridae* infect hosts ranging from microorganisms to insects to plants to vertebrates. The family *Reoviridae* is large and diverse, containing two subfamilies and numerous genera. The best-known animal reoviruses are those in the genus *Rotavirus* (major enteric pathogens of humans and other mammals), *Orbivirus* (insect-vectored animal pathogens), and *Orthoreovirus* (mild upper respiratory tract disease of mammals and poultry diseases).

The family name, *Reoviridae*, derives from the fact that the first isolates were of respiratory and enteric origin, and were not associated with known disease (thus were respiratory, enteric orphan viruses). Reoviruses have segmented, double-stranded (ds) RNA genomes. Genome

segments number from 9 to 12. Genome reassortment has likely contributed to their wide host range and great success as a family (Box 30.1). Virions are unenveloped with icosahedral symmetry and are unique in having two or three discrete capsid layers. The innermost capsid layer encloses dsRNA segments and enzymes for RNA synthesis (Fig. 30.1). Upon infection, transcriptionally active genome segments remain within cores. Newly synthesized mRNAs are released into the cytosol through pores at the fivefold axes of symmetry.

Genome organization

Reovirus genomes contain from 9 to 12 segments of linear, dsRNA. For the most part, each segment encodes a single protein (Fig. 30.2). For example, rotavirus and orbivirus genomes have 11 segments encoding 12 proteins. Genome segment sizes range from ~ 600 to 3000 bp for a total coding capacity of approximately 18,500–19,200 bp. Positive genome strands and mRNAs are capped but not polyadenylated. Negative strands have a diphosphate at their 5′ end. There are short untranslated regions at the 5′ and 3′ ends of each genome segment. Replication is cytoplasmic, but associated with discrete pseudo-organelles variously called virus factories, virus bodies, or viroplasm.

Viruses.
DOI: https://doi.org/10.1016/B978-0-323-90385-1.00032-7

© 2023 Elsevier Inc. All rights reserved.

BOX 30.1

Taxonomy

Family *Reoviridae*
Subfamily *Sedoreovirinae*
Genera include:
Genus *Orbivirus*

Orbiviruses are transmitted by insects. The genus contains over 20 named species that infect a variety of mammals. Includes blue tongue virus (BTV) and epizootic hemorrhagic disease virus (EHDV). Virions are double-layered particles containing 10 genome segments.

Genus *Rotavirus*

Rotaviruses are transmitted by the fecal–oral route. Virions are triple-layered particles containing 11 genome segments. Rotaviruses are noted for causing severe diarrhea in young animals. There are 10 species (rotavirus A–J). Most disease is associated with the so-called Group A rotaviruses (GARV).

Subfamily *Spinareovirinae* (the name refers to the large turrets at the fivefold axes of symmetry)
Genera include:
Genus *Coltivirus* (Colorado tick fever virus, CTFV)
Genus *Orthoreovirus* (hosts include birds, mammals, and reptiles)

Ubiquitous viruses, present in feces. Seldom cause disease but sometimes associated with mild upper respiratory tract disease or gastroenteritis.

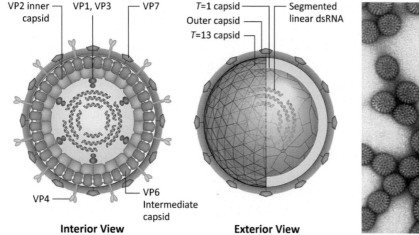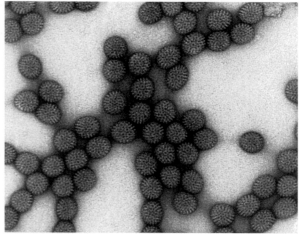

FIGURE 30.1   Virion structure of a triple-layered rotavirus (left) and electron micrograph (right). This negatively stained transmission electron microscopic (TEM) image revealed some of the ultrastructural morphology displayed by rotavirus particles. From: CDC Public Health Image Library #197. Content providers CDC/Dr. Erskine Palmer.

Virus structure

Ultrastructural studies of reoviruses reveal capsids made up of two or three discrete layers (Table 30.1). The innermost layer (core) has $T = 1$ icosahedral symmetry and is assembled from 60 dimers of single type of structural protein. The core contains genome segments, each associated with enzymes needed for transcription. It is thought that each genome segment is located at a fivefold axis of symmetry. During transcription, mRNAs exit the intact cores through channels in the structure. The core is surrounded by a capsid layer with $T = 13$ icosahedral symmetry. Some reoviruses (members of the genus *Rotavirus*) have a third protein layer, also with $T = 13$ icosahedral symmetry. Rotaviruses have glycosylated spike proteins extending from the surface giving the particles a wheel-like appearance.

The family *Reoviridae* contains two subfamilies and several genera (Box 30.2). Members of the subfamily *Sedoreovirinae* (genus *Rotavirus*, genus *Orbivirus*) have relatively smooth outer layers (or short spikes) while members of the subfamily *Spinareovirinae* (genus *Coltivirus*, genus

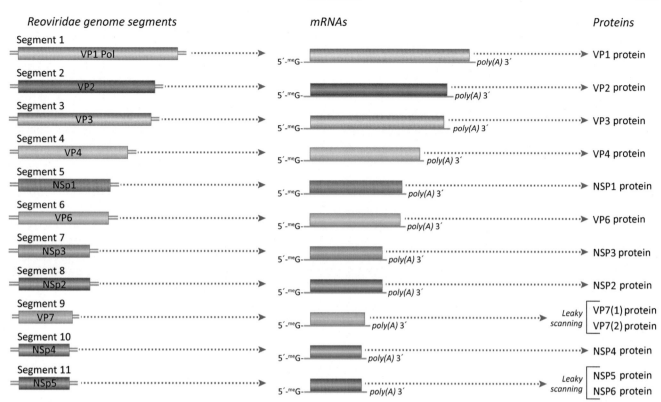

FIGURE 30.2 Depicting genome segments, mRNAs, and protein products of a rotavirus. See Table 30.2 for a comparison of the naming schemes used for different genera in the family.

TABLE 30.1   Selected members of the family *Reoviridae*.

Subfamily	*Spinareovirinae*	*Sedoreovirinae*	*Sedoreovirinae*	*Spinareovirinae*
Genus	*Orthoreovirus*	*Rotavirus*	*Orbivirus*	*Coltivirus*
RNA segments	10	11	10	12
Capsid structure	2 layers	3 layers	3 layers	2 layers
Hosts	Vertebrates	Vertebrates	Vertebrates, insects	Vertebrates, insects
Diseases	No disease or mild upper respiratory tract disease, gastroenteritis, and biliary atresia.	Gastrointestinal infections; major cause of childhood diarrhea.	Bluetongue and epizootic hemorrhagic fever.	Colorado tick fever.

BOX 30.2

General characteristics

Members of the family *Reoviridae* have segmented, dsRNA genomes. The number of genome segments ranges from 9 to 12. Reovirus replication is cytoplasmic and a unique feature of these viruses is that transcription takes place from within capsids. Virions are unenveloped, with two or three capsid layers (double- or triple-layered capsids). Middle and outer capsid layers are $T = 13$ icosahedral lattices. The inner capsids or cores are $T = 1$. As a group, reoviruses are very successful with a wide host range that includes fungi, plants, invertebrates, and vertebrates.

Orthoreovirus) have large turreted projections at the fivefold axes of symmetry.

Overview of replication

While the overall reovirus replication scheme is conserved (Fig. 30.3), it is not surprising that the details vary among genera. For example, we expect that rotaviruses (enteric viruses) and orbiviruses (transmitted by insects) might display some difference in replication, in light of their adaptation to very different host environments. The next sections highlight conserved features of reovirus replication but should be considered only a general introduction. Given that rotaviruses are a major cause of human infant mortality in some areas of the world, they have been studied in great detail. It should be noted that accepted schemes for naming/labeling reovirus genome segments and proteins vary by genus, as shown in Table 30.2. Reverse genetic systems have been developed for some reoviruses, as described in Box 30.3 and Fig. 30.4.

Attachment/entry

Reovirus attachment is mediated by capsid proteins (rotavirus VP4, orthoreovirus σ1, orbivirus VP2). A variety of cell surface molecules act as receptors. Some reoviruses use protein receptors, others use carbohydrate receptors (e.g., sialic acid) and some use a combination of the two. Rotavirus VP4 is glycosylated and is cleaved by extracellular proteases found in the digestive tract. Receptors for some rotaviruses are histo blood group antigens (HGBAs). HGBAs are complex carbohydrates found on epithelia of the respiratory, genitourinary, and digestive tracts. Intact rotavirus VP4 binds to HGBAs but VP4 cleavage products (designated VP5* and VP8*) interact with additional protein coreceptors.

Reoviruses encode proteins that interact with membranes to form pores or channels (rotavirus NSP1, orbivirus VP5, orthoreovirus μ1 protein). Penetration across plasma membrane or endosomal membranes is accompanied by disassembly of the outer capsid layers. The innermost core, containing genome segments, undergoes molecular rearrangements, but remains intact. In the case of rotaviruses, disassembly of outer layers is triggered by the low calcium concentration of the cytosol. Orthoreoviruses uncoat with the help of endosomal proteases such as cathepsins B and L.

Amplification

Penetration and uncoating activate the transcription activities of the core. Viral mRNA is synthesized within the core and newly formed (capped) transcripts are released through pores found at the vertices of the inner capsid. It is postulated that each genome segment is associated with its own pore and transcription complex. mRNAs are not polyadenylated, but form a hairpin structure at the 3′ end. In the case of rotaviruses, synthesis of viral transcripts is mediated by a complex of VP1 and VP3. VP1 contains the catalytic RNA synthesis domain while VP3 binds ssRNA and has nucleotide phosphohydrolase, guanylyltransferase, and methyltransferase activities (thus VP3 is the cap-synthesizing component of the transcription complex). Transcriptionally active cores can be prepared from purified virions (by removal of the outer capsid layers) and in vitro these can produce up to milligram quantities of mRNA, given an adequate source of precursors.

Upon exiting the cores, mRNAs are translated. In the case of rotaviruses, nonstructural protein 3 (NSP3) binds to eIF4F, the cytoplasmic cap-binding complex to facilitate translation. Thus NSP3 seems to serve the function of host polyA-binding protein in translation initiation. Some rotavirus mRNAs are translated in the cytosol but mRNAs for rotavirus VP4, VP7, and NSP4 are translated on rough endoplasmic reticulum (ER), and the proteins are glycosylated therein. Cytosolic proteins form dense viral organelle-like structures called virus factories or the viroplasm. Cores assemble in the viroplasm and these are capable of transcription to generate additional mRNAs.

Genome replication

The exact mechanism used to select and package reovirus genome segments is not well understood, but clearly the process occurs in the viroplasm. One model has mRNAs associating with pentamers of the inner core. mRNA secondary structures might provide signals for selective packaging of 9−12 different segments. It is clear that each mRNA is the template for synthesis of *single* complementary (negative) strand to form a ds genome segment.

Assembly, maturation, and release

Given their importance as worldwide human pathogens rotaviruses have been studied in great detail. Rotaviruses have triple-layered particles and their outer layer displays glycosylated proteins, synthesized in the ER. However, the virion is not enveloped, so how does it assemble and acquire its outer coat of glycoproteins?

Rotavirus double-layered particles (DLPs) are synthesized in the viroplasm concurrent with genome replication. It is thought that pentamers of VP2 dimers

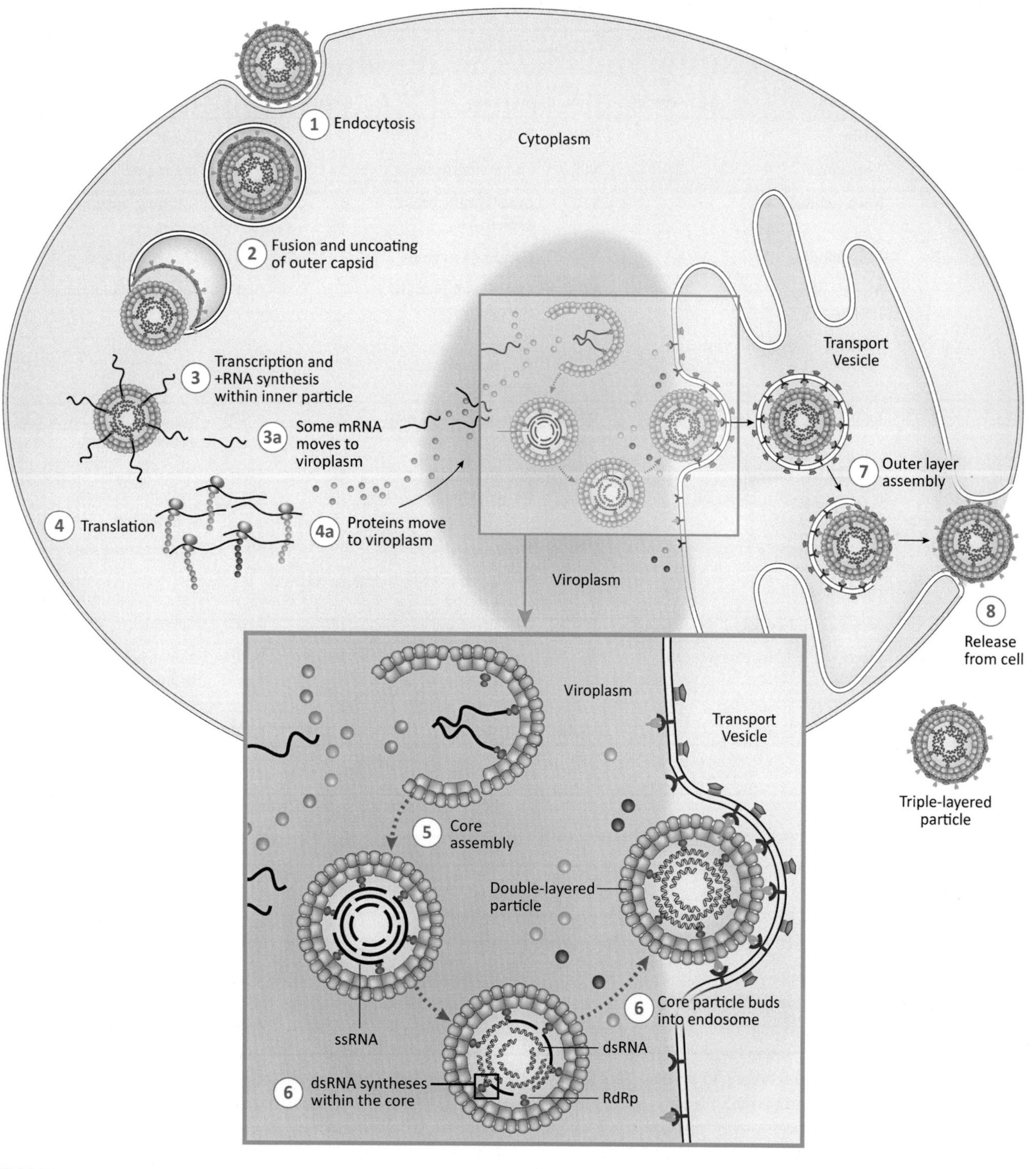

FIGURE 30.3 General overview of rotavirus replication cycle. Note that transcripts are synthesized within the core and mRNAs are released into cytoplasm through openings in the capsid. NSPs are important for development of the viroplasm, an electron dense area in the cell that contains a high concentration of viral proteins and mRNAs. Capsid assembly takes place in the viroplasm. In the model shown here, mRNAs are selected and packaged; RNA synthesis within the newly forming capsid generates the double-stranded genome segments. The outermost layer of the rotavirus triple-layered particle is obtained upon budding into ER/transport vesicles.

(the inner core protein) interact with VP1 and VP3 and a single genome segment. Assembly of pentamers forms the inner core. Subsequently VP6 is added to form the DLP. (This is very simplified description of the process.) Rotavirus DLPs then leave the viroplasm and bud into the ER lumen by virtue of interactions

TABLE 30.2 Within the family *Reoviridae*, nomenclature for genome segments and protein products differs by genus.

Genus *Orthoreovirus*			Genus *Orbivirus*			Genus *Rotavirus*		
Genome segment	Protein	Function	Genome segment	Protein	Function	Genome segment	Protein	Function
L1	λ3	RdRp	1	VP1	RdRp	1	VP1	RdRp
L2	λ2	Core spike	2	VP2	Outer shell spike	2	VP2	Inner shell
L3	λ1	RNA helicase	3	VP3	Inner core scaffold protein	3	VP3	Capping enzyme
M1	μ2	RNA binding	4	VP4	Capping enzymes	4	VP4	Outer shell spikes
M2	μ1	Major outer capsid protein	5	VP5	Outer shell protein (fusogenic)	5	NSP1	Interferon antagonist
M3	μNS	NSP, associates with viral mRNA	6	NS1	Nonstructural, enhances protein synthesis	6	VP6	Middle shell
M3	μNSC	NSP, unknown function	7	VP7	Core surface protein	7	NSP3	Nonstructural, required for systemic spread
S1	σ1	Attachment protein	8	NS2	Nonstructural, forms inclusion bodies	8	NSp2	Nonstructural forms inclusion bodies (viroplasm)
S1	σ1s	NSP, required for hematogenous dissemination	9	VP6	Helicase	9	VP7	Outer capsid shell
S2	σ2	Major capsid protein	9	NSP4	Nonstructural	10	NSP4	Enterotoxin, ER receptor
S3	σNS	Nonspecific RNA-binding protein	10	NS3	Nonstructural, virus trafficking and release	11	NSP5	Nonstructural forms inclusion bodies (viroplasm)
S4	σ3	Major outer capsid protein	10	NS3A	Unknown	11	NSP6	Nonstructural forms inclusion bodies (viroplasm)

BOX 30.3

Reverse genetics systems: all reoviruses are not created equal

Reverse genetics systems allow the production of infectious virus from cloned nucleic acids. Such systems are instrumental in allowing researchers to determine the functions of both proteins and genome regulatory regions. However, when it comes to reverse genetics systems, all reoviruses are not equal. Systems based entirely on cloned genes have been developed for some orthoreoviruses and orbiviruses. The mammalian orthoreovirus system involves transfecting cells with plasmids that encode (and express) each viral gene. The system for bluetongue virus (an orbivirus) differs only slightly. In this case, mRNAs are transcribed in vitro (in a test tube) and the single-stranded RNA products are transfected into cells. While there are important technical considerations, such as the need to produce the correct 5′ end of each mRNA, the systems are conceptually simple (as illustrated in Fig. 30.4). In contrast, similar strategies have not been successful when applied to rotaviruses. Current rotavirus reverse genetic systems require use of coinfecting helper viruses and this limits their utility. This is certainly frustrating, as the rotaviruses are widely accepted as the most important and widespread human pathogens in the family. It appears that simply expressing or introducing rotavirus mRNAs into a cell is not sufficient to produce infectious virus. The reasons remain elusive!

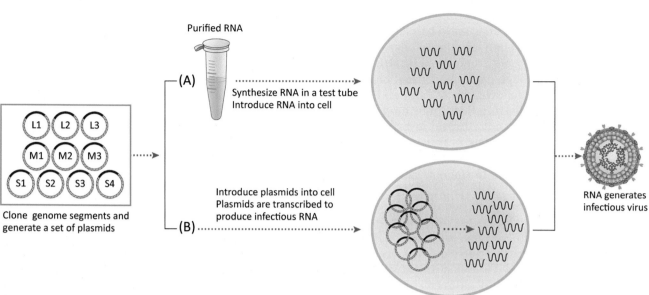

FIGURE 30.4   Reverse genetics systems have been developed for orthoreoviruses and orbiviruses. Each genome segment is cloned into a plasmid designed to produce mRNAs with defined 5′ and 3′ ends. In one strategy, the plasmids are introduced into cells and transcription of a set of viral mRNAs initiates the reovirus replication cycle. In an alternative strategy, mRNAs are synthesized in vitro and are introduced into cells. Note that the rotavirus reverse genetics system (for reasons that remain unknown) requires coinfection with helper virus.

with molecules of NSP4 located in the ER membrane. In this process, NSP4 is serving as an ER "receptor." While budding into the ER, the particle transiently acquires an envelope and VP4 and VP7. However, the lipid bilayer is subsequently lost (by an unknown biochemical mechanism). The high Ca^{2+} concentration in the ER is necessary for assembly of VP4 and VP7 into the outmost shell. Virions are released upon cell death. Rotavirus replication in gastrointestinal enterocytes is quite efficient and robust, as virions are shed in concentrations of up to 10^{10} particles per gram of feces.

An interesting feature of some orthoreoviruses is that they are fusogenic and induce formation of large syncytia (fused cells). Fusion is not mediated by capsid proteins, but by proteins that are not packaged into virions. These small transmembrane proteins are called FAST proteins (fusion-associated small transmembrane proteins). As cell fusion is proapoptotic, this may be a mechanism to enhance virion release.

Diseases

Genus rotavirus

Rotaviruses cause gastrointestinal infections of humans and other mammals. While many infections are mild to asymptomatic, some rotaviruses can cause severe gastrointestinal disease, with diarrhea and vomiting. There are 10 species of rotavirus, designated A through J. The group A rotaviruses (GARV) are those most often associated with severe disease in infants and young children. There are many different GARV, due to their propensity for genome reassortment. In agricultural settings it is not uncommon to find reassortants among from bovine and porcine rotaviruses. Human rotaviruses can also exchange gene segments with animal rotaviruses.

In humans, most severe disease (life-threatening diarrhea) is seen among children 3 months to 2 years of age. Rotaviral disease is also most severe in young animals. The exact nature of the age restriction is not clear. Virions are highly contagious and are very stable in the environment. Transmission is most often through direct contact although water-borne outbreaks have been documented. The primary treatment for rotavirus-induced dehydration is to replace fluids and electrolytes. This can be accomplished through delivery of intravenous fluids or by use of oral fluid replacement therapies.

The primary site of rotavirus replication is the small intestine where virions infect nondividing enterocytes found at the tips of villi. Rotavirus infection is cytopathic and loss of intestinal epithelial cells results in malabsorption of nutrients accompanied by mild diarrhea. In cases of severe diarrhea the culprit appears to be rotavirus NSP4. NSP4 is a multifunctional protein critical in several aspects of the rotavirus replication cycle. As mentioned above, some NSP4 is located in the ER membrane and directs budding of the newly assembled DLPs into the ER. However, some NSP4 is secreted from infected cells. Secreted NSP4 can be found at both apical and basolateral surfaces of polarized epithelial cells. NSP4 attaches to basolateral integrins to trigger mobilization of intracellular calcium (release of calcium stores from the ER), triggering release of chloride through calcium-activated channels. Release of chloride is followed by sodium and water, resulting in secretory diarrhea that can be quite severe. (Recall that the rotavirus outer capsid layer disassembles at low Ca^{2+} concentrations. Is it possible that one role of NSP4 is to trigger release of Ca^{2+} from intracellular storage compartments to stabilize newly formed virions within cells?)

Rotavirus vaccines are live-attenuated products generated by reassortment. These vaccines protect against severe disease, but do not prevent infection. In the United States, live-attenuated products that are recommended for all infants and it is estimated that vaccination prevents millions of episodes of gastroenteritis, and thousands hospitalizations, among young children each year. Worldwide the numbers are staggering, with an estimated 2 million hospitalizations and $\sim$450,000 childhood deaths from rotavirus-induced diarrheal disease per year, many of which could be prevented through vaccination.

Genus orbivirus

There are 22 named species of orbivirus that infect animals and insects. Orbiviruses are "arboviruses" as they are transmitted by insects such as ticks, mosquitoes, sand flies, and biting midges. They have a wide host range among animals and birds. Transmission requires that orbiviruses replicate in both insect and animal hosts. In animal cells they cause cytopathic infections but in insects infections are noncytopathic.

The type species in the genus *Orbivirus* is bluetongue virus (BTV), spread by biting midges of *Culicoides* sp. BTV originated in Africa but since the 1920s the virus has moved worldwide. Infections can range from entirely asymptomatic to quite severe (edema of lips, tongue and head, fever, depression, nasal discharge, excess salivation, pain, and death). In extreme cases, edema causes cyanosis of the tongue, the "blue tongue" for which the disease is named. Among domestic animals, the disease affects sheep, goats, and cattle. Sheep are usually the most severely affected.

Before 1950, BTV was reported only from parts of Africa and Cyprus. However, the virus spread rapidly across the world (in conjunction insect vectors) in a broad band between $\sim$40°N and 35°S. In 2006, BTV made news by spreading north to make its first appearance in Western Europe (the Netherlands, then Belgium, Germany, and Luxembourg). In 2007 BTV was reported in the Czech Republic and the United Kingdom. There is no evidence of transovarial transmission, and cold winters of northern Europe kill most adult *Culicoides*, but some overwintering of adults appears to occur. Some carryover of virus may also result from uterine transmission that may allow persistent infection of newborn lambs or calves. Livestock vaccines are available; however, there are multiple serotypes of BTV.

Another orbivirus of note is Epizootic hemorrhagic disease virus (EHDV), also transmitted by *Culicoides* midges. EHD is a disease of wild ruminants, particularly white-tailed deer in North America. EHD is one of the most important diseases of deer in North America. Disease onset is sudden and in extreme cases is characterized by high fever, weakness, excessive salivation, rapid pulse and respiration rate, and hemorrhage. Within only 8—36 hours of the onset of symptoms, deer become prostrate and die. African horse sickness virus is an orbivirus endemic to Africa. Hosts include horses, donkeys, mules, and zebras. The most serious disease occurs in horses and mules while zebras are often asymptomatic and are thought to the natural reservoir hosts. Transmission is through bites of *Culicoides* sp.

Genus orthoreovirus

Orthoreoviruses are in the subfamily *Spinareovirinae*. Hosts include birds, mammals, and reptiles. These viruses are ubiquitous and seldom cause severe disease. They are so common in feces that their presence is used as an indicator of fecal contamination. Diseases associated with these orthoreoviruses include mild upper respiratory tract disease, gastroenteritis, and biliary atresia.

Genus coltivirus

Members of the genus *Colitivirus* are in the subfamily *Spinareovirinae*. Their genomes contain 12 segments. In North America the vertebrate hosts of Colorado tick fever virus (CTFV) include rodents, ungulates (deer, elk, sheep), and canids (coyotes). CTFV has also been isolated from humans where infection can result in flu-like symptoms, meningitis, encephalitis, and occasionally other severe complications. CTFV disease was initially confused with a mild form of Rocky Mountain spotted fever, which is caused by *Rickettsia rickettsii*. CTFV is an arbovirus, transmitted via the bite of an infected tick (Rocky Mountain wood tick (*Dermacentor andersoni*)). Most human infections occur at 4000—10,000 feet above sea level and during the spring and summer months when ticks are most active.

In this chapter we have learned that:

- Members of the family *Reoviridae* have segmented, dsRNA genomes. Virions are unenveloped and have double- or triple-layered capsids.

- Genomes are enclosed within the innermost capsid layer (core) and transcription occurs within intact cores. mRNAs exit cores through pores at the fivefold axes of symmetry.
- The viroplasm is an electron dense region of the cell that is formed by reovirus proteins. It is the site of genome replication and core assembly.
- Human rotaviruses are a major cause of severe, debilitating diarrhea in young children. Recently developed vaccines can reduce the morbidity and mortality associated with these infections.
- Members of the genus *Orthoreovirus* are ubiquitous viruses transmitted by the fecal–oral route. Their presence in water is an indicator of fecal contamination.
- Members of the genus *Orbivirus* are transmitted by insects. Animal pathogens include BTV, EHDV, and African horse sickness virus.

31

Family *Birnaviridae*

After reading this chapter, you should be able to answer the following questions:

- What are the major characteristics of the members of the family *Birnaviridae*?
- What does the family name signify?
- In what ways do birnavirus structure and replication differ from the other dsRNA viruses such as reoviruses?

Birnaviruses are unenveloped double-stranded (ds) RNA viruses that infect birds, fish, and insects (Fig. 31.1); they have recently been detected in some mammals (Yang et al., 2017). One birnavirus, infectious bursal disease virus (IBDV), is an important pathogen of young chicks. Birnaviruses have a number of interesting molecular features. Virions contain a filamentous nucleocapsid, somewhat reminiscent of negative-strand RNA viruses. In contrast to the reoviruses, birnavirus RNA is uncoated

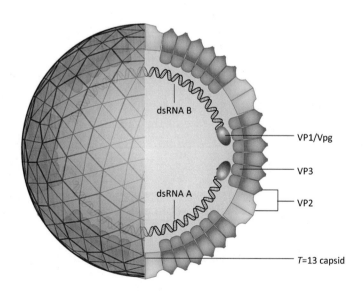
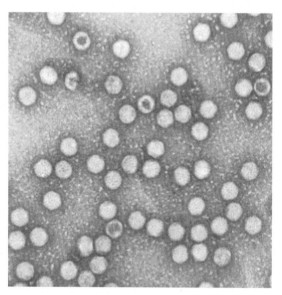

FIGURE 31.1   Virus structure (left). Electron micrograph of infectious bursal disease virus (IBDV) particles (right) (Mosley et al., 2016). Source: *(Right) Mosley, Y.Y., Wu, C.C., Lin, T.L., 2016. IBDV particles packaged with only segment A dsRNA. Virology 488, 68. https://doi.org/10.1016/j.virol.2015.11.001.*

© 2023 Elsevier Inc. All rights reserved.

BOX 31.1

Characteristics

Birnaviruses have bisegmented dsRNA genomes. Segments are ~2800–3400 bp and are named A and B. The 5′ end of each RNA strand is covalently linked to the protein primer VP1. In addition to serving as the transcription primer, VP1 contains the RdRp catalytic site. Genome segment A encodes a precursor polyprotein encoding a protease (VP4) and major structural proteins VP2 and VP3.

Segment B encodes the RdRp, the VP1 protein. Genome replication and transcription are cytoplasmic. Virions are unenveloped icosahedral capsids ~70 nm in diameter (proposed capsid structure $T = 13$) with 260 spike proteins. VP2 is the spike protein while VP3 forms the inner capsid layer and interacts with dsRNA. Virions contain a transcriptionally active polymerase (VP1).

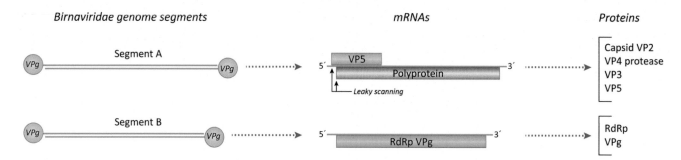

FIGURE 31.2   Genome organization and protein expression strategy of birnaviruses.

for transcription and genome replication. A precursor polyprotein encoding the major structural proteins is cleaved by a viral protease and the 5′ ends of the birnavirus genomic RNA are covalently linked to a protein primer, features similar to some positive-strand RNA viruses. While there is great interest in studying these viruses for their economic impact, many details of their molecular biology remain elusive (Box 31.1).

Genome structure

Birnavirus genomes are bisegmented (two segments) of dsRNA. (Hence the family name is *Birnaviridae*.) Segments are designated A and B and range from ~2800 to 3400 bp. The 5′ ends of each genome segment are covalently linked to viral protein 1 (VP1). VP1 is the RNA-dependent RNA polymerase (RdRp) as well as the primer for synthesis of genomic RNA. Segment A, the larger of the two, encodes an ~100 kDa precursor polyprotein (VP2-VP4-VP3). VP4 is a protease and mature VP2 and VP3 are capsid proteins. Segment A also encodes a nonstructural protein of ~17 kDa from an overlapping open reading frame. Genome segment B encodes VP1, the virion-associated RdRp. Some VP1 is free in the virion; some is linked to the 5′ ends of genomic RNA. Each genome segment contains conserved untranslated sequences at the 5′

and 3′ ends. It is likely that these regions are important for genome replication and packaging (Fig. 31.2).

Virion structure

Birnaviruses have single-layer icosahedral capsids with $T = 13$ symmetry, formed by 780 copies of the VP2 capsid protein (organized into 260 trimers). Mature virions are ~65 nm in diameter. Four small peptides (7–46 amino acids) are also associated with capsids. These are cleavage products derived from the carboxyl terminus of VP2. The longest peptide is Pep46, a 46 amino acid peptide that deforms membranes. VP3 is found on the inner face of the capsid. VP3 binds RNA and functions as scaffolding protein. Several molecules of VP1, the RdRp are found within the capsid.

General replication cycle

It is likely that attachment is mediated by the VP2 trimers that extend from the surface of the capsid and is followed by endocytosis. Within endosomes, low pH may trigger release of pep46 to produce pores in endosomal membranes. Following penetration, viral RNA is found associated with vesicular compartments in the cytoplasm. VP1, the RdRp, transcribes a single full-length mRNA from each genome segment. mRNAs are not polyadenylated but have a base-paired hairpin the 3′ end. It is not entirely clear if mRNAs have a 5′ cap, or are covalently linked to the VP1 primer. One mRNA encodes a polyprotein precursor with domains for VP2 (capsid protein), VP3 (RNA-binding and scaffolding protein), and VP4 (viral protease). The second mRNA encodes VP1, the RdRp, as shown in Fig. 31.2.

VP1 contains the catalytic site for RNA polymerization (RdRp activity) and another site that functions to self-guanylylate a site near its amino terminus. The guanylylated form of VP1 serves to prime transcription, thus generating RNAs covalently attached to a protein at the 5′ end. It is possible to generate infectious virus by transfecting cells with capped, in vitro transcribed full-length mRNAs suggesting that once they are translated, the resulting protein products are competent to complete the virus life cycle.

VP2 is the major capsid protein. Mature VP2 is the product of several cleavages that produce four peptides from its carboxyl terminal end. VP3 is a dsRNA-binding protein that serves both as a scaffold for capsid assembly and a transcriptional activator. VP3 is the major protein component of the nucleocapsid. VP3 localizes to vesicular structures bearing features of early and late endocytic compartments.

In addition to the polyprotein precursor, an addition protein (VP5) is translated from an overlapping reading frame at the 5′ end of the A segment mRNA. VP5 is a small, basic, cysteine-rich protein. VP5 is not found in cell-free virions but is present in infected cells, localized to the inner face of the plasma membrane. Among the roles ascribed VP5 are inhibitor of early apoptosis, stimulator of apoptosis, and virion release.

Disease

There are four genera in the family Birnaviridae. Primary hosts include birds, fish, and insects. Only recently have birnaviruses been reported in mammals (Yang et al., 2017).

Infectious bursal disease (Gumboro)

IBDV causes immunosuppression in young chicks as the virus targets and destroys the bursa. The bursa is a site of hematopoiesis and is critical for correct development of the avian immune system. Removal of the bursa from newly hatched chicks severely impairs the ability of the adult birds to produce antibodies, while removal of the bursa from adult chickens has little effect on immune function. An interesting feature of IBDV is that development of disease is age related. Chicks infected a less than 3 weeks of age usually have no acute signs while those infected between the ages of 3 and 6 weeks have high mortality. However, all infections of young chicks can cause immunosuppression, making birds susceptible to any number of secondary infections. Impairment of the immune system also results in poor responses to vaccines.

There are several birnaviruses that cause serious disease of natural and farmed fish. One of these is infectious pancreatic necrosis virus (IPNV). IPNV infects freshwater trout and Atlantic salmon. Highly virulent strains can cause >90% mortality in hatcheries while survivors can remain life-long carriers, serving as infectious reservoirs.

In this chapter we learned that:

- Members of the family *Birnaviridae* have unenveloped icosahedral virions ($T = 13$) that contain two segments of dsRNA. Genome replication (and possibly mRNA synthesis) is initiated with a protein primer.
- In contrast to the members of the family *Reoviridae* the birnaviruses uncoat their genomes upon entry into a cell, in a manner reminiscent of negative-strand RNA viruses. Capsid proteins are synthesized as a precursor polyprotein which is cleaved by the virus-encoded protease.
- Birnaviruses cause economically important diseases of fish and poultry.

References

Mosley, Y.Y., Wu, C.C., Lin, T.L., 2016. IBDV particles packaged with only segment A dsRNA. Virology 488, 68. Available from: https://doi.org/10.1016/j.virol.2015.11.001.

Yang, Z., He, B., Lu, Z., Mi, S., Jiang, J., Liu, Z., et al., 2017. Mammalian birnaviruses identified in pigs infected by classical swine fever virus. Virus Evol. 7, veab084. Available from: https://doi.org/10.1093/ve/veab084. 2021.

3 2

Hepatitis delta virus

After studying this chapter, you should be able to answer the following questions:

- What is the structure of the hepatitis delta virus (HDV) genome?
- What enzyme replicates the HDV genome?
- What is a ribozyme and what role do ribozymes play in HDV genome replication?
- What is the most common helper virus for HDV?

Hepatitis delta virus (HDV, family *Kolmioviridae*) is an important human pathogen that causes fulminant hepatitis in concert with hepatitis B virus (HBV, family *Hepadnaviridae*). HDV requires HBV surface proteins for packaging thus can only replicate in cells infected by HBV. HDV has a small RNA genome, but unlike other RNA viruses, host *RNA polymerase II (Pol II)* replicates its genome. HDV is the best-known member in the realm *Ribozyviria*, created in 2020 by the International Committee on Viral Taxonomy (ICTV) to contain viral agents that replicate their RNA genomes using host DNA-dependent RNA polymerases (Box 32.1).

Virion Structure

HDV virions are ∼36 nm enveloped particles with a ribonucleoprotein core (Fig. 32.1). Envelopes contain HBV surface proteins obtained from the coinfecting helper, HBV. The HDV genome is bound by two isoforms of the only virally encoded protein, the hepatitis delta antigen (HDAg). Within virions the genome is highly base paired, assuming a rod-like structure.

Genome organization

The genome of HDV is a circular ssRNA molecule of about 1700 nt. The genome is highly base paired (about 74%) and forms a rod-like structure. The genome

---

BOX 32.1

Taxonomy

The realm *Ribozyviria* contains viral agents with RNA genomes replicated by host DNA-dependent RNA polymerase. The named ribozyviria refers to the genomic and antigenomic ribozymes of the delta viruses. Additional common features include a highly base paired, rod-like genome and an RNA-binding protein.

The family *Kolmioviridae* contains eight named genera including genus *Deltavirus*. Other genera contain HDV-like viruses, recently identified by metagenomics, from a wide variety of animals, including birds, fish, snakes, toads, bats, white-tailed deer, and rodents.

Viruses.
DOI: https://doi.org/10.1016/B978-0-323-90385-1.00039-X

297

© 2023 Elsevier Inc. All rights reserved.

BOX 32.2

Viroids

Viroids are *infectious* single-stranded circular RNAs that are important plant pathogens. They are often considered subviral agents as they encode no proteins for genome encapsidation. Considered strictly, they are infectious RNA. All known viroids infect flowering plants. The first recognized viroid was the causative agent of potato spindle tuber disease. Viroids replicate their genomes using host RNA pol II via a rolling circle mechanism. Some viroids are ribozymes; their genomes contain catalytic domains that cleave and ligate RNA during genome replication. Viroid genomes range in length from about 250 to 400 nt in length. They form highly base-paired rod-like structures. Viroids are transmitted by a variety of means including through seeds and/or pollen, bulbs or tubers, grafting, and insects such as aphids. They can also be transmitted following mechanical damage to plants and by contaminated farming tools. Once a plant is infected, viroids move passively throughout the entire plant.

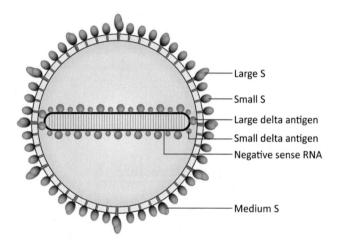

FIGURE 32.1   Hepatitis delta viruses (HDVs) are enveloped virions, about 36 nm in diameter, with a ribonucleoprotein core. In the case of HBV coinfection, three versions of HBV surface antigen are associated with the envelope.

Hepatitis delta virus genome

FIGURE 32.2   Hepatitis delta viruses (HDV) genome is a 1700 nt circular RNA. About 74% of the genome is base paired. There is a single open-reading frame, encoding S- and L- HDAgs. L-HDAg is synthesized by virtue of an RNA editing event that changes a stop codon to a tryptophan codon.

has a single open-reading frame that specifies two proteins named large and small hepatitis delta antigen (L-HDAg and S-HDAg, respectively) (Fig. 32.2). The HDV genome is structurally similar to those of plant viroids. Viroids are infectious RNAs of plants that cause severe, economically important diseases (Box 32.2). However, viroids are *infectious RNAs* that do not encode any proteins, are not packaged within a capsid or envelope, and do not require helper viruses for their replication.

Hepatitis delta antigens

L- and S-HDAg are 214 aa and 195 aa proteins. L-HDAg differs from S-HDAg in that it contains a unique carboxyl terminal extension of 19—20 amino acids. Expression of L-HDAg requires an RNA-editing event. Editing occurs at a single, specific site to convert an AUG stop codon to a codon specifying tryptophan. Editing is accomplished by host cell RNA-specific adenosine deaminase (ADAR1). ADAR1 is an interferon inducible protein that catalyzes the deamination of adenosine to yield inosine in double-stranded RNA structures. Editing occurs specifically on the antigenomic or copy RNA, which when replicated, generates an edited version of the genome strand. The editing site is informed by binding of ADAR1 to specific dsRNA sequences. The mRNAs transcribed from unedited genomes specify S-HDAg while mRNAs from edited genomes specify L-HDAg (Fig. 32.3).

L- and S-HDAgs play very different roles in viral replication. At early times after infection, S-HDAg is produced and is required for genome replication. Later, editing generates the mRNAs that specify L-HDAg, required for genome packaging. L-HDAg also inhibits genome replication. Thus HDV uses RNA editing as a mechanism to switch from genome replication to packaging. HDV S-HDAg is extensively modified posttranslationally. S-HDAg is methylated, acetylated, phosphorylated, and sumoylated. S-HDAg assembles into octamers that are wrapped by HDV RNA to form nucleosome-like structures that serve as templates for RNA synthesis by RNA Pol II. Cellular chromatin remodeling factors are involved in the process, that

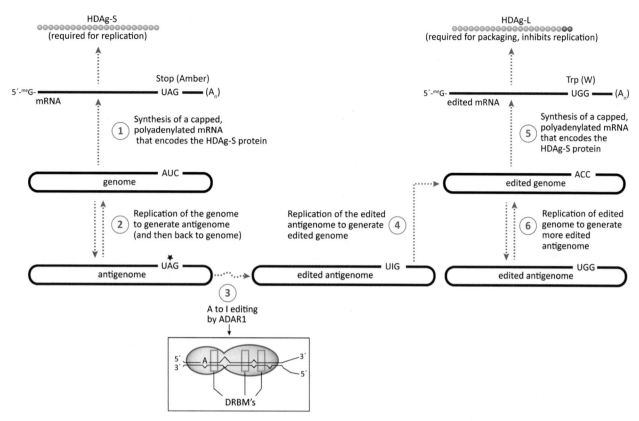

FIGURE 32.3 Overview of Hepatitis delta virus (HDV) genome replication and editing. The infecting genome is transcribed to produce a capped, polyadenylated mRNA for S-HDAg. Genome replication produces antigenome intermediates. Some antigenomes are edited by host ADAR1. As shown in the inset, ADAR1 interacts with three ds RNA-binding motifs (DRBM) that serve to direct the protein to the specific editing site. Editing converts a stop codon to a codon specifying tryptophan. The edited antigenome is copied to generate edited genomes. Edited genomes are transcribed to produce mRNAs for L-HDAg.

likely requires acetylation of S-HDAg. Phosphorylation of S-HDAg increases its affinity for a highly processive form of hyperphosphorylated host RNA Pol II. L-HDAg carries out its role in packaging via a post-translational modification. The carboxyl terminal extension is farnesylated by a cellular farnesyltransferase. This process adds hydrophobic molecules that facilitate interactions with membranes.

General replication cycle

Because the HDV particle is decorated with HBV surface proteins, it enters hepatocytes using the same receptors as HBV. Initial cell contacts are with heparin sulfate proteoglycans, followed by interaction with sodium taurocholate cotransporting polypeptide (NTCP) and epidermal growth factor receptor. Attachment is followed by endocytosis. HDV genomic RNA is transported to, and imported into the nucleus where transcription and genome replication are accomplished by RNA pol II. Transcription of genomic HDV RNA produces mRNAs that are capped and polyadenylated and are

transported out of the nucleus. S-HDAg is synthesized and transported back into the nucleus to stimulate genome replication. Genome replication takes place by a rolling circle mechanism. Concatemers of antigenomic RNA are produced and these are cleaved to unit length via a ribozyme domain in the antigenomic RNA (Box 32.3). Antigenomic RNA is the template for production of genomic RNA. Again, concatemers are cleaved to unit length by ribozyme activity and are ligated to produce circular genomes. Both genomic and antigenomic RNAs associate with HDAg to assemble into ribonuncleoproteins (RNPs).

An obvious question concerning HDV genome replication is how DNA-dependent RNA polymerase II recognizes and utilizes an RNA template. While the details are still being elucidated, recent studies are providing interesting answers. Within the eukaryotic nucleus, chromatin remodeling plays an important role in gene expression. Cellular DNA is densely packed into chromatin, and localized regions must be accessed for gene expression. This is accomplished by a process known as chromatin remodeling. Chromatin remodeling is a complicated process that requires

---

BOX 32.3

Ribozymes

Ribozymes are catalytic RNAs: They are enzymes built from RNA instead of protein. Ribozymes catalyze a variety of cellular reactions including condensation of amino acids to form polypeptides in the core of the ribosome. The ribozyme activities of HDV and viroids are nucleolytic, carrying out site-specific cleavages of the RNA backbone. Some nucleolytic ribozymes also catalyze formation of phosphodiester bonds to ligate RNA. HDV and viroid ribozymes generate unit length genomes from concatemers during rolling circle replication. Their ribozymes are encoded within their genomes. HDV encodes ribozymes on both genomic and antigenomic RNA. Ribozyme activity requires specific folding, and formation of stem loop structures to be catalytically active. The HDV ribozyme is 74 nt in length and folds to form 5 helical segments connected by a double pseudoknot.

---

histone modifications by enzymes such as histone acetyltransferases, deacetylases, methyltransferases, and kinases. Remodeling can result in histone movement, ejection of histones, or restructuring of nucleosomes. Chromatin remodeling is accomplished by assemblies of proteins called chromatin-remodeling complexes. It has recently been shown that the S-HDAg binds to a regulatory subunit of one type of remodeling complex. It is thought that this interaction is key to promoting recognition of HDV RNA by Pol II. It should also be noted that mammalian RNA pol II has recently been shown to catalyze synthesis of some cellular RNAs, in an RNA-dependent manner.

The final steps in HDV replication require assembly and release of infectious virions. In natural coinfections with HBV, HDV acquires a lipid envelope decorated with HBV surface proteins. HDV L-HDAg recruits HBV surface proteins to form infectious enveloped virions. There are three forms of glycosylated surface protein found on HBV virions, and all three forms are also found in HDV virions.

Disease

In naturally occurring infections HDV is always present with its helper virus HBV. HBV alone causes hepatitis, but HBV/HDV coinfections cause the most severe cases of hepatitis. It has been estimated that 10% of HBV-infected patients are also HDV infected, although the numbers vary widely depending on geographical region. Numbers of HDV-infected persons have been estimated to range from 12 million to 42 million worldwide. In countries where HBV vaccination is widespread, the numbers of HDV infections are lowest. HBV/HDV can be transmitted together (coinfection) or an HBV positive patient can acquire HDV infection later (superinfection). The consequences appear to be different. In a coinfection, most patients resolve the infection while the superinfection scenario leads most often to viral persistence. Chronic HDV/HBV infections often result in accelerated progression to liver cirrhosis, increased risk of liver decompensation and increased risk of hepatocellular carcinoma (HCC), compared to chronic HBV infection alone.

In 2020, bulevirtide, a 47-amino acid peptide was approved in the European Union to treat chronic HDV infection. Bulevirtide blocks viral cell entry via the NTCP receptor by mimicking the receptor binding domain of the HBV surface antigen. It thereby blocks binding of HDV and HBV to the NTCP receptor. Bulevirtide leads to a reduction in HDV RNA levels and liver enzymes in patients, however administration is daily by the subcutaneous route and treatment costs are extremely high.

In this chapter we learned that:

- HDV has a highly based paired, rod-like RNA genome that is replicated by host cell RNA polII.
- The HDV genome encodes two proteins, S-HDAg and L-HDAg. Synthesis of L-HDAg requires genome editing.
- Replication of HDV requires the helper virus, HBV. Co-infections with HDV and HBV often result in more severe disease than HBV alone. HDV RNPs are packaged into envelopes decorated with HBV glycoproteins.
- Ribozymes are catalytic RNAs. HDV encodes ribozymes in both genomic and cRNA. They cleave RNA concatamers into unit length genomic and copy RNAs.
- HDV envelopes contain HBV surface proteins thus HDV and HBV use the same cell receptors to infect hepatocytes.

33

Introduction to DNA viruses

After studying this chapter, you should be able to answer the following questions:

- What is a DNA virus?
- Which families of DNA viruses replicate in the cytoplasm?
- Which families of DNA viruses replicate in the nucleus?
- How is DNA virus replication linked to cell cycle?
- What is the general mechanism by which papillomaviruses, adenoviruses, and polyomaviruses transform cells? For these viruses, what is the relationship between cell transformation and virus replication?

Taxonomy of DNA viruses

DNA viruses have DNA genomes that are replicated by either host or virally encoded DNA polymerases. There is considerable diversity among DNA virus genomes and the relative stability of DNA allows for genomes much larger than possible for RNA viruses. Genomes of DNA viruses that infect animals range in size from less than 2 kb of single-stranded DNA to over 375 kb of double-stranded DNA. There are even larger DNA viruses that infect eukaryotic microorganisms.

The International Committee on the Taxonomy of Viruses (ICTV) has recently described three realms of DNA viruses that serve to delineate evolutionary relationships within this large and diverse group. The viruses included within each realm share basic characteristics as defined by the ICTV. Many, but not all, members of the realm *Monodnaviria* are viruses with single-stranded (ss)DNA genomes that replicate using rolling circle replication. They do not encode DNA polymerases but do encode replication (Rep) proteins that are required for genome replication. Rep proteins have endonuclease activity and nick the ssDNA genome within a base-paired region to generate a 3' OH to serve as the replication primer. Ancestors of these viruses are hypothesized to have arisen via recombination between replication initiation proteins of prokaryotic plasmids and cDNA copies of capsid protein genes of eukaryotic RNA viruses. Families containing ssDNA viruses include *Circoviridae*, *Parvoviridae*, and *Anelloviridae* among others. Some double-stranded (ds)DNA viruses are also members of the realm *Monodnaviria*. The members of the families *Polyomaviridae* and *Papillomaviridae* are hypothesized to have arisen from ssDNA ancestors as they encode proteins that direct host DNA polymerases to a viral replication origin.

The second realm containing viruses with DNA genomes is *Duplodnaviria*. As the name suggests, viruses in the realm have dsDNA genomes. But they are also defined by having a particular type of capsid protein with a so-called HK97 fold, as well as capsid portal proteins that are used in packaging DNA. The only animal DNA viruses in the realm are the herpesviruses (family *Herpesviridae*). Most members of realm are dsDNA phage.

The realm *Varidnaviria* are dsDNA viruses which encode a major capsid protein containing the vertical jelly-roll fold; these capsid proteins are noted for forming pseudohexameric capsomers. This realm also includes

© 2023 Elsevier Inc. All rights reserved.

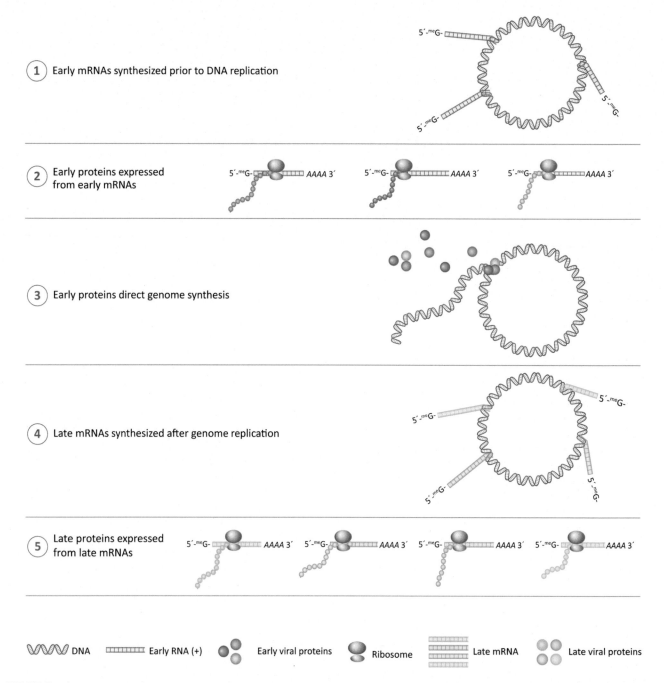

1 Early mRNAs synthesized prior to DNA replication

2 Early proteins expressed from early mRNAs

3 Early proteins direct genome synthesis

4 Late mRNAs synthesized after genome replication

5 Late proteins expressed from late mRNAs

DNA   Early RNA (+)   Early viral proteins   Ribosome   Late mRNA   Late viral proteins

FIGURE 33.1   After uncoating the genomes of DNA viruses are transcribed to produce an "early" set of mRNAs. Early mRNAs typically encode for proteins that modulate the host cell environment and/or are required for viral genome replication. After genome replication another set of mRNAs, the "late" mRNAs are expressed. Late genes encode structural proteins (and other proteins that are packaged within virions).

viruses demonstrated to have evolved from members of *Varidnaviria* that have lost the major capsid protein. Virus families discussed in this text include *Adenoviridae* (linear dsDNA genome), *Asfarviridae* (linear dsDNA genome), and *Poxviridae* (linear dsDNA genome).

A common theme among DNA viruses is separation of gene expression (transcription) into early and late phases (Fig. 33.1). Transcription of the so-called early

genes occurs before DNA synthesis to provide the protein products needed for DNA replication. After DNA replication commences the gene expression profile changes and the structural proteins needed to package DNA and form virions are produced. These structural genes are generally called "late genes." Herpesviruses encode a number of "late proteins" that are packaged into virions to support transcription upon infection of a new cell.

Viral DNA replication

The genomes of DNA viruses are quite variable in structure as well as in size: Single- or double-stranded linear molecules, single- or double-stranded circular molecules, and even highly base-paired single-stranded circular DNA that is often described as a "linear double-stranded molecule with closed ends" (family *Poxviridae*) (Fig. 33.2). As might be expected, genome structure and genome replication strategy are linked. Therefore DNA viruses employ variety of genome replication strategies. There is also considerable variation as to the source of DNA replication machinery: The smaller DNA viruses in the realm *Monodnaviria* depend almost entirely on host replication machinery while larger DNA viruses encode much of their own; some even replicating in the cytoplasm of eukaryotic cells using virally encoded RNA polymerases for mRNA synthesis.

DNA viruses have adopted a variety of strategies to prime genome synthesis: Adenoviruses use a *protein* to prime synthesis of their linear dsDNA genomes. The single-stranded linear genomes of parvoviruses adopt base-paired structures at the ends. Thus the 3′ end of the parvovirus DNA genome serves as the primer to initiate DNA replication. Other DNA viruses make use of host cell primase to synthesize RNA primers.

DNA viruses use a variety of DNA polymerases for genome synthesis. The smaller DNA viruses use host DNA replication machinery while the larger DNA viruses encode their own DNA polymerases (Table 33.1). Herpesviruses encode a viral DNA polymerase that has been exploited as a target for antiviral drugs. Of course DNA viruses that use host DNA synthetic machinery must replicate their genomes in the nucleus. Adenoviruses and herpesviruses encode their own DNA polymerases but also replicate in the nucleus as they do not encode a DNA-dependent RNA polymerase. DNA viruses that

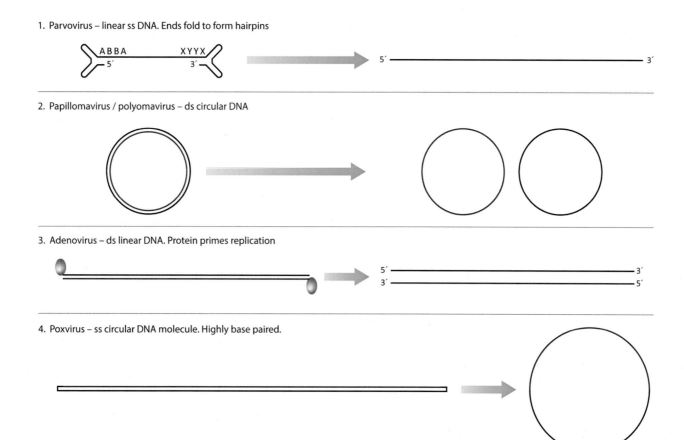

FIGURE 33.2 Examples of viral DNA genomes. Parvoviruses (top panel) have single-stranded DNA genomes that base pair to form double-stranded regions at the ends. The base-paired 3′ OH serves as the primer for DNA replication. Several virus families have circular, double-stranded DNA genomes. Adenoviruses have double-stranded linear DNA genomes. Adenoviruses use a protein to prime DNA replication. The genomes of poxviruses are often described as linear double-stranded DNA molecule. However, the ends are covalently closed, forming a single-stranded circular DNA. This is most obvious when the DNA is denatured.

TABLE 33.1   DNA viruses of animals.[a]

Virus family	Genome	Genome size (kb)	Virion morphology	Source of DNA polymerase	Source of RNA polymerase	Site of replication
Parvoviridae	ss linear	4–6	Naked, icosahedral	Host	Host	Nucleus
Circoviridae	ss circular	1.7–2.1	Naked, icosahedral	Host	Host	Nucleus
Anelloviridae	ss circular	2–4	Naked, icosahedral	Host	Host	Nucleus
Polyomaviridae	ds circular	5	Naked, icosahedral	Host	Host	Nucleus
Papillomaviridae	ds circular	8	Naked, icosahedral	Host	Host	Nucleus
Adenoviridae	ds linear	35–36	Naked, icosahedral	Viral	Host	Nucleus
Herpesviridae	ds linear	120–240	Enveloped, icosahedral capsid	Viral	Host	Nucleus
Poxviridae	ds linear[b]	130–375	Enveloped, complex	Viral	Viral	Cytoplasm
Asfarviridae	ds linear	170–190	Enveloped, icosahedral capsid	Viral	Viral	Cytoplasm

[a]*Many of these families contain members that infect a wide range of organisms.*
[b]*Ends of the molecule are linked (single-stranded circle upon denaturation).*

replicate in the nucleus have evolved mechanisms to direct viral DNA into and out of the nucleus. The process often uses both viral and host proteins.

Poxviruses and African swine fever virus (ASFV), family *Asfarviridae* have large genomes and encode DNA and RNA polymerases. They belong to a group of DNA viruses called the "nuclear and cytoplasmic large DNA viruses." Poxviruses and ASFV replicate and transcribe their genomes in the cytoplasm: In addition to polymerases, they encode topoisomerases, transcription factors, and processivity factors that link DNA polymerase to the template.

The process of actually synthesizing new DNA strands is also variable. Adenoviruses have linear dsDNA genomes and use a protein primer to initiate DNA synthesis. Synthesis begins at one end of the double-stranded molecule and the new strand remains base paired to the parental template, displacing a parental single strand. The displaced strand is then copied to generate a double-stranded molecule.

The process for replicating circular DNA genomes can be quite complex, relying on DNA nicking enzymes and methods for transitioning or resolving structures from double-stranded to single-stranded molecules. Two of the more common strategies for replication of double-stranded circular DNA genomes are illustrated in Fig. 33.3: theta replication and rolling circle replication. Herpesviruses package a molecule of linear dsDNA that circularizes in the nucleus. Herpesvirus genome replication seems to initiate with a theta replication and then switch to a rolling circle mechanism that produces long genome concatamers. The concatemers are cleaved to unit genome length as viral DNA is packaged into virions.

DNA viruses and cell cycle

A feature common to many DNA viruses is their dependence on host DNA replication to provide enzymes (e.g., DNA polymerases) and pools of deoxynucleotide triphosphate precursors. While RNA turnover is constant in metabolically active cells, DNA synthesis is tightly regulated. Cells that are not mitotically active may be poor hosts for DNA viruses. In adult animals there are many cells that are terminally differentiated and seldom, if ever divide (e.g., neurons). Some cells divide in response to damage, while others (e.g., epithelial cells) divide regularly. In contrast, the mitotic activity of a developing fetus and growing animal is much higher than in the mature adult. There are three basic strategies used by DNA viruses to ensure that they can replicate their genomes:

- Infect only mitotically active cells.
- Induce nondividing cells to enter the cell cycle by use of viral proteins.
- Encode most or all enzymes required for dNTP synthesis and DNA replication.

The smallest DNA viruses (circoviruses and parvoviruses) productively infect only mitotically active cells. As we will see in coming chapters, this impacts cell tropism and pathogenesis. The larger DNA viruses (polyomaviruses, papillomavirus, adenoviruses, and herpesviruses) encode proteins that induce resting cells to become mitotically active. Even the larger poxviruses, that encode much of the enzymatic machinery required for DNA synthesis, tend to stimulate cell proliferation.

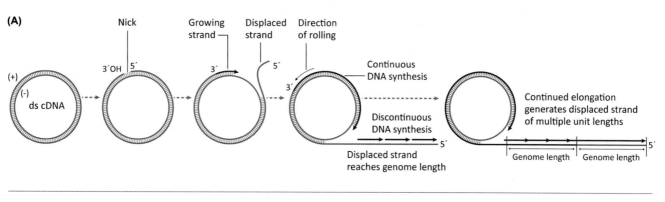

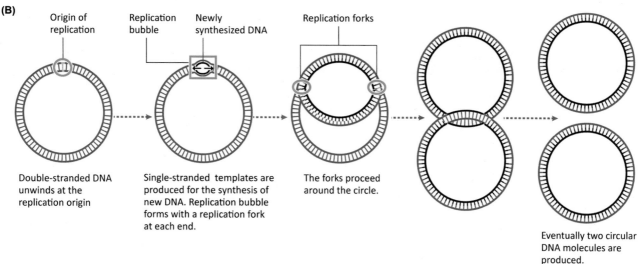

FIGURE 33.3   General models of rolling circle DNA replication (panel A) and theta replication (panel B).

Oncogenesis

Several families of DNA viruses are associated with cell transformation and oncogenesis. The polyomaviruses, papillomaviruses, and adenoviruses transform cells using a similar strategy. All encode proteins that alter cell cycle and inhibit apoptosis by inactivating key host proteins (Table 33.2). The viral proteins with these activities are called viral oncogenes. The common cellular targets are retinoblastoma protein (pRB) family members and p53. pRB and related proteins tightly control cell cycle by sequestering the transcription factor E2F. E2F activates transcription of a set of host genes that drive DNA replication and mitosis. By inactivating pRB, polyomaviruses, papillomaviruses, and adenoviruses induce host cell DNA synthesis (Fig. 33.4). Eukaryotic cells also have mechanisms to sense DNA damage, and either pause the cell cycle or activate apoptotic pathways. A key regulator of these processes is the protein p53. In order to complete their replication cycles, polyomaviruses, papillomaviruses, and adenoviruses encode proteins that inactivate p53. However, in a productive infection, neither polyomaviruses, papillomaviruses nor adenoviruses transform cells, as the end result is virion production and cell death. It is only when infection is restricted or nonproductive that cell transformation is a possible outcome. This is why adenoviruses *not* produce tumors in their normal hosts; they replicate with a resulting lytic infection. However, in nonpermissive cells the activities of viral oncogenes may be expressed and initiate the process of transformation. Some papillomaviruses and polyomaviruses *do* contribute to tumor formation in natural hosts. But tumor induction is a fairly rare event; it is not the normal outcome of infection. Good examples are the human papillomaviruses (HPVs) associated with cervical cancers. Papillomaviruses may be oncogenic (occasionally) in their natural hosts because of their overall replication strategy: Virions are only produced in differentiated epithelial cells. However, there is considerable viral DNA replication, without virion production, in nondifferentiated epithelial cells. This provides an environment in which transformation may occur. When one examines the tumors initiated by papillomavirus infection, a common finding is that tumors are monoclonal (derived from a single precursor cell) and contain part of a papillomavirus genome integrated into host

TABLE 33.2   Three families of DNA viruses use similar mechanisms to affect cell cycle.

Virus family	Proteins that interact with pRB	Proteins that interact with p53
Polyomaviridae	Large T	Large T
Papillomaviridae	E7	E6
Adenoviridae	E1A	E1B

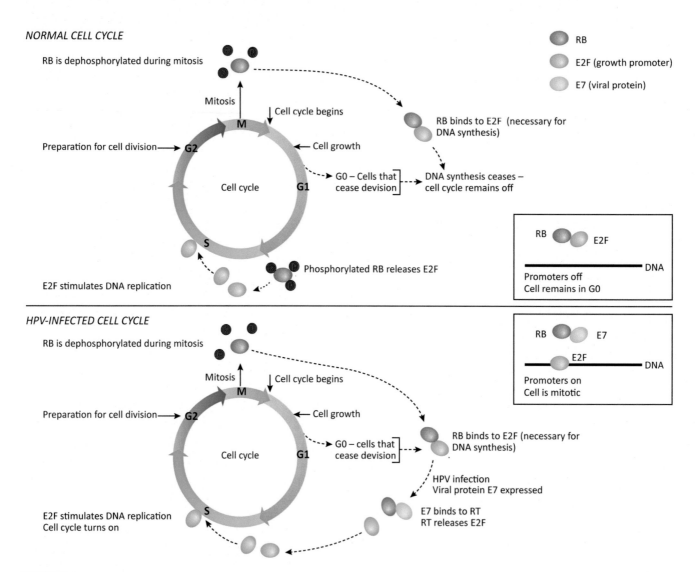

FIGURE 33.4   Comparison of normal cell cycle and HPV-infected cell cycle. Top panel: Summary of cell cycle showing contributions of pRB family proteins in regulating DNA replication. In cells that are not mitotically active (G₀), pRB is bound to the transcription factor E2F. E2F is a major regulator of genes required for DNA synthesis. In a normal cell, external signals (e.g., growth factors) may initiate mitosis, leading to the phosphorylation of pRB by cyclin-dependent kinases. Phosphorylated pRB releases E2F to turn on genes required for DNA synthesis. At the end of the cell cycle, pRB is dephosphorylated and rebinds to E2F. Bottom panel: Some DNA viruses (papillomaviruses, polyomaviruses, and adenoviruses) encode proteins that disrupt normal RB function. In this example papillomavirus E7 protein binds to RB causing the release of E2F, thereby inducing continued DNA replication in the infected cell.

DNA. The fragment of inserted DNA encodes the viral oncogenes. Thus while virally induced cancers are of great importance to the host, they are actually a "dead end" for the virus involved. It is also important to note that the viral oncogene is only an initial step in transformation. Fully transformed cells contain a variety of additional mutations that arise during cell division.

In this chapter we learned that:

- DNA viruses have DNA genomes that are replicated by a DNA-dependent DNA polymerase.
- A few of the largest DNA viruses encode a viral DNA-dependent RNA polymerase for mRNA synthesis, but most use host enzymes.
- *Poxviridae* and *Asfarviridae* are two families of large DNA viruses that replicate in the cytoplasm.
- Most families of DNA viruses replicate in the nucleus. These include *Circoviridae, Parvoviridae, Anelloviridae, Polyomaviridae, Papillomaviridae, Adenoviridae,* and *Herpesviridae.*
- Replication of several DNA viruses is tightly linked to the cell cycle. In order to replicate viral genomes, the cell must provide DNA polymerase and dNTPs. Some small DNA viruses infect only mitotically active cells; other viruses induce cells to become mitotically active.
- Papillomaviruses, adenoviruses, and polyomaviruses encode proteins that promote cell division (by inactivating pRB family members). They also encode proteins that inactivate p53, to inhibit apoptosis induced by DNA damage. In normal infection cycles virions are formed and the infected cell dies. In rare instances viral oncogenes are inserted into the host chromosome, a process that can be the first step in cancer development.

34

Family *Parvoviridae*

After reading this chapter, you should be able to answer the following questions:

- What are the major characteristics of the members of the family *Parvoviridae*?
- What cell environment does a parvovirus require for productive infection and how does this impact pathogenesis?
- What common childhood disease is caused by a parvovirus?
- What life-threatening disease of cats is caused by a parvovirus?

Parvoviruses are small, unenveloped, linear, single-stranded (ss) DNA viruses whose genome size ranges from 4 to 6 kb. Parvoviruses are unenveloped with T = 1 icosahedral symmetry; capsids assemble from the capsid protein (VP) (Fig. 34.1). The family *Parvoviridae* contains three subfamilies: *Densovirinae* (invertebrate parvoviruses), *Hamavirinae*, and *Parvovirinae* (vertebrate parvoviruses whose hosts include birds, reptiles, and mammals) (Box 34.1). There are 10 named genera in the subfamily *Parvovirinae* including *Erythroparvovirus* (human B19 parvovirus), *Dependovirus* [adeno-associated viruses (AAVs) which require a helper virus to replicate] and *Protoparvovirus* [includes feline panleukopenia virus (FPV) and canine parvovirus]. Given their small genome size, parvoviruses require much cellular assistance to replicate their genomes. For example, they require dividing cells to provide adequate pools of dNTPs. Parvoviruses do not encode proteins that activate the cell

cycle therefore many replicate only in actively dividing cells (Box 34.2). AAVs require coinfection with larger DNA viruses such as adenoviruses or herpesviruses to induce cell division. While the small size of parvovirus genomes is somewhat limiting, virions are extremely stable in the environment. Virions resist both high and low pH and are stable for an hour at 56°C (132.8 °F).

Genome organization

Parvoviruses genomes are a molecule of linear ssDNA of ~5000 nt. Two large open-reading frames (ORFs) encode a nonstructural protein (NS) and a structural protein (VP). Additional small ORFs may also be present. The number and positions of transcription promoters vary among genera. Use of variable transcription start sites, alternative splicing and alternative polyadenylation sites produce a diverse set of mRNAs from these relatively simple genomes (Fig. 34.2). At each end of the genome are noncoding regions that form stable hairpins. These hairpin structures are essential for genome replication as they serve as primers for DNA replication (Fig. 34.3).

Virion structure

Parvoviruses are small, unenveloped viruses with T = 1 icosahedral symmetry. Virions are about 25 nm in

© 2023 Elsevier Inc. All rights reserved.

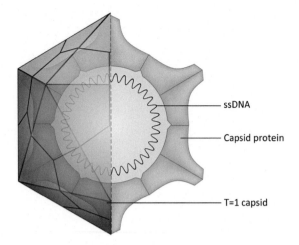
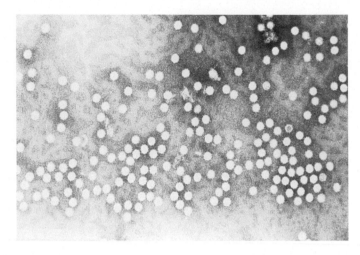

ssDNA

Capsid protein

T=1 capsid

FIGURE 34.1   Parvovirus structure (left) and electron micrograph (right). CDC Public Health Image Library Image 5618. Source: *Content providers CDC/ R. Regnery and E. L. Palmer.*

BOX 34.1

Taxonomy

Family: *Parvoviridae*
Subfamily: *Densovirinae* (11 named genera, hosts include insects, fish, crustaceans)
Subfamily *Hamaparvovirinae* (2 named genera, hosts include mammals and birds)
Subfamily: *Parvovirinae*
     Named genera including the following:

- Genus: *Amdoparvovirus* (aleutian mink disease virus)
- Genus: *Aveparvovirus* (chicken parvovirus)
- Genus: *Bocaparvovirus* (bovine parvovirus)
- Genus: *Copiparvovirus* (ungulate copiparvovirus)

- Genus: *Dependovirus* [adeno-associated viruses (AAVs) they require coinfection with an adenovirus or (less commonly) a herpesvirus. The dependoviruses are most closely related to autonomous avian parvoviruses.]
- Genus: *Erythroparvovirus* (Species *primate erythroparvovirus 1*, human parvovirus B19)
- Genus: *Protoparvovirus* [minute virus of mice (MVM), feline panleukopenia virus (FPV), canine parvovirus (CPV), and mink enteritis virus]
- Genus: *Tetraparvovirus* (primate tetraparvovirus 1)

BOX 34.2

General characteristics

Parvovirus genomes are linear, ssDNA, 4–6 kb. Most have two open reading frames (ORFs). Some have additional short ORFs. Transcription and genome replication are nuclear. Most parvoviruses replicate only in dividing cells or in the presence of a helper virus but some induce the DNA damage response and use the host DNA repair apparatus for genome replication. Virions are unenveloped with T = 1 icosahedral symmetry with spike-like protrusions, ~25 nm diameter. They assemble from 60 copies of two or three alternative forms of VP. Virions are very stable in the environment.

---

diameter with spike-like protrusions. Most autonomous parvoviruses and the AAVs assemble from 60 copies of a capsid protein. However, there are different versions of VP produced from spliced mRNAs; all share a common carboxyl-terminus. The capsid of human parvovirus B19 is assembled from two capsid proteins, VP1 and VP2.

VP2 is the major capsid component. For most or perhaps all parvoviruses, VP1 (the largest of the capsid proteins) is a minor component of the capsid. However, VP1 is critical for infectivity as it contains an essential enzymatic domain with calcium-dependent phospholipase A2 (PLA2) activity required for exit from the endosome.

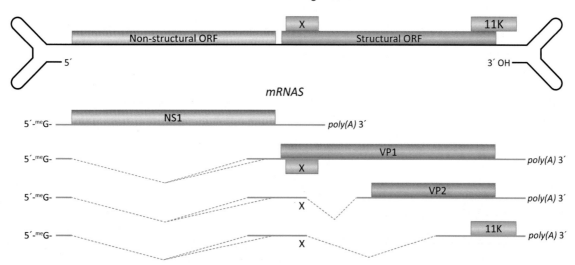

Parvoviridae genome

FIGURE 34.2   Genome organization of erythroparvovirus, B19V. Some parvoviruses produced spliced form of NS.

Overview of replication

The so-called autonomous parvoviruses replicate only in actively dividing cells while members of the genus *Dependovirus* have broader cell tropism but require coinfection with a helper virus. This section will focus primarily on replication of the autonomous (helper-independent) parvoviruses.

Parvoviruses use a variety of protein and/or carbohydrate receptors. For example, human parvovirus B19 (B19V) binds to a glycolipid receptor (called P antigen or globoside) present on primary erythroid progenitor cells. Erythroid progenitor cells are committed stem cells that give rise to red blood cells. Minute virus of mice (MVM) binds sialic acid receptors, thus hemagglutinates mouse erythrocytes. There are 12 serotypes of AAVs that are targeted to different cells by virtue of different receptor usage. Receptors for AAVs include both carbohydrate and protein receptors. The variety of receptors used by AAVs has facilitated their use as gene delivery vehicles targeting specific cell types.

Parvoviruses enter cells by endocytosis and the low pH environment of the endosome leads to critical structural changes that expose both a PLA2 domain as well as a nuclear localization signal present at the amino terminus of VP1. The PLA2 domain is a calcium-dependent phospholipase (hydrolyzes phospholipids into fatty acids) that is required to liberate the virion from the endosome. After a virion leaves the endosome, it associates with microtubules and is actively transported to the nucleus. VP1 then appears to interact with cellular proteins for translocation across the nuclear pore.

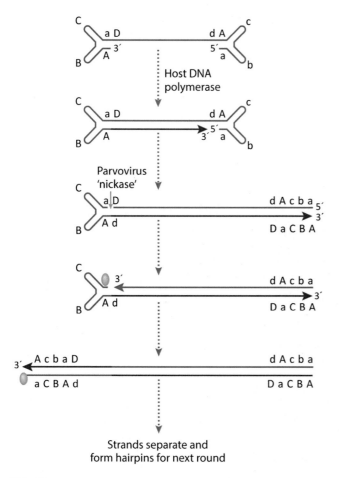

FIGURE 34.3   Parvovirus genome replication strategy.

The 3′ end (20–30 nt) of parvoviral DNA appears to be exposed on the outside of essentially intact capsids. It has been suggested that the exposed DNA could serve as the initiation site for DNA polymerase, such that the viral DNA is actively removed from the capsid without its disassembly. The genomes of autonomous parvoviruses are replicated by host cell DNA polymerases. NS1 is the key viral player in DNA replication. The current model for autonomous parvovirus DNA replication is that cellular DNA polymerase begins synthesis at the base-paired 3′ end of the genome, forming a linear duplex (Fig. 34.3). After formation, the linear duplex is nicked by the NS protein. In the process of nicking, NS is covalently linked to the 5′ end of the viral DNA and in this role, also plays a role in genome packaging. DNA synthesis continues at the new 3′ end liberated by nicking. NS is the major parvovirus regulatory protein and is a helicase, adenosine triphosphatase (ATPase), and site-specific DNA nicking enzyme (nickase). Parvovirus DNA replication takes place in discrete subnuclear structures that contain cellular DNA polymerases alpha and delta. Transcripts and proteins appear early in the course of infection. B19V NS also affects host cell transcription and induces apoptosis in some cell types.

Spliced mRNAs are used to express parvovirus structural proteins (VP1 and VP2). VP1 and VP2 share a common carboxyl-terminus; VP1 of B19V is ~84,000 kDa while VP2 is ~58,000 kDa. The majority of the B19V capsid is constructed from VP2. B19V VP1 has an additional 227 amino acids at the amino terminus and this domain contains the phospholipase activity essential for virus entry into cells. In the case of dependoviruses, the VP ORF is often called CAP, for capsid. Parvovirus capsids assemble in the nucleus. DNA incorporation occurs by insertion of viral ssDNA into the preassembled capsid and requires the helicase activity of NS to translocate ssDNA into the capsid. Capsids may be retained within the nucleus, translocated into the cytoplasm, or actively transported out of the cell.

Diseases

Human B19 parvovirus

Human B19 parvovirus (B19V) infects school-aged children and is the cause of erythema infectiosum or fifth disease (it was fifth on a list of six childhood rash-forming illnesses). It is also called slapped-cheek disease due to bright red rash on face that is a common symptom. The disease begins with general symptoms of a stuffy or runny nose, headache, and a low-grade fever. After these initial symptoms resolve a rash appears, typically on the face and then spreads to the trunk, arms, and legs.

We mentioned earlier that B19V infects erythroid progenitor cells and is found replicating in the bone marrow. Why then is the typical course of acute infection in children a mild febrile illness followed by a rash? Based on limited clinical studies, it appears that in a healthy child the virus initially replicates in the upper respiratory tract before spreading to other sites. In the bone marrow, erythroid progenitor cells are infected and produce an abundance of virus, resulting in viremia (and the clinically apparent "mild febrile illness"). By this time the immune system is responding and immune complexes (antibody and virus) are deposited in the skin, leading to the characteristic rash. Immune complex deposition could also explain symptoms of joint pain and arthritis associated with B19V infections in adult women.

It is likely that many people are infected with B19V without developing any clinical symptoms at all. However, the disease course can be very different in patients with underlying conditions that affect the life span of their red blood cells. In these patients the transient loss of erythroid progenitor cells can cause severe, life-threatening anemia. B19V is also implicated in a variety of rare, but severe blood disorders in immunocompromised individuals. Finally, there are suggested links between B19V infection and spontaneous abortion.

Metagenomics studies have led to the identification of additional human parvoviruses. In 2005 a new human parvovirus, related to bovine and canine parvoviruses (genus *Bocaparvovirus*), was identified and named human bocavirus (HBoV). HBoV is found worldwide, present in both respiratory tract samples and stool samples. The most frequently described clinical presentation of HBoV1 infection is upper respiratory tract disease. Bocaparvoviruses are interesting in that they do not require mitotically active cells for replication. Instead they induce the cellular DNA damage response and use this repair machinery to replicate their genomes.

Feline panleukopenia virus

FPV infects all felids (domestic and wild cats) and some members of related families (e.g., raccoon, mink). FPV is found worldwide and the virus is endemic in unvaccinated domestic cat populations. FPV is associated with the diseases feline panleukopenia, feline infectious enteritis, and feline distemper. These diseases are most common in recently weaned kittens who are at risk of infection as levels of maternal antibody decline. Disease occurs when virus replicates in

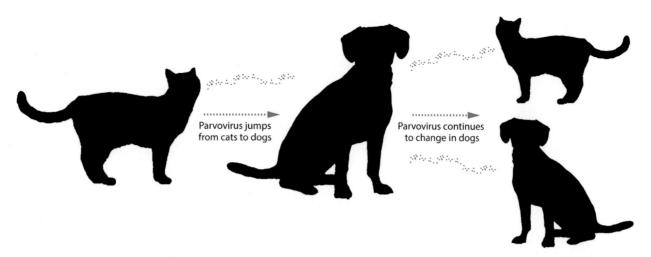

FIGURE 34.4 Canine parvovirus is closely related to feline parvovirus and may have emerged after a jump into dogs. CPV continues to spread and adapt.

mitotically active tissues including intestine, bone marrow, thymus, lymph nodes, and spleen. Numbers of circulating white blood cells may decrease by 90% resulting in panleukopenia. FPV is transmitted by respiratory and/or fecal—oral routes. Virus is shed in feces, saliva, urine, vomit, and blood. Virus can be shed for long periods (many weeks) and the virus is very stable in the environment.

Canine parvovirus

Canine parvovirus-2 (CPV-2) is closely related to FPV. While FPV does not readily infect dogs, CPV-2 presumably emerged after a jump from cats to dogs (Fig. 34.4). CPV-2 emerged as a dog pathogen in the 1970s and quickly spread worldwide. CPV-2 was initially associated with very high morbidity and mortality due to heart failure in utero or shortly after birth. However, the clinical picture changed as more dogs were exposed to the virus and acquired immunity. The virus also continued to change. Now CPV-2 infection of puppies most often leads to enteric disease.

Aleutian mink disease

Aleutian mink disease is also caused by a parvovirus genetically similar to FPV and CPV-2. Aleutian refers to the gray coat color of the farmed minks at the center of the first reported outbreak. However, the virus infects minks of all colors, as well as ferrets. The disease is worthy of mention because its pathology is quite different from the other "carnivore" parvoviruses. The infection is chronic, not acute, and ferrets can be long-term healthy carriers. When symptoms do appear, they include poor reproduction, weight loss, oral and gastrointestinal bleeding, and renal failure. The novel feature of Aleutian mink disease is that its pathogenesis is immune-mediated. Chronic infection is asymptomatic until high levels of virus and antibodies combine to form immune complexes, deposited in many organs.

In this chapter, we learned that:

- Parvoviruses are small, unenveloped viruses with linear ssDNA genomes.
- Some parvoviruses are autonomously replicating viruses that infect mitotically active cells. Others rely on coinfection with a larger DNA virus (adenovirus or herpesvirus).
- In some cases the pathogenesis of parvoviral infections is related to direct killing of mitotically active cells while in other cases disease is immune-mediated.
- The most common human parvoviral infection is B19V. It causes erythema infectiosum or fifth disease in children. The infection is usually well controlled by the immune system, but can cause chronic infection in immunosuppressed individuals. In rare cases, severe blood disorders have been associated with B19V infection.
- FPV infects wild and domesticated cats worldwide. Disease is usually associated with recently weaned kittens. The virus can cause gastrointestinal disease (enteritis) or a global reduction in white blood cells (panleukopenia).

35

Other small DNA viruses

After reading this chapter, you should be able to answer the following questions:

- What are the general characteristics of circoviruses?
- What are the general characteristics of anelloviruses?
- Why are these viruses restricted to replication in mitotically active cells?
- What is apoptin and why is it of interest to cancer researchers?

There are currently nine named families of eukaryotic single-stranded (ss)DNA viruses. With the exception of parvoviruses their genomes are circular ssDNA. Many of these viruses have been discovered by metagenomics and have never been cultured, but others are well-described and economically important pathogens of plants and animals. Two families of ssDNA viruses are described in this chapter, the circoviruses (family *Circoviridae*) and the anelloviruses (family *Anelloviridae*).

Family *Circoviridae*

Members of the family *Circoviridae* are small, unenveloped viruses ($T = 1$) that are extremely stable in the environment. There are two named genera, *Circovirus* and *Cyclovirus*. Capsids assemble from 60 copies of the capsid (C) protein (Fig. 35.1). Virions have a smooth surface appearance and a size of ~17–25 nm. Genomes are covalently closed, circular, single-stranded DNA of ~1700–1800 nt with two major open-reading frames (ORFs). The genomes of members of the genus *Circovirus* are ambisense [ORFs are transcribed in opposite directions from a single promoter site (Fig. 35.2)]. One circovirus ORF encodes the capsid (C) protein, the other encodes the nonstructural replicase-associated (Rep) protein. Circoviruses replicate in mitotically active (dividing) cells as they use host cell polymerases for genome replication and do not encode proteins that can stimulate cell division (Box 35.1). The only virally encoded proteins involved in circovirus genome replication are the replication initiation proteins, Rep and Rep'. Rep proteins have conserved endonuclease and helicase domains and the term CRESS-DNA virus (Box 35.2) has been coined to describe circular Rep-encoding ssDNA viruses.

The first step in genome replication of circoviruses is conversion of single-stranded DNA to a double-stranded molecule. The double-stranded genome is then transcribed by host RNA polymerase. Genome replication is then initiated by Rep and catalyzed by host DNA synthesis machinery. Detailed models for other processes in the circovirus life cycle (i.e., entry, assembly) are lacking but it is assumed that entry is mediated by endocytosis and the virion makes its way to the nucleus. Details of DNA packaging, assembly and release are sketchy, but will certainly be forthcoming.

Viruses.
DOI: https://doi.org/10.1016/B978-0-323-90385-1.00035-2

© 2023 Elsevier Inc. All rights reserved.

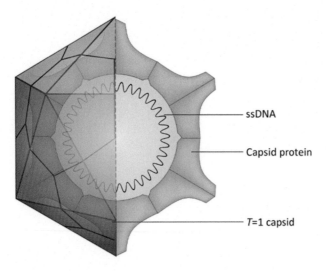

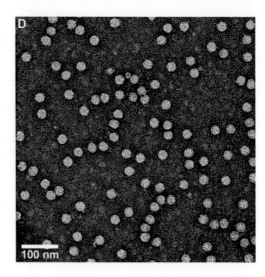

FIGURE 35.1   Circovirus virion structure (left) and cryo-electron microscopy of porcine circovirus 2d virus-like particles. Source: *From Khayat, R., Wen, K., Alimova, A., Gavrilov, B., Katz, A., Galarza, J.M., et al., 2019. Structural characterization of the PCV2d virus-like particle at 3.3 Å resolution reveals differences to PCV2a and PCV2b capsids, a tetranucleotide, and an N-terminus near the icosahedral 3-fold axes. Virol. 537, 186—197. https://doi.org/10.1016/j.virol.2019.09.001.*

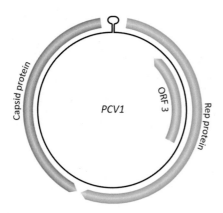

FIGURE 35.2   Circovirus genome organization. Shown is the genome of porcine circovirus 1.

Diseases

Beak and feather disease

Beak and feather disease virus (BFDV) infects psittaciform birds (parrots). The virus seems to have originated in Australia where it continues to threaten wild parrot populations (Fig. 35.3). The virus is also found worldwide in captive parrots. The virus infects mitotically active cells of the immune system as well as the cells that produce feathers and beak. Affected birds gradually lose their feathers, develop beak abnormalities that interfere with eating, and become immunocompromised, often dying of infections by a variety of microorganisms. As might be expected for a virus that

---

BOX 35.1

General description and taxonomy, family *Circoviridae*

Genomes are single-stranded circular DNA of ~1700–3800 nt encoding two major proteins: A capsid protein and a Rep protein. The Rep proteins are likely the key viral players in initiating rolling-circle replication to produce more genomes. These viruses replicate in mitotically active cells, using host DNA synthesis machinery. Virions are unenveloped, with T = 1 icosahedral symmetry.

Family: *Circoviridae*
Genus: *Circovirus*

Species include *Porcine circovirus type 1 and 2 and Beak and Feather Disease virus (BFDV)*. ICTV Taxonomy (2021) lists over 30 species, most detected in animals and birds. Some have been identified as a result of sequencing studies to characterize fecal or environmental microbiota and a few are bird pathogens.

Genus: *Cyclovirus*

ICTV Taxonomy (2021) lists over 40 species associated with animals, birds, and insects. There is no biological data regarding the infectivity, transmission, or host range of members of the genus *Cyclovirus*.

BOX 35.2

Circular Rep-encoding ssDNA (CRESS-DNA) viruses

Analysis of the global virome using unbiased nucleic acid sequencing technologies has revealed an abundance of small single-stranded (ss) DNA viruses. Many of these ssDNA viruses have circular genomes and encode replication initiator proteins (Rep) belonging to the HUH endonuclease superfamily. The term CRESS-DNA virus was coined to generally describe these circular Rep-encoding single-stranded DNA viruses. The CRESS-DNA viruses are members of the realm *Monodnaviria*. These eukaryotic ssDNA viruses may have evolved through recombination of Rep protein genes of bacterial plasmids with cDNA copies of capsid genes of eukaryotic positive sense RNA viruses. It is the presence of a Rep protein that serves to characterize and evolution-arily unite these viruses. Further, polyomaviruses and papillomaviruses (which have dsDNA genomes) may have evolved from CRESS-DNA viruses.

Rep proteins of CRESS-DNA viruses contain an endonuclease domain containing an HUH motif. This motif consists of two histidines (H) separated by a bulky hydrophobic amino acid. The endonuclease domain has a nicking function. Rep proteins also have a helicase domain which unwinds double-stranded (ds) DNA replicative intermediates during rolling-circle genome replication. Rep is actually an *initiator* of rolling-circle replication as DNA synthesis is carried out by a host-encoded polymerase. The need for host-encoded DNA polymerase limits the replication of these viruses to mitotically active cells.

FIGURE 35.3 Parrot with beak and feather disease, caused by a circovirus.

depends upon mitotically active cells for replication, young birds are much more susceptible to infection than are adult birds. There is no treatment and the disease is usually fatal.

Porcine circoviruses

Four circoviruses have been isolated from pigs to date. Porcine circovirus 1 (PCV1) is ubiquitous, and nonpathogenic in swine. Porcine circovirus 2 (PCV2) emerged as a major swine pathogen about 20 years ago and in now endemic worldwide. Little is known about the more recently discovered PCV3 and PCV4.

PCV2 is associated with postweaning multisystemic wasting syndrome, a disease characterized by jaundice, diarrhea, respiratory disease, failure to thrive, and sudden death. Other diseases associated with PCV2 infection include porcine dermatitis and nephropathy syndrome and porcine respiratory disease complex. The origins of PCV2 are not clear but it may have been a long-time ubiquitous agent in pigs. Its exact role in disease development is still somewhat murky, as PCV2 is necessary, but not sufficient, on its own, to cause disease. In experimental systems, cofactors such as infections with other agents, including porcine parvovirus, or immune stimulation, greatly exacerbate disease.

Family *Anelloviridae*

In 1997 a novel, partial viral clone was isolated from a patient (initials TT) with posttransfusion hepatitis of unknown etiology. This was the first report of TT virus (TTV) but numerous surveys since that time have revealed that TTV-like viruses comprise the most abundant component of the human blood virome. Now known as torquetenovirus, TTV is a member of the family *Anelloviridae* (Box 35.3). Sequencing studies have revealed that TTVs are ubiquitous among humans worldwide and are present in biological samples including whole blood, nasal secretions, saliva, bile, feces, tears, semen, breast-milk, and urine.

The family *Anelloviridae* currently contains 30 named genera. TTV is a member of the genus *Alphatorqevirus*.

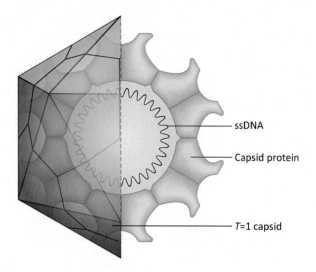

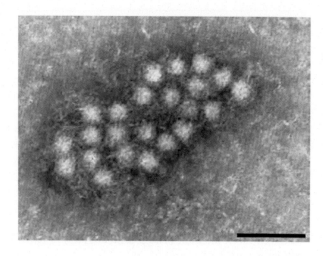

FIGURE 35.4   Anellovirus virion structure (left) and negative contrast electron microscopy (right) of particles of an isolate of Torque teno virus, stained with uranyl acetate. The bar represents 100 nm (Itoh et al., 2000). Source: *From Itoh, Y., Takahashi, M., Fukuda, M., Shibayama, T., Ishikawa, T., Tsuda, T., et al., 2000. Visualization of TT virus particles recovered from the sera and feces of infected humans. Biochem. Biophys. Res. Commun. 279, 718.*

BOX 35.3

General description and taxonomy, family *Anelloviridae*

Anelloviruses are small, unenveloped viruses with $T = 1$ icosahedral symmetry (~ 25 nm diameter). Capsids are assembled from 60 copies of the capsid protein VP1. Anellovirus genomes are a single strand of circular DNA ranging in size from ~ 2 to 4 kb. Few have been cultured. They are a major component of the human virome and are not associated with disease.

Family: *Anelloviridae*. Over 30 named genera. Chicken anemia virus (genus *Gyrovirus*) is a pathogen of poultry.

For the most part, TTVs have not been cultured so little is known about the details of their replication. Based on study of virions obtained from sera, anelloviruses are small, unenveloped viruses, ~ 30 nm diameter (Fig. 35.4). Anellovirus genomes are circular single-stranded, negative-sense DNA, range from 2 to 4 kb and encode three or four proteins (Fig. 35.5). A single untranslated region of the genome contains a promoter that directs mRNA synthesis in one direction. Two main ORFs, and additional smaller ORFs, may be deduced directly from the nucleotide sequence. These ORFs overlap partially, and their estimated sizes differ widely among isolates. Based on sequence homologies, the ORF1 product is deduced to have roles as a capsid protein and a replication protein. Introduction of cloned anellovirus genomes into cultured cells has demonstrated that at least 5—7 proteins ranging from about 12—80 kDa may be expressed from the genome.

To date, anelloviruses have not been associated with any disease in humans. Over 90% of the human

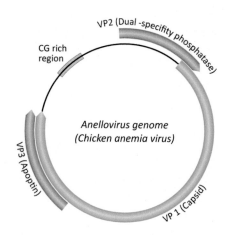

FIGURE 35.5   Anellovirus genome organization and protein expression strategy. Shown is the genome of chicken anemia virus, family *Anelloviridae*, genus *Gyrovirus*.

population is thought to be infected with these ubiquitous viruses. Studies of healthy newborns have revealed that a variety of anelloviruses appear in the gut through 12 months. Anellovirus DNA also appears in the blood of infants by 2 months of age. While we do not understand what roles, if any, anelloviruses play in human biology, it has been hypothesized that perhaps, like bacterial gut flora, they are important for development of a healthy immune system.

In this chapter, we learned that:

- Circoviruses and anelloviruses are unenveloped single-stranded DNA virus with circular genomes.
- Circoviruses replicate in mitotically active cells.
- Among the circoviruses, beak and feather disease virus (BFDV) and porcine circovirus 2 (PCV2) are significant pathogens.

- Anelloviruses are ubiquitous and most infections are not associated with disease.

References

Itoh, Y., Takahashi, M., Fukuda, M., Shibayama, T., Ishikawa, T., Tsuda, T., et al., 2000. Visualization of TT virus particles recovered from the sera and feces of infected humans. Biochem. Biophys. Res. Commun. 279, 718.

Khayat, R., Wen, K., Alimova, A., Gavrilov, B., Katz, A., Galarza, J. M., et al., 2019. Structural characterization of the PCV2d virus-like particle at 3.3 Å resolution reveals differences to PCV2a and PCV2b capsids, a tetranucleotide, and an N-terminus near the icosahedral 3-fold axes. Virol. 537, 186–197. Available from: https://doi.org/10.1016/j.virol.2019.09.001.

36

Family *Polyomaviridae*

After reading this chapter, you should be able to discuss the following topics:

- What characteristics of polyomavirus genomes made them a good model for studying processes of host-cell DNA replication and transcription?
- What characteristics of polyomaviruses made them a good model for studying cell cycle control and transformation?
- Discuss some of the functions of polyomavirus large T and small T antigens.
- What childhood disease is caused by a polyomavirus?

Polyomaviruses (PyVs) are unenveloped double-stranded (ds)DNA viruses with $T = 7$ icosahedral capsids with a diameter of 40–45 nm (Fig. 36.1). The name polyomavirus is derived from the Greek words poly (many) and oma (tumor) as the first PyV (murine PyV) was isolated from laboratory mice with multiple tumors. Simian virus 40 (SV40, species *Macaca mulatta polyomavirus 1*) was discovered as a contaminant of cell cultures used to manufacture human vaccines (from about 1955 to 1963). A few PyVs are rarely associated with disease in humans including JC polyomavirus (JCPyV), BK polyomavirus (BKPyV), and Merkel cell carcinoma polyomavirus (MCPyV). Most often however, PyVs have been studied as productive and insightful models for

understanding the processes of DNA replication, cell cycle control, and cell transformation.

Genome structure

PyVs have small (~5200 bp) circular dsDNA genomes. The PyVs genome is associated with host-cell derived histone proteins (H2A, H2B, H3, and H4) thus assumes a nucleosome structure similar to that found in cellular chromatin. PyV genomes are often called "minichromosomes" due to their shared structure with eukaryotic chromosomes (Box 36.1).

The genome organization of all PyVs is similar, although the exact number of encoded proteins can vary from 5 to 9 (Fig. 36.2). All PyV genomes contain a noncoding control region (NCCR) that serves as both the origin of replication and a bidirectional promoter controlling early and late transcription. Early transcription produces the mRNAs for the large T and small T proteins or large and small T antigen (LT-Ag, ST-Ag). The use of the term "antigen" to describe these proteins derives from "tumor antigen," originally identified as unique proteins found in tumor cells. The late transcript, produced after DNA replication commences, encodes the capsid proteins, VP1, VP2, and VP3. PyV genomes also encode micro (mi)RNAs. In some cases viral miRNAs downregulate T-Ag

Viruses.
DOI: https://doi.org/10.1016/B978-0-323-90385-1.00043-1

© 2023 Elsevier Inc. All rights reserved.

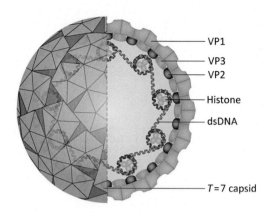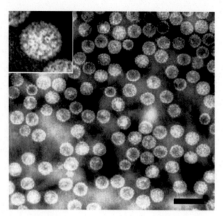

FIGURE 36.1   Virion structure (left). Transmission electron microscopic (TEM) image of polyomavirus virions (right). CDC Public Health Image Library Image ID#5629. Source: *Content providers CDC/ Dr. Erskine Palmer.*

BOX 36.1

General characteristics

PyV genomes are circular, double-stranded DNA molecules of ~5000 bp that encode 5–9 proteins. Genomes take the form of minichromosomes with the same nucleosome structure of cellular chromatin. There is a single noncoding control region (NCCR) that serves as the replication origin and a bidirectional promoter. PyVs encode T (tumor) antigens (proteins) that interact with host-cell proteins to promote DNA synthesis as PyVs rely on DNA cellular polymerases for their replication. Virions are unenveloped icosahedral structures ($T = 7$), 40–45-nm in diameter. VP1 is the major capsid protein. Capsids contain 360 molecules of VP1 arranged in 72 pentamers. The host range of PyVs includes mammals, birds, and fish.

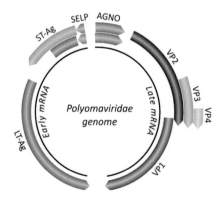

FIGURE 36.2   Genome structure of SV40, a betapolyomavirus.

expression. miRNAs may also interfere with expression of host-cell proteins.

Virion structure

PyVs have unenveloped virions with $T = 7$ icosahedral symmetry and a diameter of 40–45 nm. Capsids contain three structural proteins, VP1, VP2, and VP3. The bulk of the capsid is constructed from 360 molecules of VP1 (arranged as 72 pentamers). The C-terminal regions of VP1 are interlinked, providing capsid stability. VP1 contacts are also stabilized by calcium ions and interpentamer disulfide bonds. A single copy of VP2 or VP3 is found associated with the internal face of each pentamer. Polyomavirus virions are relatively resistant to heat or formalin inactivation.

Tumor antigens

Most PyVs encode two regulatory proteins, LT-Ag and ST-Ag that are expressed early after infection. LT-Ag has ATPase/helicase activity and binding domains for conserved motifs in the NCCR to control transcription and replication. LT-Ag also recruits cellular transcription factors required for late transcription. Additional domains in LT-Ag allow interaction with the cellular chaperone, heat shock 70 (HSc70) as well as with retinoblastoma protein family members pRb, p107, and p130. Interactions with pRb family members promote cell cycle progression. Other LT-Ag interactions block cyclin E1 degradation to further promote cell growth. Finally, some polyomavirus LT-Ags bind cellular p53, blocking apoptosis.

ST-Ag is produced by alternative splicing and shares about 80 amino acids with LT-Ag. It also has transforming properties, via binding to host proteins involved in growth control, for example, pRb family members. ST-Ags of some PyVs also have a unique domain that binds cellular phosphatase 2 A (PP2A). PP2A is a serine-threonine phosphatase that regulates cell cycle and apoptosis by desphophorylating targets such as p53. Some PyVs produce additional T-Ags that may be involved in transformation. The many roles of PyV T-Ags in promoting transformation and tumor development continue to be actively investigated.

Replication

PyV T-Ags stimulate cell to divide, so as to provide the appropriate environment for the polyomavirus genome replication. During a productive infection, virions are formed and infection is often cytopathic. However, under circumstances where cells are not fully competent to support productive replication, polyomavirus infection can transform cells (in this case virions are usually not produced). In the following sections, we will review the productive replication cycle of a polyomavirus. The virus—cell interactions that lead to transformation will be discussed later in this chapter.

Attachment and penetration

Receptors have been identified for several PyVs and they include proteins and carbohydrates. Receptor interactions are mediated by VP1. Following attachment virions are endocytosed, however, the specific endocytic pathway varies among PyVs. The endosome is only the first step in the voyage of the polyomavirus to its ultimate destination, the nucleus. Virions are delivered by endosomes to the endoplasmic reticulum (ER). In the case of SV40, uncoating begins in the ER and involves breakage and reforming of disulfide bonds that link capsid proteins. Following structural rearrangements of capsid proteins to expose hydrophobic surfaces on VP1, virions are transported across the ER membrane. Alternatively, it has been proposed that VP2, a myristoylated protein in some polyomaviruses, interacts with the ER lipid bilayer to support movement of virions out of this compartment. SV40 VP4 is a viroporin that may play a similar role.

Within the virion, PyV genomes are associated with four core histones (H2A, H2B, H3, and H4). These four core histones form a bead around which the DNA is wound. However, once the viral genome is delivered into the nucleus, histone H1 joins the process. H1 stabilizes the DNA wrapped around the core histones and binds to the linker DNA in regions between nucleosomes. Within the nucleus, the SV40 genome contains 24 nucleosomes while the NCCR remains free of nucleosomes.

Early transcription

Early gene expression begins when host transcription factors and RNA polymerase II interact with the early promoter on the infecting genome. For example, the SV40 early promoter contains a TATA-box as well as GC-rich sequences that are bound by transcription factor SP1. Thus the early promoter contributes to host range, as productive infection is restricted to cells expressing appropriate DNA-binding proteins. Early transcripts are alternatively spliced to produce LT-Ag and ST-Ag. LT-Ag is a multifunctional protein. Among the activities of LT-Ag that contribute to a productive replication cycle are: Binding to the viral origin of replication, unwinding viral dsDNA via helicase activity, inducing cell to enter S phase (by binding pRB family members), recruitment of cellular DNA replication proteins, countering cell's apoptotic response (by binding p53), and finally repressing early PyV transcription to allow transcription of late genes.

Genome replication

The process of PyV genome replication begins with binding of two molecules of LT-Ag to the origin of replication in the NCCR. Each molecule of LT-Ag recruits an additional five molecules of LT-Ag to form two hexamers that surround the origin of replication. This protein assemblage initiates melting and unwinding of the DNA. LT-Ag also recruits the cellular DNA synthesis machinery and genome replication proceeds in a bidirectional manner.

Late transcription

Late gene transcription begins with the onset of DNA replication. LT-Ag has a role in this process as well. LT-Ag stimulates late transcription via interactions with the basal transcription machinery. Late mRNAs encode the PyV capsid proteins VP1, VP2, and VP3. SV40 also encodes a VP4 protein (a viroporin) from the late transcript. VP1 is expressed from a spliced mRNA. The protein is ~350—370 aa and is the major capsid protein. VP1 functions in attachment and entry. VP1 associates with VP2 and VP3 (which are found inside the mature capsid). VP2 is myristoylated at its amino terminus. VP3 overlaps the carboxyl terminus of VP2. Another important product of late transcription is a microRNA (miRNA) whose sequence is complementary to the PyV early transcript. Thus it functions to downregulate early gene transcription.

Assembly and release

Virion assembly begins with the translocation of VP1, VP2, and VP3 into the nucleus and the formation of VP1 capsomeres. Each capsomere contains five molecules of VP1. VP1 alone is sufficient to drive capsid assembly but VP2 and VP3 are needed for production of infectious virions. In some cases PyV infections are cytopathic and virion release can accompany cell death. However, infections can be persistent and there is evidence for regulated virion release (perhaps by exocytosis).

Transformation

Murine PyV (MPyV) was the first polyomavirus identified, found to be the cause of tumors in laboratory mice. It was subsequently found that SV40 (a primate PyV) could induce tumors if injected into newborn hamsters. Thus began the study of PyV for the purpose of understanding their robust transforming potential.

Key to understanding the transforming activities of PyV is to recall that these small DNA viruses induce cells to enter S phase to facilitate viral genome replication. In a productive infection the balance favors virion production and cell death rather than transformation.

What is the process by which some cells are transformed? Transformation is associated foremost with restricted PyV replication. The general model for transformation is that T-Ags are expressed but that viral DNA replication is blocked (perhaps because of species-specific differences among DNA replication factors). Thus few or no virions are formed. Rarely the PyV genome becomes integrated into the host-cell chromosome in a manner that supports *continued expression of T-Ags*. Now the cell is maintained in a mitotically active state. The activities of LT-Ag that support transformation are its ability to bind pRb family members and to inactivate the tumor suppressor p53. The vast majority of natural human PyV infections do not result in transformation. However, as will

be discussed below, a human PyV (Merkel cell PyV) has recently been implicated in tumor formation.

Human polyomavirus and disease

Human PyV appears to be ubiquitous and infections are generally benign. In most cases the exact routes of transmission are unknown. There are several human PyVs associated with skin and direct contact is the suspected route of transmission. Other human PyVs are found in respiratory secretions (suggesting respiratory transmission) and some have been detected in urine and feces. There is strong serological evidence to support persistent infections in healthy individuals, without any evidence of disease. Serosurveys indicate that most of the world's population is infected by early to mid-childhood.

The first identified human PyV were JC polyomavirus (JCPyV) and BK polyomavirus (BKPyV) recovered from respiratory and lymphoid tissues, respectively, of diseased patients. Other human PyV includes Merkel cell carcinoma PyV (MCPyV) and human PyV 6, 7, and 8 all isolated from normal skin (Box 36.2).

PyV involvement in human disease is almost always associated with immunocompromised patients. JCPyV can cause lytic infections in the brain (progressive multifocal leukoencephalopathy) while BKPyV can cause lytic infections of the kidney, bladder, and ureter in renal transplant patients. BKPyV is often associated with kidney failure after organ transplantation. Quite simply, immunocompromised patients are unable to control replication of these viruses and lytic infection leads to tissue damage.

MCPyV is so named because of its association with a rare skin cancer, Merkel cell carcinoma (MCC). The risk of developing MCC is increased in immunocompromised patients. About 60%−80% of MCCs contain integrated MCPyV. Integration is clonal, meaning that all tumor cells have arisen from a single progenitor in which the integration event occurred. The integrated DNA encodes mutated versions of LT-Ag.

In this chapter, we have learned that:

- PyVs are small double-stranded DNA viruses whose genomes associate with histones. Thus polyomavirus genomes are often called "minichromosomes" and historically they provided a robust model for probing eukaryotic DNA replication and transcription.
- PyVs encode T-Ags that interact with cellular proteins to alter cell cycle control. PyVs drive cells into S phase in order to facilitate virus genome replication.

BOX 36.2

Taxonomy

Family *Polyomaviridae*

Genus *Alphapolyomavirus* [more than 50 named species; viruses include Merkel cell polyomavirus (MCPyV)]

Genus *Betapolyomavirus* [more than 30 named species; viruses include simian Virus 40 (SV40), BK polyomavirus (BKPyV) and JC polyomavirus (JCPyC)]

Genus *Deltapolyomavirus* (seven named species including *Human polyomavirus 6*)

Genus *Epsilonpolyomavirus* (three named species)

Genus *Etapolyomavirus* (one named species)

Genus *Gammapolyomavirus* (nine named species)

Genus *Thetapolyomavirus* (four named species)

Genus *Zetapolyomavirus* (one named species)

- In a productive PyV infection, cells are often killed. In nonpermissive or poorly permissive cells, transformation is a rare outcome. Transformed cells usually have integrated copies of genes encoding PyV T-Ags.

- T-Ags are multifunctional. Some activities include: Orchestrating DNA replication, controlling early versus late transcription, binding and inactivating pRb family members, binding and inactivating p53.

37

Family *Papillomaviridae*

After reading this chapter, you should be able to answer the following questions:

- What are the major characteristics of viruses in the family *Papillomaviridae*?
- Discuss host and tissue tropism of papillomaviruses (PVs). Why are we unable to grow these viruses in continuous cell cultures?
- What are the major functions of PV E6 and E7 proteins?
- Discuss the relationship between human PVs (HPVs) and cervical cancer. Why do some HPVs pose a higher risk of cancer development?

PVs are unenveloped icosahedral viruses with circular double-stranded (ds) DNA genomes. PVs have a strict tropism for epithelial cells and induce benign hyperplasia of skin to form common warts. PVs have been isolated from mammals, birds, and reptiles. In general, PVs are quite host specific and cross-species transmission is very rare. A notable feature of PVs is their requirement for differentiating epithelial (skin) cells in order to complete their replication cycle (Box 37.1). Thus PVs cannot be grown in transformed cells in culture, and systems for replicating PVs are limited. The study of HPVs accelerated in the 1970s when advances in molecular biology provided methods for characterizing and mapping PV genomes and transcripts from lesions. Over 100 types of HPVs have been identified. Warts rarely progress to malignant cancers although cervical and other anogenital cancers are highly associated with a small subset of HPVs.

Genome structure

PV genomes are ds circular ($\sim$8000 bp) DNA molecules. Viral DNA associates with cellular histones to form a chromatin-like complex (Fig. 37.1). PV genomes have a single untranscribed regulatory region (URR) (also called the long control region) of about 1000 bp that contains the DNA replication origin (ori) and promoters. Transcription proceeds in one direction (one strand of DNA serves as the coding strand, Fig. 37.2). mRNAs are transcribed by host cell RNA polymerase II and genome replication is catalyzed by host DNA polymerases. PV early proteins interact with the URR to direct and regulate these processes. The nucleus is the site of viral genome replication and virion assembly.

Virion structure

Virions are unenveloped and have $T = 7$ icosahedral symmetry (Fig. 37.1). Virions are $\sim$60 nm in diameter. Two structural proteins, L1 and L2, assemble to form capsids. Capsomers contain five molecules of L1 and one molecule of L2, arranged with

Viruses.
DOI: https://doi.org/10.1016/B978-0-323-90385-1.00031-5

© 2023 Elsevier Inc. All rights reserved.

BOX 37.1

General characteristics

Genomes are ds, circular DNA, ~8 kbp. Transcription and genome replication are nuclear. mRNAs are transcribed by RNA polymerase II and are extensively spliced. PV genome replication is directed by early proteins (E1 and E2) and catalyzed by cellular enzymes.

Early genes are expressed prior to genome replication (and expression continues throughout the replication cycle). Late genes are expressed after DNA replication.

Late gene expression requires epithelial cell differentiation. Products of the late genes (L1 and L2) are capsid proteins.

Virions are unenveloped icosahedral ($T = 7$) capsids containing two capsid proteins. L1 is the major capsid protein and when expressed alone, assembles to form VLPs. The production of infectious virions is tightly linked to epithelial cell differentiation, thus PVs do not replicate in transformed cells in culture.

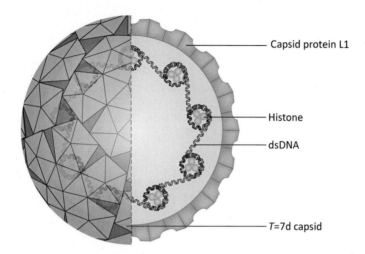

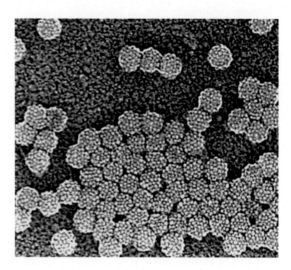

FIGURE 37.1   Virus structure (left). Electron micrograph of a human papillomavirus capsids (Belnap et al., 1996). Source: *From Belnap, D. M, Olson, N.H., Cladel, N.M., Newcomb, W.W., Brown, J.C., Kreider, J.W., et al., 1996. Conserved features in papillomavirus and polyomavirus capsids. J. Mol. Biol. 259, 249.*

L2 facing the lumen and associated with viral DNA. When expressed alone, L1 can assemble to form virus-like particles (VLPs). L2 is required for the formation of infectious virions.

Replication cycle

Key to the PV replication cycle is that productive infection is dependent on the differentiation stage of the epithelial cell. PV begins the replication cycle by infecting mitotically active basal cells where PV DNA is initially replicated. When the cell divides, daughter cells inherit copies of the PV genome. The process continues to generate a population of infected cells. However, capsid protein synthesis and production of virions must wait for cell differentiation (cells must be mitotically active to be infected, but then must

undergo differentiation). Thus the PV life cycle is intimately linked to skin development (Fig. 37.3).

Attachment/penetration

PVs infect the dividing basal cells of the skin. PVs access these cells through physical breaks in the skin. Initial attachment of HPVs is thought to involve heparan sulfate proteoglycans exposed on the surface of the basement membrane. It is likely that additional interactions are required before virions can access the basal cells themselves. Binding is followed by endocytosis. Details of endocytosis and escape from endosomes are very sketchy due to limits of productive replication systems. It appears that L2-genome complexes escape endosomes and traffic to the nucleus. A notable feature of the PV replication cycle is the length of time between surface binding and

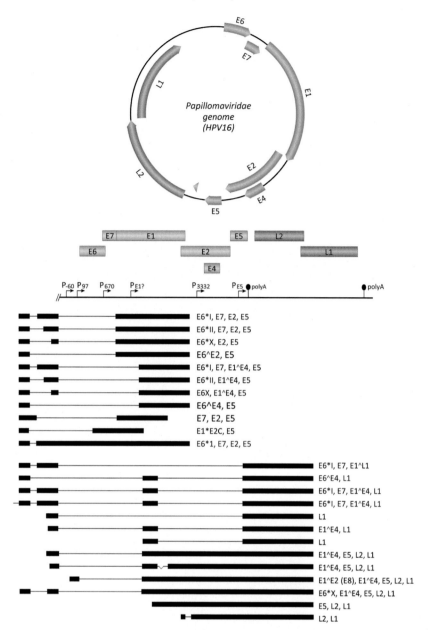

FIGURE 37.2   Genome organization and gene expression strategy of human papillomavirus 16 showing open reading frames and alternatively spliced transcripts. Note that alternative splicing generates a large number of mRNAs.

transcription, estimated to be 1–3 days (compared to minutes to hours for many other viruses).

Early transcription and genome replication

Early transcription is initiated prior to the onset of viral DNA replication. Multiple promoters and a program of alternative splicing are used to produce a family of early (E) proteins necessary for initial rounds of PV genome replication. The first viral protein products produced are E1 and E2. PV E1 is a phosphoprotein that binds the DNA replication ori.

E1 has ATPase and helicase activities. E1 unwinds the DNA at the ori and interacts with the p180 subunit of cellular primase to recruit the cellular DNA replication machinery to the ori. Binding of E1 to the ori is enhanced by the E2 protein. E2 is actually a small set of proteins that regulate viral transcription and DNA replication. E2 has sequence-specific DNA-binding and dimerization domains as well as a transactivating domain that enhances transcription. All versions of E2 bind the PV ori, but some E2 proteins lack transactivating domains. E2 proteins lacking a transactivating domain *inhibit*, rather than activate, transcription (Fig. 37.4).

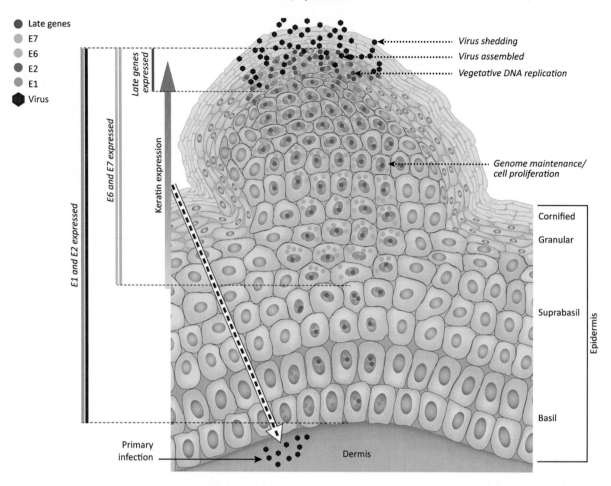

FIGURE 37.3 Papillomavirus (PV) replication is dependent upon epithelial cell differentiation. Virus infects mitotically active basal cells in the skin and genome replication begins (replicative phase) to produce 50–100 genome copies. As cells divide, genomes in each daughter cell replicate to reestablish a level of 50–100 genome copies per cell. Infected cell proliferate, but also continue on a pathway to differentiation. Expression of cellular keratin genes triggers additional PV genome replication (so-called vegetative state) and expression of late transcripts.

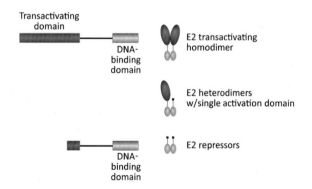

FIGURE 37.4 Papillomavirus (PV) E2 is both an activator and inhibitor of transcription. Full-length E2 (top) has an amino terminal transactivating domain followed by a linker region and a carboxyl terminal DNA-binding domain. Full-length homodimers (with transactivating domain) are transcriptional activators. Dimers of the DNA-binding domain are repressors of transcription (bottom).

E1 and E2 are produced at the earliest stages of infection. Other early proteins (E5, E6, and E7) are produced a bit later, as the basal cells begin to differentiate (Fig. 37.3). E5, E6, and E7 cooperate to stimulate replication and proliferation (hyperplasia) of the infected tissue. Later, when infected cells begin to differentiate, these early proteins continue to support viral genome synthesis. PV E6 inactivates p53 to block apoptosis while E7 interacts with pRb family proteins to prevent sequestration of the cellular transcription factor E2F. E6 and E7 proteins are conserved among all PVs, E5 is not always present.

The initial phase of PV DNA replication occurs within dividing basal skin cells. This is often referred to as the replicative phase, during which infecting PV genome is amplified to about 50–100 copies. These viral "plasmids" associate with chromatin and segregate into daughter cells during cell division. In the so-called "maintenance phase," each copy of the PV genome is replicated approximately once per cell division to maintain a stable number of PV genomes.

Late transcription and vegetative DNA replication

As skin cells move closer to the surface, they undergo major changes in gene expression. Key among these is the expression of various keratin genes. Productive PV replication is tied to this process, which triggers an increase in viral DNA replication (the so-called vegetative stage of DNA replication) and the expression capsid proteins (L1 and L2) needed to package that DNA. Virions are shed from the uppermost layers of the skin, along with dead skin cells.

In the vast majority of PV infections, the process of tumor formation does not proceed past benign hyperplasia. Normally the immune system brings the infection under control and warts are resolved. Conversely, certain types of immune impairment favor growth of large papillomas. However, it is also clear that PVs can be present in small amounts in the absence of a visible wart. Virions are quite stable and can be transmitted by procedures such as ear-tagging or tattooing livestock.

PV also infects mucosal surfaces such as the mouth, gastrointestinal, and urogenital tracts. The types of PVs that infect mucosal surfaces are distinct from those that infect other body parts. Lesions that form on mucosal surfaces are called condylomas.

Disease

Human papillomaviruses

There are at least 100 known types of HPV (Box 37.2). They are spread by direct skin contact, including sexual contact, or by surface contamination. They cause warts, epidermal, and epithelial lesions at different body sites. Most lesions are benign and rarely progress to a malignant state. In healthy individuals virus replication is controlled by T-cells, thus warts are self-limiting and will eventually resolve. However, a few HPVs are highly associated with carcinomas of cervix, anus, larynx, and lung (Box 37.3). These include HPVs 16, 18, 31, 33, 35, 39, 45, 51, 52, 56, 58, 59, and 68. Types 16 and 18 account for about 70% of cervical cancers and are targets of cervical cancer vaccines. While HPV is the most common sexually transmitted disease, most infections do not lead to cancer. Cervical and anogenital cancers may take years or decades to develop. Cancerous lesions do not produce virions but it is common to find a fragment of the HPV genome, containing genes for E6 and E7, integrated into

---

BOX 37.2

Taxonomy

Classification of PVs is based on nucleotide sequences of the L1 open reading frame. As PVs were difficult to culture, they could not be characterized by standard serologic assays, thus nucleotide sequencing was the only way to differentiate among them.

Family *Papillomaviridae*

Subfamily *Firstpapillomavirinae*. Viral genera belonging to this subfamily are named according to the Greek alphabet; there are over 50 named genera. Species names are based on the genus name and individual species are numbered. For example, within the genus *Alphapapillomavirus* (includes HPVs associated with genital and mucosal cancers) there are 14 species, numbered 1–14. There is no correspondence between the new species number designation and the historical "type" number associated. Thus HPV16, the type found most frequently in cervical cancer, is a member of *Alphapapillomavirus* species 9.

Genus *Alphapapillomavirus*. Members of this genus preferentially infect the oral or anogenital mucosa in humans and primates.

Genus *Betapapillomavirus*. Members of this genus typically cause latent infections but occasionally cause cutaneous squamous cell carcinomas (SCC) in patients with specific immune disorders. Patients suffering from the rare disease epidermodysplasia verruciformis (EV) are at a higher risk of developing SCC, and certain members of the genus *Betapapillomavirus* have been implicated.

Genus *Deltapapillomavirus*. These PVs are associated with fibropapillomas in ungulate hosts.

Genus *Gammapapillomavirus*. Typically found on the skin and oral mucosa of primates.

Subfamily *Secondpapillomavirinae*. Currently contains a single genus, *alefpapillomavirus* isolated from a fish.

BOX 37.3

Human papillomavirus (HPV) (cancer) vaccines

The history of cervical cancer as a sexually transmitted disease began in 1842 with publication of mortality statistics of women dying of cancer in the city of Verona, Italy. Italian physician, Domenico Rigoni-Stern, noted that "cancer of the uterus" was more common among married women and widows than among virgins and nuns. This was the first suggestion of a sexually transmitted cancer agent. The first links between cervical cancers and HPVs came in the middle of the 20th century with the recognition that genital warts were a sexually transmitted disease. Cytologists studying cervical cancers and precancers noted the similarities between cells associated with malignancies and those associated with cervical warts. By the early 1980s, multiple studies showed a direct link between HPV infection and cervical cancer. Over the next decade, it became increasingly clear that virtually *all* cervical cancers, and many other anogenital cancers, were the result of prior HPV infection. As molecular techniques facilitated HPV typing, it also became clear that a few "high-risk" HPV types were associated with cervical cancer.

The identification of specific viruses as the cause of cervical cancers set the stage for vaccine development. A huge hurdle was the inability to grow large amounts of virus. However, the same technologies (DNA cloning and sequencing) that facilitated studies of PVs also provided the tools for vaccine development. By the 1990s, methods were being developed to produce recombinant HPV capsid proteins and coax them to assemble into VLPs that were antigenically similar to virions but contained no HPV genetic material. In 2006 the US Food and Drug Administration licensed a Merck & Co (Kenilworth, NJ) vaccine marketed specifically for the prevention of cervical cancer. The first HPV vaccines were developed specifically for women but subsequent versions were developed to also protect against HPVs that are the major cause of genital warts in men. There are currently three licensed HPV vaccines. In the United States, Gardasil 9, a nonavalent vaccine, has been licensed for use in young women and men with recommendation for vaccination before becoming sexually active.

At over 10 years since first licensure, the efficacy of HPV vaccination in real-world settings has become obvious. Reductions in HPV 6, 11, 16, and 18 infections are up to approximately 90% in some groups of vaccinees. Reductions of up to 85% in high-grade histologically proven cervical abnormities have also been reported. Based on these types of results, the World Health Organization is pushing for increased vaccination in low-income countries. Barriers to increased vaccination rates include the high cost of these recombinant vaccines and supply limitations.

---

the chromosome. Recall that all PVs have E6 and E7 proteins that support DNA synthesis during productive HPV infection. Why are only a few types of HPV associated with development of malignant cells?

The high-risk HPVs encode E6 and E7 proteins that are more robust and varied in their ability to inactivate p53 and pRB family member proteins to promote cell growth and division. They do this by a variety of regulatory mechanisms. The E6 and E7 genes of the high-risk HPVs are also expressed differently than those of the low-risk HPVs. A single promoter drives expression of high-risk HPV E6 and E7, while the low-risk HPVs use two independent promoters. The HVP16 E6 protein has also been specifically shown to activate the enzyme telomerase in keratinocytes (recall that maintenance of telomers is important in the process of cellular immortalization and transformation).

At the cellular level, the oral mucosa is structurally similar to the vagina and cervix and most oral cancers are squamous cell carcinomas, the same type found in cervical cancers. Those facts prompted surveys of head and neck cancers for the presence of HPV genes. In one study, 25% of 253 patients diagnosed with head and neck cancers, the tissue taken from tumors was HPV positive (predominantly HPV16). Smoking and alcohol consumption can promote oral cancers and these activities may enhance to rate of transformation by high-risk HPVs.

Epidermodysplasia verruciformis

Epidermodysplasia verruciformis (EV) or "tree man illness" is an *extremely* rare genetic disorder that causes an abnormal susceptibility to HPVs. The inability to resolve HPV infections results in development of numerous scaly lesions, particularly on the hands and feet. The HPVs associated with these lesions normally cause asymptomatic infections in healthy individuals. By about 20 years of age, the lesions progress to disfiguring and incapacitating nonmelanoma skin cancers. EV is a primary immunodeficiency and the affected genes have been identified. These genes seem to regulate zinc distribution but how this affects HPV infection is not understood.

Animal papillomaviruses

Shope papillomavirus (now called cottontail rabbit papillomavirus) was discovered by Richard Shope in the 1930s. He purified virus particles from papillomas removed from wild-caught rabbits and successfully transmitted warts and tumors to laboratory rabbits. This was the first virus demonstrated to cause cancer. Bovine PVs (BoPVs) are extremely common, and they were used as early models for PV replication due to the relatively large amounts of virus present in warts. BoPVs can be spread by modern ranching techniques such as use of ear tags. The virus is difficult to inactivate and equipment must be carefully cleaned to avoid transmission of BoPV. Warts are also relatively common among young dogs but these usually resolve without treatment. Many animals have PV-associated warts; however, none are transmissible to humans.

In this chapter, we have learned that:

- PVs are unenveloped viruses with ds circular DNA genomes. Their replication cycle is tightly linked to the differentiation program of epithelial cells. Most papillomavirus infections result in self-limiting, benign warts that are formed by epithelial cell hyperplasia.
- PVs have narrow host and cell tropism.
- Papillomavirus E6 and E7 are growth-promoting/transforming proteins. E6 interferes with normal functions of p53 and E7 interferes with pRB. These interactions inhibit apoptosis and support DNA replication. The so-called "high-risk" HPVs encode E6 and E7 proteins that have stronger anti-p53 and anti-pRB activities, respectively.
- A few HPVs are associated with cancers and some of these can be prevented by vaccination.

Reference

Belnap, D.M., Olson, N.H., Cladel, N.M., Newcomb, W.W., Brown, J. C., Kreider, J.W., et al., 1996. Conserved features in papillomavirus and polyomavirus capsids. J. Mol. Biol. 259, 249.

38

Family *Adenoviridae*

After reading this chapter, you should be able to answer the following questions:

- What are the general characteristics of adenoviruses (AdVs)?
- Describe AdV genome structure and mechanism of replication.
- How do AdVs stimulate host cell DNA replication?
- Explain why human AdVs can cause tumors in rodent models, but do not cause tumors in humans.
- What are the diseases most often associated with human AdV infection?
- Explain how oncolytic AdVs are modified to kill cancer cells while not replicating in normal cells.
- Describe how AdV is used as vaccine vector.

AdVs are naked icosahedral viruses with prominent fiber proteins extending from their capsids (Fig. 38.1). AdV genomes are linear double-stranded (ds) DNA molecules ranging in size from 24 to 40 kbp. Human AdVs were first isolated from cell cultures of adenoid tissue (hence the name) and most often cause mild upper respiratory tract infections. AdVs produce large amounts of progeny and they efficiently infect cells regardless of cell cycle status (Box 38.1). Studies to physically map AdV genomes and transcripts using electron microscopy provided the first evidence of the process of RNA splicing. AdVs were valuable models of cell transformation, as human AdVs transform rodent cells (Box 38.2). AdVs also grow to high titer, making them good models for a variety of biochemical and molecular studies. Currently, AdVs are being exploited as efficient gene delivery vehicles and so-called oncolytic AdVs are being developed and tested as tumor-cell killing agents.

Genome organization

AdV genomes are a molecule of linear ds DNA. The 5' end of each strand is covalently linked to a protein called the terminal protein (TP) as a result of protein-primed DNA replication. AdV genomes range from 24 to 40 kbp. The most conserved region of AdV genomes is the central portion, which encodes the structural proteins and viral enzymes [i.e., AdV DNA polymerase (Pol)]. Regions outside the central core code for regulatory proteins that are expressed early in the replication cycle (hence are called "early" proteins) and they are variable in both sequence and size. The genomes of all AdVs have inverted terminal

Viruses.
DOI: https://doi.org/10.1016/B978-0-323-90385-1.00010-8

335

© 2023 Elsevier Inc. All rights reserved.

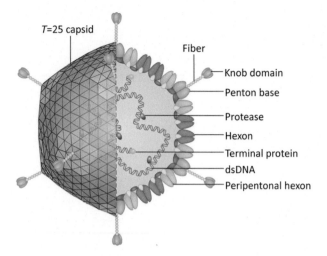

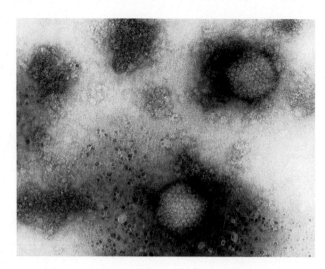

FIGURE 38.1   Virus structure (left). Transmission electron micrograph revealing some of the ultrastructural morphology exhibited by AdV-2 virions. Visible at this high magnification are the capsomeres, which in this case are hexagonally shaped, also called hexons (right). Source: *From CDC Public Health Image Library #14768. Content providers CDC/Dr. John Hierholze.*

BOX 38.1

General characteristics

AdVs are unenveloped viruses with linear ds DNA genomes. Genomes range from 24 to 40 kbp and the 5′ end of each genome strand is covalently linked to a 55 kDa viral protein (terminal protein, TP) that served as the primer for DNA replication. The best conserved region of AdV genomes is the central portion, which encodes the structural proteins and viral enzymes. Ends of the genomes encode regulatory proteins and exhibit greater variation in both sequence and size. Transcription and genome replication and assembly are nuclear. Transcripts are capped and polyadenylated and undergo extensive splicing. Most AdV infections are cytopathic.

Virions are naked with icosahedral capsids, 70—90 nm in diameter. Long fiber proteins extend from each of the 12 fivefold axes of symmetry.

BOX 38.2

Taxonomy

- Family *Adenoviridae*
- Genus *Atadenovirus* (20 species, hosts include mammals, birds, and reptiles)
- Genus *Aviadenovirus* (15 species, all infect birds)
- Genus *Ichtadenovirus* (one species, from a sturgeon)
- Genus *Mastadenovirus* (many species including human mastadenovirus A-G)
- Genus *Siadenovirus* (eight species, hosts include birds and amphibians)
- Genus *Testadenovirus* (one species, from a turtle)

---

repeat sequences ranging in size from 36 to over 200 bp. Base pairing of these regions is critical for DNA replication. AdV transcription and genome replication are nuclear. Transcripts are capped and polyadenylated and undergo extensive splicing (Fig. 38.2).

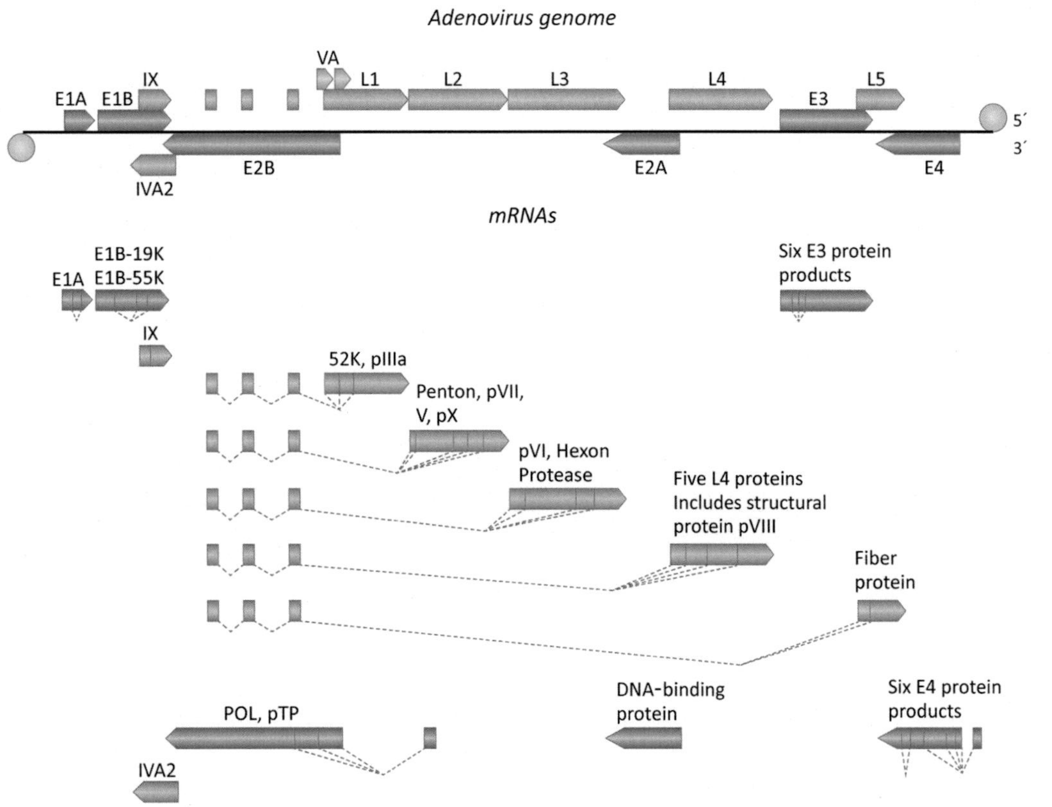

FIGURE 38.2 AdV genome organization and gene expression pattern. Note that transcripts for the major structural proteins all share a spliced, tripartite leader sequence. Alternative splicing and use of different termination signals generate the full set of mRNAs.

Virion structure

AdVs are naked icosahedral viruses (∼95 nm diameter) with long, distinct fiber proteins protruding from the 12 pentamers of the $T = 25$ icosahedral capsid. Fibers are trimers of the fiber protein. The fiber shaft is formed from a repeated structural motif and terminates with a distinct knob region at the tip (Fig. 38.1). Capsids contain at least nine polypeptides and are constructed from distinct pentamers and hexamers. Pentamers are assembled from two different proteins, the penton base protein and the fiber protein. The major capsid protein is the hexon protein. The hexon protein has two structural domains and 720 copies of hexon protein assemble into 240 trimers that further form 120 hexamers.

Adenovirus replication cycle

Attachment/penetration

AdV entry is a two-step process. The knob region of the long fiber protein makes the initial interaction to tether virions to the cell surface. This is followed by interactions of the penton base protein with cell-surface integrins, resulting in endocytosis of the virion. The process of AdV uncoating has been studied extensively. The process begins very early, with detachment of fibers, such that a fiberless virion is endocytosed. Acidification of the endosome continues the step-wise uncoating process. The final steps of uncoating occur at the nuclear pore, at which time AdV DNA is imported into the nucleus.

Early transcription

After entering the nucleus, a set of early (E) genes are transcribed by host RNA Pol II. AdV transcripts are capped, polyadenylated, and extensively spliced. The activities of the early proteins set the stage for the later processes in AdV replication. The major functions of the early protein products are described below. The activities summarized here are based on studies of human AdV-2 and human AdV-5.

E1A

A key function of E1A is to bind and inactivate cellular retinoblastoma protein (pRB) family members. A critical function of pRB is to inhibit cell cycle progression until a cell is ready to divide. Normally, phosphorylated pRB binds to the cellular transcription factor E2F and inhibits its activity. When bound by EIA, pRB cannot bind/sequester E2F, thus E2F is free to activate genes that support DNA synthesis. E1A also interacts with transcriptional modulators (i.e., histone acetyltransferases, histone deacetylases, and other chromatin remodeling factors). These interactions affect transcription of cellular genes to (among other things) inhibit the interferon response and downregulate genes for major histocompatibility complex class 1 (MHC-1) proteins. E1A protein also activates transcription of the other early viral transcription units.

E1B

The E1B transcription unit produces two mRNAs that encode proteins called E1B-19K and E1B-55K. The major role of these proteins is to block induction of apoptosis by the tumor suppressor p53. This is essential to allow time for virus to complete its replication cycle. E1B-19K and E1B-55K have discrete functions. E1B-19K inactivates two proapoptotic proteins (BAK and BAX) while E1B-55K directly inhibits p53 function by binding to its activation domain.

E2

The E2 transcription unit encodes three proteins required for AdV DNA synthesis: the precursor to the terminal protein (pTP), which primes DNA synthesis, AdV DNA Pol, and AdV ss DNA-binding protein. The E2 transcription unit is highly expressed, in part because its promoter is activated by the cellular transcription factor E2F. Recall that interaction of AdV E1A with pRB releases E2F, allowing it to activate promoters for cellular genes involved in DNA synthesis.

E3

Products of the E3 transcription unit function to antagonize host antiviral defenses. Among the characterized proteins are: E3-gp19K protein is a glycoprotein that resides in the endoplasmic reticulum where it binds to a subunit of the MHC-1. This interaction prevents trafficking of MHC-1 to the cell surface. E3-gp19K also binds to transporter associated with processing protein (TAP) and this interaction inhibits loading of peptides into MHC-1. Both of these activities prevent killing of AdV-infected cells by cytotoxic T-lymphocytes. Recall that receptors on cytotoxic T cell must bind antigen-presenting MHC-1 to mediate killing of target cells. E3−14.7K protein inhibits tumor

necrosis factor (TNF)-induced cytolysis of AdV-infected cells in culture by modulating the activity of NF-kB. E3−10.4K and E3−14.5K form a complex that internalizes receptors from the cell surface including the epidermal growth factor receptor, FAS, TRAIL receptor 1, and TNF receptor 1. Based on these activities, the proteins have been renamed receptor internalization and degradation (RID)α and RIDβ.

Viral DNA replication

AdV DNA replication is notable because the ends of the linear ds DNA genome are covalently linked to a molecule of TP that served as the primer for viral DNA synthesis. Synthesis of AdV DNA can be reconstituted in a test tube and the process requires three virally encoded proteins: pTP, AdV DNA Pol, and AdV DNA-binding protein. The addition of cellular proteins stimulates the process. Initiation of DNA synthesis is asymmetric, beginning at one end of the viral genome. As one parental strand is copied, it base pairs with the newly synthesized strand. The other parental strand is displaced and circularizes as the result of base pairing of its terminal inverted repeat regions to form a "panhandle" that serves now as a replication origin (Fig. 38.3).

Late transcription

Late transcription initiates after DNA replication. The so-called major late promoter drives expression of the late (L) genes, most of which are capsid proteins. There are five distinct families of L transcripts (L1−L5) all of which contain the same spliced, tripartite nontranscribed leader sequence at their 5′ end. L1 transcripts are produced by polyadenylation at the L1 polyadenylation site. Longer transcripts (L2−L5) are generated through use of downstream polyadenylation signals. The E4 transcription unit is also expressed after DNA replication. Functions of E4 proteins include regulation of transcription, RNA splicing, and translation. E4 products also modulate DNA replication and apoptosis.

VA RNAs

In addition to mRNAs, AdV produces short (~150−200 nt) noncoding RNAs called VA RNAs. These abundant RNAs are transcribed by cellular RNA Pol III from promoters similar to those found in tRNA genes. VA RNAs fold into very stable structures and accumulate to high levels in cells. The predominant type is VA RNAI that accumulates up to 10^8 molecules per infected cell. An important function of VA RNAI is

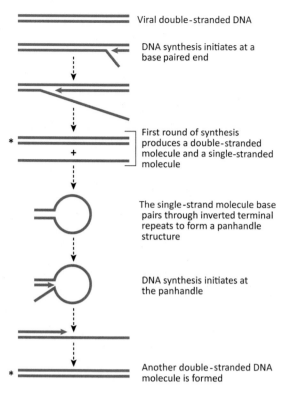

Viral double-stranded DNA

DNA synthesis initiates at a base paired end

First round of synthesis produces a double-stranded molecule and a single-stranded molecule

The single-strand molecule base pairs through inverted terminal repeats to form a panhandle structure

DNA synthesis initiates at the panhandle

Another double-stranded DNA molecule is formed

* Complete double-stranded DNA product

FIGURE 38.3 Replication of the linear ds DNA genome takes place by a strand displacement mechanism. A protein primer (not shown in this figure) primes DNA synthesis. Initiation of DNA synthesis begins at a base-paired end of the ds DNA molecule. A single strand is displaced during synthesis. Sequences at the end of the displaced strand are complementary and base pair to form an initiation site for DNA synthesis.

binding and inhibition of the ds RNA-activated protein kinase (PKR), a potent antiviral protein. Thus in the presence of VA RNAI, PKR is inhibited from doing its job as an interferon-stimulated inhibitor of viral protein synthesis. VA RNAs also interfere with other host cell processes, for example, they compete with premicro (mi)RNAs for export from the nucleus and processing by Dicer (see Box 38.3).

Assembly and release

The assembly of virions takes place the nucleus. Genome packaging requires a cis-acting domain at the left end of the AdV genome. This region contains a series of repeated sequences, termed A repeats, due to their AT-rich character. A protein from the E3 transcription unit (E3 11.6 kDa or the AdV death protein) kills cells as it accumulates during the late stage of infection and promotes cell lysis.

Adenoviruses and human disease

Many AdV infections are probably asymptomatic and persistent, and virus shedding can last for months. AdVs are spread by the fecal—oral route. When associated with disease, human AdVs usually cause only mild illnesses. They are most often associated with acute respiratory, gastrointestinal, urinary, and eye infections. It is thought that about 10% of human colds are caused by AdVs. However, disease can sometimes be severe with respiratory symptoms including pneumonia, croup, and bronchitis. Infants and people with weakened immune systems are at higher risk for severe complications. AdVs are a common cause of acute respiratory illness in military recruits, probably due to a crowded and stressful environment. The problem is significant enough that the US military has sponsored development of AdV vaccines.

Adenoviruses and transformation

Productive infections by AdVs are cytopathic; however, infection of nonpermissive cells can result in transformation. The transformation model is infection of rodent cells with human AdVs. Transformed cells always contain the E1A and E1B genes, but may not have any other viral proteins. If introduced into cells together, E1A and E1B are sufficient for transformation. The activities of E1A and E1B that result in transformation were described above. The possibility that human AdV could cause human cancers has been extensively investigated and there is no credible evidence that this occurs. When human AdV infect human cells, the complete replication cycle, which ends with cell death, is not a transforming event.

Oncolytic adenoviruses

Based on their unique properties, AdVs have been modified to enable them to kill cancer (tumor) cells while sparing nondividing cells (Fig. 38.4). Recall that the AdV E1A and E1B proteins are produced by wild-type AdVs. These proteins inactivate host cell pRb and p53 thereby stimulating host DNA replication. Without E1A and E1B, AdV cannot replicate in normal cells! But the vast majority of tumor cells lack antioncogenes (cell cycle inhibiting proteins) such as Rb and p53. Thus in constitutively dividing tumor cells AdVs lacking E1A and E1B do cause cytopathic infection. In practice, oncolytic AdVs are proving to be relatively safe and somewhat effective when injected into solid tumors. But recall that the immune system will

(A)

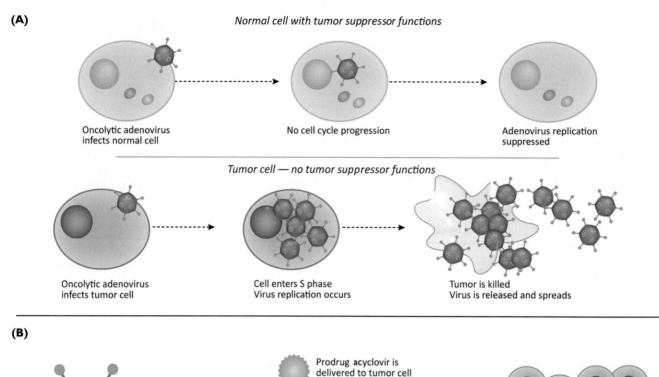

(B)

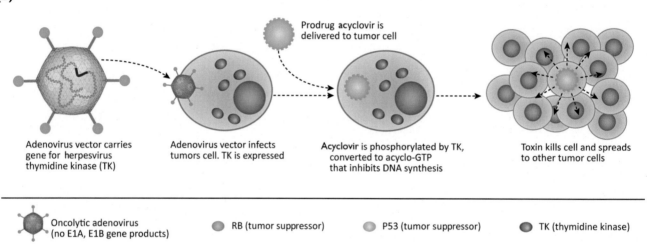

FIGURE 38.4   (A) Oncolytic AdVs lack E1A and E1B proteins therefore cannot replicate in normal, nondividing cells. However, cancer cells are mitotically active and no longer express tumor suppressors such as the p53 protein, therefore the oncolytic AdV can and does replicate, killing the tumor cells. (B) In this scenario, the AdV carries a foreign gene (e.g., herpesvirus thymidine kinase) into the cancer cells. The herpesviral TK activates the prodrug (acyclovir in this example) to produce an active drug that inhibits tumor-cell DNA replication. The activated drug can enter nearby uninfected cells.

respond to these viruses; will eventually generate an antiviral response, limiting their repeated use.

Adenoviral vectors

AdVs have proven to be useful vectors for gene delivery, with applications in basic research, gene replacement therapy, and cancer therapy. Human AdVs package about 40 kbp of DNA and the newest generation of AdV vectors can contain up to 37 kbp of foreign DNA. To make a vector, the foreign DNA is delivered to so-called "packaging" cells that contain the genes for AdV DNA Pol, TP, and capsid proteins. Within packaging cells, the foreign DNA is replicated and packaged into virions. The virions are now ready to deliver the foreign DNA to target cells. The DNA delivered by AdV vectors does not replicate in the recipient cell and it remains extrachromosomal (it is not integrated into the host genome). The foreign genes can be expressed, but as recipient cells divide, the recombinant gene is lost. This can be a

---

BOX 38.3

Competition among small RNAs in AdV-infected cells

miRNAs are small (~22 nt) cellular RNAs that inhibit translation of mRNAs to which they base pair. miRNAs are actually synthesized as long precursors that are cleaved by the cellular endoribonuclease Drosha to produce ~65 nt pre-miRNAs that are exported from the nucleus. In the cytoplasm, pre-miRNAs are further cleaved by another cellular ribonuclease, Dicer, to generate ~22 nt miRNAs that are incorporated into multiprotein RNA silencing complexes (RISC). RISC use miRNAs as a point of interaction with cellular mRNAs (via imperfect base pairing) and initiate cleavage and degradation (thus "silencing") of the bound mRNA.

Because AdV VA RNAs are synthesized at high levels in the nucleus of infected cells, they compete directly with cellular pre-miRNAs for export to the cytoplasm. Once in the cytoplasm VA RNAs also bind to Dicer, once again competitively inhibiting processing of cellular pre-miRNAs. Once incorporated into RISC, the ~22 nt VA RNA fragments do *not* target viral mRNAs (because they are transcribed from noncoding regions of the AdV genome). They may, however, bind and target a subset of cellular mRNAs. Whether or not VA RNAs directly silence cellular mRNAs, they certainly interfere with normal cellular control of translation.

---

disadvantage; however, it provides a measure of safety as well. This strategy is also used to create AdV vectors that deliver foreign viral genes, such as the SARS-CoV-2 spike protein.

In this chapter, we learned that:

- AdVs are unenveloped, icosahedral ds DNA viruses. AdV genomes are linear and covalently linked at their 5′ ends to the viral protein (TP) that served as a primer for replication.
- AdV DNA replication is asymmetric. One parental strand is copied while the other is displaced. The displaced strand forms ds hairpin structures that then serve as replication origins.
- AdV stimulates host cell DNA replication by early gene products that inactivate pRB family members and p53.

- Human AdV cannot productively infect rodent cells. Nonproductive infection occasionally results in integration of DNA encoding E1A and E1B, leading to transformation. Productive infection of human cells is cytolytic.
- Human AdV infections are often asymptomatic but can cause mild respiratory, urinary tract, and eye infections.
- Genetically modified oncolytic AdVs lack E1A and E1B, thus they cannot replicate in normal cells (they cannot stimulate DNA replication or counter the apoptotic activities of p53). As cancer cells are mitotically active, and most do not express p53, they can support the lytic growth of the genetically modified oncolytic AdVs.

39

Family *Herpesviridae*

After reading this chapter, you should be able to discuss the following:

- What characteristics are shared by all herpesviruses?
- What are the basic features of herpesvirus virions?
- What is cellular latency? How does latent infection compare with productive infection?
- What types of cells are productively infected by human herpesviruses (HHV)-1 and -2? What types of cells are latently infected by these viruses?
- What diseases are caused by HHV-1 and -2?
- What are some of the diseases caused by HHV-4 (Epstein—Barr virus or EBV) and how are they related to the cell tropism of this virus?
- What is shingles and what is the purpose of the shingles vaccine? How is the shingles virus related to the chickenpox virus?
- How does acyclovir work and why is it specific for herpesvirus-infected cells?

Members of the family *Herpesviridae* infect a wide range of vertebrates including mammals, birds, and reptiles. The family name is derived from the Greek herpein "to creep," referring to the latent, recurring infections typical of this group of viruses. After initial productive infection, these large enveloped double-stranded (ds) DNA viruses (Fig. 39.1) probably persist for the life of the organism. Infections are often completely silent, a process called latency (Box 39.1). For example, the chickenpox virus that infects a child may remain latent for decades, but may reactivate in old age to infect a new generation!

The family *Herpesviridae* contains three subfamilies. The subfamily *Alphaherpesvirinae* contains the human herpesviruses (HHV)-1 and 2, more commonly known as herpes simplex viruses. These viruses productively infect epithelial cells, producing blister-like lesions and establish latent infections in neurons. The members of the subfamily *Betaherpesvirinae* tend to replicate slowly; a productive infection cycle can take as long as a week in cultured cells. Persistently infected cultures are easy to establish and infected cells may become enlarged. Betaherpesviruses establish latency in leukocytes; human betaherpesviruses include HHV-5 (cytomegalovirus, CMV, named for the production of cytomegalia or enlarged cells), HHV-6A, HHV-6B, and HHV-7.

© 2023 Elsevier Inc. All rights reserved.

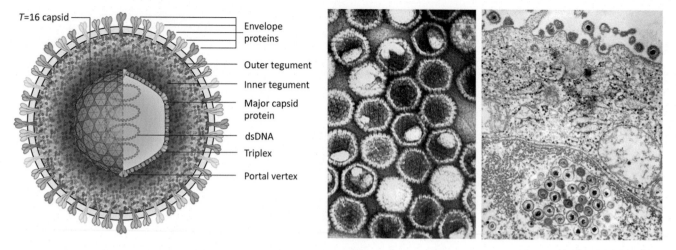

FIGURE 39.1 Herpesvirus structure (left). Negatively stained transmission electron microscopic image (right) showing numerous herpes simplex (HHV-1) virions, located both inside the nucleus, and extracellularly.
Source: CDC Public Health Image Library Image ID#10260. Content providers CDC/Dr. Fred Murphy; Sylvia Whitfield.

BOX 39.1

General characteristics

Herpesvirus genomes found within infectious virions are linear ds DNA ranging in size from ~150 to 225 kbp. However, soon after entry into the nucleus, the DNA becomes a covalently closed circle. Herpesviral genomes encode over 100 genes. Most mRNAs are unspliced but are capped and polyadenylated. There are clear patterns of early versus late gene transcription. DNA replication occurs after early transcription.

Structural proteins are synthesized from late transcripts. Transcription and genome replication are nuclear. Capsids assemble in the nucleus.

Virions are enveloped with an icosahedral capsid. Surrounding the capsid is a dense layer of viral proteins called the "tegument." There are several glycoproteins associated with the envelope.

Herpesviruses establish of lifelong latent infections.

Members of the subfamily *Gammaherpesvirinae* infect mammals. Gammaherpesviruses productively infect epithelial and/or lymphoid cells. These viruses are usually specific for replication in either T-cells or B-cells; latency is established in lymphoid cells that may proliferate in response. Human gammaherpesviruses include HHV-4 (EBV) and HHV-8 (Kaposi's sarcoma-associated herpesvirus) (Box 39.2).

episomal for the most part. Transcription and genome replication are nuclear. Transcripts are synthesized by RNA polymerase II (RNA pol II) and are capped and polyadenylated. Most viral mRNAs are not spliced and gene overlaps are common (Fig. 39.2). Genome replication requires a multisubunit herpesvirus DNA polymerase. Some herpesviruses also encode enzymes required to generate adequate pools of dNTPs. For example, HHV-1 encodes thymidine kinase (TK), dUTPase, and ribonucleotide reductase.

Genome organization

Herpesvirus virions contain a molecule of linear ds DNA. Herpesvirus genomes range in size from ~124 to 259 kbp and they encode anywhere from 70 to 200 proteins. Some types of herpesviruses have genomes with extensive repeat regions. Upon delivery to the nucleus, viral DNA circularizes prior to transcription and replication. During latency herpesvirus genomes remain

Virus structure

Virions range in size from 120 to 260 nm in diameter. Virions are complex and assembled from a large number of proteins. Estimated numbers of structural proteins range from 4 to 7 capsid proteins, 9 to 20 tegument proteins, and 4 to 19 envelope proteins. Herpesvirus capsids

BOX 39.2

Taxonomy

Order: *Herpesvirales*

The order *Herpesvirales* contains three virus families. This chapter focuses on the family *Herpesviridae*, whose members infect mammals, birds, and reptiles.

Family *Herpesviridae*
Subfamily: *Alphaherpesvirinae*

Genus: *Iltovirus* (avian herpesviruses)
Genus: *Mardivirus* (avian herpesviruses)
Genus: *Scutavirus* (turtle herpesvirus)
Genus: *Simplexvirus* [mammalian herpesviruses including human herpesviruses (HHV) -1 and -2, the herpes simplex viruses]
Genus: *Varicellovirus* (mammalian herpesviruses including HHV-3, varicella zoster virus)

Subfamily: *Betaherpesvirinae*

Genus: *Cytomegalovirus* [includes HHV-5 (cytomegalovirus, CMV)]
Genus: *Muromegalovirus* (murid betaherpesviruses)
Genus: *Proboscivirus* (elephant betaherpesvirus)
Genus: *Quwivirus* (rodent betaherpesvirus)
Genus: *Roseolovirus* (human betaherpesviruses 6 and 7)

Subfamily: *Gammaherpesvirinae*

Genus: *Lymphocryptovirus* [includes human gammaherpesvirus 4 (Epstein–Barr virus, EBV)]
Genus: *Macavirus* (mammalian gammaherpesviruses)
Genus: *Patagivirus* (bat gammaherpesvirus)
Genus: *Percavirus* (equid and mustelid gammaherpesviruses)
Genus: *Rhadinovirus* (includes human gammaherpesvirus 8)

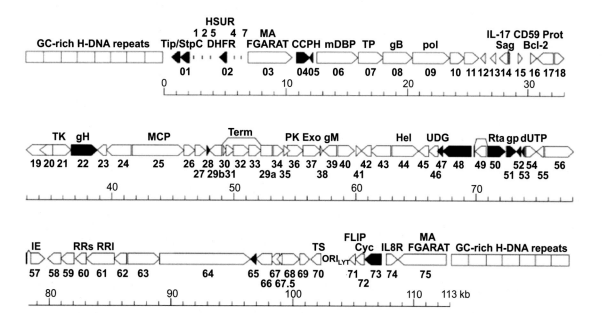

FIGURE 39.2 An example of herpesvirus genome organization showing densely packed transcripts (Ensser et al., 2003). Source: *From Ensser, A., Thurau, M., Wittmann, S., Fickenscher, H., 2003. The genome of herpesvirus saimiri C488 which is capable of transforming human T cells. Virology 314, 471–487.*

are ~125 nm in diameter. Capsids are *T* = 16 icosahedral structures assembled from 150 hexons, 11 pentons, and single portal complex. The specialized portal complex takes the place of the 12th pentamer. Hexons and pentons are assembled from the major capsid protein (MCP). The basic structure of the herpesvirus capsid protein is distinct from those of most animal viruses. Herpesviral MCP is structurally related to the capsid proteins of some DNA phages. The portal complex is assembled from 12 copies of the capsid portal protein. This is the route by which the genome enters and exits the capsid. A so-called "portal capping protein" is associated with mature DNA-containing capsids. Within the capsid, DNA is densely packed, aided by the anion spermine.

Virions are enveloped and contain several different envelope glycoproteins. Herpesvirus envelopes are densely packed with glycoproteins; the copy number of individual glycoproteins can exceed 1000 per virion. Between the capsid and envelope is a dense protein layer called the tegument. Tegument proteins are key players in the early phases of the herpesvirus replication cycle. Functions of tegument proteins include inhibition of cell defenses, shut off host protein synthesis, and stimulation of early viral gene expression.

Replication cycle

As might be expected of large, complex viruses that switch between productive and latent forms of infection, there is much variability in the details of herpesvirus replication. Add to that the importance of HHVs as pathogens, and the literature becomes extensive! The following sections address only the most basic aspects of herpesvirus replication. It should also be noted that herpesvirus protein nomenclature is variable, complex, and confusing at times. In the following sections, herpesviral proteins will be described only by their most obvious functions (e.g., DNA polymerase).

Attachment/penetration

Attachment and penetration are achieved by multiple viral glycoproteins. In the case of both HHV-1 and HHV-2 (Table 39.1), heparin sulfate proteoglycans (HSPGs) are used for initial attachment. HSPGs are glycoproteins that have a protein core to which is attached one or more heparin sulfate chains. Penetration into cells requires fusion; in some cases at

the plasma membrane, but in other cases following endocytosis. Released capsids and some tegument proteins move to nuclear pores with the help of the cell cytoskeleton. Tegument proteins may be active during the transport process, altering host cell signaling pathways to favor virus replication. DNA moves out of the capsid and through the nuclear pore in an energy requiring process. It is likely that some tegument proteins enter the nucleus as well. In the nucleus the genome is circularized.

Transcription/DNA replication

Once in the nucleus, viral DNA may be transcribed and enter a pathway to productive infection. Alternatively the cell environment may favor the establishment of latency. In the case of productive infection, the first genes transcribed are those that support DNA replication. Viral DNA replication requires many viral proteins including: viral DNA polymerase, single-stranded DNA-binding protein, Ori (replication origin)-binding protein, ds DNA-binding protein, primase/helicase, TK, ribonucleotide reductase, dUTPase, uracil DNA glycosylase among others. Enzymes involved in nucleotide biosynthesis are among those that are often dispensable for replication in cultured cells. During a productive infection, there is an ordered schedule of transcription similar to that seen for other DNA viruses. In general, transcription of structural proteins, including the tegument proteins, begins during or after DNA replication. DNA replication is accomplished through a rolling circle mechanism that results in formation of head to tail concatemers. These are cleaved into unit length genomes during packaging.

TABLE 39.1 Human herpesviruses.

ICTV[a] name	Vernacular name	Subfamily	Associated diseases[b]
HHV[c]-1	Herpes simplex 1	*Alphaherpesvirinae*	Cold sores
HHV-2	Herpes simplex 2	*Alphaherpesvirinae*	Genital herpes
HHV-3	VZV/chickenpox/shingles virus	*Alphaherpesvirinae*	Chickenpox (acute)/shingles (upon reactivation from latency)
HHV-4	EBV	*Gammaherpesvirinae*	Infectious mononucleosis; various B-cell cancers
HHV-5	CMV	*Betaherpesvirinae*	Infectious mononucleosis
HHV-6A	Roseolovirus	*Betaherpesvirinae*	Unknown
HHV-6B	Roseolovirus	*Betaherpesvirinae*	Roseola infantum (rose rash of infants, ES, or the sixth disease)
HHV-7	Roseolovirus	*Betaherpesvirinae*	Roseola infantum (rose rash of infants, ES, or the sixth disease)
HHV-8	Kaposi-associated herpesvirus	*Gammaherpesvirinae*	Usually asymptomatic; associated with KS in select populations

[a]*International Committee on the Taxonomy of Viruses.*
[b]*Many infections are asymptomatic.*
[c]*Human herpesvirus.*

Assembly/release

Capsid assembly takes place in the nucleus, the site of genome replication. Capsid assembly is complex, and occurs with the help of scaffold proteins. Nascent capsids are filled with viral DNA (through the specialized portal complex) in a process that requires energy. DNA is packaged in a "headful" mechanism whereby concatemers are cleaved at conserved sequences that define the genome ends. The DNA is tightly packed, producing a rigid capsid and the capsid is "sealed" by the portal capping protein.

DNA-containing capsids leave the nucleus and traffic through the cytoplasm. Some studies suggest that capsids bud from the nucleus and then lose their nuclear envelope. Other studies suggest that naked capsids leave the nucleus. In either case, capsids associate with tegument proteins as the capsid travels through the cytosol. Virions then obtain a final envelope by budding into a cytoplasmic organelle (e.g., recycling endosomes or Golgi). Virions are then transported, within a vesicle, to the plasma membrane and are released by exocytosis. Productive infection is usually accompanied by cell death.

Latency

Herpesviruses are noted for their ability to establish cellular latency. Cellular latency is a process during which genomes are maintained in cells, with limited expression of viral proteins and no virion production. Cellular latency is regulated at the level of individual cells. Thus one or a few cells may break from latency while other cells remain latently infected. Use of the term latency implies that virus has the *ability to reactivate* (enter a productive replication cycle) in response to certain stimuli. In this way, latency is differentiated from *abortive* infection (during which virus is never produced). The precise molecular mechanisms that lead to establishment of, and reactivation from, the latent state are not fully understood and can differ from one herpesvirus to another. Finally, it should be noted that cellular latency is distinct from clinical latency. The latter term refers to the overall state of the infected organism. Thus clinical latency is the absence of disease, not the absence of virus.

Cellular latency permits herpesviruses to remain in the infected host for long periods while avoiding detection by the immune system. Members of the subfamily *Alphaherpesvirinae* (e.g., HHV-1, -2, and -3) replicate productively in epithelial cells but establish latent infections in neurons. This process begins when virus released from epithelial cells and enters the axonal terminus of a nearby neuron. Fusion releases the capsid which uses cytoskeletal components to actively move (by retrograde axonal transport) to the nucleus in the cell body of neuron. DNA is delivered to the nucleus but transcription is limited to a set of noncoding RNAs called *latency-associated transcripts*. If circumstances permit, the HSV genome may become transcriptionally active (reactivated). Infectious virus is formed in the nerve cell body and moves by anterograde axonal transport back to the site of the original infection.

Betaherpesviruses and gammaherpesviruses establish latency in leukocytes. HHV-4 (EBV) latency has been studied extensively. HHV-4 productively infects epithelial cells and establishes latency in B-lymphocytes. HHV-4 latency is quite complex with variable numbers of proteins and small RNAs expressed. In cultured cells, EBV infection can immortalize B-cells to produce lymphoblastoid cell lines (LCL). Most LCLs are latently infected and they express a small number of HHV-4 proteins and noncoding RNAs. The proteins expressed during latency include the Epstein—Barr nuclear antigens (EBNAs) and the latent membrane proteins (LMPs). Small RNAs include EBV-encoded small RNA (EBERs) and micro (mi)-RNAs.

EBNA-1 acts as a bridge to couple the HHV-4 episome to the cellular chromosome. This enables replication and stable maintenance of the genome during cell division. EBNA-2 is a transactivating protein that works through association with cellular DNA-binding proteins to modulate expression of cellular and EBV genes. The EBNA-3 genes are required for transformation of primary B-cells. The LMPs are membrane receptors. LMP1 mimics active tumor necrosis factor receptor, TNFR. LMP2A mimics an activated B-cell receptor. Acting in concert, these receptors activate signaling pathways necessary for survival and growth of LCL.

EBERs are small noncoding RNAs. They are not expressed during lytic infection but are abundant in latently infected cells. Both EBER1 and EBER2 are transcribed by RNA Pol III; EBER1 is also transcribed by RNA Pol II, thus is quite abundant in infected cells. EBER RNAs assume stable secondary structures and localize to the nucleus where they are found in complexes with cellular proteins La and L22. La protein acts as an RNA Pol III transcription factor in the nucleus and as a translation factor in the cytoplasm. L22 is a component of the 60S ribosomal subunit. One function of EBERs is to block interferon-induced PKR phosphorylation. Phosphorylated PKR inhibits translation initiation factor eIF-2α (thereby blocking translation of host and viral mRNAs), and is normally a potent antiviral response. Therefore by blocking PKR phosphorylation, EBERs allow virus replication to proceed. EBV also encodes a group of miRNAs that likely have roles in establishing or maintaining EBV latency, targeting both viral and cellular mRNAs.

The pattern of EBV latent gene expression in LCLs is called "latency 3" to distinguish it from other patterns of gene expression that are observed in transformed cells or in B-cells of normal persons. In normal B-cells (in the immunocompetent patient) few if any, EBV genes are expressed and this state is called latency 0. In Burkitt lymphoma (BL) tumors only the EBNA-1 protein is expressed (latency 1). In other EBV-related tumors EBNA-1, LMP1 and LMP2 are expressed (latency 2). In summary, EBV latency is multifaceted and complex! The common factor is that virions are not produced and episomal DNA is maintained in dividing cells.

Diseases caused by human herpesviruses

Most herpesviruses are very host specific. They are well adapted to their hosts and their lifestyle is to remain within an infected host for life. Thus many herpesvirus infections are asymptomatic. Table 39.1 lists the human herpesviruses and their associated diseases. Occasionally relatively benign herpesviruses cause severe systemic and/or central nervous system infections. Neonates and immunocompromised individuals are the most at risk for severe infection. There are also clear associations between some herpesviruses and cancer.

HHV-1 and -2 disease

The herpes simplex viruses cause blister-like lesions. These viruses are found worldwide in the human population and are transmitted by contact. Primary infection is usually mild and localized (or asymptomatic). While virus is obviously present in lesions, it can also be present (and is transmissible) when lesions are absent.

HHV-1 and -2 are serologically related, sharing ~80% sequence identity. The two viruses usually infect different parts of the body, with HHV-2 more frequently infecting genital mucosa. However, the incidence of genital HHV-1 is rising; about 50% of primary genital infections involve HHV-1. HHV-1 and -2 establish latency in neurons. Reactivation is often associated with compromise of the immune system, for example, by the extreme stress of final exams!

Herpes simplexvirus infections of neonates can be very serious and life threatening. Infants can be infected in utero, during birth, or shortly after birth. Infections of neonates are frequently fatal due to disseminated infection and encephalitis. The incidence of such infections is estimated to be 1 in every 2000−5000

live births. To date there has been little success developing protective vaccines for either HHV-1 or- 2.

HHV-3 (varicella zoster virus)

HHV-3 causes chickenpox, a common childhood rash. HHV-3 is spread by the respiratory route with replication beginning in the upper respiratory tract. Virus then spreads throughout the body, ending up in the skin (Fig. 39.3). The virus replicates in skin cells causing a rash and itchy skin lesions due to lytic infection of the epidermis. Skin lesions contain large amounts infectious virus. Virus is shed from the skin and is passed to a new host via inhalation. In children, chickenpox is usually a mild, self-limiting infection. However, infection of adults can be more serious, resulting in pneumonia, male sterility, and/or liver failure.

HHV-3 establishes latency in dorsal root ganglia of the spinal cord (Fig. 39.3). During latency, no viral proteins are produced. Latency can last for decades. Reactivation most frequently occurs in response to stress or immunosuppression. Older adults are at risk for reactivation as their immunity to HHV-3 gradually wanes. Reactivation of HHV-3 results in a painful rash called shingles that consists of large blister-like lesions that contain infectious virus. The rash may last for weeks, and debilitating nerve pain can continue for months.

Varicella zoster virus (VZV) is the only human herpesvirus for which a vaccine is available and widely recommended. The VZV vaccine contains live attenuated virus (strain Oka, developed in Japan in 1974) and was approved for use in the United States in 1995. Current recommendations are for universal vaccination of children. Childhood vaccination is meant to prevent the more serious complication of chickenpox that can result when naïve adults are infected. There are also indications that the vaccine strain is less prone to reactivation than wild-type VZV. It is hoped, but remains to be seen, if vaccination in childhood protects against shingles decades later. A higher dose of the Oka vaccine (tradename Zostavax) was formerly licensed in the United States for use in elderly adults, however, it was not recommended for immunocompromised individuals. Zostavax was discontinued in the United States in 2020 after a safer and more effective alternative, Shingrix, was approved. Shingrix is a recombinant subunit vaccine, containing VZV glycoprotein E. It is recommended for persons over 50 years of age and for immunocompromised individuals over 19 years of age.

HHV-4 (Epstein−Barr virus)

HHV-4 infects epithelial cells and B-lymphocytes. Only infection of epithelial cells is productive. HHV-4

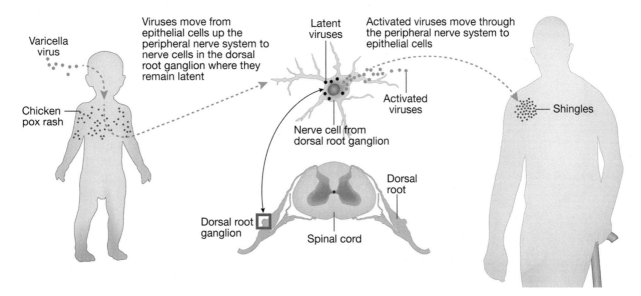

FIGURE 39.3   Disease cycle of chickenpox/shingles.

establishes latency in B-cells. When HHV-4 breaks out of latency, virions seed the oral mucosa. Studies of bone marrow transplantation patients have confirmed the "life-cycle" of HHV-4. Prior to a bone marrow transplant, the recipient's bone marrow and immune cells are destroyed. After a transplant, the HHV-4 from the *donor* is found in the transplanted B-cells and in epithelial cells of the recipient. This suggests that the only HHV-4 reservoir was in the recipient's B-cells. In the event that bone marrow from an uninfected individual is transplanted into an HHV-4 positive recipient, the recipient will now be negative for HHV-4. These findings provide good evidence that over the long term, HHV-4 is maintained in B-cells and only periodically and transiently infects epithelial cells.

During periods of productive replication, HHV-4 is produced (usually with no symptoms or lesions) and can be transmitted via saliva. HHV-4 is a very common virus and most humans are infected by age 3. Once infected, always infected, and more than 90% of the human population is infected with HHV-4. Most often there is no disease. Infection of older individuals (teenagers or older) can be symptomatic and the disease is infectious mononucleosis. Symptoms include fatigue, fever, rash, swelling of lymph nodes, and enlargement of spleen and liver. The disease is characterized by the presence of large numbers of mononuclear cells (T-lymphocytes) in the blood. These T-cells are responding to HHV-4 infection. This strong T-cell response (a cell-mediated response to the infection) is an immunopathologic event. The ability of HHV-4 to stimulate B-cell proliferation is a potentially (but not common) oncogenic event. (HHV-4 is routinely used to produce transformed B-cell cultures in the laboratory, however.)

HHV-4 and Burkitt lymphoma

BL is a childhood lymphoma (a tumor of lymphocytes) common throughout equatorial Africa. Early studies of BL-derived cell lines showed that they produced herpesviruses (HHV-4). Children with BL also have very high levels of antibody to HHV-4, and high antibody titers to HHV-4 are associated with higher risk of developing BL. However, HHV-4 infection is only *one* step in the development of BL. BL tumor cells also have a chromosomal abnormality. In fact most BL tumors have a very specific chromosomal translocation. Another strong risk factor for development of BL is coinfection with the malarial parasite *Plasmodium falciparum*.

The currently proposed scenario for the development of BL, summarized in Fig. 39.4, is as follows.

A child is infected with HHV-4, establishing an infection of B-cells. The immune system initially functions normally to suppress B-cell proliferation. However, if the child is infected with *P. falciparum*, the immune system can be suppressed, allowing for uncontrolled proliferation of HHV-4 infected B-cells. These proliferating cells are at increased risk of mutation. A chromosomal translocation that activates an oncogene is the final event: B-cells with mutated chromosomes proliferate even more, leading to the development of the B-cell tumor. BL tumors are monoclonal, that is, all tumor cells originated from a single HHV-4 infected cell that underwent a chromosomal translocation.

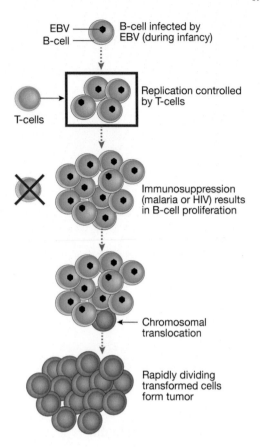

FIGURE 39.4   Model of development of Burkitt lymphoma (BL). Infection with the malarial parasite, *Plasmodium falciparum* is a strong risk factor for the development of BL. This model suggests that the link between BL and malaria is immunosuppression that allows uncontrolled replication of EBV-infected B-cells. Additional B-cell mutations occur, particularly chromosomal translocations, that impart a fully transformed phenotype.

HHV-4 and nasopharyngeal carcinoma

Nasopharyngeal carcinoma (NPC) is a malignancy of epithelial cells of the nasopharynx. NPC is rare in North America and Europe but the incidence in Asia (southern China) is as high as 1 case per 1000 persons per year. There is also a high incidence in Mediterranean, African, and Eskimo populations. Given that over 90% of humans are infected with HHV-4, how can it be associated with NPC among only certain populations? First we must consider the evidence that HHV-4 is involved in tumor development. The link is based partly on sero-epidemiologic evidence. Persons with NPC have *very high levels* of antibody to HHV-4. Healthy individuals with high levels of antibody to HHV-4 are also *more likely* than the general population to develop NPC. The NPC tumor cells are HHV-4 infected and the tumors are clonal, suggesting a rare mutation event driving cell proliferation. At this time, there is no simple model to explain the development of NPC. Disease seems to

result from a combination of environmental, genetic, and virologic factors.

HHV-4 is also suspected of causing some cases of B-cell lymphoproliferative disease (B-cell lymphoma) in immune-deficient humans. There is also an association between HHV-4 and Hodgkin disease, the most common malignant lymphoma in the Western hemisphere. Peak age of onset is 25–30 and then beyond age 45. Certain types of Hodgkin lymphomas (but not all) are strongly associated with EBV. Finally, HHV-4 in AIDS patients causes oral hairy leukoplakia, wart-like lesions of the lateral borders of the tongue that are epithelial foci of EBV replication.

HHV-5 (cytomegalovirus)

HHV-5 infects epithelial cells in many different tissues. Infection stimulates host cell DNA, RNA, and protein synthesis; infected cells become enlarged (hence the name cytomegalo or big cells). HHV-5 is transmitted by oropharyngeal secretions, breast milk, and blood. Transmission requires close contact. The virus is common, with 40%–100% infection rates by adulthood. Infection is usually asymptomatic, but may cause mononucleosis. Neonates of seronegative mothers are at highest risk of developing disease as they lack protective material IgG. Severe neurologic complications have occurred in immunocompromised patients. CMV is a common cause of AIDS retinitis.

HHV-6 and -7

HHV-6 and -7 are betaherpesviruses (genus *Roseolovirus*) associated with a febrile, rash illness called roseola infantum, exanthem subitum (ES), or sixth disease. HHV-6 and -7 are *ubiquitous viruses*, with horizontal infection commonly occurring during infancy. More than 85% of adults have antibody to both viruses. HHV-6 contains two subtypes, A and B, which are actually separate species. HHV-6B is most commonly associated with ES. HHV-7 infects CD4 + T-lymphocytes (believed to be the site of latent infection) and salivary gland epithelial cells (productive infection). HHV-6B infects lymphoid and endothelial cells. Latent infection seems to be in monocytes, macrophage, and CD34-positive progenitor cells. The HHV-6 genome is quite interesting in that it has an array of sequences similar to mammalian telomeres and these may be involved in site-specific integrations of HHV-6. Chromosomal integrations of HHV-6 (called ciHHV-6) can lead to germline inheritance. While most HHV-6 infections occur via horizontal routes, approximately 1% of the population inherits an HHV-6 genome from a parent.

ES is a common disease of infants worldwide. It is characterized by sudden fever, which lasts for a few days, followed by a rash that appears on the trunk and face and spreads to lower extremities as the fever subsides. However, it is likely that many HHV-6 primary infections of infants are asymptomatic. In adults primary infections by HHV-6 can cause mononucleosis-like disease and hemophagocytic syndrome.

HHV-8 (Kaposi sarcoma herpesvirus)

Before the HIV pandemic, Kaposi sarcoma (KS), a slowly developing endothelial cell tumor, was thought to be limited to older men in the Mediterranean region. KS can occur in various sites in the body, but skin lesions are common. In the early 1980s, KS emerged as an AIDS-defining cancer. Epidemiologic data suggested the involvement of a sexually transmitted agent, other than HIV, in the development of KS, and this lead to the identification of the KS herpesvirus (HHV-8) in 1994. Initial findings were that HHV-8 was present in cells within KS lesions, but not in normal surrounding tissues.

KS lesions are complex, containing multiple cell types. A key cell is the so-called spindle cell, of endothelial cell origin. KS herpesvirus infection, through a number of viral products (including proteins and miRNAs), contributes to the transformation of normal endothelial cells. It is thought that tumorigenesis is due to genes expressed during latency. One set of HHV-8 proteins, the Kaposins, can transform Rat-1 fibroblasts. HHV-8 is present in all KS lesions, but spindle cells loose the viral genome when placed in culture. In addition to spindle cells, KS lesions contain T-cells, B-cells, and monocytes within aberrant slit-like neovascular spaces. These spaces are lined with infected and uninfected endothelial cells and are structurally fragile. They are very susceptible to leakage and rupture, causing the hemorrhages typical of KS lesions.

Although HHV-8 was first identified in KS lesions (endothelial cell tumors), it also associated with rare B-cell lymphoproliferative diseases.

Monkey B herpesvirus

Monkey B herpesvirus infection is caused by macacine herpesvirus (formerly cercopithecine herpesvirus 1). The virus is also commonly referred to as herpes B, herpesvirus simiae, and herpesvirus B. In monkeys this virus causes a disease similar to HHV-1 in humans, that is, localized blister-like lesions. However, if a person becomes infected with the monkey B herpesvirus, it causes serious neurologic damage with high fatality rates. Persons in contact with Old World monkeys are at highest risk of contracting monkey B virus. The virus is transmitted via body fluids and most research laboratories in the United States use only animals that are antibody negative for B virus. (But according to the CDC: macaques housed in primate facilities frequently become B virus positive by the time they reach adulthood!)

Antiherpesviral drugs

Safe and effective treatments for HHV-1, -2, and -3 are widely available. Acyclovir was the first antiherpes drug developed and it provided the basis for development of additional antiherpesvirus drugs also in use today.

Acyclovir is a competitive inhibitor of dGTP. It is recognized by herpesvirus DNA polymerase and its incorporation prevents DNA chain elongation. An important feature of acyclovir is low toxicity for humans. This is due to the production of the *active* phosphorylated form of the drug *only* in herpesvirus-infected cells. Herpesviruses encode a viral TK that readily phosphorylates acyclovir while normal cells do not phosphorylate the prodrug (Fig. 39.5). In addition, the herpesvirus DNA polymerase has a higher affinity for phosphorylated acyclovir than do cellular DNA polymerases. Resistance to acyclovir is low, but virus mutants can arise. Early studies of resistant mutants were helpful in delineating the mechanism of action of the drug: the finding that resistance mapped to mutations in herpesviral TK and/or DNA polymerase led to the realization that specificity for infected cells was due to the requirement for the viral TK to produce an active drug.

Ganciclovir and the prodrug valganciclovir are competitive inhibitors of dGTP. They are often used for treating CMV infections (HHV-5), particularly in immunodeficient patients. Both drugs must first be phosphorylated to ganciclovir monophosphate by the CMV kinase. Additional antiherpesviral drugs include famciclovir and valacyclovir. Valacyclovir is a prodrug for acyclovir that can be taken orally.

In this chapter, we learned that:

- Herpesviruses are large ds DNA viruses that are notable for the establishment of latent infections.
- Virions are enveloped with an icosahedral capsid. The capsid is surrounded by a protein layer called the tegument. Within the capsid, the DNA is tightly packaged. The capsid is formed by the MCP, notable for its structural relationship to the capsid proteins of certain types of DNA phages. The capsid contains a special "portal complex" through which the DNA genome is packaged and released.

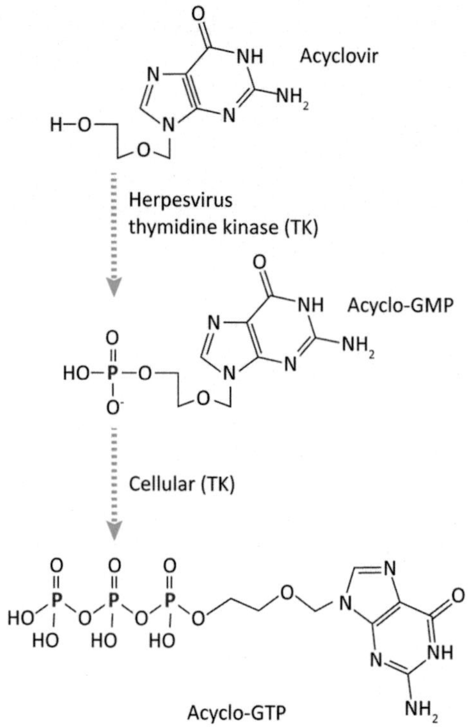

FIGURE 39.5 Mechanism of action of acyclovir. Acyclovir is an inactive "prodrug" until phosphorylated by herpesvirus TK. Monophosphorylated acyclovir is further phosphorylated by host kinases.

- Cellular latency is the maintenance of herpesviral genomes in cells, in the absence of productive infection. Few or no viral proteins are produced in the latently infected cells. Some herpesviruses (e.g., HHV-4) stimulate replication of latently infected cells. Most productive infections are cytolytic.
- HHV-1 and -2 productively infect epithelial cells but latently infect neurons.

- HHV-1 and -2 cause cold sores and genital lesions. Painful blister-like lesions are filled with infectious virus. In neonates these viruses can produce severe systemic infections. It is increasingly recognized that infectious HHV-1 and -2 can be present on skin in the absence of any obvious lesions.
- HHV-4 or EBV is one cause of infectious mononucleosis although most infections are asymptomatic. HHV-4 is a ubiquitous virus that infects most of the world's population. HHV-4 establishes latency in B-cells and induces them to proliferate. Under specific sets of conditions B-cell tumors can form.
- Shingles is the painful rash that develops when latent HHV-3 (the chickenpox virus) reactivates. The shingles vaccine is used to stimulate immune responses to the virus that tend to wane among the elderly. The same attenuated virus used for the shingles vaccine is found in the chickenpox vaccine recommended for children.
- Acyclovir is competitive inhibitor of dGTP. It competes with dGTP for incorporation into herpesviral DNA by the viral DNA polymerase. It is a DNA chain-terminator. Acyclovir is very specific to herpesvirus-infected cells because it must be phosphorylated to be active. Acyclovir is a substrate for herpesvirus TK but is not converted to the monophosphate form in uninfected cells.

References

Ensser, A., Thurau, M., Wittmann, S., Fickenscher, H., 2003. The genome of herpesvirus saimiri C488 which is capable of transforming human T cells. Virology 314, 471−487.

40

Family *Poxviridae*

After reading this chapter, you should be able to answer the following questions:

- What are the general characteristics of poxviruses?
- Where in the cell do poxviruses replicate?
- Are host or viral enzymes used for mRNA synthesis and genome replication?
- Compared to other enveloped viruses, what is unique about poxvirus morphogenesis?
- What are virokines and viroceptors and what is their function in poxvirus replication?
- What was "variolation" and why was Jenner's "vaccination" an improvement?
- Explain the outcome of the introduction of rabbit myxoma virus into Australia. What does this show about the interactions of viruses and their hosts?

Poxviruses are large, complex, DNA viruses whose various members infect animals and insects. The term "pox" derives from the English *pocks*, referring to blister-like skin lesions that are the hallmark of smallpox and other poxvirus infections. Smallpox is an ancient disease, as evidenced by scarred Egyptian mummies. By the middle ages, the disease was common in Europe and from there, it traveled to North and South America with European explorers and colonists in the 16th and 17th centuries. Smallpox was notable for many reasons. Mortality varied, depending on the particular strain, but the average was ~25%.

The disease was severe and painful, culminating in pus-filled blisters that sometimes covered the entire body. The resulting scars permanently marked smallpox survivors, sometimes causing blindness or limb deformities. Smallpox spread most readily from person to person, but could also be spread via contaminated clothes, bedding, and household items that might be infectious for weeks or months.

Smallpox is also notable for its place in the history of vaccines and vaccination. As disease survivors were clearly protected from further infection, by about the 10th century, dried scab material was being collected from mild cases of disease and used to infect naïve individuals. In China, the process involved blowing powdered material into the nostrils, while in India, infection was accomplished by inoculating scratched skin. Both methods usually resulted in mild disease with mortality significantly lower than natural infection (although these procedures were certainly not risk-free, the fatality rate was ~2%).

By the late 18th century, cutaneous variolation was a common practice in Europe and North America and this set the stage for English physician Edward Jenner to develop an even safer method to achieve protection. Jenner knew that milkmaids, who often experienced mild cowpox lesions on their hands, believed that they were resistant to both variolation and smallpox infection. In 1796 Jenner used material from a cowpox

Viruses.
DOI: https://doi.org/10.1016/B978-0-323-90385-1.00034-0

© 2023 Elsevier Inc. All rights reserved.

lesion to variolate a child. Jenner reported the resulting reaction as "barely perceptible." To test for protection, Jenner challenged the child by variolation with smallpox, and he was indeed resistant to infection. Jenner called his technique vaccination (*vacca*, Latin for cow) to distinguish it from the older procedure. Vaccination was an effective and much safer procedure than variolation. Safety was key to its widespread use, and the eventual elimination of smallpox.

While Jenner's original vaccine was cowpox, other poxviruses (i.e., horse pox and buffalo pox) were used as the source of vaccine as the procedure became widely accepted. By the late 19th century, the virus we know today as "vaccinia virus" (VACV) was widely used as the smallpox vaccine. We now know, through genome sequencing, that Jenner's cowpox was not the origin of VACV.

In 1958 the World Health Organization began the (then) unprecedented task of eliminating smallpox by launching the Global Smallpox Eradication Program. While a daunting undertaking, circumstances favoring success included the following:

- The smallpox or variola virus was exclusively a human virus.
- A safe, effective, and stable vaccine was available (lyophilized preparations were developed by 1950).
- Smallpox did not cause silent or persistent infections.
- The disease was easy to recognize, diagnose, and track.

In the United States, routine use of smallpox vaccination ended in 1972, when the small risk of serious complications came to exceed the risk of infection. The last natural case of smallpox in the world was in 1977 and the disease was declared eradicated in 1979. Routine vaccination was discontinued worldwide in 1982. Variola virus now exists only in two high-containment laboratories. However, development and production of safer smallpox vaccines continues to this day, and vaccine stockpiles serve to guard against accidental or intentional release of this deadly and disfiguring virus.

Genome organization

Poxviruses are large DNA viruses that replicate in the cell cytoplasm. Genomes are a linear, base-paired, molecule of DNA, with linked ends. Genome sizes range from 130 to 300 kbp. The genome has a unique central region flanked by inverted repeats of about 10 kbp. Very close to each end is a series of short (~70 bp) tandem repeats. Genes encoded in the central region are well conserved and essential to basic replication, while genes in the repeat regions are more virus-specific and tend to function in virulence, immune evasion, and host range. VACV is the best studied of the poxviruses. It encodes about 200 genes, densely packed on the genome. Genes have short promoters and no introns (as there is no splicing apparatus in the cell cytosol). Proteins are encoded from both strands of DNA and some genes are overlapping. Genome replication and virion assembly occur in discrete virus "factories" in the cell cytosol.

Virion structure

Poxvirions are large, enveloped, ovoid, or brick-shaped (dimensions of VACV are about $360 \times 270 \times 250$ nm^3). Virion morphology is often referred to as complex and they contain 70–80 structural proteins. Virions have multiple envelope proteins, with obvious transmembrane domains, but notably, none are glycosylated. This may be a consequence of the unusual mechanism of virion assembly described later. The VACV virion is covered with a random pattern of proteinaceous surface ridges (Fig. 40.1). Other poxviruses (e.g., parapoxviruses) have a more regular (criss-crossed) surface pattern.

Viewed in thin section, virions contain a central tubular region (often dumbbell-shaped) containing DNA surrounded by a thick core wall. Electron dense, proteinaceous lateral bodies are positioned in the curves of the core and are surrounded by the lipid envelope. The majority of poxvirions (so-called mature virions or MV) are retained within infected cells, and only released after cell death. However, some poxvirions have two envelope layers. These specialized "extracellular virions" (EVs) are released from (living) infected cells, and remain associated with cellular structures (microvilli) that actively transport them to nearby uninfected cells (Box 40.1).

Replication

The family *Poxiviridae* contains viruses that infect vertebrates (subfamily *Chordopoxvirinae*) and insects (subfamily *Entomopoxvirinae*). VACV, genus *Orthopoxvirus*, is the best studied of the poxviruses and is the focus of this chapter (Box 40.2).

Attachment/penetration

VACV has a *very* broad cell tropism (seemingly able to infect any cultured cell) and at least four viral proteins have been implicated in entry. It is likely that multiple cell receptors are used for attachment.

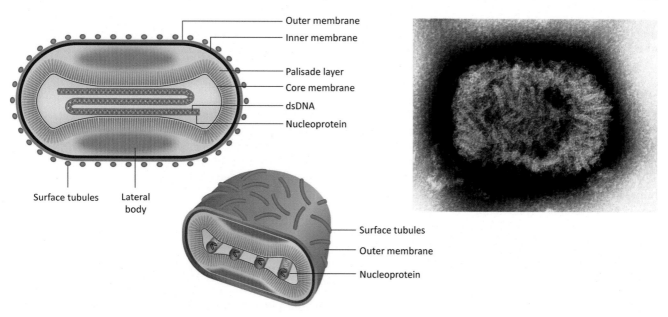

FIGURE 40.1   Virion structure (*left*). Transmission electron microscopic image depicting a smallpox (variola) virus particle (*right*). Source: *CDC Public Health Image Library Image ID#8392. Content providers: CDC/J. Nakano.*

BOX 40.1

General characteristics

Poxviruses are large DNA viruses that replicate in the cell cytoplasm. Genomes are linear, base-paired, molecules of DNA, with linked ends. Genomes range in size from 130 to 375 kbp and encode hundreds of polypeptides. DNA synthesis occurs in a discrete location in the cytoplasm to form a region called a "virus factory," an electron-dense region of cytosol devoid of cellular organelles. Poxviruses encode both RNA and DNA polymerases. They also encode many proteins that inhibit innate and adaptive immune responses. Virions are brick-shaped or ovoid, complex-enveloped structures that range in size from 140 to 260 nm in height and 220 to 450 nm in length. Although poxviruses encode all of the enzymes they require for replication in the cytoplasm, naked DNA is poorly infectious due to the need for packaged viral proteins that can initiate transcription.

BOX 40.2

Taxonomy

Family *Poxviridae*
Subfamily: *Chordopoxvirinae*. Contains 18 named genera, including the following:
Genus: *Avipoxvirus* (12 named species including *Canarypox virus*)
Genus: *Capripoxvirus* (species include Goatpox virus and Lumpy skin disease virus)
Genus: *Cervidpoxvirus* (species Mule deerpox virus)
Genus: *Leporipoxvirus* (species include Myxoma virus and Rabbit fibroma virus)

Genus: *Molluscipoxvirus* (species Moluscum contagiosum virus)
Genus: *Orthopoxvirus* (12 named species, including Variola virus, Vaccinia virus, Cowpox virus, and Monkeypox virus)
Genus: *Parapoxvirus* (species include *Orf virus* (sheep and goats), *Pseudocowpox virus*)
Subfamily: *Entomopoxvirinae* (Contains four named genera of insect poxviruses)

Transcription and genome replication

The first synthetic event in the replication cycle is transcription of a set of early genes. All proteins required for synthesis of early mRNAs are present in the lateral bodies and core. These include a nine-subunit RNA polymerase, a viral transcription factor that activates early promoters, a capping enzyme, and a poly(A) polymerase (Fig. 40.2).

Early gene transcription occurs prior to the complete uncoating of the genome and sets the stage for viral DNA replication. Functions of early gene products include the following:

- Regulate (promote) cell growth.
- Suppress various innate and adaptive immune pathways.
- Support DNA replication (including DNA polymerase and primase, enzymes needed for synthesis of dNTPs, DNA topoisomerase, helicase, and ligase, among others).

VACV DNA synthesis occurs in a discrete location in the cytoplasm to form a region called a "virus factory," an electron-dense region of cytosol devoid of cellular organelles. VACV replication is rapid, with virus factories formed by 2–3 hours postinfection. DNA synthesis begins with a nick in the one of the terminal hairpins to produce a 3′ end that serves as a primer. The overall process of DNA replication involves the formation and resolution of hairpins to produce head-to-head and tail-to-tail concatemers. The last step in the process is ligation to form a new covalently closed molecule.

Newly replicated genomes are templates for intermediate and late transcription. The products of intermediate and late transcription include structural proteins and assembly factors (scaffold and chaperone proteins) as well as enzymes and transcription factors that will be needed to initiate infection of a new cell.

Assembly/release

Poxvirus assembly is a fascinating process that can be visualized by electron microscopy (Fig. 40.3). It begins with the appearance of rigid, crescent-shaped membrane structures that contain a lipid bilayer and protein. In the case of VACV, crescents can be visualized by about 5–6 hours postinfection. The membrane fragments that form the crescents are likely derived from preexisting cellular membranes that rupture to generate these open fragments. Virion proteins are closely associated with the membrane fragments, forming an organized scaffold. Virions form during a process in which crescents close around an electron-dense material that contains viral core proteins. DNA enters the crescents before they close completely. After the envelope closes, the captured material rearranges and coalesces to form the core and lateral bodies. Finally, the ovoid or brick shape of the MV is formed.

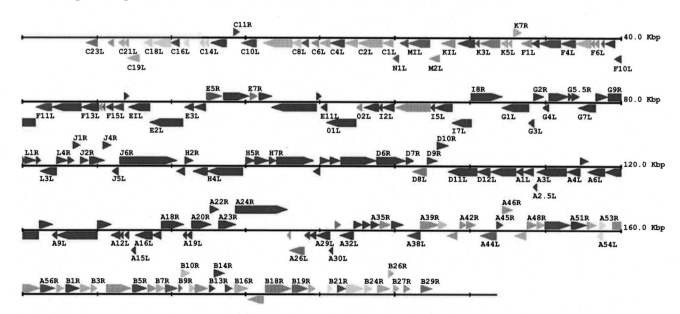

FIGURE 40.2 Diagram of the VACV strain Copenhagen genome. Genes are colored using a sliding scale, with dark blue representing those that are present in all poxviruses and yellow those that are in only one to three poxviruses (Lefkowitz et al., 2006). Source: *From Lefkowitz, E.J., Wang, C., Upton, C., 2006. Poxviruses: past, present and future. Virus Res. 117, 105–118. https://doi.org/10.1016/j.virusres.2006.01.016.*

Diseases
357

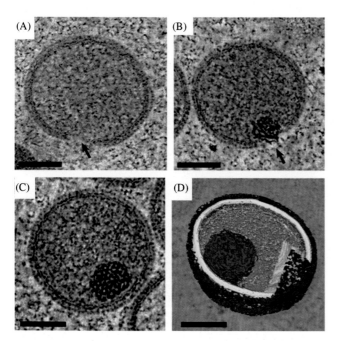

FIGURE 40.3 Poxvirus morphogenesis. After DNA uptake, the VACV crescent is sealed and forms a closed compartment. (A) Late-stage crescent. (B) Immature virus in the process of DNA uptake. (C) The membrane is sealed and the DNA is condensed. (D) Yellow indicates membrane, red indicates the outer protein scaffold, and blue indicates viral DNA (Chlanda et al., 2009). Source: *From Chlanda, P., Carbajal, M.A., Cyrklaff, M., Griffiths, G., Krijnse-Locker, J., 2009. Membrane rupture generates single open membrane sheets during vaccinia virus assembly. Cell Host Microbe 6, 81—90.*

MVs have a single envelope layer and most remain within the cell. A small number of MVs are further processed to become EVs. EVs are distinct from MVs and they play a special role in poxvirus spread. EVs are specifically designed for export from the infected cell and transport to nearby uninfected cells. Formation of EVs begins when MVs are wrapped in an additional layer of lipid (derived from the Golgi and containing a special set of viral proteins) to form the so-called wrapped virus (WV). WVs are particles with three lipid bilayers. WVs are transported to, and fuse with, the plasma membrane, releasing EVs, particles with two lipid bilayers. EVs remain associated with the cell membrane and serve as a focus for assembly of actin structures within the cell, to form EV-tipped microvilli that extend into the extracellular space. These virion-tipped microvilli contact and infect nearby cells.

Immune avoidance

Poxviruses encode a variety of proteins that interfere with the innate and adaptive immune responses. Many of these have been described for VACV (and homologs have been found among many sequenced orthopoxviruses). The number and type of immune antagonists

vary and they contribute significantly to host range and virulence. Among host defense systems compromised by VACV are the interferon system, complement system, inflammatory responses, and apoptosis.

Examples of intracellular signaling pathways affected by VACV-encoded proteins include the interleukin 1 receptor signaling pathway and multiple toll-like receptor signaling pathways. Blocking these pathways ultimately inhibits activation of nuclear factor kappa B (NF-kB) and interferon regulatory factors such as IRF3, thus inhibiting synthesis of interferon-regulated genes. VAVC also encodes proteins that interfere with mitochondrial components of the apoptotic cascade and inhibit activation of caspases. Another layer of protection is provided by a VACV protein that binds directly to double-stranded RNA to inhibit activation of dsRNA-dependent protein kinase and 2'-5'-oligoadenylate synthetase, key players in two pathways of the antiviral response.

In addition to encoding intracellular proteins that disrupt cell-signaling pathways, poxviruses encode secreted proteins and peptides in their battle with the immune system. *Virokines* are a group of secreted peptides that resemble (and thus interfere with) host cell cytokines (Fig. 40.4). Another group of secreted proteins, the *viroceptors*, bind and sequester host cell ligands. VACV produces soluble products that bind to many cytokines (including TNF, IL-1β, IFN-γ, and IL-18), chemokines, interferon receptors, and components of the complement cascade.

Diseases

Smallpox

Smallpox was an exclusively human disease. The earliest recorded descriptions of smallpox date back ~3000 years and it is likely that smallpox emerged as a human disease ~5000—10,000 years ago. One scenario has the virus moving from rodents to domesticated cattle to humans. By CE 700, smallpox is recorded in China, Japan, Europe, and Northern Africa. The virus was a relative "late-comer" to the Caribbean (CE ~1500), North America (CE ~1600), South Africa (CE ~1700), and Australia (CE 1789). Disease severity varied somewhat, but 20%—30% case fatality was very common in Europe and Asia. When the virus moved to Americas and Australia, it was even more lethal among populations newly exposed to the virus. Smallpox (and other infections, such as measles) played a critical role in the ability of Europeans to settle the new world, killing large numbers of native inhabitants with mortality rates of 90%—95%. In the early 1900s, a milder form of variola emerged (called variola minor) and eventually spread throughout the

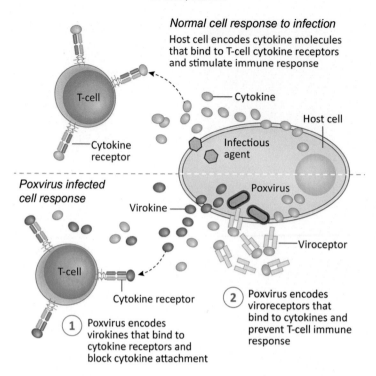

FIGURE 40.4   Poxvirus evasion of host responses to infection. Poxviruses evade immune responses by a variety of methods, including production of proteins that mimic host cytokine receptors (viroceptors) and cytokines (virokines).

United States, South America, and Europe. The cases fatality rate of variola minor was often less than 1%.

During natural infection, transmission of smallpox was by the aerosol route, inhaled into the lungs. After primary replication, the virus became systemic and caused a febrile illness typified by high fever, malaise, and prostration. The digestive tract could be affected, causing nausea, vomiting, and backache. Virus eventually moved to the skin (~14 days postinfection) and a rash developed over a 1- to 4-day period. Parts of the body involved included mucosa of the mouth and pharynx, face, arms, legs, and soles of the hands and feet. The rash progressed slowly, through characteristic stages of papules (raised spots), firm fluid-filled vesicles, pus-filled lesions (pustules), and scabs. Lesions developed deep within the skin, causing severe scarring (Fig. 40.5). The most lethal cases of smallpox (such as those seen among native Americans) occurred when deep lesions caused hemorrhage and sloughing of skin.

Molluscum contagiosum

Molluscum contagiosum is caused by another exclusively human poxvirus (Molluscum contagiosum virus). In sharp contrast to smallpox, infection results in a *mild, localized skin disease* that presents as small, raised, pearly looking lesions (called mollusca). Lesions range in size from that of a pinhead to 5 mm in diameter and may occur anywhere on the body. Transmission is via direct contact or fomites. The virus is not systemic, but can be spread to other body parts as a result of scratching the itchy lesion or by activities such as shaving.

Monkeypox

Monkeypox is an orthopoxvirus endemic to parts of Africa. Despite the name, the natural hosts appear to be rodents. Genetic analysis suggests that there are two distinct lineages of monkeypox (with different virulence for humans). Monkeypox virus infection of humans causes a smallpox-like disease and the virus was first isolated from humans in 1970 as part of a smallpox surveillance campaign. An important distinction between smallpox and monkeypox viruses is that the latter is not efficiently transmitted among humans. Most human cases involve direct contact with infected persons or animals (rodents).

In the United States, in 2003, an outbreak of monkeypox was associated with pet prairie dogs. By the end of the outbreak, the CDC reported 47 cases (37 confirmed and an additional 10 probable human cases). The cases were relatively mild (no deaths) but the outbreak prompted a significant response and investigation. How did monkeypox, an African virus, come to find its way into prairie dogs in the United States? The introduction on monkeypox was traced to importation of a large shipment of wild caught small mammals (including rodents), from West Africa. Some

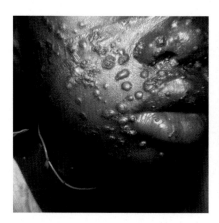 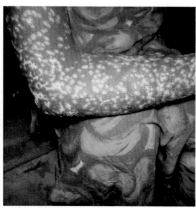

FIGURE 40.5 Smallpox lesions. *Left*: Close-up of smallpox lesions on the face of a patient living in Accra, Ghana, 1967. *Right*: Arm of an adult woman, which displays the characteristic maculopapular rash due to the smallpox virus. Source: *CDC Public Health Image Library Image ID #1994 (left) and #1990 (right). Content providers: CDC/Dr. J. Noble Jr.*

of those rodents found their way to a prairie dog breeder and all infected persons had purchased pet prairie dogs. While monkeypox is endemic in parts of Africa it was rare in other countries until 2022. In early 2022 a surprising outbreak involving dozens of countries and numbering over 5000 cases (as of July 2022) was underway. The extent of the outbreak remains to be seen.

Rabbit myxoma

Rabbit myxoma virus is a poxvirus that causes localized skin infections in American rabbits but can cause high mortality among European breeds. The virus is mechanically transmitted by mosquitoes. Rabbit myxoma virus is best known for its introduction to Australia in 1950 to control huge populations of imported European rabbits. While the virus initially produced high mortality (estimated to have reduced the rabbit population from ~600 million to ~100 million) within about 20 years, the virus became notably less virulent and Australian rabbits became much more resistant (allowing the rabbit population to rebound somewhat). This is an excellent example of host–virus coevolution. An elegant set of experiments demonstrated that viral and host genetics had both changed: Viruses from Australia were demonstrated to be less virulent when inoculated into European rabbits and Australian rabbits proved to be more resistant when infected with highly virulent myxoma virus strains.

In this chapter, we have learned that:

- Poxviruses are large, enveloped DNA viruses that encode hundreds of polypeptides. They have a complex structure.
- Poxviruses encode both RNA and DNA polymerases, which allow them to replicate in the cell cytoplasm.

- Poxviruses do not obtain their envelopes by a budding process. Instead, discrete membrane fragments appear to coalesce around viral proteins with a virus factory in the infected cells.
- In addition to encoding intracellular proteins that interfere with antiviral responses, poxviruses encode two different types of secreted protein. Virokines are peptides similar to host cytokines and they interfere with normal cytokine signaling. Viroceptors are secreted proteins that bind and sequester host ligands (e.g., cytokines). This prevents ligands from initiating signals to activate antiviral or inflammatory pathways.
- Variolation was use of dried smallpox lesions to intentionally infect people with smallpox. The process used scabs from mild cases of smallpox that were inoculated into scratches in the skin to produce a localized infection. Jenner used cowpox lesions to the same effect. However, cowpox was naturally much more attenuated for humans, and thus vaccination produced less disease, but achieved the desired protection.
- The introduction of rabbit myxoma virus into Australia was intentional, to control the rabbit population. Initially, mortality was very high; however, over time, selection resulted in changes in the genetics of both virus and host. Viruses became less virulent and the Australian rabbit population became more resistant.

References

Chlanda, P., Carbajal, M.A., Cyrklaff, M., Griffiths, G., Krijnse-Locker, J., 2009. Membrane rupture generates single open membrane sheets during vaccinia virus assembly. Cell Host Microbe 6, 81–90.

Lefkowitz, E.J., Wang, C., Upton, C., 2006. Poxviruses: past, present and future. Virus Res. 117, 105–118. Available from: https://doi.org/10.1016/j.virusres0.2006.01.016.

41

Other large DNA viruses

After reading this chapter, you should be able to answer the following questions:

- What are the general characteristics of African swine fever virus (ASFV)?
- What are the natural hosts of ASFV?
- What are the hosts of iridoviruses?
- Name one or two of the giant viruses of amoeba. Why has the discovery of these viruses prompted renewed discussion of the basic definition of a virus?

For most of the history of virology, one of the key defining features of viruses was their small size: they were smaller than bacteria, and thus were "filterable agents" that could pass through filters designed to trap bacteria. They were also too small to be seen with a light microscope (the maximum resolution of a light microscope is ~200 nm). Throughout the 20th century, the largest known animal viruses were the poxviruses and ASFV. Poxviruses range from 140 to 260 nm wide and 220 to 450 nm long. ASFV is ~175–215 nm in diameter. In comparison, mycoplasmas, among the smallest of the bacteria, are ~200–300 nm. A seminal article was published in 2003 that changed the landscape of virology. A research group from the Université de la Méditerranée in Marseille, France published an article describing a giant virus of amoeba (La Scola et al., 2003). The virus they described had first been seen in 1992 but was thought to be a bacterium, as it stained Gram positive and was clearly visible with a light microscope. The authors named the new virus mimivirus for "mimicking microbe." This

virus had a genome size of ~800 kbp of DNA and a capsid diameter of ~400 nm with filaments that projected another 100 nm from the capsid surface bringing the total diameter of the virus up to 600 nm. Since 2003, other giant viruses have been discovered, many by culturing water and soil samples with the unicellular eukaryote *Acanthamoeba polyphaga*. These large viruses are often referred to as nucleocytoplasmic large DNA viruses (NCLDV). In the 2021 taxonomy released by the International Committee on the Taxonomy of Viruses (ICTV), the phylum *Nucleocytoviricota* includes seven families of large dsDNA viruses, including *Poxviridae*, *Asfarviridae*, *Ascoviridae*, *Iridoviridae*, *Marseilleviridae*, *Mimiviridae*, and *Phycodnaviridae* (Box 41.1). The mimiviruses are particularly interesting as they are hosts to other viruses (called virophages) and have developed specific resistance mechanisms to control them. Poxviruses are described in Chapter 40. This chapter will briefly describe a few additional other large DNA viruses, beginning with ASFV.

Family *Asfarviridae*

ASFV is the sole member of the family *Asfarviridae*. ASFV is a large DNA virus with a genome structure similar to the poxviruses, having a linear dsDNA genome (170–194 kbp) with covalently closed ends. Replication is mainly cytoplasmic but the nucleus serves as a DNA replication site at early times postinfection. ASFV is endemic to Africa where it is

Viruses.
DOI: https://doi.org/10.1016/B978-0-323-90385-1.00011-X

© 2023 Elsevier Inc. All rights reserved.

BOX 41.1

Taxonomy of selected large DNA viruses

Realm *Varidnaviria*. As of 2021 the realm contains 2 kingdoms, 3 phyla, 5 classes, 12 orders, and 18 named families, which indicates the complexity and diversity of this group.

Phylum *Nucleocytoviricota*. Created to include the largest dsDNA viruses.

Family *Ascoviridae*. Hosts include insects, mainly lepidopteran larvae. They are transmitted by parasitoid wasps.

Family *Asfarviridae*. African swine fever virus is an economically important pathogen of domesticated pigs.

Family *Iridoviridae*. Hosts include fish, amphibians, and invertebrates. A few iridoviruses cause economically important fish diseases.

Family *Marseilleviridae*. Hosts include acanthamoeba and other protozoa.

Family *Mimiviridae*. Hosts include acanthamoeba and other protozoa.

Family *Phycodnaviridae*. Hosts include marine and freshwater algae.

Family *Poxviridae*. Contains 18 genera. Hosts range from insects to birds to mammals.

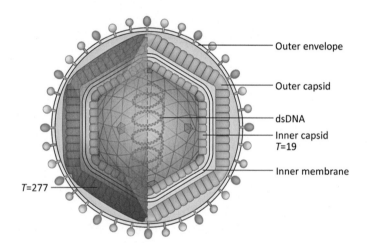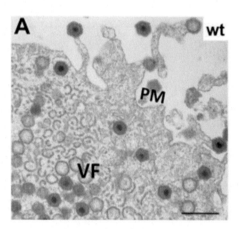

FIGURE 41.1   ASFV virion structure (*left*). Transmission electron microscopic image (Salas and Andrés, 2013). Source: *From Salas, M.L., Andrés, G., 2013. African swine fever virus morphogenesis. Virus Res. 173, 29−41. https://doi.org/10.1016/j.virusres.2012.09.016. Epub 2012 Oct 8. PMID: 23059353.*

maintained in ticks and wild pigs, causing little to no disease. However, upon transmission to domesticated pigs, the most virulent strains cause mortality rates reaching 100%. In recent years, ASFV has become a major pathogen of domesticated pigs as it has spread widely out of Africa, into both Europe and Asia.

late. The immediate early and early genes are expressed before the initiation of DNA replication. Some immediate early genes are proteins needed for DNA replication. Enzymes packaged in the virion are used to drive expression of the immediate-early genes, which takes place within partially uncoated virions.

Genome organization

ASFV genomes are a linear, dsDNA with covalently closed ends. Genome size ranges from 170 to 194 kbp, containing over 150 open reading frames (ORFs). Over 60 of these encode structural proteins. As is common for DNA viruses with large genomes, gene expression is highly regulated. Four groups of genes have been identified: Immediate early, early, intermediate, and

Virion structure

ASFV virions are complex and multilayered, ranging in size from 175 to 215 nm (Fig. 41.1). More than 50 structural proteins have been identified. At the center of the virion is an internal core containing the genome. This is surrounded by a thick protein layer named the core shell. Surrounding the core shell is an inner envelope. An icosahedral capsid forms the outermost layer of intracellular virions. Extracellular virions are surrounded by an additional lipid bilayer obtained during budding; however, this envelope is apparently not required for infectivity.

Recent cryo-electron microscopy studies have revealed more detail about ASFV capsid structure (Andrés et al., 2020; Liu et al., 2019). The $T = 277$ icosahedral capsid is assembled from more than 8000 copies of the 72-kDa major capsid protein (MCP), 60 copies of penton protein, and at least 8340 copies of minor capsid proteins. Minor capsid proteins are positioned under MCP layer, presumably to stabilize the huge shell. The core shell that surrounds the DNA is arranged as a $T = 19$ icosahedral capsid. An internal lipid membrane separates the two capsid layers. It is amazing to realize that the same underlying principals used to construct $T = 1$ capsids from 60 copies of a capsid protein are used to generate the shell of this large virus.

Overview of replication

The natural host cells for ASFV are cells of the monocyte/macrophage lineage. Receptors remain unknown but entry into cells requires endocytosis by clathrin- and dynamin-dependent mechanisms. ASFV remains in endosomes as they go through the endosomal maturation pathway. Acidification of the endosome may facilitate disassembly of the capsid, and late endosomes contain particles that have lost the outer capsid layer. This exposes the inner envelope and presumably facilitates a fusion event that releases naked cores into the cytosol.

The replication site of ASFV occurs close to the nucleus. Transcription begins using proteins packaged with the DNA genome. ASFV encodes a large number of proteins that involved in transcription and modification of mRNAs, and thus transcription is independent of host enzymes. However, there does seem to be some nuclear involvement in early DNA replication. The exact requirement for the nucleus is not understood, as at later times after infection, it is clear that the bulk of DNA replication takes place in the cytoplasm, in a viral factory. Infected cells have a single, large viral factory that is perinuclear in location. Cellular proteins are recruited to, and are required, for the structure of the factory. For example, a "cage" of the intermediate filament vimentin surrounds the factory. The factory contains viral DNA, a high concentration of viral proteins, immature and mature virions, and membrane fragments. Thus the assembly and maturation of virions occurs within this factory. Virions are then transported to the plasma membrane and are released by budding where they acquire their outer envelopes.

Epidemiology and disease

There are over 20 genotypes of ASFV and disease course and outcome vary depending on both host and viral factors. The natural hosts of ASFV are wild pigs in Africa, including warthogs and giant forest hogs. The virus appears to well-adapted to its natural hosts, causing no or mild disease. In the sylvatic cycle, transmission occurs via soft-bodied ticks of the genus *Ornithodoros*. To date, ASFV is the only known DNA virus transmitted by insects.

Transmission of ASFV to domesticated pigs first occurred in Africa in the early 20th century. The virus can cause high mortality, up to 100%, in domesticated pigs. Once introduced into a herd, transmission among animals is via respiratory secretions, with no need for a tick vector. In its most virulent and acute form, severe hemorrhagic disease develops over a 7-day period and involves vascular changes and bleeding. Less virulent genotypes can trigger persistent infections. It is likely that cytokines released by activated and necrotic macrophages are involved in disease pathogenesis.

ASFV was occasionally seen outside of Africa during the 20th century, but outbreaks were controlled and the virus did not become endemic. This changed in 2007 when the virus entered the country of Georgia and from there moved into both Europe and Asia. China reported its first case of ASFV in 2018; the virus has severely reduced pork production in that country. The virus spread rapidly and widely due to ease of transmission by fomites, food products and pigs. The virus is also present in semen, notable in that large-scale pork production is heavily dependent on artificial insemination. ASFV is highly stable, and infectious virions can persist in smoked and cured pork products for months. In Germany, ASFV is present in wild boar and therefore has become endemic. At the present time, there is no vaccine for ASFV although development is ongoing. As pigs that recover from an infection appear to be resistant to future infection against all genotypes, it is hoped that vaccines will eventually be successfully deployed.

Family *Iridoviridae*

Iridoviruses are large dsDNA viruses with a genome size ranging from 105 to 127 kbp. Virion sizes range from 120 nm for naked icosahedral particles to 200 nm for virions with an external membrane. Virions are complex; those that bud from cells have an external membrane (that is not required for infectivity) covering a viral capsid. The outer surface of the capsid is covered with flexible fibrils. A second envelope lies in between the outer capsid and an inner capsid shell. The dsDNA genome, complexed with proteins, is found at the center of the virion. The family *Iridiviridae* contains 2 subfamilies, 7 genera, and 22 species. Iridoviruses host include insects, crustaceans, fish, amphibians, and reptiles. Ranaviruses can cause significant disease, and economic loss, in cultured fish.

The giant viruses of acanthameoba

Many of the largest DNA viruses have been discovered by virtue of their ability to replicate in cultures of acanthameoba. These include viruses in two named families (*Marseilleviridae* and *Mimiviridae*) as well as several other groups that have not been formally classified at this time, including megavirus, pandoravirus, pithovirus, and medusavirus, among others. Use of other types of amoeba has revealed additional viruses. Members of the family *Phycodnaviridae* replicate in freshwater and marine algae. All of these viruses are notable for their large virions and genomes; all encode a small set of common genes, including those required for DNA replication and packaging. Many encode double-jelly roll capsid proteins. However, within their huge genomes, there is also tremendous variability in the range and scope of encoded proteins. By far, most ORFs encode proteins of unknown function, with no known homologs in the sequence data bases. However, a variety of proteins, likely captured from eukaryotes, bacteria, or archaea have been identified. These include tRNA synthases, translation factors, and enzymes needed for nucleotide biosynthesis. Surprisingly, some predicted genes have functions in basic cell metabolism such as enzymes involved in the tricarboxylic acid cycle, glycolysis, cellular fermentation, and amino acid synthesis. Additional types of identified proteins include serine and/or threonine kinases,

restriction endonucleases, ankyrin repeat-containing proteins, and histone-like proteins. The genomic content varies widely among these virus groups but it is likely that at least some of these proteins play crucial roles in manipulating the metabolism of infected cells. The large and varied gene content among the giant viruses provides fertile ground for studies of virus evolution and has stimulated renewed discussion of the basic definition of a virus. A few examples of giant viruses are provided below.

Mimiviruses were discovered in the amoeba *Acanthamoeba polyphaga* and were the first of many large DNA viruses found to replicate in these hosts. The family *Mimiviridae* currently contains three subfamilies populated by many viruses. Mimiviruses have an ~500-nm diameter icosahedral capsid (Figs. 41.2

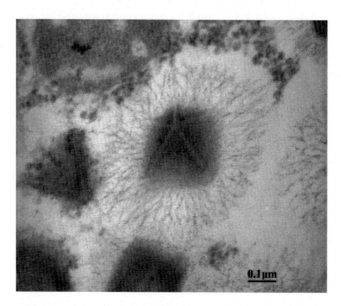

FIGURE 41.3 Electron micrograph of a mimivirus showing its stargate feature (King et al., 2012). Source: *From King, A.M.Q., Adams, M.J., Carstens, E. B., Lefkowitz, E.J. (Eds), 2012. Family Mimiviridae. In: Virus Taxonomy, vol. 2012, Elsevier, pp. 223−228, ISBN 9780123846846, https://doi.org/10.1016/B978-0-12−384684-6.00021-5.*

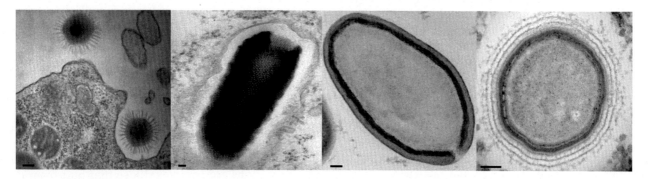

FIGURE 41.2 Giant viruses. *Left to right*: mimivirus, pithovirus, pandoravirus, and mollivirus. Scale bars: 100 nm (Abergel and Claverie, 2020). Source: *From Abergel, C., Claverie, J.M., 2020. Giant viruses. Curr. Biol. 30, R1108−R1110. https://doi.org/10.1016/j.cub.2020.08.055. PMID: 33022247. 5287090433760.*

and 41.3). Their genomes are ~1.2 Mbp linear, AT-rich (70%), dsDNA containing over 900 ORFs. They encode both DNA and RNA polymerases, as well as other transcription machinery, facilitating their replicate in the cytosol. Long fibers extend from the capsid surface. These fibers are resistant to enzymatic treatment with proteases and glycosidases and are thought to consist of a polysaccharide biopolymer. The virion is multilayered with lipid membranes lying under the capsid layer. Studies of purified virions have revealed over 100 protein products. Following phagocytosis of virions by amoeba, a region of the capsid, termed the stargate, opens to allowing fusion of the internal envelope with the host phagosome to deliver the core into the cytoplasm. By 8 h postinfection, numerous virions are assembled in virus factories, leading to a lytic infection.

Studies of mimiviruses resulted in discovery of smaller viruses that replicate within mimivirus factories in amoeba. The first of these, discovered in cultures of acanthamoeba castellanii mamavirus in 2008, was named sputnik virophage. The term virophage was used to imply a virus that used another virus as its *host*. The replication of sputnik adversely affects mimivirus replication and ability to lyse the host amoeba. Sputnik is now classified as a member of the species *Mimivirus-dependent virus Sputnik*. Sputnik is a dsDNA virus with a virion size of ~50 nm and a genome of ~18 kbp encoding over 20 putative gene products, including genes for capsid proteins. Sputnik capsids are icosahedral with a $T = 27$ organization. A number of additional viruses that replicate in concert with the giant viruses of amoebae have been discovered. Sputnik appears to bind to the long fibrils of mimiviruses as a means of transmission and uptake into new host cells. Sputnik is a member of the recently created family *Lavidaviridae*.

In 2016 a further surprise concerning the mimiviruses and their associated viruses was the description of a defense system, dubbed MIMEVIRE (Levasseur et al., 2016) that confers resistance to specific lineages of virophages. The resistance mechanism compares to CRISPR—Cas systems wherein gene sequences of virophages are acquired, and used to inhibit later infections. Short sequences of virophages can be found interspersed in the genomes of mimiviruses that resist infection, and Cas-like genes have been identified in the resistant mimiviruses. While there is still debate about its exact mechanism of action, the existence of an acquired resistance mechanism is well-established.

The family *Marseilleviridae* contains members that infect amoeba. They are smaller than mimiviruses with capsid diameters of ~250 nm. Fibers with globular ends extend from the capsid. Genomes of marseilleviruses are dsDNA ranging from 350 to 400 kbp with hundreds of predicted genes. They are ubiquitous in environmental samples worldwide.

Pandoraviruses have genomes ranging from 1.9 to 2.7 Mbp packaged in strange, elongated virions (Fig. 41.2), described as amphora shaped (amphora are ancient Greek storage vessels with pointed bottoms) of 1200 nm long by 600 nm in diameter. Genomes are GC-rich (>61%) and may encode up to 2500 proteins sharing no resemblance to those of the mimiviruses or marseilleviruses. They do not encode transcription machinery, and thus replicate their genomes in the host cell nucleus. A similarly shaped giant virus (~1500 nm in length by 500 nm in diameter) named pithovirus was found by culture of a *30,000-year-old Siberian permafrost* sample on amoeba! The name pithovirus was derived from the Greek word pithos (designating a type of large amphora) as its virion was shaped quite like the previously described pandoravirus. However, even though similar in shape, pithovirus genomes have a high AT content and are small (compared with their overall capsid size) at 610 kbp.

The aforementioned viruses represent only a few of the large DNA viruses being discovered in both current and ancient environmental samples worldwide. They represent a valuable resource for those interested in virus evolution and viral impacts on the environment.

In this chapter we have learned that:

- ASFV is a large DNA virus that causes an economically important disease of domestic pigs. In 2018 it spread to both Europe and China.
- ASFV targets monocytes and macrophages and can cause lethal hemorrhagic disease.
- The largest known viruses infect single-celled eukaryotes such as amoeba. Many have been discovered over the past 10 years. With genomes up to 1.2 Mbp, virions of over 500 nm and encoding a wide variety of genes captured from eukaryotes, bacteria, and archaea, they have stretched the definition of a virus, at least in terms of size!

References

Abergel, C., Claverie, J.M., 2020. Giant viruses. Curr. Biol. 30 R1108−R1110. Available from: https://doi.org/10.1016/j.cub.2020.08.055. PMID: 33022247. 5287090433760.

Andrés, G., Charro, D., Matamoros, T., Dillard, R.S., Abrescia, N. G.A., 2020. The cryo-EM structure of African swine fever virus unravels a unique architecture comprising two icosahedral protein capsids and two lipoprotein membranes. J. Biol. Chem. 295, 1−12. Available from: https://doi.org/10.1074/jbc. AC119.011196.

Family Mimiviridae. In: King, A.M.Q., Adams, M.J., Carstens, E.B., Lefkowitz, E.J. (Eds.), Virus Taxonomy, 2012. Elsevier, pp. 223—228. ISBN 9780123846846. Available from: https://doi.org/10.1016/B978-0-12-384684-6.00021-5.

La Scola, B., Audic, S., Robert, C., Jungang, L., de Lamballerie, X., Drancourt, M., et al., 2003. A giant virus in amoebae. Science. 299, 2033. Available from: https://doi.org/10.1126/science.1081867. PMID: 12663918.

Levasseur, A., Bekliz, M., Chabrière, E., Pontarotti, P., La Scola, B., Raoult, D., 2016. MIMIVIRE is a defence system in mimivirus that confers resistance to virophage. Nature 531, 249—252. Available from: https://doi.org/10.1038/nature17146.

Liu, S., Luo, Y., Wang, Y., Li, S., Zhao, Z., Bi, Y., et al., 2019. Cryo-EM structure of the African swine fever virus. Cell Host Microbe. Available from: https://doi.org/10.1016/j.chom.2019.11.004. 836-843.e3.

Salas, M.L., Andrés, G., 2013. African swine fever virus morphogenesis. Virus Res. 173, 29—41. Available from: https://doi.org/10.1016/j.virusres.2012.09.016. Epub 2012 Oct 8. PMID: 23059353.

42

Family *Retroviridae*

After reading this chapter, you should be able to answer the following questions:

- What are the major steps in a retrovirus replication cycle?
- Which step in the retrovirus replication cycle is unique to this virus family?
- What are the enzymatic activities of reverse transcriptase (RT), RNaseH, and integrase (IN)?
- What is the function of retroviral protease (PR) and when is it active?
- Explain the difference between exogenous and endogenous retroviruses.
- What is insertional activation and how does it lead to cell transformation?
- How are retroviral oncogenes and cellular proto-oncogenes related?

The family *Retroviridae* contains two subfamilies: *Orthoretrovirinae* and *Spumavirinea*. The main focus of this chapter is the subfamily *Orthoretrovirinae*. Retroviruses are enveloped viruses with RNA genomes. Many have icosahedral capsids or cores (Fig. 42.1). The name retrovirus was coined when it was discovered that the RNA genome is a template for synthesis of double-stranded (ds) DNA. This process was counter to the prevailing dogma that DNA was a template for RNA synthesis (the process of transcription), never the reverse. An older name for the group of viruses we now call retroviruses was "RNA tumor virus" because they were associated with transmissible cancers in mammals and birds. Retroviruses can transform cells because they *must* insert or integrate their genomes into the host cell chromosome, an activity that is inherently mutagenic. But as we see in this chapter, retroviruses are much more than just cancer agents. About 8% of the human genome is composed of retroviral sequences; retroviruses have shaped our genomes and driven our evolution. The human immunodeficiency virus (HIV), discovered only a few decades ago, infects millions of people worldwide. It causes disease primarily by damaging a subset of T-lymphocytes that are central regulators of the immune response. Modified retroviruses are commonly used for gene delivery, genetic engineering, and gene therapy; they are often used as tools to alter animal genomes. Clearly, there is a lot to know about these unique viruses!

All of the viruses discussed to this point in this textbook are transmitted from one cell to another via infectious particles (virions) or in a few cases, as nucleocapsids. However, retroviruses can be transmitted either as infectious particles or as *heritable genes*. While some viruses occasionally insert their genomes into host DNA, this is usually a "mistake" that does not benefit the virus. In contrast, retroviruses *must*

© 2023 Elsevier Inc. All rights reserved.

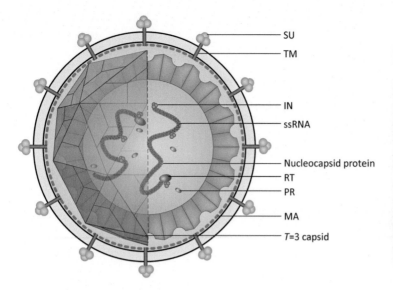

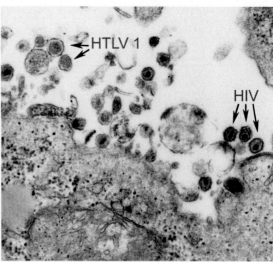

FIGURE 42.1   *Left*: Retrovirus structure. *Right*: Micrograph showing cells coinfected with human T-lymphotropic virus-1 (HTLV-1) and human immunodeficiency virus-1 (HIV-1). *Source: CDC Public Health Image Library, Image ID# 8241. Content providers: CDC/Cynthia Goldsmith.*

insert their genomes into host DNA, in order to complete a single replication cycle (Box 42.1).

There is important, but sometimes confusing, terminology used to describe retroviruses. Retroviruses transmitted as virions are called *exogenous* (think *ex*tracellular) while retroviruses transmitted exclusively via the germ-line are called endogenous retroviruses. The many retroviral sequences that populate our genomes are endogenous retroviruses. Most are defective or fragmented but some do express proteins. Endogenous retroviruses are essentially cellular genes; they rarely, if ever, are transmitted as virus particles. The presence of large numbers of endogenous retroviruses in our genomes is a reminder that we have coexisted with these viruses for hundreds of millions of years. Retroviruses are one group of a larger family of replicating molecules called retrotransposons.

HIV is an exogenous retrovirus; it is transmitted from person to person via infectious virions in blood and body fluids. But confusion often stems from the following: Exogenous retroviruses *must* integrate their genomes into host DNA for replication. We call the *integrated* versions of retroviral genomes *proviruses*. RNA transcribed from the provirus is packaged into particles for transmission (Fig. 42.2). In order for an exogenous retrovirus to become endogenous, it must integrate into *germ-line* cells and become *fixed* in the host genome. The remainder of this chapter deals almost exclusively with exogenous retroviruses.

Virion structure

Retrovirus particles are enveloped with icosahedral (or icosahedral-like) cores (Fig. 42.1). Two glycoproteins are associated with the viral envelope. A transmembrane (TM) glycoprotein is anchored into the viral envelope and a surface (SU) glycoprotein found on the exterior of the virion. SU associates with TM via noncovalent interactions. Under the viral envelope, the matrix protein (MA) associates with the cytoplasmic tails of TM, as well as with the lipid envelope. Moving further into the virion, the capsid (or core) is assembled from capsid proteins (CA). All retroviruses encode one capsid protein. Within the capsid are two identical copies of RNA (capped and polyadenylated) that serve as the retroviral genome (retroviruses therefore have diploid genomes). The RNA is associated with the positively charged (basic) nucleocapsid protein (NC). Also within the capsid are a few molecules of the enzymes RT, IN, and PR. Two molecules of tRNA are also found within the virion. They are associated with RT and are base-paired to the genomic RNA near its 5′ end, at sequence called the primer-binding site (PBS). A molecule of tRNA is a primer for DNA synthesis (Box 42.2).

Some complex retroviruses, such as HIV, package additional proteins within their virions. Proteins such as HIV-1 viral protein R (VPR), viral protein U (VPU), and virus infectivity factor (VIF) play important roles in the HIV replication cycle. They are sometimes called "accessory" proteins and they serve to counter host cell defenses to retroviral infection. They will be discussed in more detail in Chapter 43.

BOX 42.1

General characteristics of retroviruses

Members of the subfamily *Orthoretrovirinae* package two identical copies of capped, polyadenylated genomic RNA within their capsids or cores. Upon penetration/uncoating, the RNA molecules are not translated, instead they are temples for synthesis of a molecule of ds DNA that is integrated into the host chromosome. Transcription of mRNAs from the integrated DNA genome (called the provirus) is catalyzed by host RNA Pol II. Transcripts are capped and polyadenylated and some are spliced. Unspliced transcripts serve as the genomes packaged into new virions.

Retroviral genomes are 7–12 kb. All retroviral genomes encode three precursor polyproteins (GAG, GAG-POL, and ENV). Some retroviruses encode additional proteins with a variety of regulatory and auxiliary activities.

Virions are enveloped with icosahedral-shaped capsids or cores. Their envelopes contain two glycoproteins produced by cleavage of the ENV precursor. One is a transmembrane (TM) glycoprotein with a cytoplasmic tail. The second glycoprotein is the surface (SU) glycoprotein. SU is the attachment protein and TM is the fusion protein. Capsids are assembled from the capsid protein (CA). A basic, RNA-binding nucleocapsid (NC) protein binds the genomic RNA. A membrane-associated matrix (MA) protein associates with the plasma membrane and the cytoplasmic tails of TM.

Genome organization

Retroviral genomes are 5′ capped and 3′ polyadenylated mRNAs. However, upon entry into the cell, the mRNA is not translated (in contrast to the replication cycle of plus-strand RNA viruses). Instead the RNA genome serves as the template for the synthesis of a molecule of ds DNA and the RNA genome is destroyed during this process.

All retroviruses have three major genes (Fig. 42.3). *Gag* (for group-associated antigen) encodes the GAG polyprotein. The *pol* gene encodes the polymerase (POL) polyprotein and *env* encodes the envelope (ENV) polyprotein. Retroviruses encoding only *gag*, *pol*, and *env* genes are sometimes called "simple" retroviruses. The order of these genes is always *gag-pol-env*. The oncogenic retroviruses of rodents and chickens are simple retroviruses. The so-called "complex" retroviruses such as HIV and human T-cell lymphotrophic virus (HTLV) have additional genes encoding a variety of regulatory and accessory proteins (Fig. 42.4).

In addition to protein-coding regions, the retroviral genome contains a number of important regulatory sequences. We will start with a discussion of the genomic *RNA* (Fig. 42.5). At each end of genome are 30–40 nucleotide long direct repeats (R) essential for the process of reverse transcription. One copy of the R sequence is found at the very 5′ end of the genome and the R sequence at the 3′ end precedes the poly(A) tail. Directly following R at the 5′ end of the genome is a short sequence called U5 (for unique at the 5′ end). Just upstream of R, at the 3′ end of the genome, is a region called U3 (unique at the 3′ end). Two additional cis-acting sites are present near the 5′ end of the genome. The short PBS is base-paired with a molecule of tRNA. Just following the PBS is the RNA packaging site or RNA encapsidation signal (also called ψ).

The product of reverse transcription, the DNA version of the genome, is longer than the RNA genome due to the presence of duplicated sequences at each end (Fig. 42.5). This occurs during the process of reverse transcription when the U5 sequence is copied to the 3′ end of the genome and the U3 sequence is copied to the 5′ end of the genome. This process forms the long terminal repeats (LTRs) that flank the protein-coding sequences. LTRs are organized as follows: U3−R−U5. LTRs are critical elements for transcription and polyadenylation of retroviral mRNA. The promoter and enhancer elements are often found in U3 and the upstream LTR serves as the promoter for transcription of viral mRNA. A poly(A) addition site is present at the R−U5 junction of each LTR, but polyadenylation is suppressed at the upstream LTR.

Overview of the retroviral replication cycle

The retroviral replication cycle begins with binding of SU to the receptor. Some retroviruses enter cells by fusion at the plasma membrane (e.g., HIV-1) but others are endocytosed and the trigger for fusion is low pH. In either case, the hydrophobic amino terminus of the TM protein mediates membrane fusion. When the retroviral core is released into the cytosol, viral RT synthesizes a DNA copy of the RNA genome. The ds DNA molecule is then integrated into the host

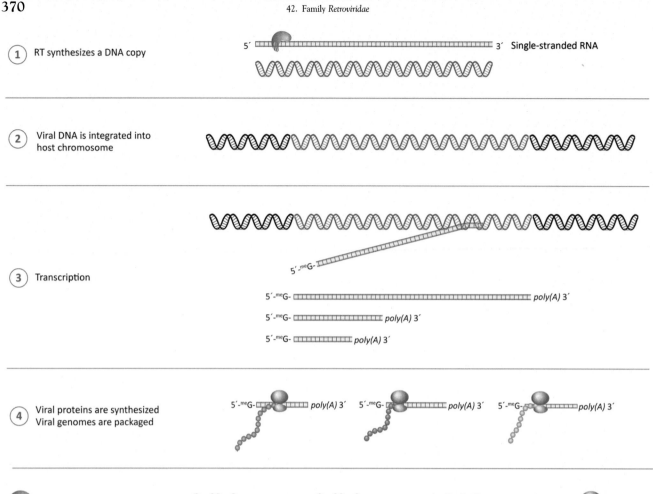

FIGURE 42.2 The major steps in retrovirus genome replication and transcription. Genome synthesis begins with reverse transcription of the infecting RNA genome. The DNA product is integrated into the host chromosome. The integrated retroviral genome is called the provirus and is transcribed by host RNA polymerase. Some viral mRNA remains unspliced and some is spliced. mRNAs are translated to produce virion proteins that package the unspliced mRNA to produce new infectious virions.

chromosome. The DNA genomes of some retroviruses (members of the genus *Lentivirus*) actively cross nuclear pores to gain access to the chromosome, but many retroviruses require a dividing cell, as their DNA genomes can only reach the host chromosome during mitosis (when the nuclear membrane breaks down). The next key player in the retrovirus replication cycle is IN. Recall that IN is present in the infecting virion. Dimers of IN interact with each end of the newly synthesized ds DNA genome. IN also interacts with host DNA and catalyzes cleavage and ligation steps that result in the covalent linkage of the viral DNA to the host DNA. The viral DNA is now, in essence, a cellular gene. Once integrated, the retroviral genome is referred to as the provirus. In order for the retroviral replication cycle to continue, the provirus must be transcribed by cellular RNA polymerase II to produce capped, polyadenylated mRNA. Spliced and unspliced viral mRNAs are transported out of the nucleus and are translated to produce a full complement of retroviral proteins. As the concentration of structural proteins increases, unspliced mRNAs are packaged to serve as the genomes of progeny virions.

The steps up to and including integration are sometimes called the early steps in the retroviral replication cycle, while those that occur after integration are called the late steps. In some cases, integration may occur months to years before virions are ever produced, as expression of the provirus may require a specific set of circumstances. In the case of HIV, the infection of a T-cell may be silent until the cell is stimulated to divide upon recognizing a foreign antigen.

Some key steps in the retroviral replication cycle are now described in greater detail.

BOX 42.2

Taxonomy of the family *Retroviridae*

The family *Retroviridae* contains seven genera within two subfamilies.

Subfamily: *Orthoretrovirinae*
Genus: *Alpharetrovirus*.
Species: *Rouse sarcoma virus* and eight others
Genus: *Betaretrovirus*
Species *Mouse mammary tumor virus* and four others
Genus: *Deltaretrovirus*. Four species recognized:
Species: *Bovine leukemia virus* (viruses include bovine leukemia virus)
Species: *Primate T-lymphotropic virus 1* (viruses include human T-lymphotropic virus 1 and simian T-lymphotropic virus 1)
Species: *Primate T-lymphotropic virus 2* (viruses include human T-lymphotropic virus 2 and simian T-lymphotropic virus 2)
Species *Primate T-lymphotropic virus 3* (viruses include human T-lymphotropic virus 3 and simian T-lymphotropic virus 3)
Genus: *Epsilonretrovirus*
Species: *Walleye dermal sarcoma virus* and two others
Genus: *Gammaretrovirus*
Species *Feline leukemia virus* and 18 other

Genus: *Lentivirus*. Nine species are recognized, including the following:
Species: *Human immunodeficiency virus 1* (human immunodeficiency virus 1; four groups, M, N, O, and P are recognized)
Species: *Human immunodeficiency virus 2* (human immunodeficiency virus 2, at least eight groups are recognized)
Species: *Simian immunodeficiency virus* (simian immunodeficiency virus)
Subfamily: *Spumavirinae*
Genus: *Bovispumavirus*
Species *Bovine foamy virus*
Genus *Equispumavirus*
Species *Equine foamy virus*
Genus *Felispumavirus*
Species: *Feline foamy virus*
Genus: *Prosimiispumavirus*
Species: *Brown greater galago prosimian foamy virus*
Genus: *Simiispumavirus*
Species: *Bornean orangutan simian foamy virus* and 14 others

---

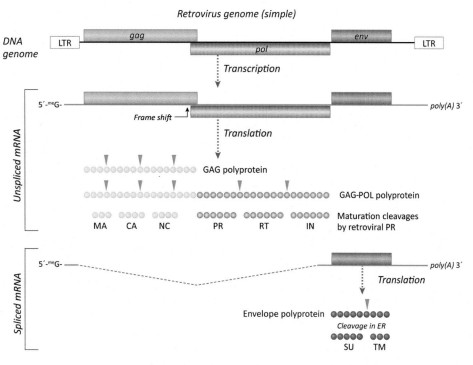

FIGURE 42.3 Genome of a simple retrovirus. Simple retroviruses have three open reading frames that encode three polyproteins that are subsequently cleaved to produce the complete array of retroviral proteins.

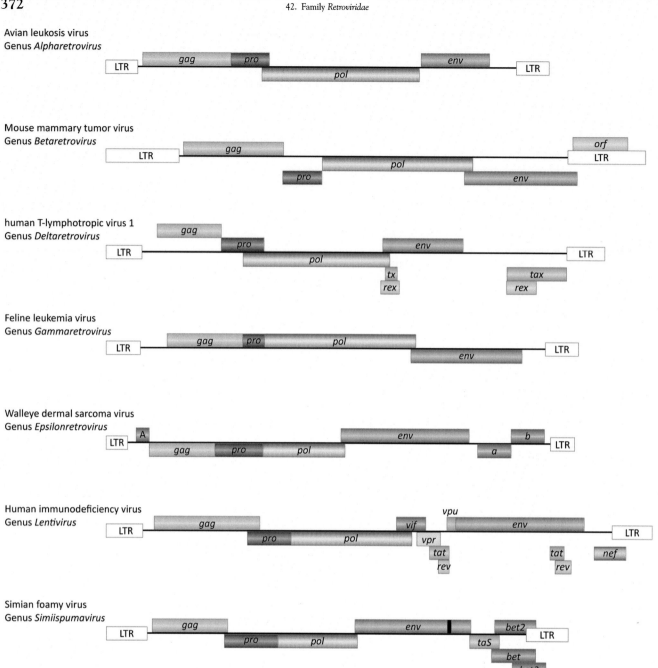

FIGURE 42.4 Comparison of retroviral genomes. Note that all retroviruses have three major open reading frames. Complex retroviruses encode up to several additional proteins.

The process of reverse transcription

The process of reverse transcription is carried out by a polymerase called RT. RT is a heterodimer. The longer polypeptide in the dimer contains the polymerization domain, a short linker domain, and an RNaseH domain. The shorter polypeptide in the dimer contains only the polymerization domain. The longer polypeptide provides the enzymatic functions of RT while the shorter polypeptide appears to play a strictly structural role. The structure of HIV-1 RT is shown in Fig. 42.6. HIV p66 folds to form the typical palm, fingers, and thumb domains of polymerases. The catalytic site is located in the palm and contains three critical aspartic acid residues and two coordinated Mg^{2+} ions.

Reverse transcription is key to the replication cycle of a retrovirus. As described earlier, retroviral genomes are 5'-capped and 3'-polyadenylated mRNAs. The RNA genome serves as the template for the synthesis of a molecule of ds DNA. Although two copies

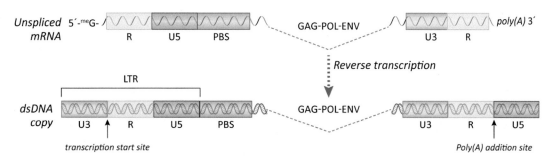

FIGURE 42.5 Comparison of RNA and DNA versions of a retroviral genome. During the process of reverse transcription, the long terminal repeat (LTR) is formed when U3 and U5 sequences are duplicated.

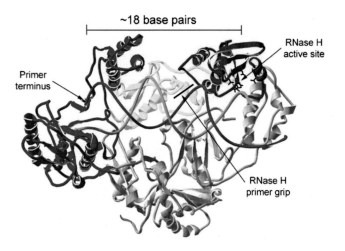

FIGURE 42.6 Molecular structure of RT. Cocrystal structure of HIV-1 RT and an RNA/DNA substrate. The polymerase and RNase H domain of p66 are shown in blue and red. The connection domain is yellow. p51 is shown in gray. The RNA template and DNA primer strands are shown in green and purple, respectively (Schultz and Champoux, 2008). Source: *From Schultz, S.J., Champoux, J.J., 2008. RNase H activity: structure, specificity, and function in reverse transcription. Virus Res. 134, 86−103.*

of RNA enter the cell, only one copy of ds DNA is synthesized.

The steps in reverse transcription are outlined in Fig. 42.7. Note that in this model, only one copy of RNA is shown. However, there is good evidence that during the process of reverse transcription, RT may jump from one of the packaged RNA strands to another. Reverse transcription begins with synthesis of a piece of DNA from the tRNA primer. The first strand of DNA synthesized is called "negative strand" because it is a copy of the mRNA (the positive strand). The first piece of DNA synthesized is rather short because RT reaches the 5′ end of the mRNA and can go no further. The RNA strand of this RNA:DNA hybrid molecule is digested away by the activity of the RNase H domain of RT. Note that the RNA:RNA duplex formed by the tRNA primer and the PBS is not digested as RNase H only degrades RNA that is part of an RNA:DNA hybrid. The short DNA product is called minus-strand strong-stop DNA.

The next step in the process is sometimes called a "jump" or a "strand transfer." In the infected cell, this step probably requires the NC protein (although this is not shown in Fig. 42.7). Key to the jump is that the newly synthesized R DNA base pairs with the R RNA at the 3′ end of the genome. The negative strand of DNA is then synthesized through to the 5′ end of the RNA. Not shown in Fig. 42.7 is that most of the RNA template is degraded *during* the process of DNA synthesis. One piece of RNA remains undigested, however: the so-called polypurine tract (PPT). The PPT is the RNA primer for synthesis of plus-strand DNA (Fig. 42.7, step 7). As we saw with minus-strand DNA synthesis, the first product of plus-strand DNA synthesis is a short fragment. As shown in step 8, part of the tRNA is copied, generating the PBS. Now the tRNA is digested away by RNase H and a second jump occurs: The PBS region on the plus strand is base-paired to its complement on the minus strand. The ds DNA molecule is completed by extension of the two partial DNA strands to form the completed linear ds DNA. The LTRs generated during reverse transcription are key regulatory regions of the provirus. It is important to note that RT is not the only viral player in the RT process. NC and other virion proteins (CA and IN) remain associated with the DNA as it is synthesized, forming the so-called preintegration complex (PIC). The PIC traffics to the nucleus. Many retroviruses replicate only in mitotically active cells as they require breakdown of the nuclear membrane to reach the host chromosome. However, lentiviruses, such as HIV, can replicate in nondividing cells because the PIC actively crosses the nuclear pore.

Integration

IN is bound to the very ends of the linear ds DNA molecule (Fig. 42.8); this complex is called the *intasome*. IN polypeptides interact to form dimers and tetramers, bringing the two ends of the linear provirus together. The number of IN proteins in the intasome ranges from 4 to 16

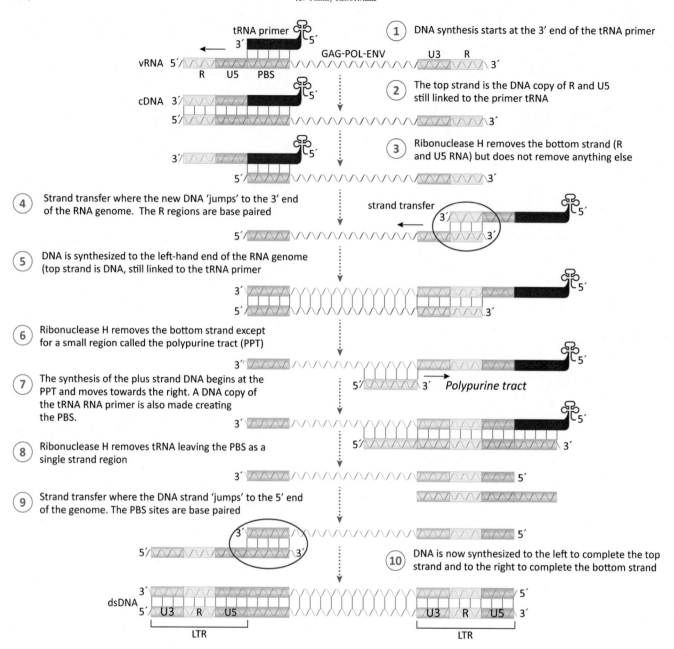

FIGURE 42.7 Steps in the reverse transcription process. Reverse transcription begins with synthesis from the tRNA primer to produce a short piece of minus strand strong stop DNA. R DNA is transferred to base pair with R RNA at the 3′ end of the genome and synthesis of minus strand DNA continues. RNA, with the exception of the PPT, is degraded. PPT primes synthesis of plus-strand strong stop DNA. Base pairing of PBS sequences occurs during a second strand transfer, followed by completion of both DNA strands.

(varying by genus). The intasome carries out three enzymatic activities: First it is an exonuclease that cleaves 2—4 nt from the 3′ ends of the linear, unintegrated ds DNA molecule. Second, it is an endonuclease that makes a staggered cut in host cell DNA. Third, it has ligase activity to join viral DNA to host cell DNA (ligation). Successful integration also requires host cell enzymes to repair gaps at the ends of the newly integrated molecule. The sites in the chromosome that serve as targets for integration are generally considered to be random. However, some types of retroviruses preferentially integrate in region of active gene transcription, while others do not.

Integration can be considered the last step in the *early phase* of the retrovirus replication cycle. The viral genome is now a host cell gene that can be passed to daughter cells. The provirus may remain silent for some time, but successful completion of the replication cycle requires that it eventually be transcribed.

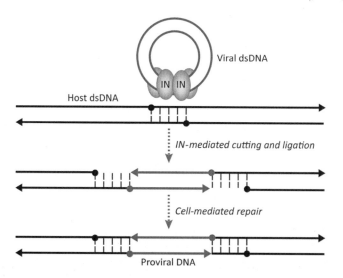

FIGURE 42.8 Integration. The ends of linear retroviral DNA are associated with IN protein. IN removes 2 nt from the 3′ end of each strand of viral DNA and makes a staggered cut in target chromosomal DNA and ligates each strand. DNA repair (to fill gaps) is performed by host enzymes. This process generates short direct repeats at each end of the insertion.

Transcription of retroviral mRNA

The retroviral LTR controls transcription of the provirus. U3 contains a core promoter as well as enhancer elements. Transcripts are synthesized by the host RNA Pol II complex. The efficiency of transcription is regulated by sequences in the LTR. Some retroviral promoters are active in many cell types while other promoters are only active in specific cells at specific times. An instructive example is the retrovirus mouse mammary tumor virus (MMTV). The MMTV promoter is active only in the mammary glands of a lactating mouse. Virus is produced during lactation, when it can be successfully transmitted to the suckling mouse pups. Some retroviral promoters, for example, HIV (a lentivirus) and HTLV (a deltaretrovirus), are relatively weak and require virally encoded transcription factors to increase transcription to sufficient levels for successful completion of the retrovirus replication cycle. In contrast, some oncogenic retroviruses often have strong promoters that drive transcription of not only the provirus, but of nearby cellular genes, a process called insertional activation.

All retroviral transcripts begin at the R region of the upstream LTR and are cleaved and polyadenylated at a site in the downstream LTR (Fig. 42.5). This process produces a capped and polyadenylated mRNA identical in structure to the infecting genome, with R sequences at each end. Some of the transcripts remain unspliced but others are spliced, to produce mRNAs that encode the ENV polyprotein. Complex retroviruses may produce multiple spliced mRNAs

that encode transcriptional activators and other accessory proteins.

Both spliced and unspliced mRNAs exit the nucleus. The GAG polyprotein is translated from unspliced mRNA (Fig. 42.3). The POL polyprotein is also synthesized from the unspliced mRNA, but this requires either a ribosomal frame shift or suppression of a stop codon to produce the GAG−POL polyprotein.

The ENV precursor is synthesized from a singly spliced mRNA. ENV is translated on rough ER and the precursor is processed in the ER and Golgi. Processing includes cleavage to produce SU and TM, as well as glycosylation. Complex retroviruses also encode accessory and regulatory proteins from a variety of singly and multiply spliced mRNAs. In the case of HIV and other lentiviruses, two key regulatory proteins are TAT and REV. TAT is a small protein that increases transcription from the HIV LTR. REV is a small protein that regulates transport of spliced and unspliced mRNAs from the nucleus.

Structural proteins

The full-length (unspliced) viral transcript is transported to the cytoplasm where it is translated to produce two precursor polyproteins, GAG and GAG−POL. GAG contains the domains for MA−CA−NC while the GAG−POL polyprotein also contains the RT and IN domains. As shown in Fig. 42.4, the PR domain is sometimes part of the GAG precursor and sometimes part of GAG−POL. Thus retroviruses have evolved a variety of strategies for producing PR. The GAG and GAG−POL precursors remain uncleaved within the infected cell. PR is not active until virions assemble.

The ENV polyprotein is translated from a singly spliced mRNA that lacks *gag* and *pol* sequences. The ENV polyprotein is synthesized on rough ER. Within the ER lumen, ENV is cleaved (by furin-like proteases) and the cleavage products are glycosylated as they travel through the ER and Golgi. For many retroviruses, the final destination of the ENV glycoproteins is the plasma membrane. Complex retroviruses like HIV use multiply spliced mRNAs to produce additional accessory proteins. Some of these are packaged into new virions to exert their effects during the next cycle of replication.

Assembly, release, and maturation

In the cytosol, the RT domain of the GAG−POL precursor binds the tRNA that will be used to prime DNA synthesis in the next cell. Full-length retroviral mRNAs have a "packaging signal" near the 5′ end that

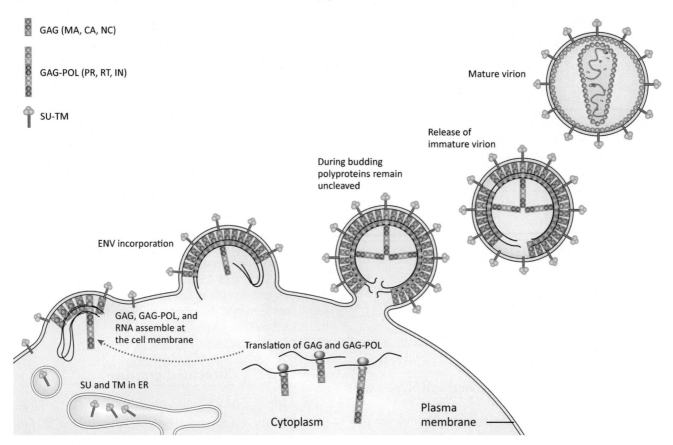

GAG (MA, CA, NC)

GAG-POL (PR, RT, IN)

SU-TM

Mature virion

Release of
immature virion

During budding
polyproteins remain
uncleaved

ENV incorporation

GAG, GAG-POL, and
RNA assemble at
the cell membrane

Translation of GAG and GAG-POL

SU and TM in ER

Cytoplasm

Plasma
membrane

FIGURE 42.9   Assembly of a retrovirus. This figure illustrates assembly of HIV-1 at the plasma membrane of the infected cell. The immature virion assembles from polyprotein precursors that are cleaved to form infectious virions after release from the cell.

allows specific incorporation of genomes into particles. GAG and GAG–POL polyproteins associate with a cell membrane (the PM in the case of HIV). That interaction is facilitated by a fatty acid molecule that is linked to the 5′ end of the MA domain of the GAG and GAG–POL polyproteins. These polyproteins also contain interaction domains that allow them to specifically interact with the cytoplasmic tail of TM (Fig. 42.9).

Mutational analysis of retroviral *gag* genes has revealed the presence of short amino acid motifs that promote the release of the budding particles from the PM. These motifs are called "late domains" because they are required in the very last steps of release from the cell. Three types of retroviral late domains have been identified: Pro–Thr/Ser–Ala–Pro (PTAP or PSAP), Pro–Pro–Pro–Tyr (PPPY), and Tyr–Pro–Xn–Leu (YPXnL, where X is any amino acid and n = 1–3 residues). The motifs are found in various places in GAG: The HIV late domain is in p6 (near the carboxyl terminus of GAG) while the HTLV-1 late domain is in MA. Mutation of these motifs disrupts virion budding and release. Retroviral late domains function by interacting with components of the cellular endosomal sorting machinery. The so-called ESCRT complexes (endosomal sorting complexes required for transport) are multiprotein complexes that sort cellular proteins into vesicles; in addition to their "sorting" function, they promote budding and membrane scission. There are four different ESCRT complexes that function together in cells; only a subset of ESCRT machinery is required for retrovirus budding.

HIV is an example of a retrovirus that buds from the plasma membrane. During budding, GAG and GAG–POL polyproteins become densely packed, triggering activation of the PR domain. PR cleaves the precursor proteins into their final products. MA remains associated with the inside of the viral envelope, and CA rearranges and assembles to form a capsid that surrounds the genome, NC, RT, and IN. The virion is now ready to infect a new cell (Fig. 42.9). During an HIV infection, the virus infects several types of cell that are key players in the immune response. Disruption of the activities of these cells leads to immune suppression and disease. These processes will be discussed in more detail in Chapter 43.

Unique aspects of spumaretrovirus replication

Members of the subfamily *Spumavirinae* differ in some noteworthy ways from the members of the subfamily *Orthoretrovirinae*. The name spumavirus derives from the "foamy" appearance of infected cells; hence, these viruses are also frequently referred to as the foamy viruses.

One major unique feature of the spumaviruses is the presence of DNA in the virion. *This occurs because the process of reverse transcription begins as the viruses are assembling.* The synthesis of the complete ds DNA probably is completed after entry, followed by integration. Transcription of the provirus generates a full-length mRNA that is used to produce the GAG polyprotein, but unlike the orthoretroviruses, there is a spliced mRNA used to produce the POL polyprotein. Thus there is no GAG–POL polyprotein produced. A third spliced mRNA is used to produce the ENV precursor protein.

While the orthoretroviruses produce all mRNAs from promoter sequences in the 5′ LTR, the spumaretroviruses also use a second promoter near the 3′ end of the genome. This internal promoter is used to produce mRNAs for a transactivating protein (TAS) and an accessory protein (BET). Protein processing is also unique for spumaretroviruses. The GAG polyprotein is not highly processed: There is only one cleavage of 71 dDa GAG, producing a 68 kDa protein and a 3 kDa peptide. Virions contain a mixture of p71/p68. Finally, the *pol* mRNA produces a precursor with domains for PR, POL, RNaseH, and IN. However, the polyprotein is cleaved only once, between the RNaseH and IN domain.

Mechanisms of retroviral oncogenesis

Retroviruses *must* integrate (insert) their genomes into host DNA for the purposes of replication, and disruption of host DNA is an inherently mutagenic event (Box 42.3). Early studies of avian retroviruses revealed discrete mechanisms of transformation: Some retroviruses [e.g., Rous sarcoma virus (RSV)] are *acutely* transforming and rapidly give rise to polyclonal tumors (each tumor is derived from a different infected cell). These so-called "acute" transforming retroviruses carry an oncogene into every infected cell (Fig. 42.10). In fact, these retroviral oncogenes were originally "captured" from cells.

BOX 42.3

A brief history of retroviral oncogenesis

Studies of oncogenic retroviruses date back to the early 1900s with descriptions of "filterable agents" capable of causing avian leucosis and avian sarcoma. Over the next 20 years, many distinct avian cancers were attributed to transmissible agents (viruses). By the 1950s, a group of related, tumor-causing RNA viruses had been identified from birds, rodents, and other mammals. They shared biochemical features and were morphologically similar, but displayed different phenotypes as regarded host specificity, tumor types, and transforming ability. For many years, these viruses were collectively called "RNA tumor viruses." In general, they caused noncytopathic infections, required dividing cells for establishment of productive infection, and were tumorigenic; once tumor cells were formed, the transformed phenotype was stably inherited by daughter cells.

In the 1960s Howard Temin (working at the McArdle Laboratory for Cancer Research at the University of Wisconsin-Madison) began to suspect that after infection, the genetic information of RNA tumor viruses became *stably associated* with the cell genome. Temin thought that process might be similar to DNA bacteriophage such as lambda phage, known to integrate into bacterial genomes. But the RNA tumor viruses had *RNA genomes*, and synthesis of DNA from an RNA template was unheard of! Temin performed a variety of studies to support his idea, but the breakthrough came when he demonstrated that RNA tumor viruses package a DNA polymerase. Temin and Mizutani (1970) published this seminal work, side-by-side with a paper from Baltimore (1970). Thus in 1970 two laboratories independently and convincingly demonstrated that RNA tumor viruses package an enzymatically active DNA polymerase. Discovery of the polymerase we now call *RT* revolutionized the study of RNA tumor viruses. In 1975 the Nobel Prize in Physiology or Medicine was awarded jointly to David Baltimore, Renato Dulbecco, and Howard Temin for their discoveries concerning interactions between cells and tumor viruses. In 1980 the first human oncogenic retrovirus (HTLV-1) was described (Poiesz et al., 1980). HTLV-1 was discovered in the laboratory of Robert Gallo at the US National Cancer Institute. Also in the early 1980s, a retroviral etiology was ascribed to lung tumors in sheep (Safran et al., 1985).

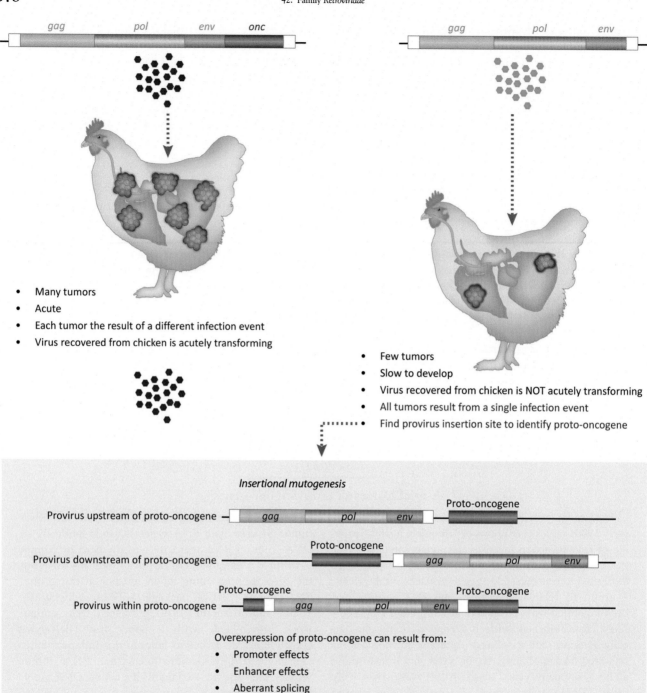

FIGURE 42.10   Two methods of retroviral oncogenesis. On the *left*, a chicken is infected with an acutely transforming retrovirus (e.g., RSV). The viral genome encodes an oncogene that is delivered to every infected cell. Many cells become transformed to generate tumors. The tumors in the chicken on the *left* are polyclonal (each tumor originated from a different cell). The chicken on the *right* is infected with a retrovirus that has the potential to insertionally activate a chicken proto-oncogene. During infection, the retrovirus inserts into chicken chromosomal DNA. In rare cases, insertion will activate a growth-promoting gene and the cell may be transformed. There are a variety of mechanisms by which a proto-oncogene can be activated. Often the gene is overexpressed by the strong retroviral promoter. The tumors in the chicken on the right at monoclonal. One cell was initially transformed.

Other oncogenic retroviruses are less "potent": Tumors are slower to develop and those that do are usually monoclonal. These tumors are caused by a process called *insertional* activation or mutagenesis (Fig. 42.10). Many cells are infected, but only rarely does chromosomal insertion provide a trigger for cell transformation. Studying tumors caused by insertional activation led to the discovery of dozens of "growth-promoting" cellular genes (so-called proto-oncogenes). These genes fall into two broad classes: Cell *receptors* that respond to extracellular signals to promote growth and *kinases* that activate signaling pathways to transmit messages to the cell nucleus.

Insertion of a provirus within a proto-oncogene may result in production of an aberrant protein: perhaps one with unregulated growth-promoting potential. But many insertions are near, not within, a proto-oncogene. These types of insertions may increase the *amount* of normal protein product. The expression of growth-promoting proteins is tightly regulated and insertion of a provirus, with its own promoter, enhancer, splicing, and polyadenylation signals can disrupt that regulation. The overexpression of a growth-promoting protein is often the first step in cell transformation. Retroviral insertions may also promote cell transformation by *inactivating* a growth-*suppressing* gene. However, this is much less common, as the inactivation of one allele leaves a second, intact allele to produce an active protein product. Finally, it is important to understand that cell transformation is a readily observable event: *most* insertional events are silent.

Jaagsiekte sheep retrovirus (JSRV) is an acutely transforming betaretrovirus that causes a contagious lung cancer in sheep called ovine pulmonary adenocarcinoma. JSRV brings an oncogene into every infected cell. However, the JSRV "oncogene" is not a captured cellular gene; it is the viral TM protein. JSRV TM directly activates cell-signaling cascades to promote cellular proliferation; additional mutations in proliferating cells lead to malignant transformation. Experimental infection of day-old lambs with JSRV induces tumor formation within 3–6 weeks.

Feline leukemia virus (FeLV) (genus *Gammaretrovirus*) causes a variety of diseases in infected cats, including tumors induced by insertional mutagenesis. The pathogenesis of FeLV is complex because the virus *mutates during infection* and can *recombine with endogenous FeLVs* to generate variants with different cell tropisms. FeLV infection of adult cats is well controlled by the immune system (Fig. 42.11) and very little, if any, replicating virus is detected after the acute infection; these cats do not transmit virus, nor do they develop FeLV-associated tumors. Infection of kittens often has a very different outcome. Kittens fail to control virus replication and become persistently infected (PI) cats. They shed virus and most will eventually become sick. Some will develop tumors, initiated by insertional mutagenesis. However, the most interesting feature of PI cats (from the perspective of viral pathogenesis) is the emergence of mutated viruses with altered cell tropism. These variants arise by point mutations in SU and/or by recombination of the *env* gene of exogenous FeLV with *env* genes of endogenous copies of FeLV-like viruses. One variety of mutated FeLV is called the T subtype. FeLV-T is tropic for, and causes cytopathic infection of T cells, leading to immunosuppression. Another variety, FeLV-C, causes fatal anemia. FeLV-T and C are rarely transmitted, but instead, arise from virus mutations within PI cats.

Some retroviruses that cause cancers do so by a completely different mechanism. They encode proteins that modify cell transcription and signaling pathways to promote cell growth. HTLV-1 is an oncogenic human retrovirus. HTLVs cause life-long infections that are largely asymptomatic but can increase risk of developing leukemia. Oncogenesis of HTLV-1 is thought to require two small viral proteins, TAX and bZIP (HBZ) (Fig. 42.4). Both are multifunctional and interact with a variety of host cell proteins. TAX is thought to be a key player in tumor initiation but by the time cells are transformed, TAX is silenced, and HBZ is expressed (see Box 42.4).

Walleye dermal sarcoma virus (WDSV) is a member of the genus *Epsionretrovirus*. Infection is associated with proliferative lesions in walleye, a freshwater fish native to Canada and the Northern United States. In infected fish, proliferative lesions are seasonal. The highest incidence of disease occurs in the Fall/Winter seasons and lesions regress as water temperatures warm up in the Spring/Summer seasons. WDSV is a complex retrovirus that encodes auxiliary proteins believed to be responsible for tumor development. One of these is a cyclin-like protein [retrovirus (rv)-cyclin] that may stimulate cell growth. The other (ORF B) protects cells from apoptotic stimuli. The specific factors involved in the seasonality of tumor formation are unclear and this interesting retrovirus has yet to reveal all of its secrets.

In this chapter, we learned that:

- The major steps in the retrovirus replication cycle are attachment, penetration, reverse transcription, integration, transcription, translation, assembly, release, and maturation.
- Integration of the retroviral genome (the ds DNA version) is unique to the family *Retroviridae*. Reverse transcription is *almost* unique to this family but the hepadnaviruses also use reverse transcription for genome synthesis.
- Retroviral PR cleaves the GAG and GAG—POL precursors to generate MA, CA, NC, PR, RT, and IN. PR is active after budding. Budded virions are initially immature and become mature after PR cleavages are complete.

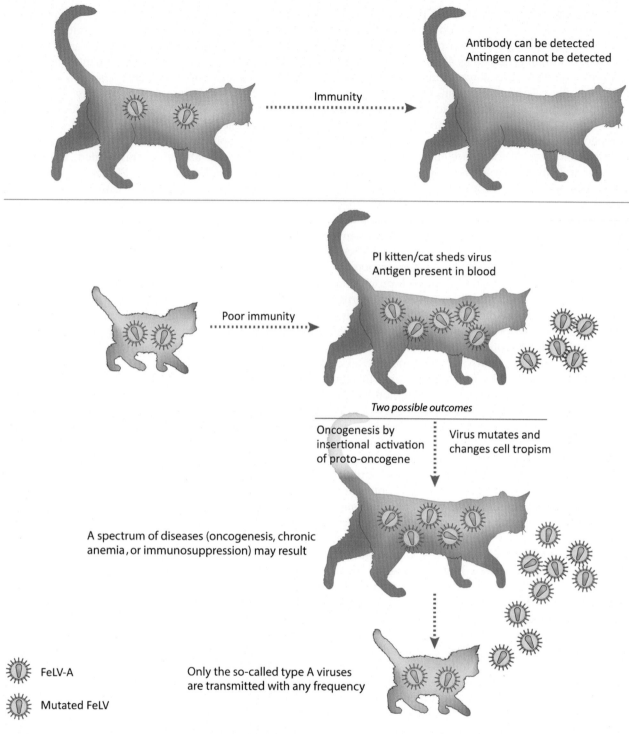

Antibody can be detected
Antingen cannot be detected

Immunity

PI kitten/cat sheds virus
Antigen present in blood

Poor immunity

Two possible outcomes

Oncogenesis by
insertional activation
of proto-oncogene

Virus mutates and
changes cell tropism

A spectrum of diseases (oncogenesis, chronic
anemia, or immunosuppression) may result

FeLV-A

Mutated FeLV

Only the so-called type A viruses
are transmitted with any frequency

FIGURE 42.11   Pathogenesis of FeLV is complex. Infection of an adult cat often results in development of protective immunity. The cat is antibody positive but not viremic. If a kitten is infected with FeLV, it becomes persistently infected (PI). The PI kitten or cat sheds virus and is identified by the presence of viral antigen in the blood. The antigen-positive cat may develop any number of diseases, including cancers or acute (fatal) anemia. The viruses found in the sick cat often contain mutations.

- Exogenous retroviruses are transmitted horizontally, by infectious particles, from one animal or person to another. Endogenous retroviruses are transmitted through the germ-line (sperm and eggs) from parent to offspring. They are seldom found as particles and many are defective. About 8% of the human genome consists of endogenous retroviruses (or their remains).

BOX 42.4

Human T-lymphotropic virus 1

Oncogenic animal and bird retroviruses were discovered and characterized several decades before the discovery of HTLV-1 in 1980. HTLV-1 and its close relative HTLV-2 are complex retroviruses in the genus *Deltaretrovirus*.

HTLV-1 is endemic in parts of Japan, the Caribbean Islands, Africa, and South America. Mechanisms of virus transmission include sexual contact, breast-feeding, blood transfusion, and contaminated needles. It is estimated that 10−20 million people worldwide are infected. Most infections are asymptomatic; however, 3%−5% of infected persons develop HTLV-1-associated T-cell tumors. A smaller percentage of infected individuals develop a demyelinating disease called HTLV-1-associated myelopathy (HAM) or tropical spastic paraparesis (TSP).

In vivo HTLV-1 infects primarily T-lymphocytes but in vitro tropism is broader and the virus infects a variety of other cell types, including B-lymphocytes, monocytes, endothelial cells, and fibroblasts. HTLV-1 is a unique retrovirus in several respects. First, HTLV-1 is a highly cell-associated virus. Whether in vitro or in vivo, few cell-free virions are released; for the most part, virus is transmitted directly from cell to cell via a virological synapse. Second, within an HTLV-1-infected individual, there is very little virus expression. Polymerase chain reaction (PCR) is required to detect very low levels of viral RNA; viral proteins are undetectable. (Note the contrast with HIV, another T-cell tropic human retrovirus.) When peripheral blood mononuclear cells from an infected patient are cultured, levels of viral RNA and protein increase significantly. This suggests that chronically infected patients maintain a pool of "latently" infected cells and HTLV-1 is passed to daughter cells via mitosis.

The HTLV-1 genome is shown in Fig. 42.4. *Gag, pro, pol,* and *env* genes overlap and their arrangement is reminiscent of simple retroviruses. However, an open reading frame called X, found at the 3′ end of the genome, encodes several additional proteins. The best studied of these are TAX and REX. TAX is a transactivating protein while REX increases nuclear export of unspliced and singly spliced viral mRNAs. TAX interacts with cellular proteins and forms a complex on the LTR, thereby increasing transcription. The mechanism of transcription activation seems to involve changes to chromatin structure.

HTLV-1 bZIP factor (HBZ) is also encoded in the X region but is transcribed in the opposite direction to all other transcripts. The promoter for HBZ transcription is the 3′ LTR. HBZ is a Basic Leucine Zipper Domain (bZIP) protein that interacts with many cellular proteins, including several important transcription factors. HBZ activities are complex: (1) HBZ is required for efficient viral infectivity and persistence in a rabbit model; (2) HBZ is dispensable for HTLV-1-mediated cellular transformation in cultured cells but (3) HBZ by itself promotes proliferation of T-cell lines; and (4) HBZ suppresses TAX-mediated viral gene transcription.

Adult T-cell leukemia (ATL) is a very aggressive T-cell cancer that may arise decades after infection with HTLV-1. The leukemic cells of ATL are monoclonal and harbor HTLV-1 proviral DNA at random chromosomal integration sites. This suggests an important role for HTLV-1 in tumor development but also suggests that insertional activation of a cellular proto-oncogene does not drive cell proliferation. Models of ATL focus on the activities of TAX and HBZ. The process is clearly complex and multistep, but some common findings are as follows: (1) ATL cells are genetically unstable and TAX promotes chromosomal damage. This suggests an early role for TAX in tumor development. (2) TAX expression is largely silenced in ATL cells. As TAX is a target of cytotoxic T cells, lack of expression would protect proliferating tumor cells. (3) HBZ mRNA and protein are found in most ATL tumors. (iv) Continued expression of HBZ is necessary to maintain tumor cell proliferation.

It is estimated that between 0.25% and 3% of HTLV-1-infected persons may eventually develop HAM/TSP, a slowly progressive, chronic disease of the spinal cord. Spinal cord damage results from inflammatory responses directed against infected cells. Unfortunately, nerve cells are also damaged in the process. The main symptoms are painful stiffness and weakness of the legs.

• Insertional activation describes the process by which integration of a retrovirus activates a cellular proto-oncogene. Proto-oncogene is a general term to describe any growth-promoting gene. Retroviral integration can directly mutate a proto-oncogene or it can increase the level of expression of a normal protein product.

Expression of growth-promoting genes is tightly regulated. Insertion of a provirus may provide a strong promoter, may provide enhancer elements, may change splicing patters, and change use of a polyadenylation signal or RNA stability element. Insertional activation of a proto-oncogene is often the first step in the transformation process. Often additional mutations are required to confer a fully transformed phenotype.

References

Baltimore, D., 1970. RNA-dependent DNA polymerase in virions of RNA tumour viruses. Nature 226, 1209–1211.

Poiesz, B.J., Ruscetti, F.W., Gazdar, A.F., Bunn, P.A., Minna, J.D., Gallo, R.C., 1980. Detection and isolation of type C retrovirus particles from fresh and cultured lymphocytes of a patient with cutaneous T-cell lymphoma. Proc. Natl. Acad. Sci. USA 77, 7415–7419.

Safran, N., Zimber, A., Irving, S.G., Perk, K., 1985. Transforming potential of a retrovirus isolated from lung carcinoma of sheep. Int. J. Cancer 35, 499–504.

Schultz, S.J., Champoux, J.J., 2008. RNase H activity: structure, specificity, and function in reverse transcription. Virus Res. 134, 86. Available from: https://doi.org/10.1016/j.virusres.2007.12.007.

Temin, H.M., Mizutani, S., 1970. RNA-dependent DNA polymerase in virions of Rous sarcoma virus. Nature 226, 1211–1213.

43

Replication and pathogenesis of human immunodeficiency virus

After reading this chapter, you should be able to answer the following questions:

- What are the most common routes of transmission of human immunodeficiency virus (HIV)?
- How can transmission of HIV be avoided?
- HIV infects what types of cells?
- How does the cell tropism of HIV relate to its pathogenesis?
- What is the cell receptor for HIV?
- What proteins serve as coreceptors for HIV?
- What is the function of HIV TAT?
- What are some common features of the auxiliary proteins encoded by HIV and simian immunodeficiency virus (SIV)?
- What is a viral "restriction factor"?
- What classes of antiretroviral drugs are available for treatment of HIV infection?

DOI: https://doi.org/10.1016/B978-0-323-90385-1.00030-3

© 2023 Elsevier Inc. All rights reserved.

- Explain why it is important to use drug "cocktails" to treat HIV infection.
- What are some of the obstacles to development of an effective HIV vaccine?

History of HIV and acquired immunodeficiency syndrome

The modern history of HIV-AIDS began in June 1981 with a paper in *Morbidity and Mortality Reports Weekly* published by the United States Centers for Disease Control. The paper described an unusual cluster of cases of severe immunodeficiency in previously healthy men. In the weeks and months that followed, additional cases were reported. All involved the development of severe immunodeficiency in previously healthy individuals. By 1982 over 800 cases of severe immunodeficiency had been reported in the United States (Box 43.1). Persons at highest risk were intravenous drug users, hemophiliacs, blood transfusion recipients, homosexual men, and children born to mothers at risk. The epidemiology was consistent with an *infectious agent* transmitted via *blood and body fluids* as the cause of this new disease. The name given to the syndrome was acquired immunodeficiency syndrome (AIDS). By 1984 a lymphotropic retrovirus had been isolated by a number of laboratories (Fig. 43.1). Blood screening tests were quickly developed to help eliminate HIV from the blood supply. Blood screening tests were also used to identify healthy persons in high-risk groups, with antibodies to HIV. These patients were then closely followed in an effort to understand the progression of the disease.

BOX 43.1

The origins of HIV/AIDS

AIDS was first recognized in the United States and other developed countries in the early 1980s. The origins of the disease were unknown at that time. Once it was shown to be caused by HIV, a retrovirus, it was possible to trace back its origins. For example, it was soon shown that similar retroviruses were present in numerous species of African monkeys. Genomic sequence analysis showed that some were much more closely related to HIV than others. Chimpanzee retroviruses were the most closely related, suggesting that the human virus originated from chimpanzees. Within African chimpanzee populations, there are multiple subpopulations separated from each other by broad rivers. (Chimpanzees do not like to swim and little mixing occurs.) The most closely related virus to HIV could thus be tracked down to a forest in central Africa by the Congo river. By estimating the mutation rate of HIV, and comparing chimpanzee strains with human strains, it has been calculated that the chimpanzee virus entered humans between 1900 and 1920.

Chimpanzees are quite good to eat and are a significant source of bush meat for some African villagers. They are, however, fierce and a bow-and-arrow is not the best way to kill them. Only when guns became available to villagers around 1900 did chimpanzees become a significant food source. Once killed, the meat had to be carried back to the village. It was therefore necessary to butcher the carcass. This was a bloody business and inevitably some chimpanzee blood could contaminate cuts or scratches on a hunter. Once infected, the hunter would return to the village and perhaps infect his wife. Their deaths would have attracted little attention, and provided they were monogamous, the infection would not have spread.

However, both French and Belgian colonists in the Congo employed forced labor and many villagers would have been compelled to work in rubber plantations and other activities. To ensure the health of their workers, the colonial authorities made widespread use of antimalarial drugs, usually administered by injection. It is highly likely that HIV was thus spread from isolated villages into a wider work force. Nevertheless, the number of HIV cases would have remained low and unremarkable against a background of many other tropical diseases.

Beginning in the 1950s, both the Belgian and French Congos gained their independence but were politically highly unstable. Huge numbers of villagers moved to the cities in search of better opportunities. With so many single men in the cities, prostitution thrived, and as a result, the prevalence of HIV in the urban population of cities like Kinshasa grew rapidly in the 1950s and 1960s. Because of instability and chronic warfare, few Western aid workers entered these countries and they relied on development aid and training from countries such as Haiti. Sometime in the 1970s, the virus probably infected Haitian workers and was introduced into the Haitian sex-trade. From Haiti, it spread to the United States and Western Europe by way of the trade in blood products as well as sex-tourism. Eventually it spread across the globe. It should be pointed out, however, that 75% of AIDS cases are still occurring in Sub-Saharan Africa.

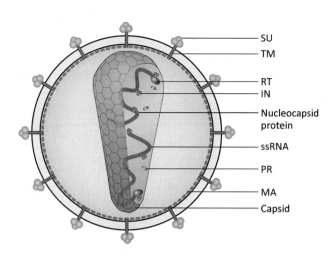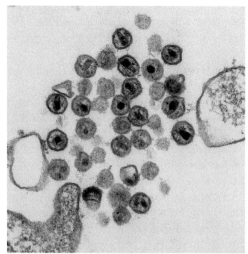

FIGURE 43.1 Retrovirus structure (*left*). Thin section transmission electron microscopic image depicts numerous HIV virions (*right*). Source: *CDC Public Health Image Library Image ID# 13472. Content providers: CDC/Maureen Metcalfe, Tom Hodge.*

In the 1980s–90s a diagnosis of AIDS or HIV infection was considered a death sentence. However, development of novel antiviral drug therapies changed the prognosis drastically. While it would be a mistake to minimize the costs and potential side effects of drug treatment, HIV infection is now considered a chronic disease. HIV has taken a huge toll on human life and many billions of dollars have been spent developing vaccines (mostly unsuccessful to date) and treatments, as well as to probe the human immune system and unravel the disease process.

As treatments have improved, the number of persons living with HIV/AIDS has increased and the number of deaths has decreased. In 2020 the World Health Organization estimated over 37 million HIV infections worldwide, with approximately 680,000 deaths. Estimates in the United States were ~1.9 million infected persons in 2019. Unfortunately, about one in seven infected people do *not* know that they carry the virus and they pose a significant risk for transmitting the disease. Although the earliest reports of HIV/AIDS in the United States focused on homosexual men, it is quite clear that HIV does not discriminate: It is an equal opportunity infection for those who ignore the hazards of unsafe sexual practices and sharing needles for illicit drug use.

Where did HIV come from and how did it emerge so rapidly? The closest relatives to HIV are the SIVs that infect a variety of Old World monkeys. The virus we call HIV-1 jumped from simian to human hosts at least four independent times. These cross-species jumps likely occurred through hunting and consuming monkeys. The vast majority of human HIV-1 infections are caused by the so-called M group viruses. M group viruses emerged in Central Africa and are now found worldwide. The spread of HIV-1 was certainly facilitated by the long asymptomatic period during which the virus can be silently and unknowingly transmitted.

HIV is a lentivirus

When HIV was identified in the early 1980s, it was clear that it was an unusual retrovirus. It was soon determined that the closest known relatives of HIV were an obscure group of animal retroviruses called lentiviruses (lenti = slow) that cause chronic life-long infections of horses, sheep, and goats. The family *Retroviridae* contains seven genera within two subfamilies. Members of the genus *Lentivirus* include HIV, SIV, equine infectious anemia virus (EIAV), visna-maedi viruses, feline immunodeficiency virus (FIV), and bovine immunodeficiency virus. Lentiviruses are among the so-called complex retroviruses because they encode a number of gene products in addition to the GAG, POL, and ENV polyproteins. All known lentiviruses have a preference for infecting cells of the immune system. HIV and SIV infect T-helper cells, monocytes, macrophages, and dendritic cells (DCs). FIV infects feline T- and B-cells while EIAV is more restricted, infecting only equine monocytes and macrophages. One characteristic shared by these cell types is that they are often found in resting state, although they can be activated by immune stimulation. Lentiviruses are unique among the retroviruses in their ability to infect nondividing cells. Lentiviruses encode proteins that actively move a preintegration complex (PIC) across a nuclear pore to gain access to host DNA. Another common trait of lentiviruses is their high mutation rate. Even before the discovery of HIV, studies of horse, sheep, and goat lentiviruses suggested that these viruses persisted for years in their infected hosts, in part, because they mutated to escape immune responses.

Today we know that HIV's closest relatives are other primate lentiviruses. Understanding the taxonomy of the primate lentiviruses provides an understanding of the origins of HIV. All primate lentiviruses have a common ancestor. It appears that these viruses evolved with their hosts, such that today, each species of monkey has its own SIV. SIVs are known to naturally infect at least 40 different species of nonhuman primates in Africa. Most SIVs cause minimal disease in their natural host, but a cross-species infection (naturally or experimentally) may cause high morbidity and mortality in the new host.

Humans can be infected with two different viruses, HIV-1 and HIV-2. An examination of their genomes reveals differences in their auxiliary proteins (Fig. 43.2). HIV-1 has spread widely while HIV-2 has remained more restricted to countries in West Africa. The closest relative of HIV-1 is chimpanzee SIV (SIV$_{CPZ}$) while the closest relative of HIV-2 is the sooty mangabey virus, SIV$_{SM}$. In addition to genomic differences, the human epicenters of HIV-1 and HIV-2 are discrete. Thus HIV-1 and HIV-2 are two different viruses that jumped independently into the human population.

In fact, both HIV-1 and HIV-2 have jumped into the human population *multiple* times. In the case of HIV-1, four distinct "groups" have been identified (Fig. 43.3). Each group is the result of an *independent* introduction into the human population. The M (main) group of HIV-1 is without a doubt the *main* group. M group viruses emerged in Central Africa and have spread worldwide, infecting many millions of people. In addition, sublineages or clades have developed within HIV-1 M during worldwide spread. Sublineages are

important because these are genetically distinct enough that vaccines developed for a single subgroup are not likely to be effective across all subgroups. M group viruses are readily transmitted from person to person (both horizontal and vertical transmission). While highly pathogenic (in untreated patients), the long asymptomatic period (clinical latency) facilitates

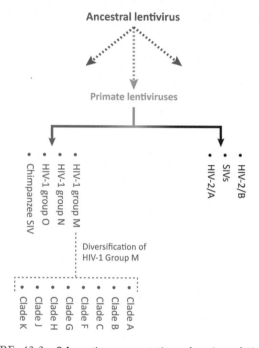

FIGURE 43.3 Schematic representation of major relationships among primate lentiviruses and diversification. Nonhuman primate lentiviruses have entered the human population at least five times. HIV-1 groups M, N, and O are related to chimpanzee simian immunodeficiency virus (SIV) and represent three independent jumps into humans. HIV-1 group M has spread worldwide and has diversified extensively to produce groups called clades. HIV-2/A and HIV-2/B are not as common as HIV-1. Their closest relatives are SIVs that infect sooty mangabeys (*Cercocebus atys*).

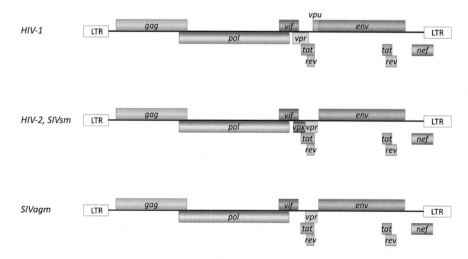

FIGURE 43.2   Comparison of primate lentivirus genomes showing differences in auxiliary proteins.

their transmission. However, use of highly effective antiretroviral therapies seems to be slowing their transmission.

The HIV-1 O (outlier) group viruses diverge in sequence from M group strains by as much as 50% in the envelope gene. The O group viruses are closely related to chimpanzee and gorilla SIVs. A third genetically distinct group of HIV-1 was discovered in 1998; only a handful of human infections have been identified to date. This group is called N, for non-M and non-O.

HIV-2 entered the human population in West Africa. Both HIV-1 and HIV-2 have the same modes of transmission and are associated with similar opportunistic infections and development of AIDS. However, in persons infected with HIV-2, immunodeficiency seems to develop more slowly and to be milder. HIV-2 may be less transmissible than HIV-1. Currently, two groups of HIV-2 have been identified, Groups A and B. The identification of genetically distinct groups of HIV-1 and HIV-2 is medically important. Current tests to screen patients and blood may not identify HIV-2 or non-M group HIV-1 infections. In addition, drugs developed to treat HIV-1 may be ineffective against HIV-2.

Sequencing thousands of HIV genomes confirmed that genome diversity is indeed a common trait of lentiviruses. It has been estimated that during an infection, viral genetic diversity in HIV patients increases by ~1% per year from the initial founder virus strain (the infecting virus). Several factors contribute to the genetic heterogeneity of HIV:

- Reverse transcriptase (RT) does not have a proofreading/correction function. Estimates of mutation rate in cell culture are $1-2 \times 10^{-5}$ mutations per nucleotide synthesized per replication cycle. Estimates of mutation rates in patients are up to five times higher.
- Within an infected patient, there are millions of HIV replication cycles per year.
- Virus production in vivo can be an astounding 10^9 particles per day.
- Recombination by template switching (during reverse transcription) increases viral genetic diversity.

HIV Structural proteins

HIV and other lentiviruses are complex retroviruses, but their basic genome organization and replication cycle have much in common with other retroviruses. The basic steps in lentivirus replication are attachment/entry, reverse transcription, integration, transcription, protein synthesis, assembly, budding, and maturation. These processes were described in the previous chapter, family *Retroviridae*. In this section, we review HIV proteins and their functions. The three longest open reading frames of all retroviruses encode precursor polyproteins (Fig. 43.4). HIV uses an unspliced mRNA to produce the GAG precursor polyprotein and the GAG–POL precursor polyprotein. The ENV precursor is expressed from a singly spliced mRNA. ENV is processed in the endoplasmic reticulum to produce the HIV surface (SU) and transmembrane (TM) glycoproteins.

GAG precursor (Pr55^{Gag})

Pr55^{Gag} is translated from unspliced mRNA. During virion maturation, Pr55^{Gag} is cleaved by the vial protease (PR) to produce four structural proteins (MA, CA, NC, and p6) and two spacer peptides (SP1 and SP2). The positions of the structural proteins in the precursor reflect their final locations in the mature virion. MA, at the amino-terminus of Pr55^{Gag}, is found in association with the lipid envelope. CA is the capsid protein that forms the elongated icosahedral shell around the genome (this is positioned in the center of the mature virion). The positively charged NC protein binds to RNA.

The MA domain of Pr55^{Gag} functions during the late phase of the HIV replication cycle and is a key player in assembly. The MA domain directs and localizes Pr55^{Gag} to the plasma membrane. MA is linked to a molecule of myristic acid (a 14-carbon fatty acid) via a glycine near its amino-terminus. The fatty acid moiety inserts into the plasma membrane to form an anchor for Pr55^{Gag}.

The CA domain has several discrete functions that can be identified by mutagenesis of different regions of the protein. One region of CA promotes oligomerization of Pr55^{Gag}, and mutations in the oligomerization domain block production of immature particles. Other CA mutations interfere with formation of mature capsids. Still other CA mutants produce viruses that bud and form capsids, but are noninfectious (these virions are probably deficient in a function needed for successful entry into a new cell).

Another important biological activity of HIV-1 CA is that it binds to the host CypA protein during virus assembly. CypA is a member of the cyclophilin family of peptidyl-prolyl *cis-trans* isomerases. HIV CA binds CypA via a proline-rich loop and the interaction leads to incorporation of CypA into virions. It appears that CypA may counteract a host restriction factor (TRIM5α) early in the infection process.

The NC domain interacts with the retroviral genome (unspliced mRNA) and is critical for its encapsidation. HIV-1 NC has two zinc finger motifs that specifically

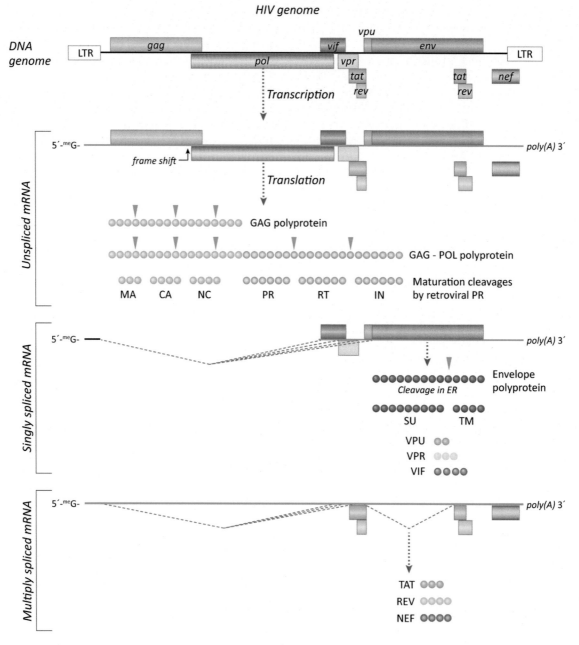

FIGURE 43.4   HIV genome and gene expression strategy.

interact with HIV-1 RNA at the packaging signal. NC also assists in placement of the tRNA primer on the primer-binding site and participates in oligomerization of Pr55^{Gag}. After particle budding and maturation, NC remains bound to the genome. NC also functions during the early stage of infection when its activity as a nucleic acid chaperone greatly facilitates successful reverse transcription. NC appears to remain associated with newly synthesized DNA and plays a role in integration as well.

The p6 domain is located at the very carboxyl terminus of Pr55^{Gag}. The p6 domain is a key player in release of immature virions from the budding site. Some mutations in p6 yield particles that are tethered at the PM. The critical site for particle release is a highly conserved amino acid motif called PTAP (so named because its amino acid sequence is Pro—Thr—Ala—Pro). PTAP is a so-called *late* domain signifying its role in the very last stages of virion production. The p6 domain interacts with host proteins of the cellular endosomal sorting machinery (ESCRT machinery). P6 also interacts with HIV-1 viral protein R (VPR) (described later) facilitating its incorporation into virions.

GAG–POL polyprotein precursor (Pr160^Gag-pol^)

Pr160^Gag-pol^ is a 160-kDa polyprotein synthesized as the result of a ribosomal frame-shift in GAG. It is estimated that frame-shifting occurs about 20% of the time. HIV Pr160^Gag-pol^ contains domains for protease (PR), reverse transcriptase (RT), and integrase (IN).

PR is an aspartic protease that cleaves pr55^Gag^ and pr160^Gag-pol^. PR is active as a dimer and the active site contains two opposed aspartic acids that coordinate a molecule of water to catalyze hydrolysis of a peptide bond. The regions flanking the active site contribute to substrate recognition. PR catalytic activity is activated upon budding and is absolutely required for production of mature, infectious virions. HIV-1 PR inhibitors are effective antivirals that are incorporated into immature particles and inhibit processing of precursor polyproteins.

HIV RT is a heterodimer. One subunit is 66 kDa (p66) and the other is 51 kDa (p51). P66 contains an RNase H domain missing from p51. In the mature RT, p66 appears to contribute catalytic activity while p51 plays a structural role. HIV IN is located at the carboxy terminus of Pr160^Gag-pol^. IN is 32 kDa and is the major player in integration. Upon infection, IN remains associated with the core during reverse transcription and is part of the PIC. Enzymatic activities of IN include exonuclease, endonuclease, and ligase. IN forms tetramers that bind to the ends of the linear double-stranded DNA.

ENV precursor

The ENV precursor polyprotein is a 160-kDa glycoprotein called gp160. It is cleaved to produce the gp120 SU and gp41 TM proteins. Gp160 is translated from a singly spliced mRNA. The ENV precursor is translated on rough ER and within the ER gp160 is glycosylated and forms trimers. Gp160 is cleaved by furin or a furin-like PR in the Golgi. SU and TM remain associated during transit through the Golgi to the PM. TM is anchored in the PM by a single TM domain, but SU is maintained on the surface of cells or virions by relatively weak (noncovalent) interactions with TM.

Gp120 is highly glycosylated. In fact, ~50% of the total mass of SU is carbohydrate. The carbohydrate shields the polypeptide backbone, limiting immune recognition. The gp120 molecule has alternating regions of conserved (C) and variable (V) amino acids. The so-called variable domains are regions that form flexible loops on the more conserved backbone of the molecule. SU is the attachment protein and the CD4-binding site maps to highly conserved regions within SU. CD4 binding induces conformational changes in gp120 that allows binding to coreceptors (Fig. 43.5). The coreceptors for HIV are members of the G protein-coupled receptor superfamily of seven-transmembrane domain proteins. The two major

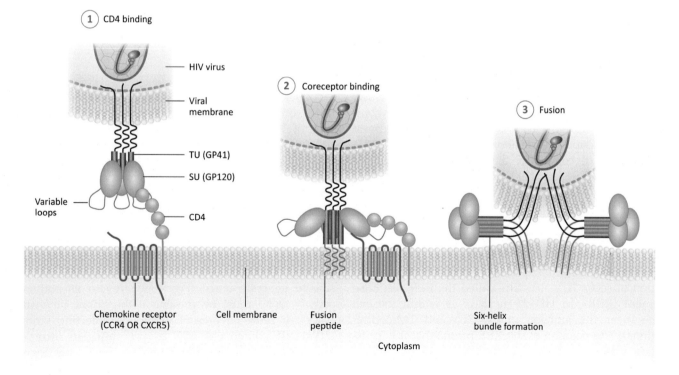

FIGURE 43.5 HIV attachment and fusion. Attachment starts with interactions between surface (SU) and the CD4 protein receptor. This is followed by additional interactions between SU and a chemokine receptor that allow the fusion domain of transmembrane (TM) to interact with the plasma membrane.

coreceptors for HIV are CCR5 and CXCR4. CCR5 is found on monocytes and macrophages; HIVs that use CCR5 are called R5 isolates. Strains that bind CXCR4 are referred to as X4 viruses. Other chemokine receptors can also serve as HIV coreceptors. CCR5 appears to play a major role in establishment of infection as individuals with a mutant version of CCR5 (called CCR5/Δ32) do not express the protein on the surface of cells. Individuals homozygous for this mutation are only rarely infected with HIV-1.

Gp41 or TM is the HIV fusion protein. Upon binding to CD4 and a chemokine coreceptor, conformational changes in SU activate the membrane fusion potential of TM. A stretch of hydrophobic amino acids at the N-terminus of TM initiate fusion by inserting into the PM (Fig. 43.5). Mutational analyses have also identified additional regions of the TM ectodomain that are critical for membrane fusion. Identification of regions critical for fusion provides targets for development of novel drugs.

TAT

Complex retroviruses encode a variable of number of proteins in addition to those produced from GAG, GAG-POL, and ENV precursors. Two of these, TAT and REV, are encoded by all lentiviruses. TAT is a small protein (101 aa) with a critical role as an activator of transcription (Fig. 43.6). In the absence of TAT, RNA pol II is poorly processive; while transcripts are initiated, most terminate close to a stem-loop region called the transactivating response element (TAR). However, a few transcripts will be completed and these are spliced and exported from the nucleus. As a result, TAT is synthesized, imported back into the nucleus, and interacts with TAR. TAT also binds to cellular proteins, thus forming a protein complex that modifies host RNA pol II to significantly increase transcription elongation. In the presence of TAT, the number of completed transcripts is increased by ~100-fold. TAT is essential for HIV-1 replication.

REV

REV is a small (116 aa) RNA-binding protein produced from a multiply spliced mRNA. REV binds to a region of secondary structure called the REV response element (RRE). In HIV-1, the RRE is within the *env* gene (near the junction of the SU- and TM-coding regions). The position of the RRE means that it is present in unspliced mRNA as well as in singly spliced mRNA but is absent from multiply spliced mRNAs. When REV is present in the nucleus, it binds to the RRE, oligomerizes, and facilitates export of unspliced and singly spliced mRNA from the nucleus (Fig. 43.7).

In the absence of REV, all viral mRNAs are multiply spliced before leaving the nucleus. Therefore without REV, no structural proteins can be synthesized. REV disassembles from mRNA in the cytoplasm and moves back into the nucleus to bind additional mRNA targets. Like TAT, REV is essential for HIV replication. Neither TAT nor REV is a structural protein, and neither is found in virus particles.

Other HIV-1 proteins

HIV-1 and other primate lentiviruses encode additional proteins required for efficient replication and a comparison of primate lentiviruses genomes (Fig. 43.2) reveals some variation. A common trait of these auxiliary genes and proteins is their small size and lack of enzymatic activity. They tend to interact with multiple cellular proteins. All auxiliary proteins appear to be required for optimal replication in the natural host, but all are expendable in some types of cultured cells (although virus infectivity may be affected up to several thousand-fold, depending on the accessory gene and the type of infected cell). A primary (although not exclusive) feature of primate lentivirus auxiliary proteins is their role in antagonizing host antiviral proteins. Among the primate lentiviruses, HIV-1 encodes virus infectivity factor (VIF), VPR, viral protein U (VPU), and NEF proteins; HIV-2 lacks VPU but encodes viral protein X (VPX); SIVagm lacks VPU and VPX.

Virus infectivity factor

VIF is required for production of infectious virions from some cells types. The cell-type-dependent phenotype of VIF mutants made it challenging to determine VIF function. A model of VIF activity emerged whereby *VIF was hypothesized to inhibit inclusion of an antiviral protein into virions.* The packaged antiviral protein then functioned to prevent infection of a new cell (Fig. 43.8).

The antiviral factor antagonized by HIV-1 VIF is a single-stranded DNA cytidine deaminase called APOBEC3G (A3G). A3G blocks retrovirus replication by damaging newly synthesized DNA during reverse transcription. The damage is caused by deamination of cytosine to generate uracil (Fig. 43.9). Key to its antiviral activity, A3G must be present at the "right time and place": It must be present within the infecting virion. HIV-1 VIF counteracts A3G by preventing its incorporation into virions by targeting it for proteosomal degradation. The reason that VIF is cell-type-dependent is that some cells do not express A3G (and some species do not encode A3G). If a cell does not express A3G, there is no need for VIF. Thus we have a host antiviral factor (A3G)

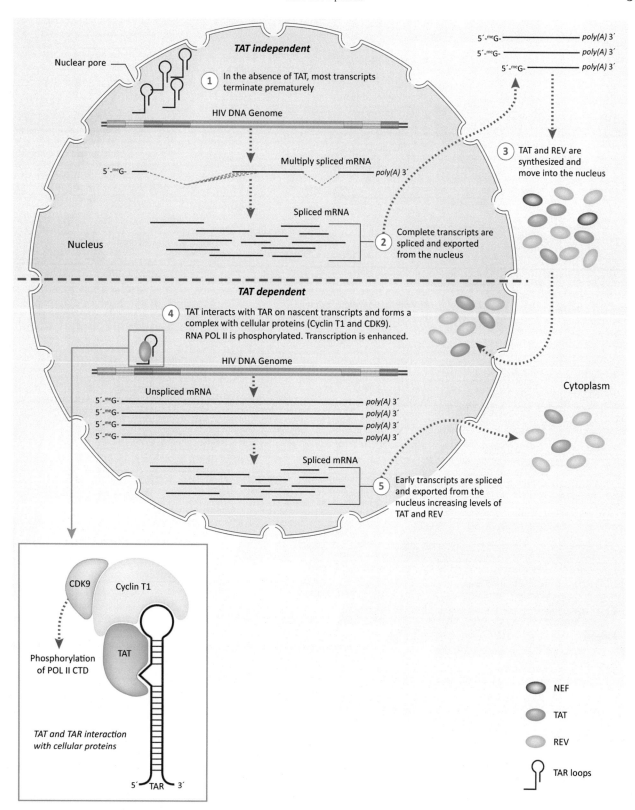

FIGURE 43.6 TAT transactivation. TAT is one of the first proteins expressed from the integrated HIV genome. In the absence of TAT, transcription initiates but is poorly processive and most transcripts terminate close to the promoter. The few full-length transcripts are multiply spliced to produce mRNAs that encode TAT. After translation, TAT moves into the nucleus and binds the TAR RNA stem-loop. Cellular proteins cyclin T1 and CDK9 join the complex and activate (by phosphorylation) RNA pol II. Transcription becomes highly processive.

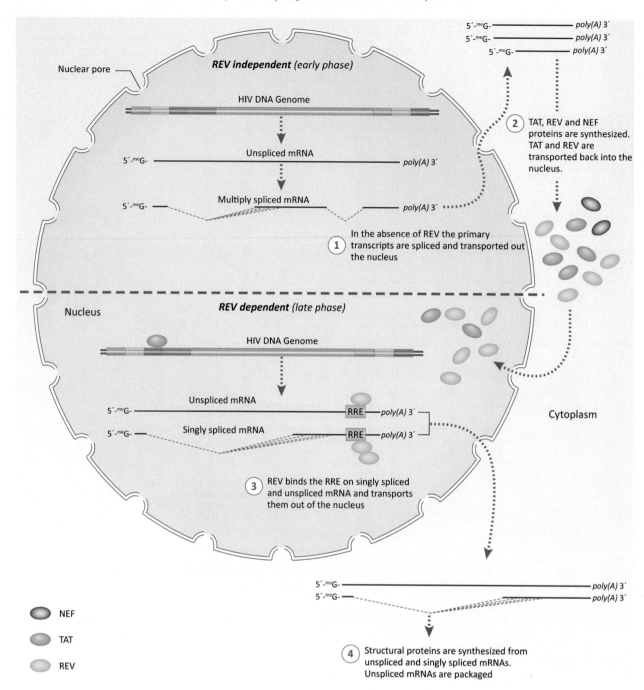

FIGURE 43.7   HIV REV mediates export of unspliced and singly spliced mRNA from the nucleus. REV protein is expressed from spliced mRNAs, early in the replication cycle. REV moves into the nucleus and binds to viral mRNAs containing an RNA element called the RRE. The RRE is found on unspliced and singly spliced mRNAs. Multiple copies of REV bind to the RRE. In the cytoplasm, REV is released from the RRE and cycles back to the nucleus. REV is required for the production of structural proteins.

and viral protein (VIF) to counteract it. Defenses and counter-defenses in action!

Viral protein U

VPU is encoded by HIV-1 but is absent from HIV-2 and most SIVs. This small (81 amino acid) protein is a

key player in the evolution of HIV-1, because of its ability to *antagonize the interferon-inducible antiviral protein, tetherin, in human cells*. VPU is synthesized at high levels in the infected cell and forms membrane-associated multimers. VPU has not been detected in virus particles. HIV-1 VPU mutants are defective late in the replication cycle: Particles fail to complete budding and maturation.

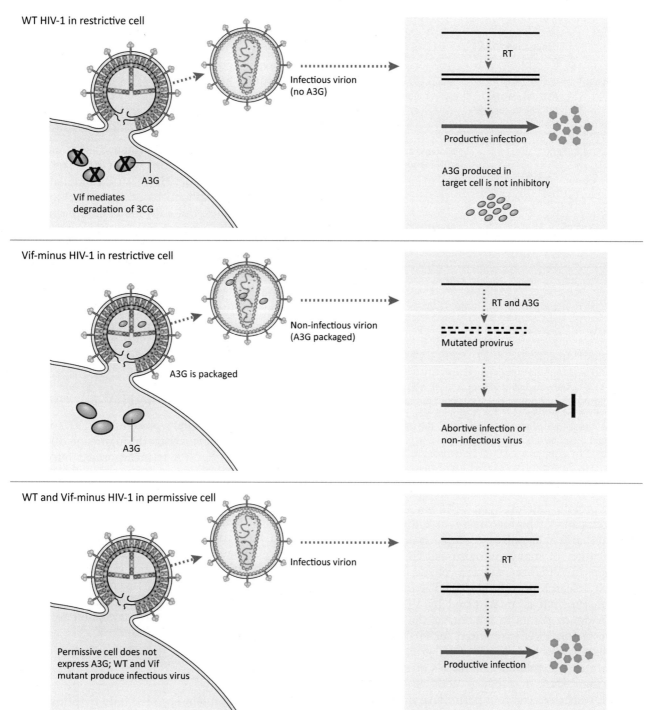

FIGURE 43.8 Interactions of HIV virus infectivity factor (VIF) and host protein apolipoprotein B mRNA editing enzyme, catalytic polypeptide-like (APOBEC) 3G (A3G). A3G is a single-stranded DNA cytidine deaminase that restricts retroviral replication. If A3G is present within an infecting virion, it deaminates newly synthesized retroviral DNA, a process that is mutagenic. In the top panel, HIV VIF mediates degradation of A3G, inhibiting its incorporation into virions. In the middle panel, A3G is incorporated into virions and deaminates cytidine during reverse transcription. In the bottom panel, the cell producing virions does not express A3G, thus wild-type or mutant viruses lacking VIF produce infectious virions.

Early studies of VPU mutants revealed a cell-type-dependent phenotype, suggesting that it interacts with a cellular factor expressed in some, but not all human cells. Treatment of so-called "permissive" cells with interferon changes them to a restrictive type, an indication that the inhibitory cell factor is an interferon-

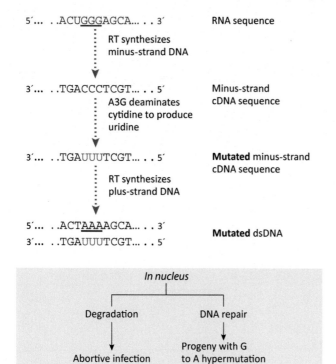

FIGURE 43.9   Antiretroviral activity of A3G. If A3G is present during reverse transcription, it deaminates cytosine (C) to generate uracil (U) in newly synthesized first-strand cDNA. During synthesis of the second DNA strand, uracil base pairs with adenine (A). The ds DNA product containing uracil may be degraded by repair enzymes in the nucleus. If the retroviral DNA is integrated, U will be removed and replaced with thymine (T). Thus the original bases GGG become mutated to AAA.

inducible protein. Interestingly, VPU can also increase virus production of highly divergent retroviruses, an indication that it antagonizes a cellular factor that inhibits replication of many retroviruses. The molecular mechanism of VPU remained a puzzle until 2008, when two research groups identified a protein, now called tetherin, that is antagonized by VPU. A great deal of evidence points to tetherin as the mysterious restriction factor: VPU is required for HIV replication when cells express high levels of tetherin, but is dispensable in cells expressing low levels of tetherin; use of silencing RNA to reduce tetherin expression changes a restrictive cell to a permissive one; the ability of VPU to counteract tetherin is species-specific. HIV-1 VPU specifically counteracts human tetherin. Tetherin from some nonhuman species can restrict HIV-1 release but cannot be counteracted by HIV-1 VPU. Finally, tetherin restricts the replication of many retroviruses as well as other enveloped viruses, including ebolaviruses, arenaviruses, and herpesviruses, and these viruses have evolved their own tetherin antagonists. There is also evidence, from genetic studies of primates, that tetherin itself is undergoing *rapid positive selection*, possibly in response to lentivirus infections.

Another activity of VPU is its ability to bind to the CD4 receptor in the ER, mediating its degradation. This activity is important to the HIV replication cycle, as the presence of CD4 in the ER sequesters the envelope precursor gp160 therein.

Viral protein R

VPR is a 96-amino acid protein incorporated into virus particles by interactions with GAG. The need for VPR is cell-type specific. HIV-1 VPR mutants replicate in proliferating peripheral blood mononuclear cells or cultured T-cell lines. But VPR mutants replicate poorly in nondividing monocyte-derived macrophages (MDM). VPR is part of the PIC and this may explain some of its activity in nondividing cells. However, VPR also acts by usurping the host ubiquitin—proteasome-mediated protein degradation pathway, modulating the activities of multiple host proteins. Targets of VPR-mediated degradation appear to include proteins involved in DNA repair, although how this facilitates HIV-1 replication in MDM is unclear.

Viral protein X is unique to HIV-2

HIV-2 encodes auxiliary protein VPX. VPX is related to VPR (and may have arisen through gene duplication of VPR). Like VPR, VPX is incorporated into virions. HIV-2 VPX is required for efficient infection of nondividing cells (e.g., MDM and DCs). Like VPR, VPX activates/modulates a ubiquitin ligase complex. One target of VPX appears to be an interferon-inducible host cell restriction factor called Sterile Alpha Motif and HD Domain Protein 1 (SAMHD1), an enzyme with important roles in nucleic acid metabolism. SAMHD1 specifically blocks reverse transcription of retroviruses in MDM and DCs. In fact, expression of VPX in such cells renders them susceptible to infection by retroviruses other than HIV-2.

NEF

NEF is small protein encoded only by primate lentiviruses. HIV NEF was named after early studies that suggested it was a negative factor for HIV-1 replication. Although subsequent studies did not support the early data, the name stuck. The *nef* gene is located at the 3' end of the HIV genome and partially overlaps the 3' LTR. One domain important for NEF activity is its amino-terminus. NEF is myristylated and contains a cluster of basic amino acids that mediate membrane interactions. Another important NEF domain is a cluster of amino acids (Pro—X—X—Pro; a so-called Src homology 3 domain) that mediates binding to several important cellular kinases (e.g., Lyn, Hck, and Lck); NEF binding modulates the catalytic activities of these kinases, leading to global changes in cell signaling and gene expression. NEF also disrupts endosomal trafficking in infected cells, resulting in downregulation of important cell surface molecules, including CD4, MHC-I, CD8, and the CD3 T-cell receptor complex.

If NEF modulates signaling and protein expression patterns of immune cells, we might expect that NEF mutants would differ from wild-type virus as regards pathogenesis, and this is indeed the case. An intact *nef* gene is crucial for pathogenesis of HIV and SIV. In studies of SIV, a *nef* deletion was used to generate an attenuated vaccine. The vaccine induced strong antibody responses and the attenuated virus was cleared. Direct evidence for a role of NEF in HIV-1 pathogenesis comes from studies of long-term nonprogressors, a small cohort of HIV-1-infected patients that develop disease very slowly. The HIVs infecting these patients have defective *nef* genes. The ability of NEF to modulate cell signaling and receptor proteins appears to activate key cells in the immune system. Continuous activation of immune cells is detrimental and is linked to the pathogenesis of HIV-1.

Another activity of the NEF proteins of many SIVs (*but not HIV*) is to antagonize tetherin. HIV NEF is unable to antagonize tetherin as it lacks a "tetherin-interacting" domain. However, recall that HIV VPU does antagonize tetherin. This points to the importance of tetherin as an antiviral factor, and the flexibility of viruses to counter its inhibition.

Summary of HIV replication

Before discussing HIV-1 pathogenesis, a brief review of the major steps in the HIV replication cycle is warranted. Details of these processes have been described in Chapter 42.

Binding/entry

SU is the HIV attachment protein. It binds to its primary receptor, the immunoglobulin family member protein, CD4. CD4 expression restricts HIV replication to CD4+ T-lymphocytes, monocytes, and DCs. Following interaction with CD4, SU undergoes conformational changes that unmask binding sites for coreceptors CCR5 and CXCR4. Coreceptor interaction is required for entry and probably serves to place the virion in close proximity to the PM. The next step in the process is insertion of the fusion domain of TM into the PM. Fusion takes place at the PM, releasing the core into the cytoplasm (Fig. 43.5).

Reverse transcription, nuclear entry, and integration

The first viral product to be synthesized is the double-stranded DNA copy of the genomic RNA. The process is similar to other retroviruses, as described in Chapter 42. Key proteins in reverse transcription include heterodimeric RT and NC. However, the process takes place in association with other proteins in the infecting core.

Lentiviruses are unique among retroviruses in their ability to replicate in nondividing cells. This is a result of the ability of the ~56-nm-diameter PIC to actively move across nuclear pores to gain access to cellular chromatin. It is not entirely clear which viral components are involved the process. Suggested players include IN, VPR, and CA.

The IN protein is the key player in integration of all retroviruses. HIV IN enters the nucleus as part of the PIC, in association with ds DNA. IN catalyzes removal of two nucleotides from the 3′ end of each viral DNA strand and makes a staggered double-stranded cut in the target host cell DNA. Finally, IN mediates DNA strand ligation. Integration is generally considered to be the final step in the *early* phase of the retrovirus life cycle. In the case of HIV-1, entry into quiescent cells may result in a long "latent" period before the replication cycle continues. This is a problem in terms of treatment and "elimination" of HIV infection, as cells may harbor transcriptionally inactive proviruses for many years.

Transcription and translation

Retroviral transcription is dependent on recognition of the retroviral promoter by host cell transcription machinery. The HIV promoter is relatively weak (compared to the strong promoters of some oncogenic retroviruses). Transcripts are initiated but not efficiently extended in the absence of the HIV-1 TAT protein. The level of transcription in the absence of TAT is not sufficient to support virus replication. When present, TAT binds to the TAR region of the nascent viral transcripts and promotes assembly of host proteins that activate transcription.

Early HIV transcripts are multiply spliced to produce mRNAs that encode a variety of regulatory and auxiliary proteins. But unspliced and single-spliced HIV-1 transcripts are required for production of virions. HIV REV acts by binding to an RNA sequence (the RRE) present in unspliced and singly spliced viral mRNAs. REV binding enhances nuclear export of these mRNAs, allowing expression of HIV-1 structural proteins.

Assembly/release/maturation

HIV-1 assembly takes place at the plasma membrane. Budding particles contain uncleaved pr55^{Gag} and pr160$^{Gag-pol}$ at membrane sites decorated by SU and TM. Other viral and cellular proteins, for example, CypA and VPR associate with pr55^{Gag}. Two copies of unspliced, capped, and polyadenylated HIV-1 mRNA are brought to the assembly site by interaction with the NC domain. HIV-1 release requires cooperation of cellular and viral proteins. HIV-1 p6 contains a highly conserved amino acid motif PTAP called a "late domain" because it functions at the very end of the replication cycle. Without a late domain, immature HIV particles remain attached to the plasma membrane. Late domains function by interacting with a subset of components of the cellular ESCRT machinery.

Assembly and release of HIV-1 clearly takes place at the PM of T-cells. However, the process is not as clear in MDM. In these cells, HIV-1 appears to assemble at internal sites as well as at the PM. However, it has recently been determined that the "intracellular" locations of HIV assembly in MDM are actually deep invaginations of the PM. HIV virions become infectious upon maturation, a process that occurs after release. The PR domain of pr160$^{Gag-pol}$ is activated, generating mature structural proteins. MA remains associated with the envelope, and CA forms the capsid. Within the capsid, RNA is bound by NC. RT and IN are also present within the capsid.

Overview of HIV pathogenesis

HIV pathogenesis is complex and is impacted by both host and viral genetics. Some untreated, infected individuals remain clinically healthy for decades while others develop AIDS within months. What is very clear is that inhibiting HIV replication, by use of highly effective drug cocktails, is profoundly successful in preventing both disease and transmission! In the following sections, we outline some of the basic interactions between HIV and host cells that contribute to the disease process in untreated individuals.

HIV-1 infection of CD4 + T-lymphocytes

T-lymphocytes are major players in the immune response. They are called "T" cells because they develop in the thymus. There are different types of T-cell; those most important in HIV replication and pathogenesis are the so-called T-helper cells (Th cells). Th cells express the CD4 protein on their cell surface (thus another name for a Th cell is CD4 + T-cell).

Key to their role in immunity, Th cells display receptors that recognize foreign antigens. A naïve CD4 + T-cell is one that has not yet encountered its cognate foreign antigen. When a naïve Th cell binds to its specific antigen, it proliferates. Proliferating Th cells are in an activated state; they secrete cytokines and productively interact with other cells. After many generations, activated T-cells differentiate into memory T-cells. Memory Th cells are long-lived cells that recognize the same antigen as their progenitors.

All Th cells express the CD4 protein on their surface, thus all can bind HIV. However, the results of infection vary, depending on whether or not the Th cell is a naïve or memory cell. Memory CD4 + T-cells express the chemokine receptors CXCR4 and CCR5. Memory CD4 + T-cells are present at high levels in the gastrointestinal tract and lung (where they are likely to contact microbial invaders). Most memory CD4 + T-cells are activated and highly susceptible to HIV-1 infection. The result is production of large numbers of virions and death of the infected cells. Within a few days of initial infection, there is wholesale depletion of gut-associated T-cells.

Some memory Th cells are resting cells. They can also be infected by HIV-1 but the result is integration, followed by a latent infection (no viral proteins are expressed). These cells serve as important long-term reservoirs for the virus. These latently infected cells are directly responsible for the inability to "cure" most HIV-1 infections even after prolonged treatment with antiretroviral drugs.

Naïve Th cells are the predominant type found in peripheral blood. Naïve Th cells express the chemokine receptor CXCR4. Thus at early times postinfection, most naïve Th cells are resistant to HIV-1 infection. However, in some infected individuals, X4 variants of HIV-1 will emerge, a process called coreceptor switching. These viruses now use the CXCR4 coreceptor to infect and deplete the naïve T-cell population; this is associated with accelerated progression to AIDS.

What about the innate immune responses that should control early HIV-1 replication? Unfortunately, early, innate responses induce production of inflammatory molecules that increase the numbers and activation status of Th cells at the site of infection.

HIV-1 infection of monocytes and macrophages

Monocytes and macrophages express CD4 and CCR5 on their surface. However, not all HIV isolates replicate well in these cells. Isolates from the blood usually replicate poorly in cultured macrophages; in contrast, isolates from tissue or the central nervous system replicate much more productively in these cells. In cultured cells, vigorous replication of HIV requires fully differentiated macrophages; these are generally considered to be analogous to tissue macrophages in vivo. Brain-resident macrophages are called microglia cells and their infection may lead to HIV-associated encephalitis.

HIV-1 infection of dendritic cells

DCs capture, transport, and present antigens to CD4 + and CD8 + T-lymphocytes. There are two main types of DCs, myeloid DCs (MDCs) and plasmacytoid DCs (PDCs). MDCs are found in tissues, including skin and intestinal and genital tract mucosa. (MDCs in the skin are called Langerhans cells.) MDCs at mucosal surfaces are the first line of defense against infectious agents encountered at these sites. Their job is to engage infectious agents, degrade them, and carry processed antigens to draining lymph nodes rich in T-lymphocytes. Unfortunately, MDCs do not degrade HIV-1 efficiently and thus bring infectious virus to CD + T-cells in the lymph nodes.

Disruption of immune function

Regulation of the immune system is a very complex process. Not only are many different cell types involved, but gene expression patterns change in response to various stimuli. Th cells, monocytes, macrophages, and DCs are all susceptible to HIV infection. But depending on the activation state of the cell, infection may be productive or latent. A confounding factor in truly understanding HIV pathogenesis is that even in the absence of direct cell killing, infection can change the ability of immune cells to do their jobs. For example, CD8 + T-cells may be less able to kill their targets, even though they are not directly infected by HIV-1.

Immune activation

Another factor in HIV-1 pathogenesis involves its ability to persistently activate the immune system. As mentioned earlier, activated Th cells are targets for productive infection. But persistent immune activation is damaging, even in the absence of direct infection. There is evidence that uninfected, but persistently activated CD4 + T lymphocytes undergo enhanced apoptosis, contributing to overall depletion of T-cells.

There are several lines of evidence-linking HIV-1 NEF protein to immune activation and pathogenesis. It was shown early in the AIDS epidemic that a few patients were so-called long-term nonprogressors (infected individuals that did not develop immunodeficiency). Analysis of their HIVs revealed naturally occurring NEF mutations. Another line of evidence for the role of immune activation in disease development

comes from studies of SIV infections of natural hosts. Where SIV and monkey host are well adapted to one another, there is little evidence of generalized immune activation and little disease.

HIV transmission

HIV is transmitted by infected body fluids. Thus HIV is transmitted most often by sexual contact, sharing needles, or from mother to child (during birth or through breast milk). Sexual transmission can be avoided by abstinence or greatly reduced by practicing safe sex (i.e., using condoms). Risk of sexual transmission increases between persons with other sexually transmitted diseases. Treatment of HIV-infected women can greatly reduce risk of transmission to their babies. Prophylactic treatment with antiretroviral drugs is often recommended after a possible exposure. There is also strong evidence that antiviral treatments that reduce viral loads also reduce risk of transmission. Thus knowing ones HIV status, and seeking prompt treatment, is an effective way to prevent spread.

Disease course

The disease course of HIV-infected individuals is variable, but for the most part, untreated HIV infection leads to AIDS and death. Within 2−4 weeks after infection with HIV, a person will usually develop a high-titer viremia and a generalized febrile or flu-like illness that may be accompanied by lymphadenopathy. However, some patients will be viremic, but experience no symptoms. During acute infection, much of the gut-associated lymphoid tissue is lost and circulating CD4 + T-cell numbers are reduced. Immune responses develop but do not clear the infection; in particular, levels of neutralizing antibody are slow to develop and titers remain quite low. T-cell numbers rebound, but usually remain below preinfection levels. Patients recover from the acute illness but virus continues to replicate and can be readily transmitted.

The best predictors of the course of infection are numbers of circulating CD4 + T-cells and virus load. Lower CD4 + T-cell counts and higher virus loads are predictors of a more rapid onset of AIDS-defining opportunistic infections or cancers. In terms of T-cell counts, the definition of AIDS is a CD4 + T-cell count of less than 200 cells/μL of blood. The time to development of AIDS may be as short as 6 months or as long as several decades (even in some untreated patients). Treatment with antiretroviral drugs drastically changes the progress of the infection. As highly effective drugs, with fewer side effects, have been developed, the trend is to treat patients earlier. At one time, the trigger for treatment was a CD4 + T-cell count of <350 cells/μL blood; however, some patients may choose to begin drug treatment when their T-cell counts are higher. It is clear that antiretroviral drugs drastically reduce viral load, preserving healthy levels of CD4 + T-cells.

Antiretroviral drugs

Development of drugs to treat HIV infection is an ongoing process. Currently, there are six classes of drugs actively used to treat HIV infection. Most are used in multidrug cocktails. Many of the newest drugs used to treat HIV were developed in a targeted manner, based on understanding the specific molecular interactions required for each step in the replication process.

Entry/fusion inhibitors

Maraviroc binds to the CCR5 coreceptor on CD4 + cells and macrophages, thereby preventing an interaction with HIV gp120. It should be noted that maraviroc does not inhibit strains of HIV that utilize the CXCR4 coreceptor but these viruses usually emerge late in an infection. Enfuvirtide (brand name Fuzeon) is a 36-amino acid peptide based on the sequence of HIV-1 TM. It was the first peptide inhibitor approved by the US Food and Drug Administration (FDA). The peptide interacts with a region of TM inhibiting formation of the specific structure required for fusion. Enfuvirtide has low toxicity and high efficacy but must be injected subcutaneously on a daily basis, limiting its utility for long-term therapy.

Nucleoside/nucleotide reverse transcriptase inhibitors

There are several approved nucleoside reverse transcriptase inhibitors (NRTIs), including azidothymidine (AZT), dideoxyinosine (ddI), and dideoxycytidine (ddC). All NRTIs are 2′3′-dideoxynucleoside analogs that must be phosphorylated by the host, after which they act as chain terminators, inhibiting elongation of the DNA chain by RT. Tenofovir disoproxil fumarate (TDF) is a nucleotide analog RT inhibitor. In vivo TDF is converted to tenofovir, an acyclic analog of deoxyadenosine 5′-monophosphate (dAMP). Tenofovir lacks a hydroxyl group in the position corresponding to the 3′ carbon of the dAMP. All of the aforementioned drugs can prevent establishment of an infection by interfering with reverse transcription. Unfortunately, once an integrated provirus exists, they are not able to prevent production of virions from an infected cell. Also, when used alone, resistance to these drugs develops quickly; therefore they are recommended to be used in combination therapies.

Nonnucleoside reverse transcriptase inhibitors

There are five approved nonnucleoside reverse transcriptase inhibitors (NNRTIs), including nevirapine and delavirdine. These compounds are noncompetitive inhibitors of RT, binding to the catalytic site, to inhibit activity. They do not require any metabolic processing by the host and are not incorporated into DNA. The approved NNRTIs inhibit HIV-1 but not HIV-2 RT. NNRTIs are frequently used as first-line drugs and are often combined with other drugs.

Integrase inhibitors

There are at least three FDA-approved IN inhibitors, including raltegravir, dolutegravir, and elvitegravir. These drugs block insertion of HIV DNA into the host chromosome. Recall that integration is an obligatory step in the HIV replication cycle. These drugs target HIV-1 IN to block the covalent linkage of the 3′ ends of viral DNA to host chromosomal DNA.

Protease inhibitors

PR inhibitors were the first successful "designer" antiviral drugs. PR inhibitors were developed based upon a clear understanding of the *molecular structure and catalytic activity of PR*. They are competitive inhibitors of HIV-1 PR and function by inserting into the substrate-binding cleft of PR. The first PR inhibitors were derivatives of small peptides. There are at least eight approved PR inhibitors, and in the presence of these drugs, noninfectious virions are produced from cells containing integrated copies of the HIV genome. PR inhibitors are extremely effective when used in combination with NRTIs, perhaps because they target a different step in the virus replication cycle.

Combination drugs

Combination HIV medicines contain two or more HIV medicines from one or more of the drug classes listed earlier. Some currently approved combinations include up to four active compounds. Use of drug combinations to target HIV is often called highly active antiretroviral therapy (HAART) and was initially used to avoid the development of drug resistance, which was common when NRTIs were used alone. Drug combinations that target different points in the HIV replication cycle are quite effective in limiting virus replication and spread within a patient. HAART often results in pushing down HIV replication to undetectable levels. This both limits damage to the immune system and prevents spread.

Prophylaxis

HIV-negative individuals with potential exposures to HIV can be treated with drugs to avoid infection. Postexposure prophylaxis (PEP) is a short course (usually 28 days) of anti-HIV drugs designed to prevent infection. PEP is recommended in emergency situations and treatment should begin as soon as possible, within a 72-hour window postexposure. PEP might be recommended if you are HIV-negative and experienced a needle stick in the workplace, were a victim of sexual assault, you shared needles or syringes to inject drugs, or experienced a condom break during sex with a partner of unknown HIV status (or a partner whose HIV is not well controlled).

Another type of prophylaxis is also available to protect against HIV infection. Called preexposure prophylaxix (PrEP), this is a drug treatment designed for persons at continued risk of contracting HIV through sex or injection drug use. PrEP is also an option for women with HIV-infected partners, wishing to get pregnant. Before taking PrEP, an individual must test negative for HIV, as the drug regimen is designed to prevent infection, not treat ongoing infection. Persons taking PrEP must have follow-up health care and get tested for HIV regularly.

Vaccines

It has been over 30 years since the discovery of HIV and a vaccine remains elusive, despite considerable effort and investment by vaccine companies, governments, and a variety of nonprofit agencies. Challenges to development of an efficacious vaccine include:

- *Correlates of protection remain undefined.* Because the healthy immune system does not control HIV during a natural infection, we have no road map to the types of immune responses that might be protective. There are small groups of individuals that resist HIV infection or remain healthy without treatment. These individuals are intensively studied for clues about natural mechanisms of resistance that could be "mimicked" by a vaccine.
- *Genome variation.* HIV is genetically variable and evolves within patients and on a worldwide population scale. It is likely that no single vaccine will provide protection against all groups or clades

of HIV. RT has a high intrinsic mutation rate as it does not have a proofreading mechanism. RNA recombination is frequent (perhaps because the process of reverse transcription requires translocation of newly synthesized products from one position to another on the template). Additional mutations can be generated by the antiviral activities of cellular cytidine deaminases and cellular DNA repair machinery.

- *Immune escape.* Genome variation gives rise to mutants that escape established immune responses in infected individuals. Viruses that escape vaccine-induced immune responses might emerge if protection is not complete.
- *Primate models are expensive, ethically challenging,* and have not yet provided a clear path to an effective vaccine. However, they have provided a wealth of information about pathogenesis of, and immune responses to, lentiviral infection.
- *Regulatory hurdles for human vaccine trials are high and trials are expensive.* Over 200 vaccines have been tested for safety and immunogenicity but only five have been tested for efficacy, and for the most part, the results have been disappointing. To determine efficacy, half of the individuals in a trial will receive a placebo, not the candidate vaccine. Given the life-long consequences of HIV infection, it is imperative that all individuals enrolled in such trials be provided with education and support aimed at avoiding exposure to HIV. Volunteers also need access to follow-up care and treatment should they become infected.

In this chapter we have learned that:

- The most common routes of HIV transmission are sexual contact and shared use of contaminated needles. HIV can also be transmitted from mother to child (during birth or by breast feeding). HIV can also be transmitted through nonsterile medical instruments or blood transfusions (screening of blood and blood products has virtually eliminated this mode of transmission).
- Virus transmission can be avoided by practicing "safe sex" (using condoms) and not sharing needles. Being aware of one's infection status (by getting tested) and seeking treatment can be very effective in reducing spread. Strict attention to infection control in medical settings is important to avoid spread of many infectious agents, including HIV. Blood and blood products should be tested.
- HIV infects cells that display the CD4 receptor on their surface (a coreceptor is also needed). Cells that express CD4 include a subset of T-lymphocytes, monocytes, macrophage, and DCs.
- Coreceptors for HIV are chemokine and cytokine receptors, including CCR5 and CXCR4.

- HIV is tropic for cells that are major players in the immune response. HIV kills some infected cells and alters the function of others. HIV causes disease by damaging the immune system; this leads to immunodeficiency (AIDS). Immunodeficient persons are at high risk for, and cannot control, many viral, bacterial, and fungal diseases.
- HIV TAT is a protein that is required for replication. It is a transcriptional activator that increases the production of HIV transcripts by $\sim$1000-fold.
- HIV and SIV encode a variety of auxiliary proteins. While they have different activities, they also share some features: They are small proteins that lack enzymatic activity. They work by interacting with various cellular proteins. Interaction with cellular proteins can result in antagonizing antiviral defenses, altering cell growth or signaling pathways, and inhibiting certain proteins from being expressed on the cell surface.
- Viral restriction factors are host-encoded proteins that have evolved to inhibit virus replication. Some examples include tetherin, A3 family proteins, and SAMHD1. Retroviruses can be inhibited by restriction factors but some have evolved specific proteins (i.e., HIV VIF and VPU) to antagonize them. Retroviral restriction factors are important in limiting cross-species transmission.
- Different classes of drugs target different events in the HIV replication cycle. Entry inhibitors interfere with attachment, fusion inhibitors interfere with fusion, and inhibitors of reverse transcription include nucleoside/nucleotide RT inhibitors. PR inhibitors block virion maturation.
- It is important to use drug cocktails to treat HIV infection because of the combination of high mutation rate and high virus load. It is less likely that drug-resistant mutants will arise when two or three different viral proteins are targeted at the same time.

44

Family *Hepadnaviridae*

After reading this chapter, you should be able to answer the following questions:

- What are the major characteristics of the members of the family *Hepadnaviridae*?
- By what process is the hepatitis B virus (HBV) genome synthesized?
- Describe the three particle types found in the blood of persons with chronic HBV infection.
- What are the major host cells for HBV replication? How does this impact the ability to study HBV in cultured cells?
- What animal models are used to study hepadnavirus pathogenesis?
- Why is HBV vaccination recommended for infants?
- What groups of adults are at highest risk for acquiring HBV?

HBV, a member of the family *Hepadnaviridae*, is one of several important human hepatitis viruses. HBV is the etiologic agent of hepatitis B, or blood-borne hepatitis. HBV is highly infectious and is transmitted by blood and body fluids. HBV pathogenesis is variable and the outcome of infection largely depends on host immune responses. A robust immune response may transiently damage the liver, but clears the infection in a process mediated by cytotoxic T cells. Weaker immune responses may fail to clear the virus, resulting in a long-term, chronic infection. Chronic infection is often asymptomatic as HBV is noncytopathic for hepatocytes. Thus infected individuals remain healthy carriers for

years or decades, but eventually develop liver damage or cancer. The World Health Organization (WHO) estimates that nearly 300 million people were living with chronic HBV infection in 2019, with 1.5 million new infections each year. WHO further estimates that HBV caused an estimated 820,000 deaths in 2019, mostly from liver cirrhosis and hepatocellular carcinoma.

Hepadnaviruses are enveloped viruses with $T = 4$ icosahedral capsids (Fig. 44.1). Genomes are relaxed circles of partially double-stranded DNA (Box 44.1). Only a handful of hepadnaviruses have been identified, and known hosts include birds, mammals, fish, and frogs. A unique aspect of hepadnavirus replication is the synthesis of genomes by reverse transcription. Hepadnaviruses have a definite preference for replication in differentiated hepatocytes thus it was a long and somewhat difficult process to determine even the most basic aspects of the hepadnavirus replication cycle. HBV does not infect animals (other than apes) so natural infections of woodchucks, ground squirrels, and ducks, with their respective hepadnaviruses have provided important insights into replication and pathogenesis (Box 44.2).

Genome organization

The HBV genome is quite small, only ~3.2 kbp. Within virions, hepadnavirus genomes have an unusual form: They are partially double-stranded,

© 2023 Elsevier Inc. All rights reserved.

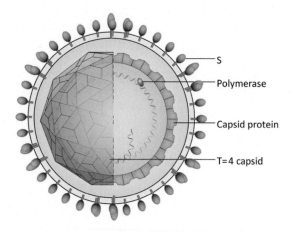

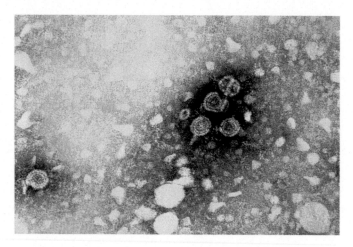

FIGURE 44.1   Virus structure (left). The transmission electron microscopic image (right) shows hepatitis B virions. The large round virions are known as Dane particles. Source: *CDC Public Health Image Library Image ID 270. Content providers: CDC/Dr. Erskine Palmer.*

BOX 44.1

General characteristics

Hepadnaviruses are small enveloped viruses with icosahedral capsids/cores. The HBV virion is ~ 42 nm in diameter with a core of ~ 34 nm. A single capsid (C) protein assembles to make a structure with $T = 4$ icosahedral symmetry. A set of overlapping envelope glycoproteins (large, medium, and small S) are associated with the envelope.

Genomes are relaxed circles of partially double-stranded DNA (~ 3.2 kbp) that are synthesized by reverse transcription. Early after infection genomes are converted to cccDNA in the nucleus. There are four ORFs. The largest encodes the polymerase (P). P has three domains: The amino terminus encodes a protein primer (terminal protein) used for priming minus-strand DNA synthesis. P also has a reverse transcriptase domain and an RNase H domain. The compact nature of the genome is achieved by extensive use of overlapping reading frames. Hepadnavirus mRNAs are capped and polyadenylated. The genome has multiple promoters and a single poly(A) addition site.

BOX 44.2

Taxonomy

Family *Hepdanaviridae* (five named genera)

Genus *Avihepadnavirus* (host are birds; three named species)

Genus *Herpetohepadnavirus* (hosts are reptiles and frogs; one named species)

Genus *Metahepadnavirus* (hosts are fish; one named species)

Genus *Orthohepadnavirus* (12 named species including *hepatitis B virus*; hosts are mammals)

Genus *Parahepadnavirus* (hosts are fish; one named species)

---

relaxed circles of DNA (Fig. 44.2) and the two stands are of unequal length. All viral proteins are encoded one strand. The longer DNA strand is the noncoding or negative (−) strand; it contains short direct repeats (r) at its ends. The negative strand of DNA is covalently linked to a protein at the 5′ end, the result of protein-primed DNA synthesis. The coding or plus (+) strand of DNA is of variable length, but is always incomplete and shorter than the negative strand. The coding strand is linked to an RNA oligonucleotide at

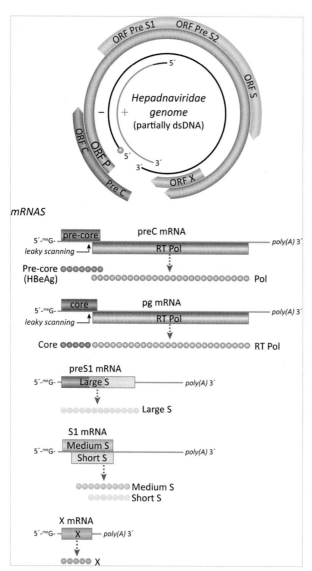

FIGURE 44.2 Genome organization. The genome is partially dsDNA. The longer, (−) strand is covalently linked to a protein primer. The shorter, plus-strand is primed by an RNA oligonucleotide. Note the extensive use of overlapping ORFs.

its 5′ end, as plus-strand DNA synthesis is primed with an RNA primer.

Upon delivery to the nucleus, the relaxed circular DNA genome is converted to a covalently closed circle (ccc) of DNA that associates with histones and other cell proteins. This form of the genome is called cccDNA. It serves as a template for transcription, but is *not* replicated by DNA polymerases. The HBV genome has four open-reading frames (ORFs) and all of the ORFs are overlapping, thus all three possible reading frames are used to encode proteins!

The longest ORF encodes the polymerase (P), a reverse transcriptase (Fig. 44.2). A set of surface (S) glycoproteins is encoded by the S ORF. The C ORF

encodes the core protein and a longer "preC" ORF encodes a secreted protein (E) not present in the virion. The small X ORF encodes a nonstructural protein with regulatory functions.

An examination of the genome reveals the large extent of gene overlap characteristic of the hepadnaviruses. The S ORF is contained completely within the P ORF and all ORFs share some overlap with P. Table 44.1 lists HBV ORFs and their protein products and functions. As a number of naming schemes have been used over the years, alternative names are provided in Table 44.1. Hepadnavirus mRNAs are capped and polyadenylated. The genome has multiple promoters but only a single poly(A) addition site. One mRNA, called pregenomic RNA (pgRNA), is longer than the cccDNA, thus contains repeated regions at the ends. The repeats are critical for the process of reverse transcription. pgRNA is the template for reverse transcription (by P) to form a new molecule of relaxed circular DNA. Reverse transcription occurs within a newly assembled capsid.

Virion structure

Hepadnaviruses are enveloped particles with a diameter of ~42 nm. The envelope contains three related glycoproteins, the large, medium, and small surface (S) proteins. (In this chapter, they are referred to as large S, medium S, and small S.) The envelope surrounds an icosahedral, $T = 4$ capsid that is formed by the capsid (C) protein. Capsids are ~34 nm in diameter. The partially double-stranded DNA genome is associated with a molecule of reverse transcriptase, often referred to as polymerase (P). In addition to infectious virions, two other types of particles are seen in the blood of persistently infected patients. These can be present in very large amounts. The most abundant (up to 10^{12} particles per mL serum) are 20 nm particles that are largely comprised of surface proteins. Longer filaments or rods are also ~20 nm in diameter and contain lipid and surface proteins. Neither of these particle types contains cores or DNA, thus they are noninfectious.

Replication cycle

The general steps in the hepadnavirus replication are shown in Fig. 44.3.

Attachment/penetration

HBV infects differentiated liver cells (hepatocytes) to establish a persistent infection. Initial interactions

TABLE 44.1   Hepadnavirus open-reading frame (ORF) and protein products.

ORF	Protein products	Functions	Alternative names
P	Polymerase. Contains functional domains for the protein primer (terminal protein, TP), reverse transcriptase and RNaseH	Terminal protein domain primes DNA synthesis; RT synthesizes DNA; RNAse H cleaves RNA during reverse transcription process.	Reverse transcriptase, pol
preS/S	Small S	Envelope glycoprotein. Shares a common carboxyl terminus with Medium S and Large S proteins.	Small HB surface protein (SHBs), S, HBs, S-HBsAg
	Medium S	Envelope glycoprotein.	Medium HB surface protein (MHBs), preS1, M, M-HBsAg
	Large S	Envelope glycoprotein. Large S contains the receptor-binding domain.	Large HB surface protein (LHBs), preS2, L, L-HBsAg
C	Core	Core/capsid formation	HBcAg, C
preC	E	Secreted from cells, not found in the virion; has immunomodulary functions.	HBe-protein, HBeAg
X	X	Regulatory protein, important in development of hepatocellular carcinoma.	

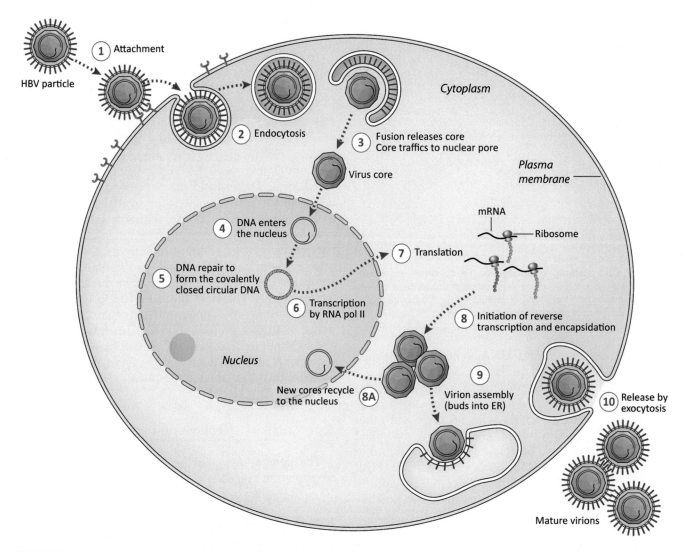

FIGURE 44.3   Overview of the hepadnavirus replication cycle.

are based upon weak interactions with heparan sulfate proteoglycans. Stronger binding results from interaction of large S with the liver-specific hepatic bile acid transporter: sodium-taurocholate cotransporting polypeptide. HBV uptake is not well characterized but probably occurs via an endocytic pathway followed by fusion to release the DNA-containing core into the cytosol. The core traffics to the nuclear pore and the genome enter the nucleus.

Transcription

Within the nucleus the HBV genome is converted to a covalently closed circular molecule (cccDNA) that associates with histones and other cellular proteins. While the details are not known, it is thought that cellular DNA repair enzymes are responsible for generating cccDNA.

cccDNA is maintained as stable episome within the nucleus. Unlike the reverse transcribing retroviruses, hepadnaviruses genomes are not normally integrated into chromosomal DNA. The episomal dsDNA genome is transcribed by host-cell RNA polymerase II to generate a set of capped, polyadenylated mRNAs. Four promoters have been identified on the HBV genome but transcription start sites are variable, forming a more complex set of mRNAs. There is a single polyadenylation site. Transcription depends, in part, on liver-specific transcription factors. The major transcripts seen by Northern blot analysis are 3.5, 2.4, and 2.1 kb and are termed pre-C/C, pre-S, and S messenger RNAs (mRNAs), respectively. There is also small amount of a 0.7 kb transcript, encoding the X protein. The 3.5 kb mRNA band seen on Northern blots is actually heterogeneous and contains mRNAs with three different 5′ ends. The two longer mRNAs are called preC mRNAs while the shorter mRNA is called the pregenomic (pg) mRNA. Together these mRNAs are used to produce the E, C, and P proteins. Pg mRNA is also the template for production of the genome. preS1 and S mRNAs overlap each other and completely overlap the P ORF. PreS1 mRNA is used to produce the large S protein. S mRNA is used to produce medium and small S proteins. All S proteins share the same C-terminal amino acid sequences but medium and large S have additional N terminal sequences. The promoter for X mRNA is within the P ORF. X mRNA is ~ 0.8 kb and is used to produce the X protein.

Protein products

The capsid protein, C, is a 21 kda phosphoprotein that assembles to form the $T = 4$ viral capsid. Hepadnaviral C proteins lack the basic structure of antiparallel beta sheets of amino acids shared by many capsid proteins. Instead C folds into alpha-helices and assumes a T-like shape. C dimers form via interactions in the stem of the "T". The top of the "T" faces inward and the stem faces outward, forming spikes on the capsid surface. The carboxyl terminal region of C protein is arginine rich, and may form a platform for DNA synthesis.

HBV e-antigen or E protein is produced from preC mRNA. Although as initially translated E is larger than C, the N-terminal domain contains a signal sequence that directs the protein to the endoplasmic reticulum (ER). Within the ER the E is processed by removal of the signal sequence as well as 34 aa from the carboxyl terminus to produce a 15 kd product that is secreted from the cell. In animal models of hepadnaviral infection E is not required for infectivity and some patients with chronic HBV harbor mutants that are defective for E production. The exact function(s) for E are not understood, but it appears to be immunomodulatory and may play a role in establishment of chronic infections.

HBV produces three envelope proteins: small, medium, and large S proteins. They share a common carboxyl terminal region. HBV large and medium S proteins contain the small S region but have amino terminal extensions. All versions share the same membrane spanning hydrophobic domains and are glycosylated; they assemble to form homodimers and heterodimers in the envelope. Large S is modified by N-terminal myristoylation while medium S is modified by N-terminal acetylation. Large S exists in two different conformations within the cell. In one conformation the N-terminal domain is cytosolic where it binds capsids and participates in virion assembly. In its second conformation the N-terminus of is present in the ER lumen, thus becoming exposed on the virion surface after particle release. In this conformation large S serves as the viral attachment protein. Some hepadnaviruses, notably those of birds, fish, and reptiles, do not appear to produce the medium S product.

The P protein is 90 kda and has reverse transcriptase and RNaseH domains. It also has an amino terminal domain, called TP for terminal protein, the primer for DNA synthesis. The RT domain is a target for several HBV antiviral drugs. To transcribe DNA, P must bind a specific genomic sequence, called epsilon, located at the 5′ end of pg RNA. Binding also requires host cell chaperone proteins. A tyrosine in the TP domain serves as a primer for DNA replication thus P remains covalently linked to the DNA product.

X is small nonstructural protein required for efficient virus replication. It can be expressed at very high levels and appears have a variety of cell-signaling and regulatory functions. Its functions have been studied in a variety of cultured cells (as well as in transgenic animal models), but none of these models accurately reflects the natural infection of differentiated hepatocytes. X interacts with many different cellular proteins, explaining the wide variety of activities that have been ascribed to it. It appears to affect viral and host gene expression and it signals through proteins that regulate cell proliferation and apoptosis.

Genome replication

Hepadnavirus genomes are produced by reverse transcription (Fig. 44.4). Thus genome synthesis actually begins with transcription of capped, polyadenylated pg mRNA. The process of reverse transcription occurs in the cytoplasm and is tightly linked to capsid formation. Genomes may be synthesized within newly formed capsids and it has been proposed that they are incomplete due to size limits of the capsid. HBV P, a reverse transcriptase, is key to the process of genome synthesis. The terminal domain of P serves as the primer for minus-strand DNA synthesis. Hepadnavirus RT and RNaseH function in similar manners to their retroviral counterparts.

Cellular chaperones participate in the initial association between P and pg mRNA. The priming reaction occurs at the epsilon site, near direct repeat 1 (Dr1) on the pg mRNA. The initial priming reaction leads to the formation of a covalent bond between a tyrosine residue and dGMP, followed by synthesis of just a few nucleotides. The primer then translocates to the downstream Dr1 sequence and the minus-strand is elongated toward the 5' end of the pg mRNA. RNaseH cleaves most of the RNA from the DNA/RNA hybrid molecule, but a small, capped mRNA fragment (~18 nt) remains uncleaved, to prime synthesis of the plus-strand of DNA. The plus-strand is synthesized following circularization of the complete minus-strand and translocation of the mRNA primer from Dr1 to Dr2. Synthesis of the plus-strand is not complete.

There are two fates for the new relaxed circular DNA genome (Fig. 44.3). Some capsids deliver genomes back to the nucleus where they will be repaired, to serve as templates for transcription, thus "amplifying" the infection. Other DNA-containing capsids will associate with S proteins in the ER, to form new infectious virions.

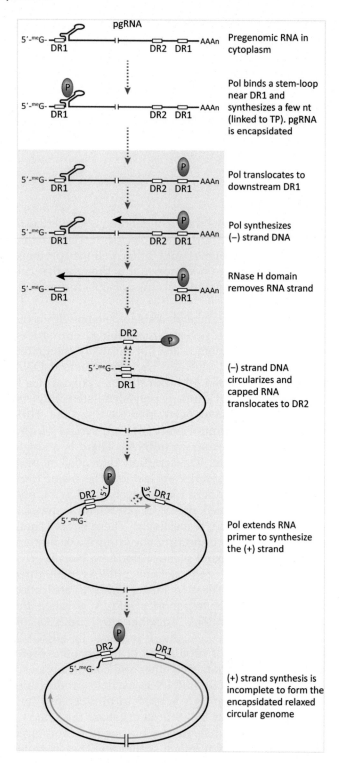

FIGURE 44.4 Process of reverse transcription to generate HBV genomes. The gray-shaded box indicates events that occur in association with capsid proteins.

Assembly/release of infectious particles

Formation of the HBV capsid is linked to reverse transcription. The capsid is not simply an inert "shell" as mutations in C affect activities as diverse as DNA synthesis, transport of capsids back to the nucleus, and

envelopment of capsids to generate infectious virions. Assembly of infectious particles requires interactions between the capsid and large S in the cytosol. Envelopment seems to occur by budding into the ER, with accumulation of virions in multivesicular bodies (MVB) of the late endosomal compartment. Virion release would then occur upon fusion of the MVB with the plasma membrane.

Hepatitis B virus and disease

The course of HBV infection is highly dependent on the age of infection. Among adults, ~90% will clear the infection within 6 months. Virus clearance is accomplished by the immune system. Disease is usually mild, and acute infection may be asymptomatic. However, approximately 10% of infected adults do not clear the infection and have chronic (usually lifelong) HBV infection. In contrast, infection of infants usually results in development of chronic HBV infection. Only ~10% of infected infants will clear the infection. The rate and extent of liver damage among individuals with chronic HBV infection varies widely; some will experience liver cirrhosis after a few decades while others will remain healthy. Chronic HBV infection is a significant risk factor for development of hepatocellular carcinoma. It is estimated that among chronically infected patients, 25% will develop liver cirrhosis and 5% will develop HCC.

In the United States, there are an estimated 1 million HBV carriers. The WHO estimates 200 million carriers worldwide, and asymptomatic carriers serve as a large virus reservoir. However, vaccination is reversing the trend. Since 1991 the number of new infections among infants in the United States has declined by over 80% as vaccinating infants has become a common practice. If this trend continues, the overall burden of HBV infection in the United States will continue to decline, as chronic infection becomes a rare event. However, there is still a significant risk of contracting HBV infection among certain groups, and vaccination is highly recommended for high-risk adults (including unvaccinated health care workers or any individuals whose job involved potential contact with human blood). According to the CDC, adults at highest risk for catching HBV:

- have sex with an HBV-infected partner,
- inject drugs,
- live in the same house with someone who has lifelong HBV infection,
- have a job that involves possible exposure to human blood,
- travel to areas where hepatitis B is common.

The first-generation HBV vaccine was prepared from pooled plasma from chronically infected HBV patients; taking advantage of the large quantities of noninfectious 20 nm particles and filaments (that contain abundant S protein) found therein. Although this vaccine was highly effective it was a blood-derived product and as such, was not widely used. The second-generation HBV vaccine is a recombinant product. When expressed in yeast, the S protein assembles into particles that are antigenically similar to those found in blood. In the United States, the recombinant vaccine is recommended for all infants and anyone under 18 years of age who has not been vaccinated, persons of any age whose behavior puts them at high risk for HBV infection and persons whose jobs expose them to human blood. The WHO is committed to reducing the burden of HBV worldwide and recommends vaccination for all infants.

In this chapter, we have learned that:

- Members of the family *Hepadnaviridae* are small, enveloped viruses with dsDNA genomes. They infect differentiated hepatocytes, causing persistent, noncytopathic infection.
- Hepadnavirus genomes are synthesized by reverse transcription. Thus every dsDNA genome is produced by reverse transcription of a molecule of capped, polyadenylated viral mRNA. DNA synthesis is tightly linked to encapsidation. The products of reverse transcription are relaxed circles of DNA. One strand (the plus strand) is usually incomplete.
- HBV P is a reverse transcriptase. The amino terminal domain is noncatalytic and serves as the primer for minus-strand DNA synthesis. The catalytic RT and RNaseH domains are similar to those found among retroviruses.
- Blood of persons with chronic HBV infection contains large numbers of noninfectious 20 nm particles, mostly consisting of S protein. There are also longer 20 nm diameter filaments. Least abundant is the Dane particle, the 42 nm infectious particle.
- HBV replicates in differentiated hepatocytes, which cannot be maintained in culture. Thus cell culture systems for HBV are of limited utility. HBV infection is also limited to apes (expensive and morally questionable animal models). Thus pathogenesis is often studied in woodchucks (woodchuck hepatitis virus), ground squirrels (ground squirrel hepatitis virus), and ducks (duck hepatitis virus).
- HBV vaccination is recommended for infants as they are at highest risk of developing chronic infection if exposed to HBV. At particular risk are infants of HBV-positive mothers.
- Adults at highest risk for acquiring HBV are nonvaccinated persons with HBV-infected sexual partners, i.v. drug users, and those with jobs involving possible contact with human blood.

Index

Note: Page numbers followed by "*b*," "*f*," and "*t*" refer to boxes, figures, and tables, respectively.

Printed in the United States
by Baker & Taylor Publisher Services